# THE SPIRAL STRUCTURE OF OUR GALAXY

INTERNATIONAL ASTRONOMICAL UNION
UNION ASTRONOMIQUE INTERNATIONALE

SYMPOSIUM No. 38
HELD IN BASEL, SWITZERLAND, AUGUST 29–SEPTEMBER 4, 1969

# THE SPIRAL STRUCTURE
# OF OUR GALAXY

EDITED BY

W. BECKER

*Astronomisch-Meteorologische Anstalt, Binningen, Switzerland*

AND

G. CONTOPOULOS

*University of Thessaloniki, Thessaloniki, Greece*

D. REIDEL PUBLISHING COMPANY

DORDRECHT-HOLLAND

1970

*Published on behalf of
the International Astronomical Union
by
D. Reidel Publishing Company, Dordrecht, Holland*

Library of Congress Catalog Card Number 75–115886

ISBN-13:978-94-010-3277-3     e-ISBN-13:978-94-010-3275-9
DOI: 10.1007/978-94-010-3275-9

*50th Anniversary of the IAU*

*1919–1969*

# INTRODUCTION

The idea of the organization of a Symposium on Spiral Structure came at a special meeting of Commission 33 on Spiral Structure during the 12th General Assembly of the IAU in Prague, 1967.

So much interest was shown during this meeting that one of us proposed a special Symposium on the 'Spiral Structure of Our Galaxy' for 1969. The response was immediate and it was finally agreed upon holding the Symposium in Basel, a center of galactic research in the center of Europe.

During the next months a special 'List of Problems', related to this Symposium, was sent to many prospective participants by the president of Commission 33. This stimulated an increase of interest in problems of galactic spiral structure and a concentrated effort on some problems.

The organizing Committee of the Symposium was composed of Drs. L. Woltjer (president), W. Becker, A. Blaauw, B. J. Bok, G. Contopoulos, F. J. Kerr, C. C. Lin, S. W. McCuskey and S. B. Pikel'ner. Most of the work for the organization of the Symposium was carried by Dr. L. Woltjer.

The Local Committee, composed of Drs. W. Becker, U. W. Steinlin, R. P. Fenkart, and G. A. Tammann, made every effort to secure the success of the Symposium. Most of the credit goes to Dr. Steinlin.

The Symposium was supported financially by the IAU and by the Swiss National Science Foundation.

The meetings took place at the University of Basel, which provided also secretarial help and many other facilities.

The Basel Symposium marked the 50th anniversary of the IAU. A special celebration took place at the Opening of the Symposium, during which the President and the General Secretary of the IAU, Drs. O. Heckmann and L. Perek, as well as Dr. J. H. Oort spoke about this happy event.

The scientific meetings lasted from August 29 to September 4, 1969, and were attended by 145 participants from 23 countries.

Eighty-five papers were presented, followed by discussions. Some papers served as Introduction to general subjects. At the beginning there was a general 'Survey of spiral structure problems' by Dr. J. H. Oort, and at the end a comprehensive 'Summary and outlook' by Dr. B. J. Bok. The final discussion of 'Desiderata for future work' was introduced by Drs. Oort, Becker, Burke and Lin.

We have decided to avoid the publication of discussions, except in cases where the participants wanted their remarks to be inserted. This simplifies the edition of the Proceedings; on the other hand many of the most interesting discussions took place outside the meetings, and this is one of the most important functions of any Symposium.

In preparing this Volume for the press we have been helped considerably by Dr. J. Hadjidemetriou in Thessaloniki and by Dr. R. Fenkart in Basel.

In closing this Introduction we would like to thank all those that have contributed to the success of the Symposium.

*The Editors:*

W. BECKER

G. CONTOPOULOS

# CONTENTS

## PART II /
## OBSERVATIONS OF SPIRAL STRUCTURE IN OUR GALAXY

A. *Radio Observations*

## PART III / THEORY OF SPIRAL STRUCTURE

### A. *Normal Spirals*

### B. *Barred Spirals*

### C. *Numerical Experiments*

## PART IV / COMPARISON OF THEORY AND OBSERVATIONS

### A. *Gravitational Effects*

# 1. SURVEY OF SPIRAL STRUCTURE PROBLEMS

J. H. OORT

*Leiden Observatory, Leiden, The Netherlands*

If one tries to look at the structure of spiral galaxies with the eyes of an unbiased newcomer, and if one then pages through the *Hubble Atlas of Galaxies* or Arp's *Atlas of Peculiar Galaxies* the first reaction is that these objects must be *young*, and could hardly have lived over much more than one revolution. Consideration of data on the strong differential rotation that is characteristic for spiral galaxies further strengthens this impression.

And yet we know, from convincing age determinations of clusters, that at least our own Galaxy is old. There is, furthermore, so much similarity between the spectral characteristics and general composition of other spirals and so much resemblance with our Galaxy that it seems unreasonable to believe that these other spirals would be much younger. We may assume, therefore, that the ages of most spiral galaxies count several dozens of revolutions. Let us adopt 50 revolutions as an order-of-magnitude for the average age.

These high ages are for the systems as a whole. They do not necessarily apply to the spiral patterns, which might be more ephemeral. But if we would suppose the spiral arms to be short-lived there has to be a mechanism by which they are revived every few revolutions. Otherwise we could not explain why at present the majority of non-elliptical stellar systems show a well-defined arm structure. A particular difficulty is that they would have to be revived as more or less continuous structures throughout the entire system. Up to the present no plausible mechanism has been worked out which could explain such a periodical renewal of the large-scale spiral structures.*

It was suggested many years ago by Bertil Lindblad that spiral patterns could be maintained for a long period by a sectorial wave system. This idea has in recent years been further worked out by Lin, Shu and Yuan, who showed among other things how such a density wave causing a spiral pattern could be sustained by its own spiral gravitational field superposed on the general axisymmetrical field of the galaxy. The density excess in the arms needed for this spiral field is entirely reasonable in view of the observational data on the distribution of gas and young stars in the Galaxy. As will be indicated in some reports at the Symposium there is considerable direct evidence, especially in the humps of the rotation curve found from 21-cm observations in our Galaxy, but likewise in other 21-cm observations, that streamings of the kind and amplitude required by the wave theory, actually exist. Moreover there is evidence that formation of stars occurs preferentially when the gas in the regions concerned is compressed by the wave.

---

* In the course of the Symposium this problem was extensively discussed. It appeared that several promising theories on the re-excitation of spiral structure had already been partly worked out.

*Becker and Contopoulos (eds.), The Spiral Structure of Our Galaxy, 1–5. All Rights Reserved.*
Copyright © 1970 by the I.A.U.

A very important step forward thus appears to have been made in understanding how spiral patterns can persist for a much longer time than the particular arms we see at a given instant.

The theory explains the maintenance but *not* the *origin* of spiral structure. I do not think this is an important shortcoming, for it is easy to conceive of processes which would *start* a spiral structure. Indeed, any mechanism of formation of a rotating galaxy is likely to produce a system with large scale initial asymmetry in the plane of rotation, which will almost automatically develop into a two-armed spiral structure.

A more serious problem seems that of the *long*-term permanence of the spiral waves. Can they continue to run round during 50 revolutions without fatal damage to their regularity? Looking at the irregularities in the actual spiral galaxies one wonders whether the present spirals could *continue* to exist for such a large number of revolutions. The problem seems particularly acute for the outer parts of Sc spirals, like M 101 or NGC 628, and still more for some SBc systems, like NGC 3187 in the Leo group. In the latter case interaction with other members of the group might possibly have been a factor in determining the non-equilibrium shape of arms. But in most cases we cannot invoke interaction. Moreover, if there had been interaction in the past it might well have contributed to *disturb* the regular wave pattern, but less likely to rebuild it.

The foregoing considerations give one the feeling that, though the principal part of the spiral phenomenon may be explained by wave motions of the type described, there is something else besides which influences, or perhaps incites, the structure.

I have not yet mentioned the important class of spirals which beside spiral arms have a bar structure. The difficult dynamical problems which these systems present will be reported on by Freeman. The wave theory cannot be directly applied to barred spirals. Interesting observations of velocity patterns made by Mrs. Burbidge indicate that the phenomena are in any case quite complicated.

There are, of course, also the irregular galaxies to be considered. In addition, some rather regular galaxies exist whose forms do not fit in any of the categories considered (cf. the ring galaxy reproduced as No. 147 in Arp's *Atlas of Peculiar Galaxies*).

At this point I want to say a few words about the general appearance of spiral galaxies. Dr. Lin has sometime quoted me as having stated that the important characteristic of spiral galaxies is not that they have spiral-like features – because spiral shapes are the natural shape into which any larger feature will automatically be drawn out by differential rotation – but that in so many cases spiral arms can be followed more or less continuously through the entire galaxy. I do not want to withdraw this statement, but I must point out that it should be supplemented by two essential additions. First, that in about half of the spirals the structure is either unclear, or there are more than two arms. Second, that even in the half that can be classed among the two-armed spirals there are invariably important additional features *between* the two principal arms, while the latter have often a number of secondary branches coming off their outer rims. In a fair number of cases these secondary features make it difficult to follow the main arms throughout the system.

What is already difficult in extraneous systems is an order of magnitude more difficult in the galaxy which is the title subject of this symposium. Our own Galaxy is seen edge-on, and almost the only way to unravel the general structure is through observation of radial velocities. But these are affected as much as the structures themselves by the irregularities described, so that we get two classes of irregularities superimposed. Nevertheless it has proved possible by means of the 21-cm observations, recently aided also by velocities of H II regions derived from radio recombination lines, to trace spiral arms through at least one half of the Galactic System. A full discussion of this will be given in the course of this symposium by Kerr, Mezger, Westerhout, Weaver, Varsavsky, and, for the inner arms, by Burton and Shane.

Though for a *general* study of the spiral structure the Galactic System is less accessible than other galaxies it *does* permit the study of a number of phenomena connected with the spiral arms which, so far, it has not been possible to investigate in other galaxies. I mention the detailed study of the stellar populations of spiral arms, the distribution of individual motions of stars in the arms, the thickness of the arms perpendicular to the galactic plane, the connection between the arms and large-scale streamings of the gas, the character of large-scale irregularities like the splitting of the Perseus arm discovered by Münch and extensively studied by Rickard, and, particularly, the large radial motions of the arms in the central region.

As the latter will not be discussed in other parts of the symposium, except in connection with the Scutum arm, I want to say a few words about the possible significance of these phenomena.

In the region within 3 or 4 kpc from the centre we find an arm which at the point where it passes between the centre and us has a velocity component of 53 km s$^{-1}$ directed away from the centre. This so-called 3-kpc arm has all the characteristics of the spiral arms observed in the outer parts of the Galaxy except for its large radial motion. It contains a similar order of mass per unit length, has a similar linear thickness, and is a *major* feature, that can be followed over at least a full quadrant of galactocentric longitude. On the *far* side of the nucleus there is an approximately equal amount of hydrogen, apparently also moving away from the centre, but with velocities which are two to three times higher. Part of this matter may be rather close to the centre. The arm structures on that side are not so well-defined as the 3-kpc arm.

The problem how these expanding arms have acquired their outward motions is still unsolved. It is tempting to think, as several astronomers have done, that the outward momentum comes from violent ejections from the galactic nucleus. A realistic basis for this idea has been furnished by recent detailed investigations of Seyfert nuclei, showing that violent motions involving large and massive individual clouds exist in the nuclei of these otherwise normal spirals. Clouds or streams with masses of the order of $10^4$ or $10^5$ solar masses are apparently escaping from their nuclei at velocities of the order of 1000 km s$^{-1}$. If a comparable activity had happened in our Galaxy it might well have caused the observed outward motions of the arms in the central region. It is interesting that rather direct evidence of activity of a similar kind has now been found in the vicinity of the nucleus of the Galactic System. But the velocities are an order of a magnitude lower than in the Seyfert nuclei. The evidence for the ejection of large masses of gas from the nuclear region was first discovered in 1966 by W. W. Shane from a survey for high-velocity features in the central region outside the galactic plane. Van der Kruit, who has followed up this discovery, finds that within a region extending to 5° in latitude and 10° in longitude from the centre there exists a mass of roughly $5 \times 10^6 M_\odot$ of neutral hydrogen well outside the galactic disk. His observations suggest an ejection of clouds from the nucleus in two roughly opposite directions making large angles with the plane. The velocities range from about 100 km s$^{-1}$ near the centre to 50 km s$^{-1}$ at 5 to 10° distance. Though the momentum involved is insufficient to have yielded the motion of the 3-kpc arm, the observed part of which has a mass of $4 \times 10^7 M_\odot$, we might imagine that strong nuclear activity is a thing that comes in bursts, that we are now in a relatively quiescent period, and that the motion of the 3-kpc arm is due to much stronger activity at an epoch a few times $10^7$ years ago. The circumstance that the streams appear to be ejected at a considerable angle with the galactic plane makes it understandable that the small rotating nuclear disk found by Rougoor and Oort might have survived the period of violent activity.

The preferential direction might well be connected with the structure of the magnetic field. It is suggestive, in this connection, that Kerr and Sinclair have found that the emission at 20 cm near the centre shows two separate ridges, one practically along the galactic circle, the other along a line going through the centre but inclined at an angle of 47° to this circle. The latter ridges not only lie in the same quadrants as the high-velocity streams, but also the inclination to the galactic equator is comparable to what the streams indicate in the region close to the centre. The inclined radiation ridge has not been followed for more than 1°.4 from the centre. Kerr and Sinclair suggest that it is due to jets from the nucleus. No data have yet been published on the spectrum of this most interesting radiation.

If the motion of the 3-kpc arm is ascribed to ejections from the nucleus one should ask also what consequences such spasmodic massive violence would have on the more outlying parts of the disk. Could this possibly enhance the spiral waves in these outer parts, and thereby keep them going for longer periods than if left entirely to themselves?

In this connection a special interest attaches to the observations of the Scutum arm, which Shane will discuss. The observations are indicative of an expanding motion also in this arm. Likewise in this connection I must mention the extremely interesting observations by Mrs Rubin and Ford on the central region of M 31. The almost zero velocity of rotation which they find for the gas at 2 kpc from the centre may well have been caused by matter with little or no angular momentum expelled from the nucleus under an angle with the galactic plane, and having again collected in the plane around a distance of 2 kpc, leaving the rapidly rotating gas which is observed inside this distance more or less undisturbed.

We are clearly not yet in a position to answer the question of a possible excitation of spiral structure by the nuclear activity – at any rate not in a positive sense. The idea itself that the entire structure of a galaxy is strongly influenced by phenomena originating in its nucleus has been expressed on several occasions by Ambartsumian. It has very recently been suggested anew by Arp.

There are two more points I want to mention before concluding this survey.

The first is the extension of the arms perpendicular to the plane. Their average thickness is about 250 parsecs. But they have long 'wings' which, at least for the outer arms, extend to distances of 1 to 2 kpc from the plane. These phenomena will be discussed in a later communication. A remarkable thing about these wings is that they have no tendency to merge into a more or less continuous halo, but seem to have remained connected with the arms from which they have originated. They must be highly unstable, and, consequently, should be continually renewed. An important question to be answered – also in connection with the general problem of the spiral arms – is how this repopulation of the very high $z$-levels is done, whether by violent events, or by some kind of magnetic field instability.

The second and last point is the influence on the dynamics of spiral arms of a possible inflow of gas from intergalactic space. There is, in my opinion, strong evidence that gas is flowing into the Galaxy from outside. This subject will not be discussed in the present conference. It is useful, however, to give a few numbers indicating the magnitude of the interaction with the disk gas as inferred from the high- and intermediate-velocity clouds observed in Groningen and Leiden. From these I have estimated a flow into the Galaxy of roughly $3 \times 10^{17}$ atoms per cm$^2$ of the galactic plane per $10^6$ years. This is a considerable amount. In one revolution of the Galaxy it would increase the amount of gas in the disk by roughly 10%, and might sensibly decrease the rotational velocity of this gas.

# PART I

# SPIRAL STRUCTURE IN GALAXIES

# 2. SPIRAL STRUCTURE IN EXTERNAL GALAXIES

W. W. MORGAN

*Yerkes Observatory, Williams Bay, Wisc., U.S.A.*

**Abstract.** The need for observing at least four different categories of optical objects for a satisfactory definition of optical spiral structure is emphasized. A sequence of optical form-types for the description of spiral structure is outlined. A pronounced difference in the nature of the spiral structure of the outer and inner parts of Messier 33 is illustrated. Striking local differences between spiral arms defined by differing criteria are noted.

## 1. Introduction

There are at least four different categories of optical objects which can be considered to define spiral arms: (a) H$\text{II}$ regions, (b) dust lanes, (c) blue supergiants (with a certain admixture of h-$\chi$ Persei-type red supergiants), and (d) a non-blue stellar population which can be observed on plates exposed to the yellow spectral region. The appearance – and even location – of the arms depicted varies somewhat on passing from one category to another.

The H$\text{II}$ regions define arm segments which are narrow, somewhat irregular in shape, and which may be relatively short in length. The arms as defined by dust segments tend to be narrow also, may be quite regular in the inner parts of the main body, and sometimes coincide in position with the arm segments as defined by the H$\text{II}$ regions. The brightest resolved blue stars (and red supergiants of the h-$\chi$ Persei-type) lie in arm-like segments somewhat less sharply defined than those of the first two categories. The 'non-blue star-arms' tend to have the smoothest and most regular appearance in the inner parts of spirals. The local width of the arms increases progressively from the H$\text{II}$ region type (the narrowest), to the non-blue star-arms (the widest).

From these circumstances it can be seen that if we wish to discuss the general phenomenon of spiral structure we must describe in detail, and describe separately, the spiral arms as shown by the various categories of objects. Only in this way will we be able to bring full observational weight to bear on the definition and evolution of extragalactic spiral structures. In addition, it is only by such a complex procedure that we will be able to compare with precision the spiral structure of our own Galaxy with that of external galaxies. There is no single indicator capable of defining completely the general phenomenon called 'spiral structure'; it is only by a collation of the results from the various categories of objects (including those in radio astronomy) that a satisfactory description of spiral structure can be arrived at.

## 2. The Frame of Reference

The optical form-classification of Hubble in its 1935 state is described in Hubble's

*Becker and Contopoulos (eds.), The Spiral Structure of Our Galaxy, 9–14. All Rights Reserved.*

(1936) *The Realm of the Nebulae*. The normal spirals (Sa, Sb, Sc) are illustrated in the famous tuning fork diagram; the two principal criteria for their classification were: (1) the 'relative luminosity of nuclear region and spiral arms', and (2) the 'openness of the arms'. An inspection of Hubble's (1936) diagram suggests that the principal discriminant between classes Sa and Sb was the openness of the arms, since the relative luminosity of the nuclear region to the spiral arms in the sketches of classes Sa and Sb is not very different.

The practical result of this weighting of the two criteria on later classification work on the Hubble system (Humason *et al.*, 1956; Sandage, 1961) has been to create an Sa class which includes objects exhibiting a wide range in the nuclear concentration of luminosity – and a pronounced overlap between classes Sa and Sb in this same characteristic. From this it is suggested that the sequence Sa-Sb-Sc is not a completely one-dimensional sequence, since the two principal classification criteria can give conflicting Hubble types. The Hubble Sc type includes some Yerkes types in the range aI-gS; the Hubble type Sb includes some Yerkes types in the range fgS-kS. As shown in Figure 1, the Hubble Sa class may parallel class Sb with respect to the Yerkes one-dimensional stellar-population form-sequence. Six standard galaxies are reproduced in Figures 2 and 3; they are from Mount Wilson 60-inch and 100-inch plates.

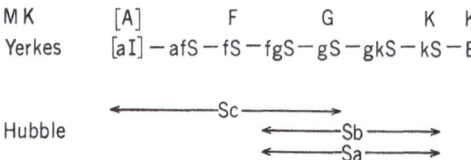

Fig. 1.   A comparison of Yerkes form classification with blueviolet spectral types, and Hubble types for spirals. The Sb and Sa classes overlap with respect to spectral type of the nuclear region. A relationship somewhat similar to Figure 1 can be set up for the barred spirals.

The Yerkes form types (based on the relative luminosity of the nuclear region to the main body) are fairly well correlated with the integrated spectral type of the nuclear region. The greatest stellar contribution in the blue-violet to the luminosity of the nuclear region progresses from A stars (4490, Sc, aI) to K giants (2841, Sb, kS). No spectrogram was available for 4535. The populations in the later parts of this sequence appear to be of the ordinary strong metallic-line type. *We may label the above Yerkes sequence as an evolutionary sequence for the stars contributing most to the luminosity of the inner region of the galaxies concerned.*

There is a wide variation in the detailed optical aspects of galaxies classified in each category of the Yerkes system; the criterion used has been the relative luminosity of the nuclear region to that of the remainder of the main body; the only added criterion was that of visible spiral structure. NGC 4490, classified on the Hubble system as Sc, has been considered to belong to the Irregular class on the Yerkes system; however, this is a delicate distinction; and the galaxy may become a prototype for a category described by G. R. Burbidge and by Morgan, and labelled by them

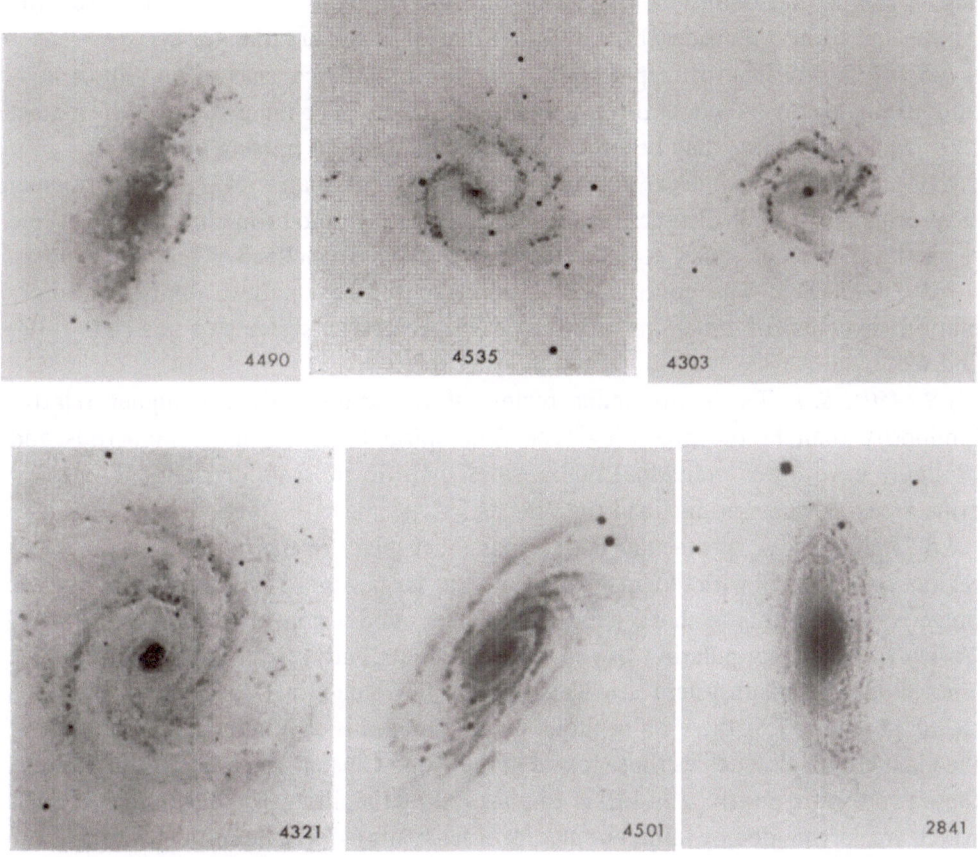

Fig. 2 and 3. A series of six standard galaxies illustrating Yerkes form types for spirals. NGC 4490 (aI; Sc) has an A-type absorption spectrum. The sequence 4535, 4303, 4321, and 4501 is in order of increasing luminosity of the inner region, relative to that of the main body as a whole. NGC 2841 possesses a large, luminous, nuclear subsystem; the latter is entirely different in appearance from the other galaxies illustrated; the light of the inner region in the blue-violet is due principally to yellow giants. The inner spiral arms (dust or stars) tend to be rather highly inclined for all except 2841.
Mount Wilson 60-100-inch photographs.

'integral signs'; such objects may turn out to be closely related in nature to such systems as 5247, and objects Nos. 31 and 33 in the Arp *Atlas of Peculiar Galaxies*.

The earliest-type object with well-developed spiral structure has been classified as afS (4535). In the series illustrated, the spirals 4303, 4321, 4501 and 2841 form a sequence of progressively more luminous inner regions – and also of progressively later spectral type in blue-violet wavelengths.

## 3. The Phenomena of Spiral Structure as a Function of Yerkes Form-Type

*aI (4490, Sc)*. This might be considered as a galaxy consisting of a long, distorted, very broad spiral arm. [OII] 3727 is outstanding in strength in the central region;

and N 1, N 2, and Balmer emission lines are present in some regions. However, the A-type spectrum is considerably later than that of irregulars like 4214.

*afS (4535, Sc).* The principal spiral structure in 4535 consists of an almost continuous single arm extending through the small nucleus. The inner star arms lie almost 180° from each other; they become less highly inclined on passing outward.

*fS (4303, Sc).* The nuclear region is more luminous than in 4535. The principal arms originate at some distance from the nucleus and tend to lie in almost straight-line segments. The outer, faint, spiral structure is less highly inclined.

*fgS (4321, Sc).* The contrast (or visibility with the smallest number of information elements) of the spiral arms is at its greatest in systems of this class (5194 and 628).

*gS (4501, Sc).* The entire inner region of the main body has higher relative luminosity than in the preceding type. The spiral arms are more numerous and probably have lower individual luminosities than do those in class fgS. The spiral arms are somewhat less inclined than in the fS systems.

*kS (2841, Sb).* A spectacular change has taken place in the nature of the nuclear region, as compared with the preceding spirals: a large, amorphous, luminous, nuclear bulge is present; this is characteristic of Yerkes subdivision kS (M 31, 4594, M 81, 2841). The stellar population of the principal contributors to the luminosity of the amorphous inner subsystem are K giants (in the blue-violet spectral region). The stellar population of these amorphous nuclear bulges is strikingly similar to that of the giant ellipticals, such as those found in the Virgo Cluster; the relationship between these two form types (kS and kE) is emphasized by the discovery of certain ellipticals which are surrounded by faint, circular, arm segments (474). The spiral arms observed in 2841 are of low inclination and of considerably lower contrast and luminosity than in the fg spirals.

## 4. The General Large-Scale Characteristics of Spiral Arms

The inclination of the inner arms of fS and fgS galaxies tends to be high; that of the spiral arms of the kS galaxies tends to be considerably lower, and not far from circular. The integrated MK spectral types of the nuclei of fS and fgS galaxies are of class F in the blue-violet region; the nuclear bulges of kS galaxies are of MK spectral class gK. A relationship between the average inclination of the inner arms and the average evolutionary state of the most luminous fraction of the nuclear stellar population is therefore suggested.

Figure 4 shows the outer part of M 33 from the blue print of the National Geographic Society-Palomar Sky Survey. The positions of the principal two inner star arms are marked from Yerkes photographs by Dr. Robert Garrison. The complexity in number of the outer arms is in contrast to the pair of principal inner arms.

Figure 5 illustrates two photographs of M 101 obtained by Garrison at Yerkes. A section of an H II arm that appears to be separated from the nearest strong star arm is marked.

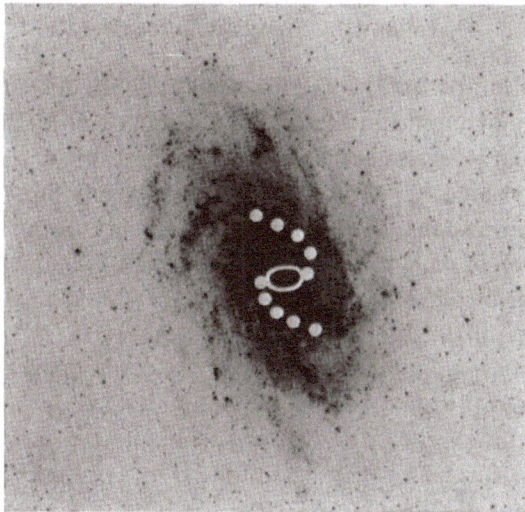

Fig. 4. The spiral M 33 (fS; Sc) reproduced from the blue print of the National Geographic Society-Palomar Sky Survey. The two principal inner star arms are located approximately by the white dots. The outer spiral structure can be seen to be complex and to a certain extent independent of the two principal arms. Copyrighted by the National Geographic Society-Palomar Observatory Sky Survey.

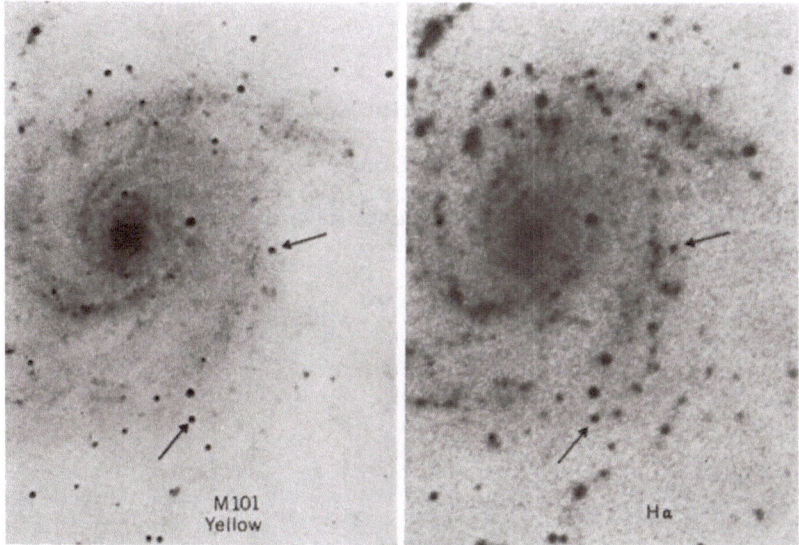

Fig. 5. Yellow and Hα photographs of M 101. On the yellow plate, one of the principal star arms can be seen extending slightly to the left of the two stars marked. On the Hα plate a string of HII regions can be seen slightly to the right of a line joining the two marked stars. This can be described as a 'splitting' of an arm segment, with the two segments differing in physical constitution and width. Yerkes photographs by Robert Garrison.

## 5. Some Directions for Future Work

The detailed delineation of spiral structure separately for the four optical indicators listed above is of major importance. A careful comparison of Hα photographs with exposures in the yellow region allows a clean separation to be made between the H II arms and the non-blue star-arms. Positive-negative combinations of plates taken with differing spectral response give spectacular evidence of differences in arm structure; the interpretation of such results requires considerable care, if unique conclusions are to be drawn. A near-infrared survey in which the Hα line is filtered out could give useful evidence of the inner non-blue star-arm structure.

Additional information along the above lines could increase considerably our knowledge of the local optical spiral structure of our own Galaxy – when spectroscopic parallax programs now in progress on the blue-giant stars are completed. Such an approach would also furnish new information necessary for the collation of optical and radio observations within three kiloparsecs of the sun.

## Acknowledgments

We are indebted to Dr. H. W. Babcock, Director, for permission to reproduce the six Mount Wilson photographs of Figures 2 and 3, from 60-inch and 100-inch originals – and for permission to reproduce Figure 4 from the National Geographic Society-Palomar Sky Survey. The photographic work was carried out by Mr. J. W. Tapscott.

We wish to acknowledge support from a grant on Stellar Classification from the National Science Foundation.

## References

Hubble, E.: 1936, *The Realm of the Nebulae*, Yale University Press, New Haven.
Humason, M. L., Mayall, N. U., and Sandage, A. R.: 1956, *Astron. J.* **61**, 97.
Sandage, A.: 1961, *The Hubble Atlas of Galaxies*, Carnegie Institution of Washington, Washington.

# 3. SPIRAL STRUCTURE OF OUR GALAXY AND
## OF OTHER GALAXIES

B. A. VORONTSOV-VELYAMINOV

*Moscow University, Moscow, U.S.S.R.*

Time and again Dr. B. Bok publishes his reviews on the state of optical searches of the spiral arms in our Galaxy. A great optimist at the beginning, concerning the agreement of different results, he became later more critical in his conclusions. But in his paper of 1967 he claims anew, as well established, that we can trace the spiral arms up to a distance of 14000 pc from the centre.

I could never agree that the spirals made of H I, or rather the gaseous rings, did confirm the optical spirals (then outlined), which make an acute angle with the radius.

In fact the two patterns, optical and radio, could not, and still cannot be reconciled at all.

Formerly the analogy with the most popular photographs of galaxies possessing only two arms was expected. Consequently only two arms in our Galaxy were searched. Nobody worried that a large number of complete revolutions was necessary in order to draw the spiral arms up to the sun or to a greater distance, though hardly there could be shown a galaxy with two complete revolutions of its spirals. At the same time, not so long ago it was suggested that there is a possibility that our Galaxy is a barred spiral. Recently some models with many arms were advanced. These divergences in the localisation of the spiral arms in our Galaxy can hardly allow for a satisfactory model.

The efforts made to construct a model of our Galaxy with many arms are fruitless because there exist no such numerous arms that emerge from the nucleus. If there are many 'arms', they represent in this case only bits of spirals and can be traced as continuous formations at most for $\frac{1}{4}$ of a complete revolution. Such galaxies are very different from the ideal symmetric models with many arms so far proposed. On the other hand the many arms model, especially with fragmentary arms, is easier to reconcile with the observational data.

The same is true with the dark matter. The scheme of a homogeneous layer was first proposed, then the contrary view of a multitude of average dark nebulae. Finally the topographic method of presentation was established. The same must be done with the spiral structure. We shall see what it is like, after we shall have good distance determinations, without the drastic efforts to adapt the observational data to abstract schemes.

To obtain a reliable idea of spiral arms in a giant galaxy, from observations made within a radius of 2000–2500 parsecs, is very difficult. The reason is that in real galaxies the structure of the spiral arms is complicated as a rule. We may compare M 33 and NGC 2403 with their broad coalescent arms and NGC 5364, 210. I may also mention the complicated patterns, which remind of the lace of the lines of a

*Becker and Contopoulos (eds.), The Spiral Structure of Our Galaxy, 15–17. All Rights Reserved.*
*Copyright © 1970 by the I.A.U.*

magnetic field. There are straight bands connecting the spiral arms, as in M 51, etc.

The simple geometrical models are needed by theoreticians, but the nature is much more complicated. Take, for instance, two volumes 1–2 kpc in radius (3 kpc for a giant galaxy) in different places of one and the same galaxy. Try to obtain concordant tracings of spiral arms reconstructed from these two volumes.

I will stress that taking simple patterns and adjusting them to conform to a scanty number of objects, small as compared to their separation, we can deceive ourselves. Especially because the impression of our finding depends in a striking way on the way we sketch our objects and join them by lines (Figure 1).

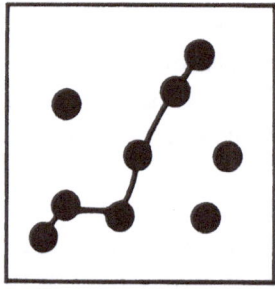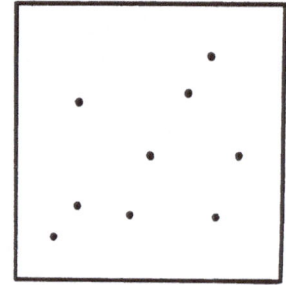

Fig. 1.

We are 8000 pc distant from the centre of our Galaxy and we look for the spiral arms still farther from the centre, beyond the solar position. We know the distribution of hot giants only approximately and at most up to a distance of 2–3 kpc from us. Even here some of them are hidden behind dark clouds, especially the more distant ones.

The nearest galaxy comparable to ours according to its dimensions, is M 31, a spiral considered to be even somewhat larger than our Galaxy. At a distance of 8000 pc from its centre the surface brightness is very low. The outermost borders of its spiral arms, very faint features, barely touch this limit. Further on no traces of spiral arms can be recognized. The features which Baade called the spiral arms V–VIII are but the loose groups of hot stars intersected by the prolonged great axis of M 31. They cannot be traced as real *arms*, even along some fraction of a revolution. Of course Baade did no attempt to trace them. So even in M 31, which apparently is larger than our Galaxy even the borders of unmistakable spiral arms are closer than 8000 pc from the centre. It appears that we do not know a galaxy with spiral arms extending beyond 8000 pc. (Of course under 'optical spiral arms' we understand always the distinct features above the general galactic light background, and not the small detached groups of stars. The spiral arms are more or less continuous.)

No spiral arms can be expected in the direction of anticentre in our Galaxy and less so up to 14 000 pc.

Some concentration of objects called the Orion arm does not represent a regular arm.

My doubt of the reality of the spiral structures in the solar vicinity is supported by the existence there of differential rotation. Radio observations of other galaxies made far from their centres supported my former conclusion that the distinctly visible spiral arms are situated in the region of a rigid body rotation. So they can hardly exist farther than the solar distance from the centre. In M 31 the differential rotation is observed in the region of the spiral arms. But this is an exception and Sandage holds even M 31 as a multiarm galaxy as NGC 5055 and 2841. Long ago I departed from the notion that the differential rotation destroys the regular shape of logarithmic spirals. However Marochnik and Suchkov claim that in their theory differential rotation and arms may coexist.

It is not proper to base the spiral arms on the 'coincidence' of localisation of hot stars and of H II regions. In fact we locate H II regions just where we locate the hot stars tentatively responsible for their luminosity.

The discussion of the 21-cm observations for the location of spiral arms uses a law of rotation which is postulated beforehand. Therefore the locations mentioned are not independent. On the other hand the broad ring of H I (this is just what really is seen on the map of densities) corresponds probably to the rings of H I described by M. Roberts. In many galaxies (not in M 31) the optical spiral arms are inside the H I ring.

The thin spiral arms, or rather the rings observed optically in NGC 488 are very rare. But when they are present they never make so many revolutions as are needed to extend them from the nucleus to a distance of 12–14 kpc.

I suggest not to be in a haste to construct a model of our Galaxy but to search the real patterns without bias.

# 4. STATISTICS OF SPIRAL PATTERNS AND COMPARISON OF OUR GALAXY WITH OTHER GALAXIES

## G. DE VAUCOULEURS

*University of Texas, Austin, Tex., U.S.A.*

**Abstract.** Apparent relative frequencies of various types of spirals are given for 900 spirals with the best revised Hubble classifications. Mean diameters of inner ring structures vary from 2.1 kpc in ordinary spirals (SA) to 4.4 kpc in barred spirals (SB) with a total range of 10 to 1 within each type. The probable morphological type of our galaxy is estimated from 6 criteria (multiplicity of spiral pattern, inner ring diameter, broken ring structure, radio structure of nucleus, Yerkes type, HI diagram); arguments advanced in 1963 for an SAB(rs) structure of the inner regions of the galactic system are strengthened by this analysis. Several examples of galaxies in this area of the classification plane are discussed.

1. Spiral structure is observed in about 60% of the brighter galaxies for which classification is available on the revised Hubble system (de Vaucouleurs 1959, 1963; Sandage 1961). In this 3-dimensional system two *families* A, B and two *varieties* r, s (with transition types AB, rs) are distinguished at each *stage* of the sequence from early (S0/a) to late (Sm) through a, ab, b, ...d, dm for a total of 10 stages. The multiplicity of the spiral pattern is correlated with the absence or presence of a bar (A vs. B) and with the presence or absence of an inner ring structure (r vs. s), being highest at SA(r) and lowest at SB(s) as illustrated in Figure 1. Examples of each type have been described and illustrated elsewhere (de Vaucouleurs 1959, 1963; Sandage 1961) and revised types given on this system for 2300 objects (G. and A. de Vaucouleurs, 1964). Among the 1500 objects with the best revised classification (de Vaucouleurs, 1963), about 900 are spirals and among them the relative frequencies of the main types (all stages combined) are listed in Table I. A more detailed tabulation of relative frequencies as a function of stage along the sequence also appears in Table I. These apparent frequencies are probably affected by selection effects and bias; in particular it is almost certain that the abundance of ordinary spirals SA(s) is overestimated by inclusion of objects which with higher resolution or more favorable orientation would have been recognized as A(rs), AB(s) or AB(rs). Similarly some objects of type AB(s) or B(rs) must have been lumped with the regular barred spirals SB(s). The classification space is, of course, a continuum related to dynamic and physical evolution, but the classification forces an arbitrary division into discrete cells and in the process some types, more easily recognized, tend to be favored. Similar problems arise in spectral classification work without vitiating its significance. Again the abundance of ringed types (r), from A(rs) to A(r) and AB(r), must be underestimated because the inner ring structure in these types has a rather small diameter and may not be detected if the orientation is unfavorable or the resolution insufficient. The average diameters of the inner ring structures in spiral galaxies of different types are given in Table II (de Vaucouleurs and Schultz, 1970) for 212 galaxies whose distances may be estimated from membership in groups (de Vaucouleurs, 1966).

*Becker and Contopoulos (eds.), The Spiral Structure of Our Galaxy, 18–25. All Rights Reserved.*

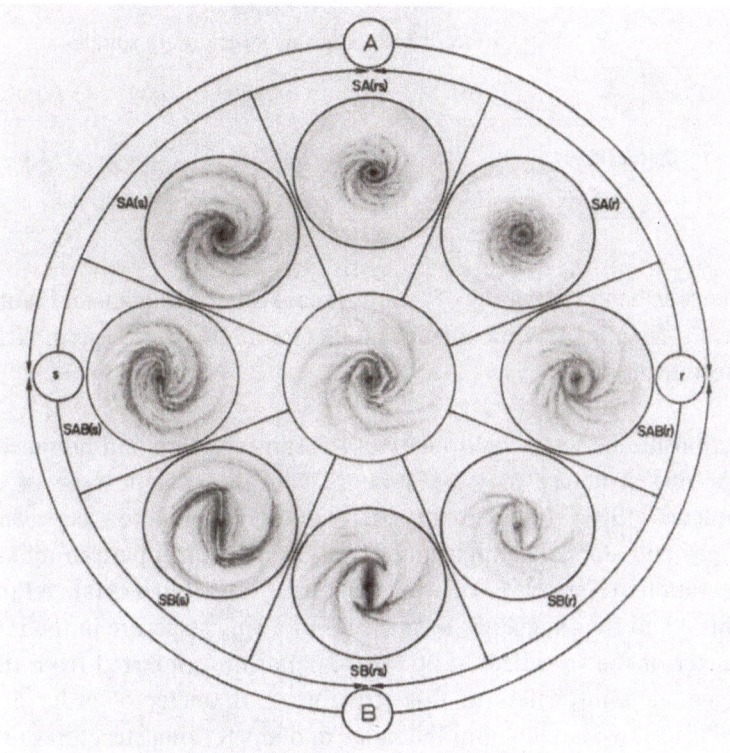

Fig. 1. Cross-section of classification volume at stage Sb–Sc showing main families (A, B) and varieties (r, s) of spiral pattern with transistion types (AB, rs).

TABLE I

Apparent relative frequencies of spiral types (%)

| Type | Stage | | | | | |
|------|-------|-----|-----|-----|-----|-----|
| | a | b | c | d | m | All |
| | 0, 1 | 2, 3 | 4, 5 | 6, 7 | 8, 9 | |
| A(s) | 21.1 | 13.6 | 22.7 | 23.0 | 12.0 | 19.8 |
| AB(s) | 8.0 | 9.0 | 5.3 | 8.9 | 16.0 | 8.2 |
| B(s) | 28.5 | 16.0 | 10.1 | 23.7 | 56.0 | 19.9 |
| B(sr) | 11.0 | 10.9 | 7.7 | 12.6 | 14.0 | 10.6 |
| B(r) | 11.0 | 16.5 | 11.0 | 5.9 | 0 | 8.6 |
| AB(r) | 5.1 | 8.6 | 5.7 | 2.2 | 0 | 5.7 |
| A(r) | 5.1 | 7.4 | 3.6 | 0.7 | 0 | 4.4 |
| A(rs) | 2.9 | 9.3 | 9.2 | 0 | 0 | 6.6 |
| AB(rs) | 7.3 | 8.6 | 24.7 | 23.0 | 2.0 | 16.3 |

### TABLE II

Mean diameters of ring structures in spirals

| Type | A(rs) | A(r) | AB(r) | B(r) | B(rs) | AB(rs) |
|---|---|---|---|---|---|---|
| (r) diam. (kpc) | 2.2 | 2.1 | 3.4 | 4.4 | 4.3 | 3.5 |
| n | 26 | 17 | 41 | 50 | 38 | 40 |

The distance scale in the survey of groups is consistent with a mean Hubble constant $H = 110$ km s$^{-1}$ Mpc$^{-1}$ (de Vaucouleurs and Peters, 1968), but was established without reference to redshifts.

2. The relationship between multiplicity of the spiral pattern and presence or absence of an inner ring structure may be used to infer the galactic type of our Galaxy (de Vaucouleurs, 1964) provided there is a general correlation between the distributions of gas and stars. The high multiplicity of the spiral pattern indicated by the radio observations (Kerr, 1967) is not consistent with types A(s), AB(s), B(s) and probably B(rs) and thus suggests the presence of a ring structure in the inner regions. If the diameter of the so-called 3-kpc (or 4-kpc) arm, as inferred from the longitude of the tangential point, refers to this structure, a diameter of order 6 to 8 kpc is indicated. This value is well within the range of diameter ring structures in types B(rs), B(r), AB(r) and AB(rs), which vary by a factor 3 either way from the mean values in Table II, but is far in excess of the observed diameters in types A(rs) and A(r). Further, if the galactic ring is incomplete, as seems to be the case from the radio data, the structure is consistent with types B(rs), AB(rs) and possibly A(rs) (if not excluded by diameter considerations), but not with A(r), AB(r) or B(r) – always with the assumption that gas and stars do not form entirely different patterns.

3. The outward velocity of the gas observed in absorption at $-53$ km s$^{-1}$ in the 21 cm line profile of Sgr A has been interpreted as evidence for gas streaming along a bar tilted by about 30 to 45° to the sun-center line (de Vaucouleurs, 1964), a phenomenon that has been observed in several barred spirals (G. and A. de Vaucouleurs, 1963; de Vaucouleurs et al., 1968). The basic structure of a broken inner ring and bar structure for the inner regions of our Galaxy (Figure 2a) was first described at the Canberra symposium in 1963 (de Vaucouleurs, 1964). Recently the same basic scheme was adopted by Kerr (1967) to interpret related, but relatively independent data on gas motions slightly above or below the galactic plane (Figure 2b). Then again, from a comparison between radio observations of the galactic center and optical appearance of the nuclei of other galaxies Cameron (1968) has concluded that "our Galaxy is likely to be of type SAB or SB" and that "it is clearly inappropriate to accept M 31 as a counterpart for our Galaxy", in full agreement with the earlier conclusions of Arp and the writer (de Vaucouleurs, 1964). Finally Morgan (1962) and Morgan and Osterbrock (1969) used his classification scheme to evaluate spectral and other proper-

ties of our Galaxy and suggested that both NGC 4501 and NGC 4216 closely match our Galaxy; NGC 4501 was classified type SA(rs) b and NGC 4216 type SAB(s) b in the revised system. This may be compared with my proposed classification of our Galaxy as intermediate between NGC 4303, type SAB(rs) b and NGC 6744, type SA(r) bc (de Vaucouleurs, 1964). All these comparisons and proposed identifications are in the same general area of the central cross-section of the classification volume at stage Sb or bc and definitely exclude the ordinary two-armed spirals of the SA(s) type exemplified by M 31, M 51 and M 81.

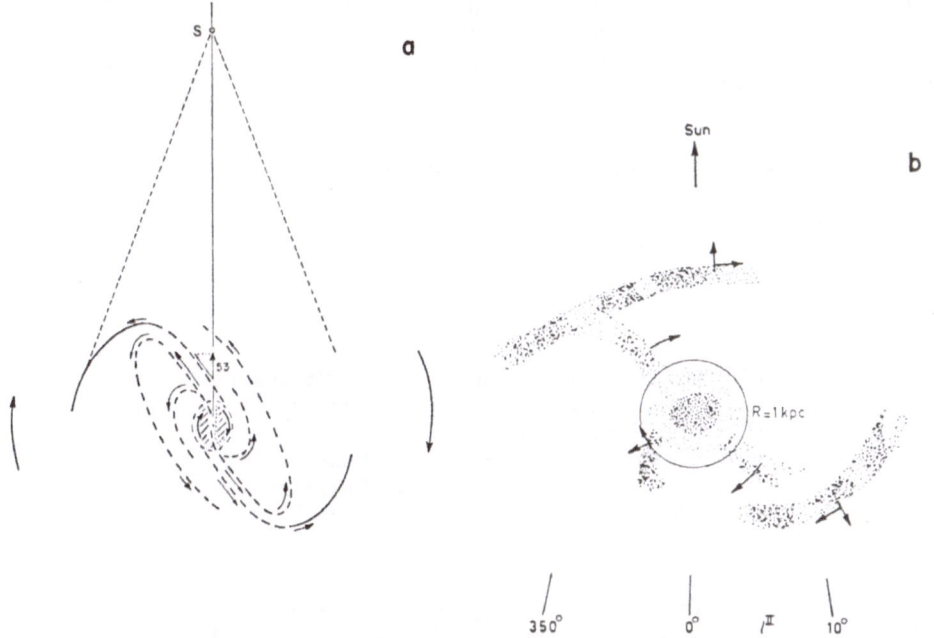

Fig. 2.    (a) Schematic SAB structure in inner regions of Galaxy inferred from radio and optical data (de Vaucouleurs, 1964). (b) Sketch of galactic hydrogen spiral structure (Kerr, 1967) showing SB(rs) structure in inner regions.

Table III summarizes in matrix form the probability that our Galaxy belongs to a given type according to a given criterion, i.e. a 'rating' of each of the main galaxy types with respect to each of the several criteria. The total 'scores' of each type, given by the sums in the last column, define contours of equal probability in the classification plane (Figure 3) and restrict to a fairly small domain of this plane the possible morphological types of our Galaxy which agree best with all classification criteria. The highest scores in Table III are at types SAB(rs) and SB(rs) and confirm that on present evidence the most probable identification of our Galaxy is with type SAB(rs) bc. This classification corresponds to definite morphological features that have been described earlier (de Vaucouleurs, 1958, 1959, 1964) and it should not be construed as being merely an expression of total uncertainty (Burbidge, 1967); short of deliberate misrepresentation it is difficult to see how such gross misinterpretation can arise.

## TABLE III

### Classification of Galaxy from several criteria*

| Criterion<br>Source | (1)<br>[a] | (2)<br>[a, b] | (3)<br>[a, b] | (4)<br>[c] | (5)<br>[d] | (6)<br>[e] | Sum |
|---|---|---|---|---|---|---|---|
| A(s)   | −1 | −1 | −1 | −1 | 0  | −1 | −5 |
| AB(s)  | −1 | −1 | −1 | −1 | +1 | −1 | −4 |
| B(s)   | −1 | −1 | −1 | 0  | 0  | −1 | −4 |
| B(rs)  | 0  | +1 | +1 | 0  | 0  | +1 | +3 |
| B(r)   | 0  | +1 | 0  | +1 | 0  | 0  | +2 |
| AB(r)  | +1 | +1 | 0  | 0  | 0  | 0  | +2 |
| A(r)   | 0  | 0  | 0  | −1 | 0  | −1 | −2 |
| A(rs)  | 0  | 0  | +1 | −1 | +1 | −1 | 0  |
| AB(rs) | +1 | +1 | +1 | 0  | 0  | +1 | +4 |

* Probability rating that type is consistent with criterion:

−1: Excluded or improbable,
  0: Possible, but not probable,
+1: Possible and probable.

Criterion:

(1) Multiplicity of spiral pattern
(2) Inner ring diameter
(3) Broken ring structure
(4) Radio structure of nucleus
(5) Yerkes type
(6) HI diagram

Source:

[a] de Vaucouleurs (1964)
[b] this paper
[a, b]
[c] Cameron (1968)
[d] Morgan (1962, 1969)
[e] Kerr (1968)

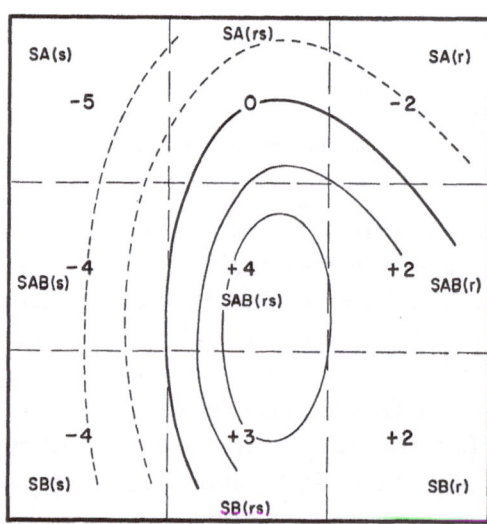

Fig. 3. Equi-probability contours in classification plane from total scores in Table III. Most likely classification of our Galaxy is in area within contour + 3 including types SAB(rs) and SB(rs); possible but less likely types on present evidence are inside contour 0. Most unlikely is ordinary spiral type SA(s).

Examples of galaxy types in the area of interest of the classification plane are illustrated in Figure 4. NGC 6744 is the type example for SAB(r) bc, it has a broad diffuse nucleus, weak bar, relatively faint inner ring and many filamentary arms, but the pattern has lower multiplicity than in type SA(r) such as NGC 7217. NGC 5921 is a good example of type SB(r) bc, it has a very bright, sharp elliptical nucleus, a strong bar and complete elliptical ring (not a projection effect); two main arms emerge

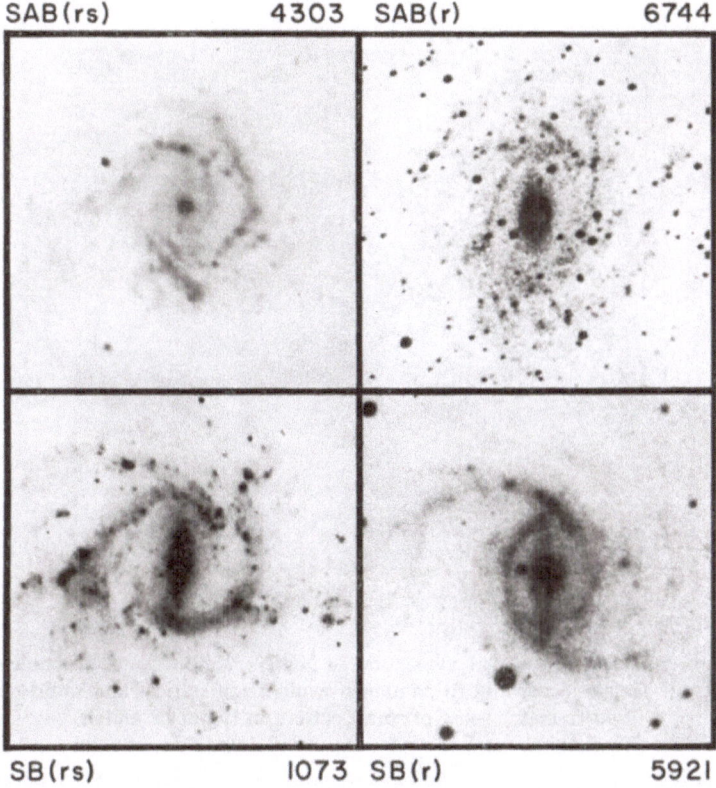

Fig. 4.   Typical examples of Sb or bc spiral galaxies in the positive region of classification plane (Figure 3) reproduced on uniform scales of ring structure. Orientation was reversed where necessary to make sense of spiral pattern consistent with Figure 2. Photographs are from McDonald 82-inch (4303), Mt. Stromlo 30-inch (6744), Palomar 200-inch (1073), Haute Provence 48-inch (5921).

from the ring near the ends of the bar and two weaker ones are branching out near the minor axis of the ring. NGC 1073, type SB(rs) c, has also a strong bar, but the ring is now broken and opens near the minor axis while the arcs near the ends of the bar or major axis of the ring are asymmetrical; one forms the origin of one arm, the other of two arms of which one is weaker. NGC 4303 is the type example of SAB(rs) b, it has a small bright nucleus in the center of a broad bar with complex dark lanes (running mainly parallel to the bar as in barred spirals but less closely so than in a typical SB system); the inner pseudo-ring – almost an hexagon – is formed by straight segments of three main spiral arms branching out into the fainter spiral structure of the outer regions. NGC 1232, type SAB(rs) c, illustrated in the Hubble Atlas (Sandage, 1961) and in which three main arms emerge from a short bar is a transition type between SA(s) and SAB(rs). NGC 7424, type SAB(s), (see photograph in G. and A. de Vaucouleurs, 1961) which has also a short bar but only two main arms is a transition type between SA(s) and SAB(s), the latter exemplified by NGC 5236 (also in Hubble Atlas). A sketch of the central regions of NGC 4303 which illustrate best the most probable structure of our Galaxy is shown in Figure 5.

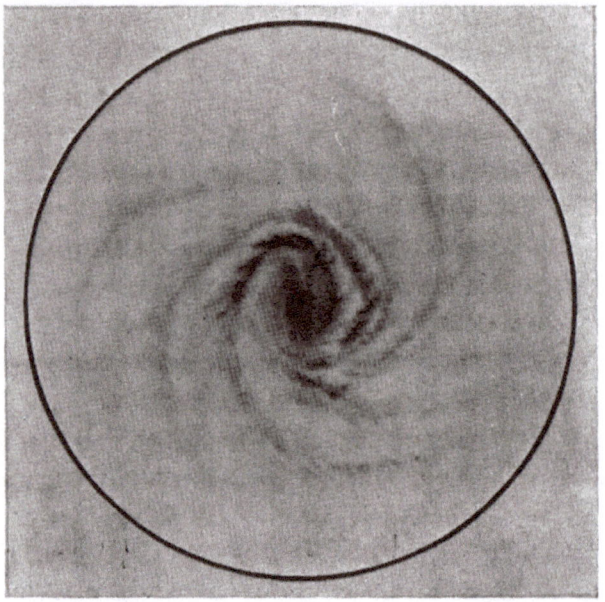

Fig. 5.   Sketch of typical SAB(rs)bc galaxy, similar to NGC 4303, illustrates probable structure of inner regions of our Galaxy according to combined evidence of optical and radio data. Print is reversed to match sense of spiral pattern in Figures 2 and 4.

This study is part of the McDonald Observatory extragalactic research program supported by the National Science Foundation and the Research Institute of the University of Texas.

### References

Burbidge, E. M.: 1967, IAU Symposium No. 31, p. 209.
Cameron, M. J.: 1968, *Observatory* **88**, 254.
Kerr, F. J.: 1967, IAU Symposium No. 31, p. 239.
Morgan, W. W.: 1962, *Astron. J.* **67**, 788.
Morgan, W. W. and Osterbrock, D. E.: 1969, *Astron. J.* **74**, 515.
Sandage, A.: 1961, *The Hubble Atlas of Galaxies*, Carnegie Institution of Washington, Washington.
Vaucouleurs, G. de: 1958, *Rev. Mod. Phys.* **30**, 926.
Vaucouleurs, G. de: 1959, *Handbuch der Physik* **53**, 275.
Vaucouleurs, G. de: 1963, *Astrophys. J. Suppl. Ser.* **8**, 31.
Vaucouleurs, G. de: 1964, IAU-URSI Symposium No. 20, pp. 88 and 195.

Vaucouleurs, G. de: 1966, 'Nearby Groups of Galaxies', to be published in *Stars and Stellar Systems*, **9**.

Vaucouleurs, G. and A. de: 1961, *Mem. Roy. Astron. Soc.* **68**, 69.

Vaucouleurs, G. and A. de: 1963, *Astron. J.* **68**, 278.

Vaucouleurs, G. and A. de: 1964, *Reference Catalogue of Bright Galaxies*, Univ. Texas Press, Austin.

Vaucouleurs, G. de and Peters, W. L.: 1968, *Nature* **220**, 868.

Vaucouleurs, G. de and Schultz, P.: 1970, 'Ring Structures in Galaxies as Distance Indicators', unpublished.

Vaucouleurs, G. and A. de and Freeman, K. C.: 1968, *Monthly Notices Roy. Astron. Soc.* **139**, 425.

# 5. THE DISTRIBUTION OF DARK NEBULAE IN LATE-TYPE SPIRALS

B. T. LYNDS

*Steward Observatory, University of Arizona, Tucson, Ariz., U.S.A., and*
*Institute of Theoretical Astronomy, Cambridge, England*

**Abstract.** Seventeen Sc- galaxies have been studied in order to determine the distribution of dark nebulae within them. Hα plates were used to compare the distribution of obscuring material with the location of H II regions, and it was found that the H II regions are always tangent to or imbedded in regions of high obscuration. Sandage's conclusions are confirmed that strong, regular dust lanes exist in the central regions of these galaxies and wind outward on the inside edge of the two prominent arms. Characteristic dimensions of the dust lanes are noted, and secondary dust characteristics are described.

The most detailed notes on the distribution of interstellar dust in galaxies have been given by Sandage and are to be found in the description of the Hubble classification and in the captions for the illustrations of the Hubble Atlas (Sandage, 1961). In this publication, Sandage points out that the spiral pattern for Sc-galaxies first becomes apparent near the center of such systems as two principal dust lanes; luminous spiral arms appear only at larger distances from the nucleus, at which distances the two principal dust lanes wind out along the inside of the two most luminous outer arms. After winding out for nearly half a revolution, these two principal arms branch into segments which continue to spiral outward to form the multiple-arm structure. Furthermore, Sandage points out that the dust is not confined to these lanes but rather is seen in spiral patterns over the entire face of such systems.

Several detailed studies of individual galaxies have been made in which the distribution of the interstellar dust is discussed, but the emphasis has usually been on the determination of the sense of direction of the motion of spiral arms. Other than the report of Sandage, no detailed studies of the sizes and locations of dark nebulae in galaxies are known to the author. Because of the recent revival of interest in the role played by the dust, a systematic analysis of the sizes and distributions of the dark material in external galaxies should be of interest.

Through the courtesy of the Mount Wilson and Palomar Observatories, the plate collection of 100- and 200-inch photographs of galaxies constitutes the raw material for this study. All galaxies in the Shapley-Ames Catalogue were visually inspected, and a sample of these objects was selected for detailed measurement. The $x$-, $y$-coordinate measuring machine of the Mount Wilson Observatory was used to scan across each sample galaxy and to record the positions of the dark and bright spiral features. This report is confined to results obtained from a study of seventeen late-type spirals, classified as Sc- by Holmberg (1958).

Although interstellar dust is most easily detected in spiral systems whose plane of symmetry lies nearly in the line of sight, it is very difficult to determine the exact location of the dust in such systems. Therefore, for this study, the galaxies selected

*Becker and Contopoulos (eds.), The Spiral Structure of Our Galaxy, 26–34. All Rights Reserved.*
*Copyright © 1970 by the I.A.U.*

are those seen nearly face-on for which there is some degree of certainty of identification of the dust clouds. The results given here therefore represent data for those Sc-galaxies in which the dust is readily apparent and relatively easily measured. It is a happy coincidence that such Sc galaxies appear to be the most commonly observed and may reflect average properties of Sc-systems in general in spite of the fact that the sample of 17 studied was in no sense unbiased.

The usual criteria for identification of a dark nebula in any stellar system are firstly, the absence of stars relative to the surrounding area, and secondly, the ability to detect a relatively well-defined boundary of the nebula. In the case of an external spiral galaxy, one can often use these two criteria with as great a certainty as one can in our own Milky Way System. The identification becomes more difficult in the inter-arm regions of spirals and in the outer zones of a galaxy beyond its luminous arms. An effort was made to confine the data to dark regions whose presence would be difficult to interpret other than to say that they are true obscuring clouds.

It was initially hoped that the brightest H II regions of a galaxy would define the spiral pattern against which the positions of the dark nebulae could be compared. Therefore, in this preliminary report, those galaxies (which met the first requirement of containing easily-detected dark clouds) which have available Hα interference-filter photographs in the Mt. Wilson and Palomar plate collection were selected. Perhaps the most interesting result of this study is the confirmation of the oft-quoted statement that there is a 'one-to-one correspondence' between the presence of dust and the presence of young stars. On the basis of the sample studied here, it is concluded that the brightest H II regions of a galaxy are always found either next to or imbedded in regions of high obscuration. The converse of the statement is not true; many regions of equally dense obscuration contain no H II regions. It is also the conclusion of this study that the primary dust lanes of a galaxy better define the spiral pattern than do the H II regions, which often are found on the inside edge of a bright spiral arm but may be found on the outside edge or somewhere within the luminous arm where a 'feather' of a dark lane cuts across the arm. Occasionally, H II regions are even found in heavily-obscured inter-arm areas (if the arms are defined by the regions of relatively uniform luminosity).

Figures 1–5 contain the distribution of the dark nebulae within those galaxies for which Hα photographs were available. The lightly-shaded areas represent the dust lanes of the galaxy; the heavy black marks are the locations of the brightest H II regions.

When one studies the sketches of Figures 1–5; or, better, when one examines the photographs themselves (see, for example the prints of the Hubble Atlas), certain regular patterns of the distribution of dust become apparent. The two most striking features are those pointed out by Sandage: the existence of very strong dust lanes in the central regions and the presence of dark lanes along the inside edge of the two luminous arms. These latter primary dust lanes can occasionally be traced through more than 360°, that is, through a 'first' and 'second' winding of the arms. In addition to these characteristics, thin dust lanes are also found cutting their way across a bright arm with pitch angles of about 50° relative to the primary lanes.

Once such 'feathers' have been traced across a luminous arm, they are no longer visible against a bright background. There is, however, some evidence that an extensive area beyond a luminous arm also contains dust condensations. Another, smaller scale, characteristic of dark nebulae in these systems is the appearance of nearly

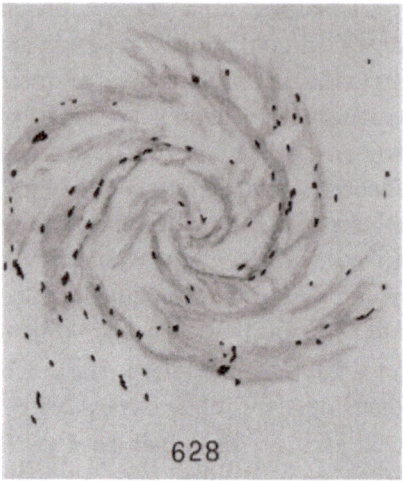

Fig. 1.   NGC 628. This is the most regular of the Sc-galaxies studied. The two primary dust arms can be relatively easily traced through a first and second winding. The luminous arms are found sandwiched between the dark nebulae of the sketch.

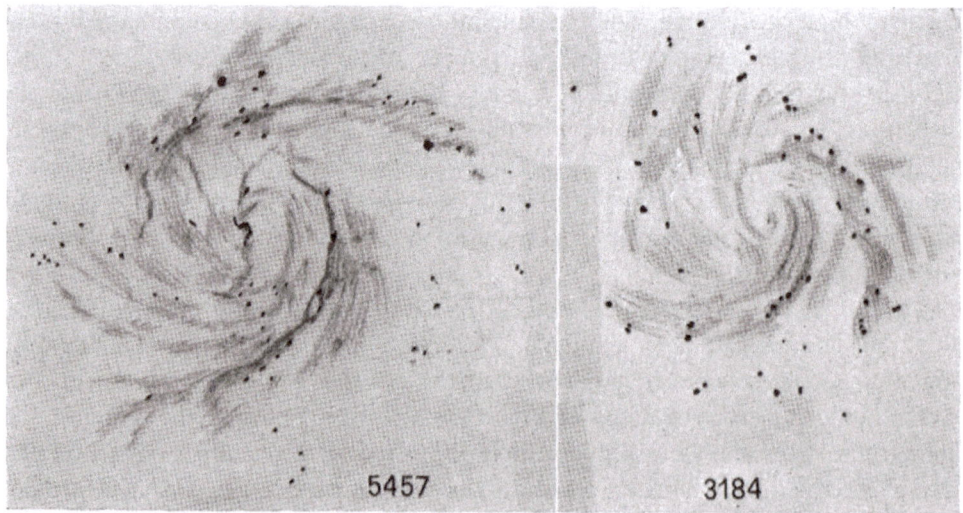

Fig. 2.   NGC 5457. M 101 is the prototype of the special subgroup of Sc-galaxies which shows the most regular dust patterns. The luminous arms are thinner than those of NGC 628, which belongs to the same subgroup. Sandage has pointed out that in M 101 the dust lanes appear as separate segments which combine to form a spiral pattern. NGC 3184: No Hα interference filter photograph was available for this galaxy, but the H II regions listed by Hodge (1969) were identified on the 100-inch 103aO photograph. Hodge's data were based on 48-inch Schmidt photograph, so that the central regions are incomplete in the H II identifications. The dust in this galaxy is more difficult to detect.

circular clouds which often contain a very bright central knot. These bright knots are frequently not H II regions but may be one or more bright blue stars.

Table I lists the representative sizes of the characteristic interstellar features. The distances $(m-M)$ and radii $(R)$ were taken from Holmberg (1964). The galaxies are

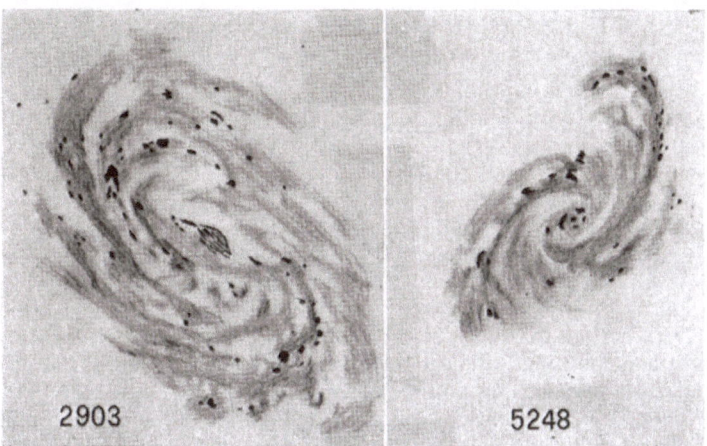

Fig. 3. NGC 2903. These two galaxies are viewed more highly inclined to the line of sight. NGC 2903 belongs to the NGC 253 subgroup of Sc-galaxies which has ill-defined spiral arms and more chaotic dust. For the subgroup, spiraling dark lanes are still quite apparent over the entire galaxy. – NGC 5248. This galaxy is of the NGC 1637 type, whose luminous arms are not as well-defined as those of M 101 subgroup. Nevertheless, this subgroup shows the same characteristic dust pattern as does the M 101 type.

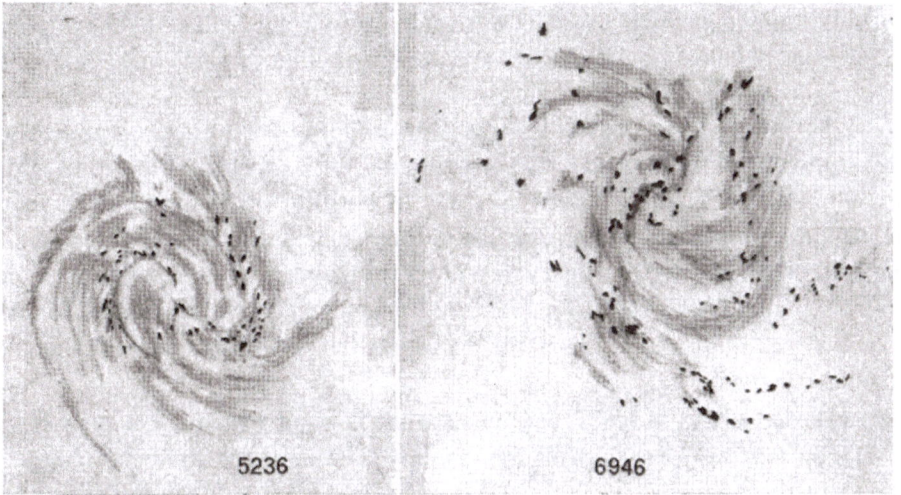

Fig. 4. NGC 5236. The spiraling dust segments in this very open galaxy are its most dominant feature. The arms appear to be exceptionally rich in H II regions and are quite obviously closely associated with the dust. – NGC 6946. The dust pattern in this galaxy is similar to NGC 5236 but more difficult to measure and the luminous arms are more difficult to trace. The 48-inch Schmidt print of this galaxy strongly suggests that obscuration exists in the outer regions beyond the luminous primary lanes, as sketched in the figure.

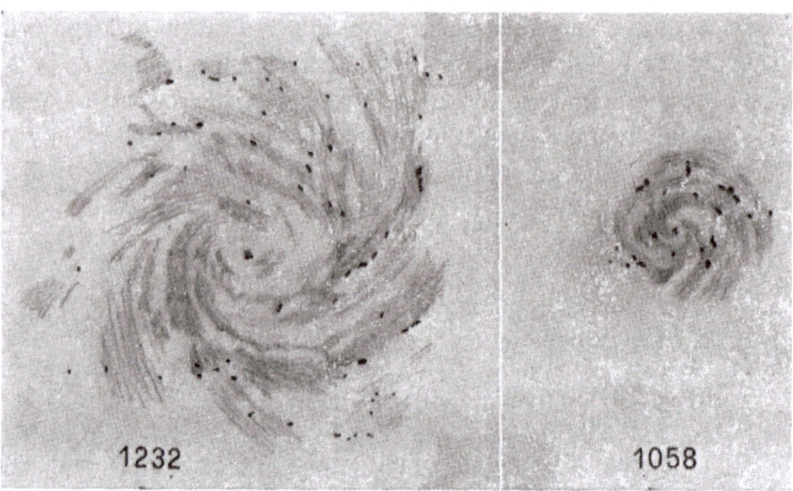

Fig. 5.   NGC 1232. This galaxy belongs to the NGC 5364 type in which the primary arms begin
tangent to an inner luminous ring. Although there is no complete inner ring in this object, the arms
begin tangent to the central region, as do the primary dust lanes. The dust lanes in this galaxy are
difficult to trace but appear to lie both on the inside and outside of the easily-seen bright arms. –
NGC 1058. Holmberg's radius of this galaxy is 6.6 kpc, which makes it the smallest galaxy in this
study. The primary dust lanes produce the spiral appearance of this galaxy.

divided into the Sc subgroups defined by Sandage (1961). Five of his six major
subgroups are represented in the table; only the NGC 4395-type (having very chaotic
and ill-defined spiral arms) is not included. NGC 2403, like M 33, has obvious dust
patches, but the dominant spiraling dark lanes of the other types of Sc systems are
not readily apparent in this type. The remaining four subgroups all show spiraling
dark lanes as the most obvious pattern of the distribution of dust, with the regularity
increasing from type NGC 253 to NGC 1637 to M 101. The sixth subgroup, having
NGC 5364 as its prototype, has its two primary spiral arms starting tangent to an
external circular luminous ring. Here again, the dust appears in the form of dark
lanes either on the inside or on the outside (or both) of the luminous arms, but the
dark nebulae are more difficult to detect in these galaxies. Columns 4, 5, and 7 give
the measured widths of the primary dust lanes for the nuclear regions ($n$), 1st winding
and 2nd winding distances, respectively; columns 6 and 8 give the respective nuclear
distances ($d$). Columns 8, 9, and 10 list the widths ($w$), lengths ($l$), and pitch angles ($i$)
of the dark feathers.

Ten galaxies of the M 101 subgroup were included in this study. Figure 6 shows
the relation between the width of the primary dust lane and its distance from the
nucleus for these galaxies. Both quantities are expressed in units of the radius of the
galaxy. It appears that there is a general increase in thickness of the primary dust lane
as it winds its way out from the nucleus; the overlapping of the points representing
first (filled circles) and second (open circles) winding positions indicates that the width
depends more on the nuclear distance than on the tightness of the winding of the arms.
The lower portion of Figure 6 shows that the measurements of the width of the primary

TABLE I

Dust measurements in Sc-galaxies

| NGC | m–M | R | Primary dust lanes | | | Feathers | | | | | Plates[a] |
|---|---|---|---|---|---|---|---|---|---|---|---|
| | | | n | 1st | d | 2nd | d | w | l | i | |
| | | kpc | pc | pc | kpc | pc | kpc | pc | kpc | | |
| **M 101-Type** | | | | | | | | | | | |
| 157 | 31.7 | 18 | – | 350 | 2.8 | 900 | 7 | 160 | 1.0 | 75° | 103aO |
| 628 | 29.7 | 15 | 60 | 100 | 1.1 | 530 | 3 | 60 | 0.5 | 50 | 103aO, D, E; Hα |
| 2532 | 34.0 | 35 | – | 300 | 2.0 | – | – | – | – | – | 103aO |
| 3184 | 29.3 | 10 | – | 80 | 0.4 | 190 | 2 | – | – | – | 103aO (100-inch) |
| 4254 | 30.5 | 13 | – | 250 | 1.2 | 320 | 2 | 80 | 0.8 | 45 | 103aE |
| 4303 | 30.5 | 20 | 130 | 100 | 2.4 | – | – | – | – | – | 103aO |
| 4321 | 30.5 | 18 | 120 | 270 | 2.4 | 270 | 5 | 130 | 0.9 | 45 | 103aO, D, E |
| 5236 | 28.2 | 8 | 70 | 200 | 1.4 | 300 | 3 | 40 | 0.5 | – | E40 (100-inch), Hα |
| 5457 | 27.7 | 14 | 30 | 50 | 0.6 | 200 | 2 | 40 | 0.2 | 60 | 103aD, E; Hα |
| 6946 | 28.4[b] | 9 | 40 | 150 | 1.2 | 540 | 3 | 50 | 0.4 | – | 103aO (100-inch), 103aE; Hα |
| **NGC 1637-Type** | | | | | | | | | | | |
| 1637 | 29.5 | 9 | | 100 | 0.8 | 130 | 1.3 | | | | 103aO |
| 5248 | 30.9 | 18 | | | | 210 | 4 | | | 55 | 103aO, Hα |
| 1058 | 29.4 | 7 | | 80 | 0.2 | 240 | 0.5 | | | | 103aO, D, E; Hα |
| **NGC 5364-Type** | | | | | | | | | | | |
| 864 | 31.5 | 13 | | 560 | 2. | 1000 | 4 | | | | 103aO |
| 1232 | 30.7 | 19 | | | | 640 | 2 | | | 40 | 103aO, Hα |
| 5364 | 30.1 | 12 | | 330 | 1.9 | 550 | 4 | | | | 103aO |
| **NGC 253-Type** | | | | | | | | | | | |
| 2903 | 29.5 | 16 | | ~100 | 3 | | | | | | 103aO, Hα |
| **M 33-Type** | | | | | | | | | | | |
| 2403 | 27.6 | 14 | dust patches ~ 100 pc segments of lanes of width ~ 30 pc | | | | | | | | 103aO, D, E |

[a] Plates are 200-inch photographs unless otherwise stated.
[b] Distance from redshift; diameter measured by Lynds.

dust lane are not distance-dependent for this sample. Most of the plates measured were 200-inch direct photographs, with a scale of 11.06″ per mm. It is possible to be fairly certain in the identification and measurement of a dark lane which has a thickness equal to or greater than 0.1 mm on the plate; the observational lower limit to the detection of a dark lane thus corresponds to about one second of arc. The measured thicknesses of the thin feathers of dark nebulae branching from the primary lanes do show a distance effect, as illustrated in Figure 7. The width of the feathers appears to be about one-half that of the primary lane, if only the upper part of Figure 7 is considered. However, the lower part of the figure suggests that although the measured feather widths appear to be greater than the observational lower limit for galaxies whose distance moduli are less than 30, beyond this distance of 10 Mpc the measurements represent only the selected wider lanes. The dashed line in the lower part of

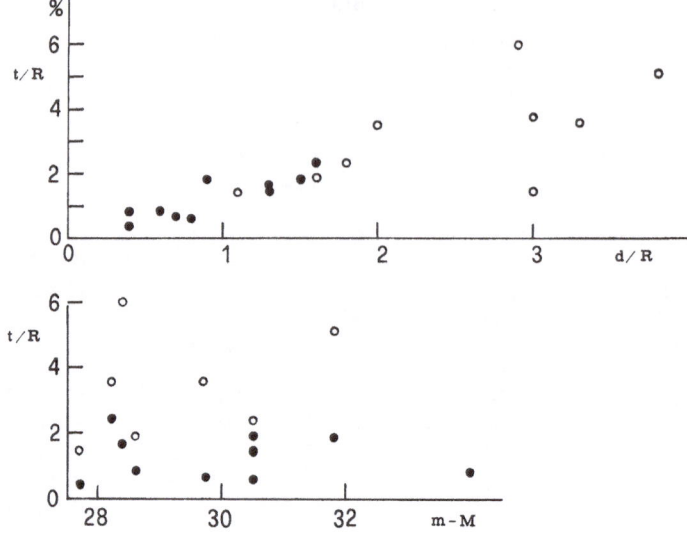

Fig. 6.   Variation in width of primary dust lanes with distance from the nucleus (upper) and with distance modulus (lower). The units are in terms of the radius of the galaxy. The ordinate is the percentage width of the primary lane; the upper abscissa is the fractional distance from the center of the galaxy. The lower array of points shows that the measurements are independent of distance for the ten M 101 galaxies studied. The filled circles are measurements of the first winding thicknesses; the open circles are the second winding measurements.

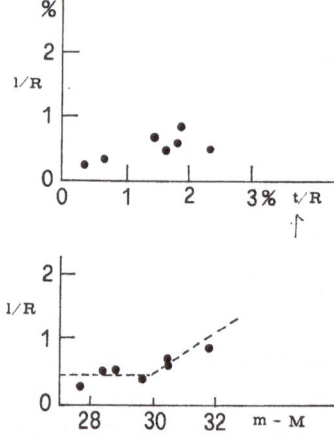

Fig. 7.   Variation in width of dust feathers with width of primary lanes (upper) and with distance (lower). The Units are percentage of radius of the galaxy. The dashed curve of the lower diagram represents the expected variation in the measured feather widths if a galaxy of 11 kpc had dark feathers of widths 50 pc and greater.

Figure 7 illustrates this resolution effect and defines the locus of observational points to be expected if a galaxy of radius 11 kpc had feathers of widths 50 pc and greater.

Table I also lists the emulsion sensitivities of the plates measured in this study. For some spirals, scans across the galaxy were made on O, D, and E plates. The accuracy of positional registration of these plates for intercomparison is about 30 microns, and to this accuracy no difference was found in the sizes of the dark nebulae over the wavelength regions represented by the red and the blue plates. This must mean that the nebulae selected to be measured are of very large optical thickness. Furthermore, in many regions of these galaxies, dark lanes can be seen occulting a luminous spiral arm which itself is several magnitudes above the plate limit. It is therefore concluded that the extinction through the dark clouds whose characteristic sizes are listed in Table I must be greater than about 3 magnitudes.

If 3 magnitudes is adopted as the lower limit to the optical extinction through the dark nebulae and if Holmberg's distances are assumed, then the actual physical sizes of the nebulae may be estimated. If we adopt 15 kpc as an average radius of a M 101-type Sc-galaxy, then Figure 7 may be used to estimate the width of the primary dust lanes. Near the nucleus the lane has a thickness of about 50 pc; the width of the lane gradually increases, reaching a value of about 240 pc at a distance of 2 kpc from the center. Unfortunately, we have no means of estimating the $z$ thickness of the cloud, which is the pathlength needed to correspond to the 3 magnitude extinction estimate. If we use the simple model in which the lane is as 'thick' as it is wide, the pathlength may be set equal to the measured width of the lane. In this case we can estimate the lower limit to the grain density within the primary lane. According to Solomon (1969), the grain density of a cloud of 3 magnitudes absorption is given by

$$\varrho_g = 10^{-4}/L \text{ g cm}^{-3},$$

where $L$ is the pathlength in cm through the cloud. For our model lane the grain densities must be greater than $7 \times 10^{-25} \text{ g cm}^{-3}$ for the nuclear regions and $4 \times 10^{-25} \text{ g cm}^{-3}$ at a distance of 2 kpc from the center. This assumes that the extinction is produced by the same type of particle which exists in our galaxy. If we further assume the usual 100:1 ratio of gas to dust, we find an interstellar density greater than about $10^{-23} \text{ g cm}^{-3}$ throughout the winding of the primary dust lane. In an analogous manner, the mass of a typical feather of obscuration must be greater than $10^6$ solar masses. Such a lane of dark nebulosity crossing a luminous arm meets the requirements of Spitzer and Schwarzschild (1953) who predicted similar interstellar clouds on dynamical arguments related to stellar velocity dispersions.

In summary, the seventeen Sc-galaxies studied here indicate that the spiral pattern of these galaxies is revealed by the distribution of the interstellar dust. The dominant lanes of obscuration evident in most of these systems probably have densities comparable to those of the luminous arms themselves. Although each galaxy on close inspection has many individual features peculiar to itself, the overall pattern of dust as defined in this study seems to be characteristic of such systems.

## Acknowledgements

The author is greatly indebted to Dr. Allan Sandage for his generous assistance in making the Palomar plate collection available for this study, and to Dr. Horace Babcock for permission to use the Mt. Wilson and Palomar Observatories equipment. The manuscript was prepared while the author was a visitor at the Institute of Theoretical Astronomy, and it is a pleasure to acknowledge the hospitality of Prof. F. Hoyle and his associates.

## References

Hodge, P. W.: 1969, *Astrophys. J. Suppl. Ser.* **18**, 73.
Holmberg, E.: 1958, *Medd. Lunds Astron. Obs.*, Ser. II, No. 136.
Holmberg, E.: 1964, *Ark. Astron.* **3**, 387.
Sandage, A.: 1961, *The Hubble Atlas of Galaxies*, Carnegie Institution of Washington, Washington.
Solomon, P.: 1969, private communication.
Spitzer, L. and Schwarzschild, M.: 1953, *Astrophys. J.* **118**, 106.

# 6. ANGULAR MOMENTA OF LATE-TYPE SPIRAL GALAXIES

N. HEIDMANN

*Observatoire de Meudon, Meudon, France*

**Abstract.** The angular momenta of galaxies may be evaluated from their photometry. The mass to light ratio in NGC 224 is discussed and the calculation is applied to two late-type spirals.

## 1. Introduction

In a previous study an attempt was made towards the investigation of specific angular momenta of spirals by the use of *indicative* specific angular momenta defined by

$$A_i = 0.10aV_m \tag{1}$$

where $A_i$ is in kpc km s$^{-1}$, $a$ is the 26.5 $m_{pg}$ per square arcsec photometric diameter in kpc and $V_m$ the maximum rotational velocity in km sec$^{-1}$ (Heidmann, 1969). The largest $A_i$ values are obtained mostly for the Sbc type, suggesting that there may be a relation between the angular momentum and the degree of development of spiral arms.

However other factors may be of importance and the relation between indicative specific angular momentum $A_i$ and real specific angular momentum $A$ has first to be closely investigated.

The evaluation of the angular momentum of a galaxy from the rotation curve is very sensitive to its shape after the turnover point. If $r_m$ is the turnover radius and $V_m$ the maximum rotational velocity, the specific angular momentum for a flat disk is:

$$A = kr_mV_m \tag{2}$$

where $k$ is a coefficient dependent on the shape of the rotation curve $V(r)$. For example, Toomre's (1963) calculations show that in two cases in which $V(2r_m)=0.9\ V_m$ or $0.7\ V_m$, $k$ differs by a factor 4. Similar results are obtained from Takase and Kinoshita's (1967) calculations.

As it is difficult to measure radial velocities after the turnover radius, we attempt to derive angular momenta from photometric profiles. With values of the mass to luminosity ratio, the brightness distribution gives the density distribution, which in turn gives the potential, then the force, the velocity and the angular momentum.

With respect to the axis of cylindrical coordinates $(r, \theta, z)$, the specific angular momentum of a galaxy is:

$$A = \frac{1}{M_T} \sum_{\text{particles}} r\Theta m, \tag{3}$$

where $M_T$ is the total mass, $\Theta$ the tangential velocity and $m$ the mass of a particle.

*Becker and Contopoulos (eds.), The Spiral Structure of Our Galaxy, 35–40. All Rights Reserved.*
*Copyright © 1970 by the I.A.U.*

A first summation on the particles in a small volume element changes $\Theta$ into $V_0$, the centroid velocity. For a stationary system with rotational symmetry, Jeans' (1922) equations may be used. The calculation is quite easy in case the galaxy is flat enough to assume that:

 (i) the velocity dispersions are negligeable,

 (ii) the circular velocity $V_c(r, z)$ is nearly independent of $z$ inside the system.

If the system has an equatorial plane of symmetry, the momentum is then:

$$A = \frac{4\pi}{M_T} \int_0^\infty r^{5/2} [K_r(r, 0)]^{1/2} \int_0^\infty \varrho(r, z)\, dz\, dr, \tag{4}$$

where $K_r(r, 0)$ is the radial force in the equatorial plane and $\varrho$ the density. Usually the main contribution to $A$ will arise from intermediate $r$ values, so that the two above assumptions have to be fulfilled especially for these $r$ values. Formula (4) will be applied to late type spirals.

## 2. Photometric Profiles

De Vaucouleurs (1962) has shown that the photometric profile of elliptical galaxies follows a '$r^{1/4}$' law and that the profile of late-type spirals follows an exponential law. He has also shown (1958) that the profile of an intermediate type spiral such as NGC 224 can be represented by the sum of a '$r^{1/4}$' profile due to the bulge (spheroidal component) and an exponential profile due to the disk (flat component). We may assume that in a first approximation the profile of a galaxy may be represented by the sum of two such profiles.

## 3. Mass to Luminosity Ratio

The global mass to luminosity ratio for ellipticals is 20 (Holmberg, 1964) and for late type spirals it is 4 (Heidmann, 1969); here the luminosities are corrected for absorption according to Holmberg (1964).

The mass to luminosity ratio $f(r)$ at various distances $r$ from the center of a galaxy has been worked out for the Large Magellanic Cloud by De Vaucouleurs (1960) using two Perek spheroids. It is constant for $r > 2°$; for $r < 2°$ it is smaller and decreases by two thirds for $r = 0$ according to this method; however, in this analysis, there was no rotation observation available for $r < 1.5°$. It seems that for this galaxy, made up practically of a flat component, $f(r)$ can be taken to be a constant in most of the object.

The value of $f(r)$ has also been given by Gottesman et al. (1966) for NGC 224 using a Brandt curve (Brandt and Scheer, 1965) with index $\frac{3}{2}$ to fit their 21-cm observations. It is reproduced in Figure 1. In a dynamical study of the central parts of NGC 224 up to $r = 10'$, Kinman (1965) obtained $f \sim 50$. When comparing his rotation curves with the one observed by Lallemand et al. (1960), we obtain $f = 25$. These values are plotted in Figure 1.

Though the quasistellar nucleus of NGC 224 might play a special role, Kinman's values suggest that $f$ is indeed large in the central parts of NGC 224, as for elliptical galaxies. It is possible that the $f(r)$ values of Gottesman *et al.* in the central regions are smaller because Brandt's formalism is not valid for a mass distribution with large axis ratios and the radiotelescope beam has the effect to lower the slope of the rotation curve.

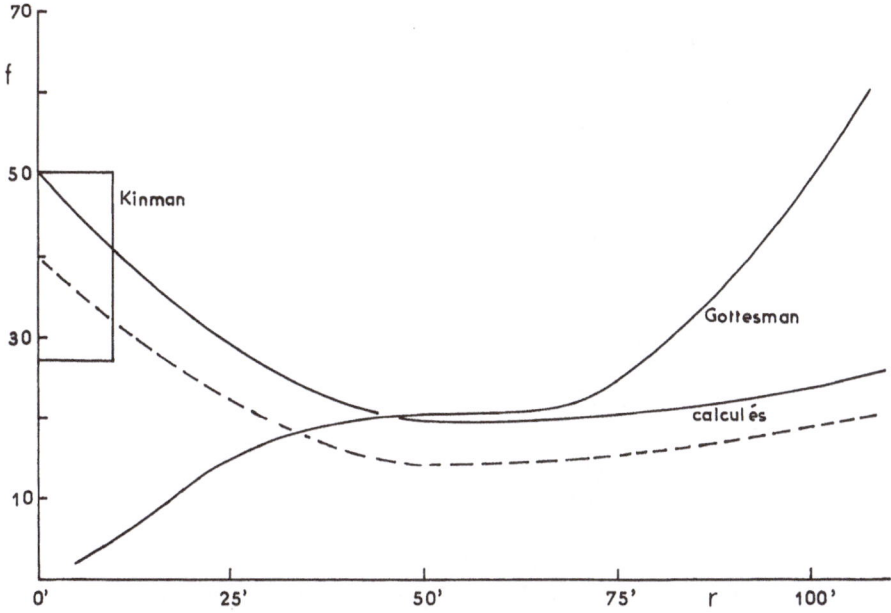

Fig. 1. Mass to luminosity ratio $f$, in solar units, vs. the distance from the center $r$ in arc min for NGC 224. The values from Gottesman *et al.* and from Kinman are plotted. The lines *calculés* are calculated for $f_s = 50$, $f_t = 15$ (full line) and for $f_s = 40$, $f_t = 10$ (broken line).

Using the two profiles given by De Vaucouleurs (1958) for the spheroidal and for the flat components of NGC 224 with the $f$ values $f_s = 50$ and $f_f = 15$ for each component respectively, we obtain the $f(r)$ full line curve labelled *calculés* in Figure 1. Note that the $f$ values here are not corrected for absorption. For a total absorption of 1 magnitude, assumed to be the same for the bulge and for the disk (cf. De Vaucouleurs 1958 for discussion), the corrected values would be 20 and 6. The broken line is for $f_s = 40$ and $f_f = 10$.

The agreement with Kinman's values for small $r$ and with Gottesman's values for intermediate $r$ is satisfactory. For large $r$, Gottesman's values are larger by about a factor 2. This may be due to an overestimation of the surface density as it appears from comparison with the two best solutions which Roberts (1966) obtained from his observations. Figure 2 is a plot of the surface density $\sigma(r)$ and shows that Roberts' values are 2 or 4 times smaller than Gottesman's in the region 70–110′ (and are larger for small $r$ values). This difference may be due to beamwidth and to bandwidth effects and to the fact that Gottesman *et al.* use mean velocities while Roberts uses corrected peak velocities which are closer to rotational velocities. Roberts' values for a Brandt curve with index $\frac{3}{2}$ are in agreement with the $f(r)$ curve calculated with the hypothesis of constant values for $f_s$ and $f_f$.

It may then be assumed that the mass to luminosity ratio $f$ of each of the spheroidal and flat components of a galaxy is the same at all point of each component.

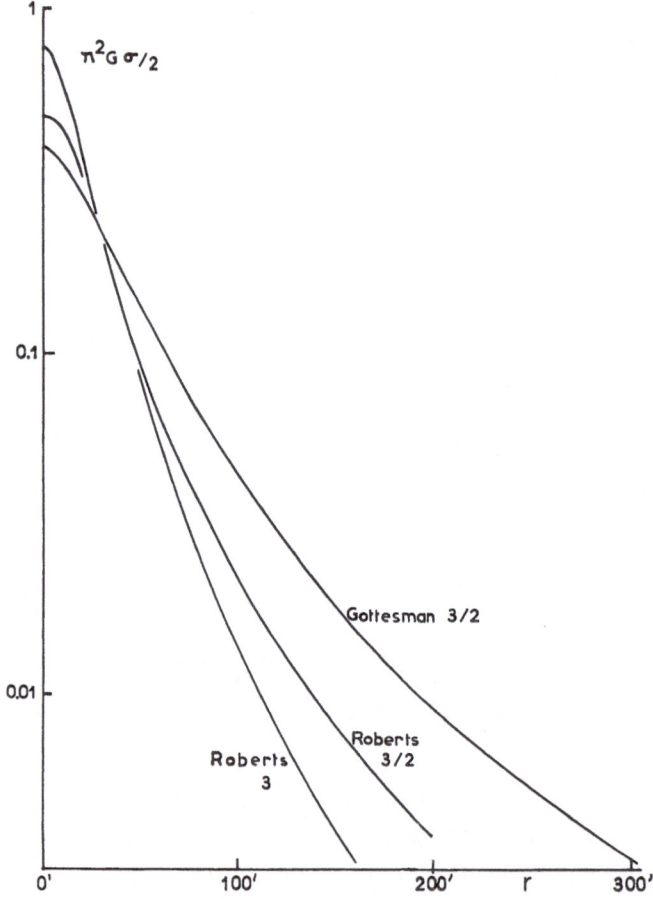

Fig. 2. Surface density $\sigma$ times $\pi^2 G/2$, in $10^7$ cgs units, for NGC 224 vs. the distance from the center $r$ in arcmin. The curves are from Gottesman *et al.* and from Roberts with the indicated Brandt indices.

## 4. Momentum of Late Type Spirals

The specific angular momentum of late type spirals is the easiest to evaluate by the photometric method. There is only a flat component and Equation (4) may be used. Integration along $z$ introduces the surface density, which is exponential since $f$ is constant inside the component and the brightness is exponential. The calculation of $K_r$ has been explicitly made in this case by Wyse and Mayall (1942). Using it, the specific angular momentum in kpc km s$^{-1}$ is:

$$A = 0.325 \left( D^3 R^3 f \left( b/a \right) I_0 \right)^{1/2} \tag{5}$$

where $D$ is the distance in Mpc, $f$ the mass to luminosity ratio in solar units corrected for absorption, $b/a$ the axis ratio, $I_0$ the central brightness corrected for absorption in units of 25 magnitude per square arcsec and $R$ the distance in arcmin at which the brightness has fallen to the value $I_0 \, e^{-4}$.

This calculation is applied to two late type spirals for which photometric profiles follow the exponential law: NGC 300 and LMC. The profiles are corrected for absorption in our Galaxy ($\Delta m_g$) and for internal absorption ($\Delta m_i$) according to Holmberg (1964); $B$ magnitudes are reduced to the $pg$ system by subtracting 0.1. Values and references are given in Table I.

TABLE I

Data for galaxies for which the angular momentum $A$ is evaluated from the photometry

|            | NGC 300    | LMC       | Units                     |
|------------|------------|-----------|---------------------------|
| type       | Sd (d)     | Sm (d)    |                           |
| $I_0$ obs. | 21.7 (a)   | 21.5 (b)  | $m_{pg}$ arcsec$^{-2}$     |
| $R$        | 13 (a)     | 370 (b)   | arcmin                    |
| $\Delta m_g$ | 0.2 (a)  | 0.4 (b)   | $m_{pg}$                  |
| $\Delta m_i$ | 0.3 (c)  | 0.3 (c)   | $m_{pg}$                  |
| $I_0$ corr. | 33        | 48        | 25 $m_{pg}$ arcsec$^{-2}$  |
| $b/a$      | 0.70 (d)   | 0.89 (b)  |                           |
| $D$        | 2.5 (e)    | 0.048 (f) | Mpc                       |
| $A$        | 580        | 320       | kpc km s$^{-1}$           |
| $A_i$      | 340        | 170       | kpc km s$^{-1}$           |

(a) G. de Vaucouleurs and Page (1962); (b) G. de Vaucouleurs (1960); (c) Holmberg (1964); (d) G. and A. de Vaucouleurs (1964); (e) Bottinelli *et al.* (1968); (f) Sandage (1962).

We hope to extend to more late type and to other type galaxies the calculation of momenta using the photometric data. The case of the intermediate types Sc-Scd might be difficult to handle because a small spheroidal component may exist which barely affects the luminosity profile but which may give an important contribution to the density. For example, NGC 598 has a small spheroidal component with diameter about 4′ (De Vaucouleurs, private communication) of which the photometry has not yet been done and which is probably responsible of the parabolic shape of the central part of the rotation curve obtained by Courtès' group (Baudel, private communication). The spatial flatness of the galaxy allows use of Equation (4) but the decomposition into two components may be difficult.

For comparison, the indicative specific angular momenta $A_i$ are also listed in the Table I. A further comparison for an earlier type spiral is obtained from data for our

Galaxy. Integration of the momentum distributions of Innanen's (1966) models yields $A = 900-1300$ kpc km s$^{-1}$; Kinman's (1959) study of globular clusters gives $A = 800$ kpc km s$^{-1}$. To obtain the indicative specific momentum the photometric diameter of our Galaxy may be estimated from the neutral hydrogen diameter through a statistical relation found by Bottinelli (private communication) between type and hydrogen diameter to photometric diameter ratio: $a \sim 40$ kpc. With $V_m = 260$ km s$^{-1}$ (Kerr, private communication), the indicative momentum is $A_i \sim 1040$ kpc km s$^{-1}$, which is also close to $A$.

## Acknowledgements

It is a pleasure to thank Drs. J. Heidmann and G. de Vaucouleurs for advice and discussions.

## References

Bottinelli, L., Gouguenheim, L., Heidmann, J., and Heidmann, N.: 1968, *Ann. Astrophys.* **31**, 205.
Brandt, J. C. and Scheer, L. S.: 1965, *Astron. J.* **70**, 471.
Gottesman, S. T., Davies, R. D., and Reddish, V. C.: 1966, *Monthly Notices Roy. Astron. Soc.* **133**, 359.
Heidmann, N.: 1969, *Astrophys. Letters* **3**, 153.
Holmberg, E.: 1964, *Ark. Astron.* **3**, 387.
Innanen, K. A.: 1966, *Astrophys. J.* **143**, 153.
Jeans, J. H.: 1922, *Monthly Notices Roy. Astron. Soc.* **82**, 122.
Kinman, T. D.: 1959, *Monthly Notices Roy. Astron. Soc.* **119**, 559.
Kinman, T. D.: 1965, *Astrophys. J.* **142**, 1376.
Lallemand, A., Duchesne, M., and Walker, M. F.: 1960, *Publ. Astron. Soc. Pacific* **72**, 76.
Roberts, M. S.: 1966, *Astrophys. J.* **144**, 639.
Sandage, A.: 1962, IAU Symposium No. 15, p. 359.
Takase, B. and Kinoshita, H.: 1967, *Publ. Astron. Soc. Japan* **19**, 409.
Toomre, A.: 1963, *Astrophys. J.* **138**, 385.
Vaucouleurs, G. de: 1958, *Astrophys. J.* **128**, 465.
Vaucouleurs, G. de: 1960, *Astrophys. J.* **131**, 265.
Vaucouleurs, G. de: 1962, IAU Symposium No. 15, p. 3.
Vaucouleurs, G. de and Page, J.: 1962, *Astrophys. J.* **136**, 107.
Vaucouleurs, G. and A. de: 1964, *Reference Catalogue of Bright Galaxies*, Univ. Texas Press, Austin.
Wyse, A. B. and Mayall, N. U.: 1942, *Astrophys. J.* **95**, 24.

# 7. APPLYING THE MODEL OF A NORMAL LOGARITHMIC SPIRAL TO GALAXIES

R. M. DZIGVASHVILI and T. M. BORCHKHADZE

*Abastumani Astrophysical Observatory, Abastumani, Georgia, U.S.S.R.*

**Abstract.** The characteristic $\mu$ angles for the spirals of 6 multi-arm galaxies (NGC 1232, 5247, 4303, 4321, 3938, 3184) have been determined. It has been found that they vary along spiral arms within a rather large interval, with mean square deviations ranging from 4° to 20°. The mean $\mu$ for each arm deviates in a relatively small degree from the mean $\mu$ of the whole galaxy.

The authors apply the normal logarithmic spiral model in estimating the number of arms in our Galaxy. On the basis of this model the number of spiral arms for the above galaxies have been determined and compared with the actual number. It is shown that the normal logarithmic spiral model is unsuitable for our purposes. The number of spiral arms calculated in this way is not reliable.

# 8. DENSITY DISTRIBUTION AND THE RADIAL VELOCITY
# FIELD IN THE SPIRAL ARMS OF M 31

J. EINASTO and U. RÜMMEL

*W. Struve Astrophysical Observatory, Tartu, Estonian S.S.R.*

**Abstract.** The density distribution and the radial velocity field in the Andromeda galaxy, M 31, have been studied on the basis of the 21-cm radio-line data from Jodrell Bank and Green Bank. The true density has been obtained from the observed one by solving a two-dimensional integral equation. As the resolving power of the radio telescopes is too low to locate all spiral arms separately, optical data on the distribution of ionized hydrogen clouds have been also used. The mean radial velocities have been derived by solving a two-dimensional non-linear integral equation with the help of hydrogen densities, and a model radial velocity field.

The inner concentrations of hydrogen form two patchy ringlike structures with mean radii 30′ and 50′, the outer concentrations can be represented as fragments of two *leading* spiral arms.

The rotational velocity, derived from the radial velocity field, in the central region differs considerably from the velocity curves obtained by earlier authors. The difference can be explained by the fact that in this region the correction for the antenna beam width is much greater than adopted by previous investigators.

## 1. Introduction

In the present paper the density distribution and the radial velocity field of neutral hydrogen in the Andromeda galaxy, M 31, have been studied. The investigation is based on the 21-cm radio-line data from Jodrell Bank and Green Bank Observatories, kindly sent to us by Dr. R. D. Davies and Dr. M. S. Roberts. Optical data on the distribution and motion of ionized hydrogen are also used.

When studying the distribution and motion of hydrogen in external galaxies it is necessary to take into consideration the angular resolving power of radio telescopes. Gottesman *et al.*, (1966) found that the correction for the beam width of the 250-feet Jodrell Bank telescope both in the density and the velocity does not exceed 10%. Our calculations, however, have shown that in some cases the correction needed is much greater. This indicates that the Jodrell Bank investigators have used too simplified reduction method. The Green Bank data have been reduced neglecting the antenna smearing effect (Roberts, 1966). For that reason the available radio data are to be reduced once again. At present the program is not finished. In this paper the preliminary results are reported.

## 2. The Integral Equations for the Density and the Mean Radial Velocity

Let $X$, $Y$ be the rectangular galactocentric coordinates in minutes of arc, the $Y$-axis being directed to the NE side of the major axis of the galaxy; $V$ the true radial velocity; $D(X, Y)$ the true projected density of neutral hydrogen; $E(V - \bar{V})$ the distribution function of residual radial velocities in the direction $X, Y$; $\bar{V} = \bar{V}(X, Y)$ is the mean radial velocity in this direction.

*Becker and Contopoulos (eds.), The Spiral Structure of Our Galaxy, 42–50. All Rights Reserved.*

The radio telescope, directed to the point $X_p$, $Y_p$ and disposed to the frequence, corresponding to the radial velocity $V_k$, will record the flux

$$T(X_p, Y_p, V_k) = \int\!\!\!\int\!\!\!\int_{-\infty}^{+\infty} D(X, Y) F(X - X_p, Y - Y_p)$$

$$\times E[V - \bar{V}(X, Y)] G(V - V_k)\, dX\, dY\, dV, \tag{1}$$

where $F(X-X_p, Y-Y_p)$, is the angular sensitivity function of the telescope and $G(V-V_k)$ is the corresponding frequency sensitivity function.

Integrating (1) over all observed velocities $V_k$ we obtain the observed projected density of hydrogen $\bar{D}(X_p, Y_p)$, which is connected with the true density $D(X, Y)$ by means of the equation

$$\bar{D}(X_p, Y_p) = \int\!\!\!\int_{-\infty}^{+\infty} D(X, Y) F(X - X_p, Y - Y_p)\, dX\, dY. \tag{2}$$

This is a two-dimensional homogeneous Fredholm integral equation of the first kind for the determination of the true density $D(X, Y)$. If the density is known the Equation (1) can be considered as a non-linear integral equation for the determination of the mean radial velocity $\bar{V}(X, Y)$.

The observations of point radio sources indicate that the function $F$ can be fairly well approximated by a two-dimensional Gaussian with half-intensity diameters 15' and 10' in the case of the Jodrell Bank and Green Bank telescopes respectively (Davies, 1969; Roberts, 1969). The function $G$ has in the case of the Jodrell Bank telescope also a Gaussian shape with half-intensity width 200 kHz, which correspond to a velocity dispersion of 17 km s$^{-1}$. The Green Bank telescope has a rectangular shaped function $G$ of 95 kHz $=20$ km s$^{-1}$ wide.

### 3. The Density Distribution

From the analogy with our Galaxy we may expect that the neutral hydrogen in M 31 is concentrated in the spiral arms. The optical observations of ionized hydrogen (Baade, 1963; Arp, 1964) indicate that the Andromeda galaxy has 4 or 5 spiral arms in both sides of the galaxy. The mean distance between every two arms is 20' $=4$ kpc, in projection only 4'–8', except the region around the major axis. The ionized hydrogen arms coincide with the neutral hydrogen arms within the actual distance of 5'; the neutral hydrogen arms are situated closer to the centre of the galaxy (Roberts, 1967).

The resolving power of the radio telescopes used is not sufficient to separate all spiral arms in the Andromeda galaxy; only the most dense arms N 4, S 4, and S 5 (designated after Baade, 1963) can be 'seen' individually (Roberts, 1967). To locate the other neutral hydrogen arms the optical data on the distribution of the ionized hydrogen clouds (Baade and Arp, 1964) can be used.

The true density distribution has been determined from the integral Equation (2) by two methods. Near the minor axis the equidensity lines are almost parallel to the major axis, and the two-dimensional equation can be reduced to the one-dimensional one. Representing the observed density distribution by a sum of Gaussian functions we get the solution of the equation also in the form of a sum of Gaussian functions.

For points far off from the minor axis the solution of the Equation (2) has been found by successive approximations. The arms have been located by combining optical and radio data, the corrected densities have been derived from the observed radio densities by a trial-and-error procedure. The densities have been found for a network of points, placed in $X$ and $Y$ at intervals 2′ and 10′ respectively.

The observed (Green Bank) and corrected density profiles (first approximation) along the major and minor axes of the Andromeda galaxy are shown in Figures 1 and 2 respectively. The picture is quite similar to the neutral hydrogen density profiles found for our Galaxy; an example of them, drawn on the basis of the Dutch survey (Westerhout, 1957; Schmidt, 1957), is given in Figure 3.

The $X, Y$-distribution of ionized hydrogen clouds (Baade and Arp, 1964) is given in Figure 4. The map of equidensity contours of neutral hydrogen is presented in Figure 5. The $R$-distribution (integrated over all position angles $\theta$) of the neutral and ionized hydrogen, as well as of the stellar associations (Van den Bergh, 1964) is plotted in Figure 6. The original distributions are reduced to an equal total number of objects, $N = 1000$.

The inspection of the data obtained leads us to the following conclusions:

(a) the spatial distribution of neutral hydrogen is similar to the distribution of

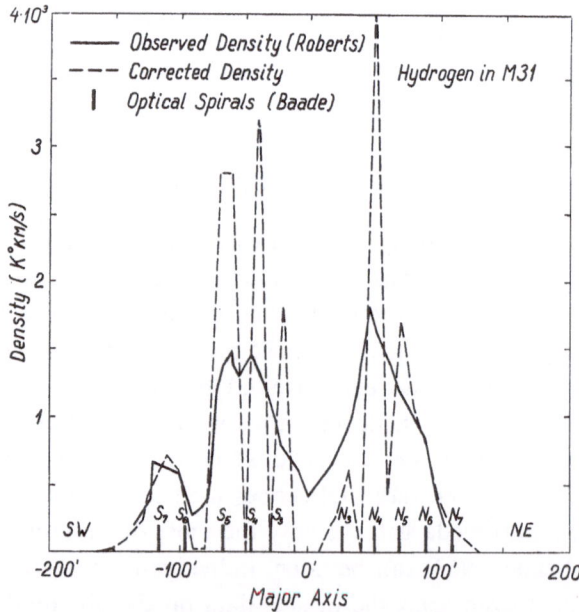

Fig. 1.   The observed (Roberts, 1967) and corrected surface densities of neutral hydrogen along the major axis of M 31. The location of optical arms according to Baade (1963) is also indicated.

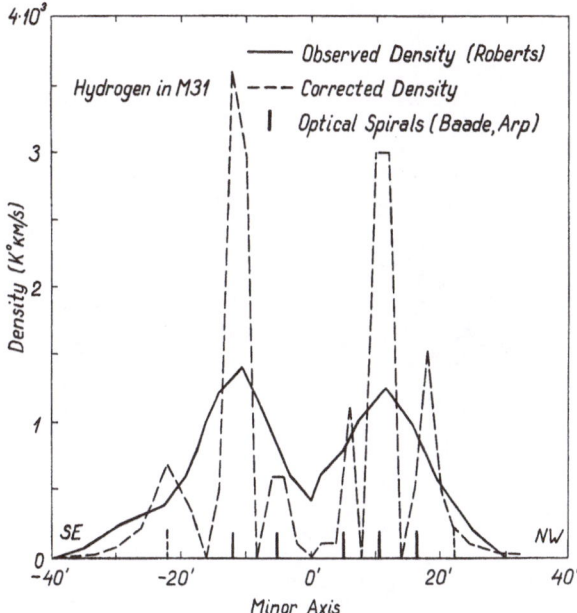

Fig. 2.    The observed and corrected surface densities of neutral hydrogen along the minor axis of M 31.

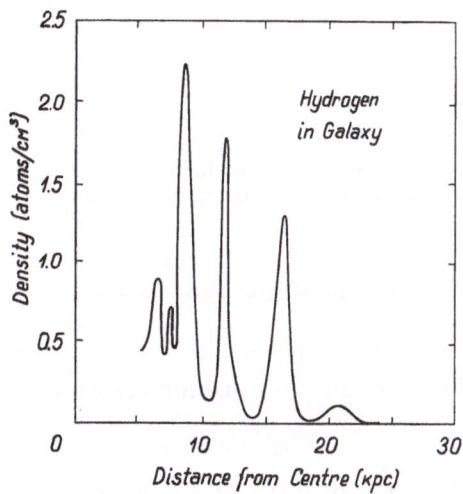

Fig. 3.    The space density of neutral hydrogen in the plane of our Galaxy.

ionized hydrogen and stellar associations; at great distances from the centre the relative density of neutral hydrogen is higher than that of the ionized hydrogen;

(b) the inner concentrations of hydrogen form two patchy ring-like structures with the mean radii 30' (the arms N 3, S 3 after Baade) and 50' (the arms N 4, S 4);

(c) the outer hydrogen concentrations can be fairly well represented as fragments of two *leading* spiral arms S 5–N 6, N 5–S 6.

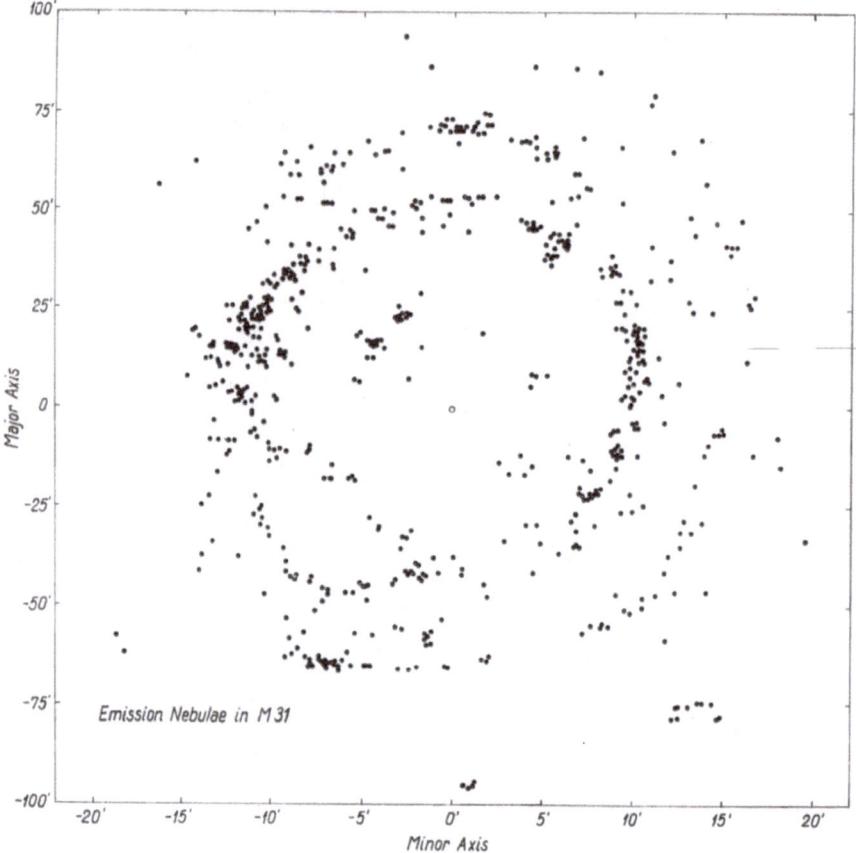

**Fig. 4.** The distribution of ionized hydrogen clouds in M 31 according to Baade and Arp (1964) data. The scale in $X$ (minor axis) is enlarged 4.5 times, corresponding to a tilt angle 12°.8 of M 31.

## 4. The Radial Velocity Field

The density distribution function $D$, and the angular sensitivity function $F$ are independent of the velocity $V$, and in Equation (1) we can integrate first over the velocity

$$T(X_p, Y_p, V_k) = \int\!\!\!\int_{-\infty}^{+\infty} D(X, Y) F(X - X_p, Y - Y_p)$$
$$\times H[V_k - \bar{V}(X, Y)] \, dX \, dY, \tag{3}$$

where

$$H[V_k - \bar{V}(X, Y)] = \int_{-\infty}^{+\infty} G(V - V_k) E[V - \bar{V}(X, Y)] \, dV. \tag{4}$$

If the velocity dispersion is independent of the position $X, Y$, the Formula (3) can be made more suitable for numerical computations. Let us use instead of $X, Y$ the

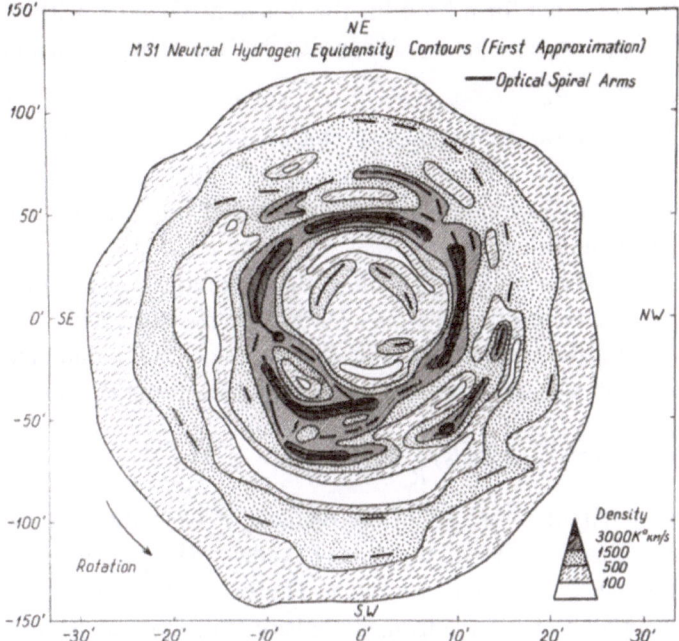

Fig. 5.   The preliminary equidensity contours of neutral hydrogen in M 31. Main spiral optical arms are indicated by dark lines.

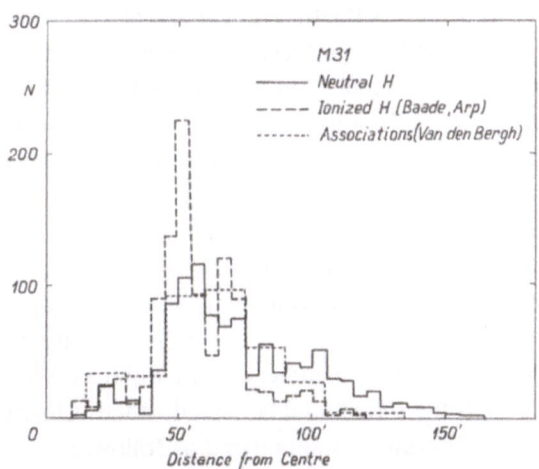

Fig. 6.   The distribution of neutral hydrogen, ionized hydrogen clouds (according to Baade and Arp, 1964), and of stellar associations (Van den Bergh, 1964) in M 31.

variables $S$, $\bar{V}$, where $S$ is the length along the line $\bar{V}(X, Y) = \text{const.}$ We have

$$T(X_p, Y_p, V_k) = \int\limits_{-\infty}^{+\infty} H(V_k - \bar{V})$$

$$\times \left[ \int\limits_{S} D(X, Y) F(X - X_p, Y - Y_p) J\left(\frac{X, Y}{S, \bar{V}}\right) dS \right] d\bar{V} . \qquad (5)$$

Assuming the Gaussian form both for the functions $G$ and $E$ with the dispersions $\sigma_G$, $\sigma_E$, respectively, then the function $H$ has also the Gaussian form with the dispersion

$$\sigma_H^2 = \sigma_G^2 + \sigma_E^2. \tag{6}$$

Interferometric observations show (Deharveng and Pellet, 1969) that the radial velocity dispersion has practically a constant value $\sigma_E = 17$ km s$^{-1}$ (due to the projection effect the dispersion $\sigma_E$ is greater than the true radial velocity dispersion in a small volume element of the galaxy).

Formula (5) has been used to calculate the theoretical 21-cm line profiles. An effective radial velocity dispersion $\sigma_H = 24$ km s$^{-1}$, the corrected hydrogen density field, and a model radial velocity field have been used. The radial velocity field has been calculated from a plane disc pure rotation model, using the obvious formula

$$\bar{V}(X, Y) = V_0 + V(R)\frac{Y}{R}\cos i, \tag{7}$$

where $V(R)$ is the circular velocity at the distance $R$ from the centre of the galaxy, $V_0$ – the mean radial velocity of the galaxy, and $i$ – the tilt angle of the plane of symmetry of the galaxy to the line of sight. The velocity $V(R)$ was taken from our four-component model of the Andromeda galaxy (Einasto and Rümmel, 1969), the constants are chosen as follows: $i = 12°.8$, $V_0 = -300$ km s$^{-1}$.

Gottesman et al., (1966) have derived for 231 points $X_p$, $Y_p$ the line profiles (spectra) $T(V_k \mid X_p, Y_p)$. For all these points the theoretical profiles have been calculated. These are quite similar to the observed profiles, but, in general, shifted in the velocity. The comparison of the profiles enables us to correct the model radial velocity field.

In this way we have found a solution to the integral Equation (1). From the corrected radial velocities near the major axis points a new improved rotation velocity curve has been derived.

The results are presented graphically. In Figure 7 the 21-cm line profiles for a major axis point are given. The theoretical profiles are calculated by using both the corrected and the uncorrected (observed) hydrogen densities, the model velocity field being identical. Mean radial velocities and the point velocity $V_p$ (the model radial velocity at the point $X_p$, $Y_p$) are also indicated. In Figure 8 the rotation curves are presented, and in Figure 9 the model and observed radial velocity field.

The analysis of the results can be summarized as follows:

(a) the change of the density causes both vertical and horizontal shifts in the line profiles, therefore an unbiased radial velocity field can be derived only by using carefully corrected densities;

(b) when the radio telescope is directed to a point of low hydrogen density or large density gradient, the mean radial velocity of the profile does not coincide with the point velocity; in extreme cases near the major axis the difference exceeds 100 km s$^{-1}$. This effect has caused large systematic errors in the previous reductions of radio-data (Argyle, 1965; Gottesman et al., 1966; Roberts, 1966);

(c) the corrected radial velocity field has great irregularities in respect of the model field.

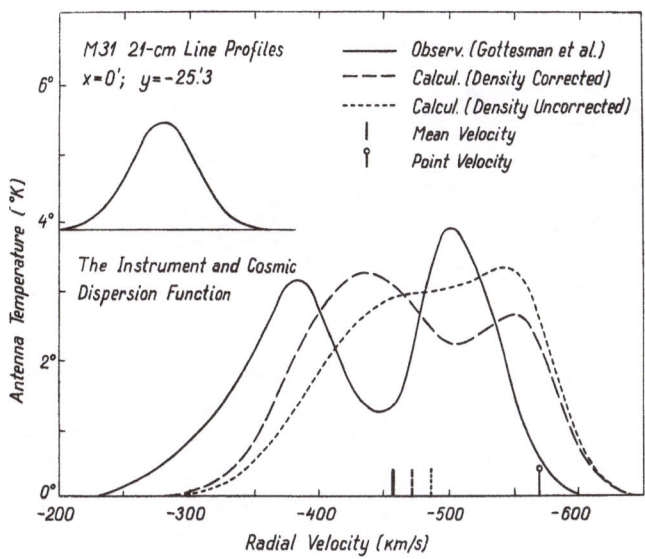

Fig. 7.   The 21-cm radio line profiles for a major axis point of M 31. The instrumental and cosmic
dispersion function is indicated.

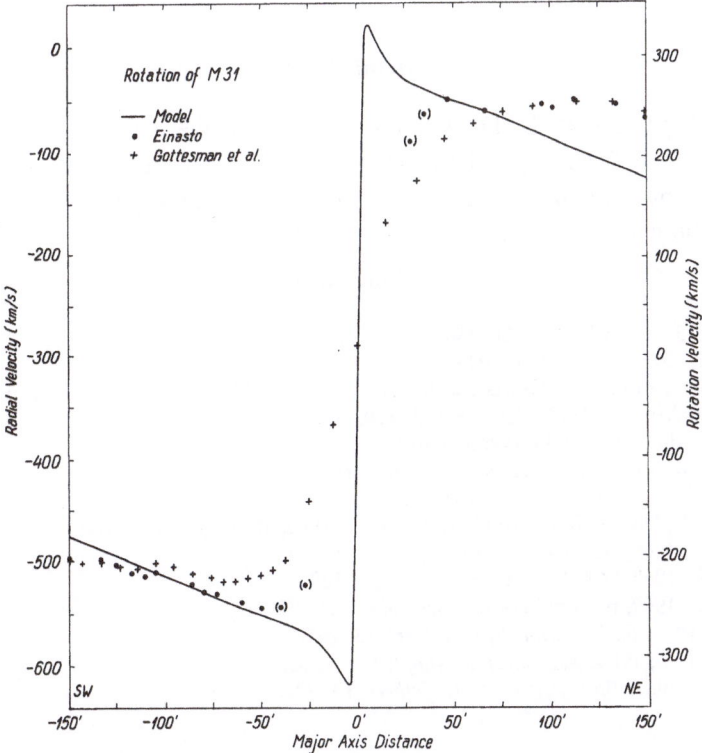

Fig. 8.   The model circular velocity and rotational velocities according to Gottesman *et al.*, (1966)
and our present data.

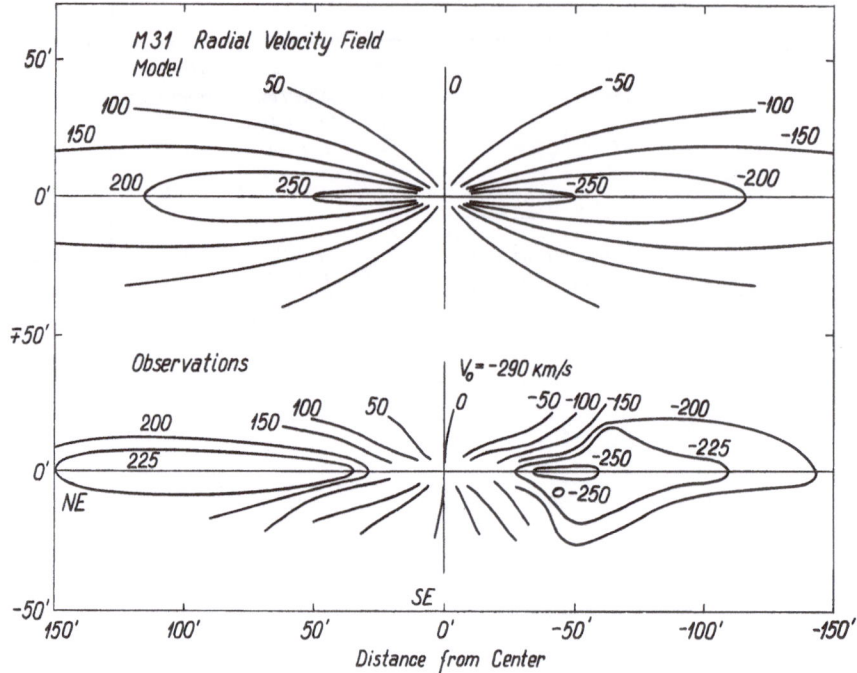

Fig. 9.   The observed and model radial velocity fields in M 31.

## Acknowledgements

We wish to thank Dr. R. D. Davies, and Dr. M. S. Roberts for the copies of original radio data, Dr. I. Petersen, Chief of the Computer Centre of the Estonian Academy of Sciences, for providing necessary computer time, and Dr. G. Kuzmin for discussion of the paper.

## References

Argyle, E.: 1965, *Astrophys. J.* **141**, 750.
Arp, H.: 1964, *Astrophys. J.* **139**, 1045.
Baade, W.: 1963, *Evolution of Stars and Galaxies,* Harvard University Press, Cambridge.
Baade, W. and Arp, H.: 1964, *Astrophys. J.* **139**, 1027.
Davies, R. D.: 1969, personal communication.
Deharveng, J. M. and Pellet, A.: 1969, *Astron. Astrophys.* **1**, 208.
Einasto, J. and Rümmel, U.: 1969, *Astrofiz.,* in press.
Gottesman, S. T., Davies, R. D., and Reddish, V. C.: 1966, *Monthly Notices Roy. Astron. Soc.* **133**, 359.
Roberts, M. S.: 1966, *Astrophys. J.* **144**, 639.
Roberts, M. S.: 1967, IAU Symposium No. 31, p. 189.
Roberts, M. S.: 1969, personal communication.
Schmidt, M.: 1957, *Bull. Astron. Inst. Netherl.* **13**, 247.
Van den Bergh, S.: 1964, *Astrophys. J. Suppl. Ser.* **9**, 65.
Westerhout, G. 1957, *Bull. Astron. Inst. Netherl.* **13**, 201.

# 9. THE ROTATION CURVE, MASS, LIGHT, AND VELOCITY DISTRIBUTION OF M 31

J. EINASTO and U. RÜMMEL

*W. Struve Astrophysical Observatory, Tartu, Estonian S.S.R.*

**Abstract.** A model for the Andromeda galaxy, M 31, has been derived from the available radio, photometric, and spectroscopic data. The model consists of four components – the nucleus, the bulge, the disc, and the flat component.

For all components the following functions have been found: the mass density; the mass-to-light ratio; the velocity dispersions in three perpendicular directions (for the plane of symmetry and the axis of the galaxy); the deviation angle of the major axis of the velocity ellipsoid from the plane of symmetry; the centroid velocity (for the plane of symmetry).

Our model differs in two points from the models obtained by other authors: the central concentration of mass is higher (in the nucleus the mass-to-light ratio is about 170), and the total mass of the galaxy ($200 \times 10^9$ solar masses) is smaller. The differences can be explained by different rotation curves adopted, and by attributing more weight to photometric and spectroscopic data in the case of our model.

## 1. Introduction

The most convenient way to express in condensed and mutually consistent form the various observational data on galaxies is the construction of their models. In the case of bright galaxies there are available the following data: the photometric data for the galaxy as a whole, and for some subsystems (neutral and ionized hydrogen, young bright stars, novae, cepheids), spectrophotometric data (mean spectral type, stellar content) for the nucleus and the bulge, and kinematical data (the systematic radial motion and velocity dispersion) for the gaseous component, the nucleus, and the bulge.

On the basis of these data, using the necessary dynamical and geometric equations it is possible to construct a composite hydrodynamic model of the galaxy. The method has been described earlier (Einasto, 1968a, 1968b, 1969a, 1969c) and applied to the Andromeda galaxy (Einasto, 1969b; Einasto and Rümmel, 1969). In the present paper both the method and the model have been improved. For the first time it is possible to derive reasonable values for all hydrodynamical descriptive functions of the main components of a galaxy.

## 2. Theory

A. ASSUMPTIONS AND DESCRIPTIVE FUNCTIONS

We assume that the galaxy has an axis and a plane of symmetry, common for all subsystems, that the galaxy is in the steady state, and consists of a number of physically homogeneous subsystems. The equidensity surfaces of the subsystems are similar concentric ellipsoids.

The hydrodynamic descriptive functions, determining the space density of matter and the velocity dispersion tensor, are designated as follows:

$\varrho(a)$ – the space density of matter, $a$ being the major semiaxis of the equidensity ellipsoid with the axial ratio $\varepsilon = b/a$;

$\sigma_R$, $\sigma_\theta$, $\sigma_z$ – the velocity dispersions in a galactocentric cylindrical coordinate system $(a^2 = R^2 + \varepsilon^{-2} z^2)$;

$V_\theta$ – the rotation velocity;

$\alpha$ – the inclination angle of the major axis of the velocity ellipsoid in respect to the plane of symmetry of the galaxy.

## B. GEOMETRIC EQUATIONS

The space density of matter can be found from the observed projected luminosity density $L(A)$, where $A$ is the major semiaxis of the projected equidensity ellipse with the apparent axial ratio $E$, $E^2 = \sin^2 i + \varepsilon^2 \cos^2 i$, where $i$ is the angle between the axis of the system and the line of sight. From geometric considerations, neglecting the absorption of light, we have (Einasto, 1969a)

$$L(A) = \frac{2\varepsilon}{Ef} \int\limits_A^{A^0} \frac{\varrho(a)\, a\, \mathrm{d}a}{\sqrt{a^2 - A^2}}, \tag{1}$$

where $f$ is the mass-to-light ratio of the subsystem, considered as a constant, and $A^0$ the major semiaxis of the limiting ellipsoid of the subsystem.

## C. HYDRODYNAMIC EQUATIONS

In a steady state galaxy the gravitational attraction of the galaxy is counterbalanced by the pressure (velocity dispersion) and the rotation. In cylindrical coordinates the hydrodynamic equilibrium equations are

$$\frac{1}{R}(\sigma_R^2 - \sigma_\theta^2) + \frac{1}{\varrho}\frac{\partial}{\partial R}(\varrho\sigma_R^2) + \frac{1}{\varrho}\frac{\partial}{\partial z}\left[\varrho\gamma(\sigma_R^2 - \sigma_z^2)\right] - \frac{V_\theta^2}{R} = -K_R, \tag{2}$$

$$\frac{1}{R}\gamma(\sigma_R^2 - \sigma_z^2) - \frac{1}{\varrho}\frac{\partial}{\partial R}\left[\varrho\gamma(\sigma_R^2 - \sigma_z^2)\right] + \frac{1}{\varrho}\frac{\partial}{\partial z}(\varrho\sigma_z^2) = -K_z, \tag{3}$$

where

$$\gamma = \tfrac{1}{2}\mathrm{tg}\,\alpha, \tag{4}$$

and $K_R$, $K_z$ – the radial and vertical components of the gravitational acceleration of the whole galaxy. The latter quantities can be derived from the mass density distribution function (Einasto, 1969a). In the steady state galaxy the functions $\sigma_R$, $\sigma_\theta$, $\sigma_z$, $V_\theta$, $\gamma$ fully determine the velocity ellipsoid as two axes of the ellipsoid lie in the meridional plane of the galaxy, and the radial and vertical components of the centroid motion are equal to zero.

## D. ADDITIONAL EQUATIONS, CLOSING THE SYSTEM OF EQUATIONS

In order to obtain composite models of galaxies, the mass and light distribution of subsystems is first to be determined from photometric and spectroscopic data. Then the gravitational acceleration of the whole galaxy can be found. Finally the kinematical functions of the subsystems can be derived. Given the density and the acceleration the Equations (2) and (3) involve 5 unknown kinematical functions. As we have only two equations the problem is not closed: to solve the system of equations three additional equations are needed.

It is convenient to give the additional equations for the functions, which determine the orientation of the velocity ellipsoid, $\gamma$, and its shape

$$k_\theta(R, z) = \sigma_\theta^2/\sigma_R^2, \quad k_z(R, z) = \sigma_z^2/\sigma_R^2. \tag{5}$$

From the theory of the third integral of motion of stars follows (Kuzmin, 1953)

$$\gamma = Rz/(R^2 + z_0^2 - z^2), \tag{6}$$

where $z_0$ is a constant, depending on the gravitational potential of the whole galaxy.

The equations for $k_\theta$ and $k_z$ are in general case complicated (Einasto, 1969c). In the present paper we have computed the kinematical functions for the plane and the axis of the galaxy only. The theory of the steady state galaxy gives (cf. Einasto, 1969a)

$$k_\theta(R, 0) = \tfrac{1}{2}\left[1 + \frac{\partial \ln V_\theta}{\partial \ln R}\right]. \tag{7}$$

We assume that in the first approximation the centroid velocity $V_\theta$ is proportional to the circular velocity $V_c$. In this case $k_\theta(R, 0)$ are identical for all subsystems. From the symmetry condition on the axis of the galaxy we have

$$k_\theta(0, z) = 1. \tag{8}$$

For flat subsystems the ratio $k_z(R, 0)$ can be found from the theory of irregular gravitational forces. Kuzmin (1961) has derived the following approximate relation

$$[k_z(R, 0)]^{-1} = 1 + [k_\theta(R, 0)]^{-1}. \tag{9}$$

On the other hand from the theory of the third integral we have for the axis $R=0$, supposing the ellipsoidal distribution of velocities (Einasto, 1969c)

$$k_z(0, z) = k_z(0, 0)/k_z(\sqrt{z^2 - z_0^2}, 0). \tag{10}$$

Formulae (9) and (10) can be used, if $R^2 \gg z_0^2$, and $z^2 \geqslant z_0^2$ correspondingly. For small $R$ and $z$, $k_z(R, z)$ is to be interpolated, using the value $k_z(0, 0)$, derived from the virial theorem.

### E. APPLICATION OF THE VIRIAL THEOREM

The nucleus of a galaxy can be considered in a good approximation to be an isolated dynamical system. In this case we may apply the tensor virial theorem (Kuzmin, 1964).

Assuming a rigid body rotation and ellipsoidal shape for the nucleus we have

$$\overline{\sigma_R^2} + \tfrac{1}{3}\omega^2\overline{a^2} = \tfrac{1}{2}\beta_R G \mathcal{M} \bar{a}^{-1}, \tag{11}$$

$$\overline{\sigma_z^2} \qquad = \tfrac{1}{2}\beta_z G \mathcal{M} \bar{a}^{-1}. \tag{12}$$

In these formulae $\omega$ is the constant angular velocity, $G$ the gravitational constant, $\mathcal{M}$, the mass of the nucleus, and

$$\overline{a^2} = \frac{1}{\mathcal{M}} \int\limits_0^\infty \mu(a)\, a^2\, \mathrm{d}a, \tag{13}$$

$$\bar{a}^{-1} = \frac{2}{\mathcal{M}^2} \int\limits_0^{\mathcal{M}} \frac{M(a)\, \mathrm{d}M(a)}{a}, \tag{14}$$

where

$$\mu(a) = 4\pi\varepsilon\varrho(a)\, a^2 \tag{15}$$

is the mass distribution function, and

$$M(a) = \int\limits_0^a \mu(a)\, \mathrm{d}a \tag{16}$$

the integral mass distribution function.

The constants $\beta_R$ and $\beta_z$ depend on the shape of the system. Denoting $e^2 = 1 - \varepsilon^2$ we have

$$\beta_R = \frac{1}{2e^2}\left[\frac{\arcsin e}{e} - \varepsilon\right], \tag{17}$$

$$\beta_z = \frac{\varepsilon^2}{e^2}\left[\frac{1}{\varepsilon} - \frac{\arcsin e}{e}\right]. \tag{18}$$

From (11)–(12) we obtain

$$\bar{k}_z = \frac{\overline{\sigma_z^2}}{\overline{\sigma_R^2}} = \frac{\beta_z}{\beta_R}\left(1 + \frac{\omega^2\overline{a^2}}{3\,\overline{\sigma_R^2}}\right). \tag{19}$$

As in the nucleus of the Andromeda galaxy $\omega^2\bar{a}^2 \ll \bar{\sigma}_R^2$, the mean axial ratio of the velocity ellipsoid depends sufficiently only on the axial ratio of the system itself.

The value of $\bar{k}_z$, found for the nucleus of the galaxy, can be adopted for $k_z(0, 0)$.

## 3. The Model

The theory outlined has been applied to a model of the Andromeda galaxy, consisting of four components: the nucleus, the bulge, the disc, and the flat component.

The observational data used have been described in a published paper (Einasto, 1969b).

The distance 692 kpc of the galaxy is accepted, corresponding to the true distance modulus $(m - M)_0 = 24^m.2$ (Baade and Swope, 1963).

The inclination of the galaxy has been estimated by combining the data on the axial ratio of isophotes in the outer region of the galaxy, and the distribution of emission nebulae (Baade and Arp, 1964). The value $i = 12°.8$ has been found. It is in good agreement with an earlier estimate by Baade $i = 12°.7$, quoted by Schmidt (1957). Somewhat larger values found by Arp (1964), and by some other authors cannot be accepted, as in this case the true axial ratio of the equidensity surfaces of the disc population will be too small, of the order of 0.01. The disc component of a galaxy consists of the old population I stars. Their vertical dispersion of velocities at the distance $R = 10$ kpc from the centre is of the order of 20 km s$^{-1}$. From these data we can estimate the thickness and the axial ratio of equidensity surfaces; the latter quantity becomes of the order of 0.1.

The parameter $z_0$ was derived from the gravitational potential of the system. An effective value $z_0 = 0.5$ kpc has been found.

The principal descriptive function, the space density of matter, has been chosen in the form of a generalized exponential function

$$\varrho(a) = \varrho_0 \exp\left[-(a/a_0 k)^\nu\right], \tag{20}$$

where $\varrho_0$ is the central density of the component, $a_0$ – the effective (harmonic mean) radius of the component, $\nu$ – the structural parameter of the model, and $k$ – a dimensionless parameter depending on $\nu$. The central density depends on the mass, effective radius, and the axial ratio of the component:

$$\varrho_0 = \frac{h}{4\pi\varepsilon} \frac{\mathcal{M}}{a_0^3}, \tag{21}$$

where $h$ is a dimensionless parameter depending on $\nu$.

The derived parameters of the components are given in Table I.

To obtain better agreement with observations, the flat component of the galaxy has been represented by a sum of two functions (20), one of them being negative.

TABLE I

Parameters of the components of M 31

| Quantity | Unit | Total | Nucleus | Bulge | Disc | Flat component | |
|---|---|---|---|---|---|---|---|
| | | | | | | + | − |
| $\nu$ | | | 1 | 1/4 | 1 | 1 | 1 |
| $k$ | | | 0.5 | $1.263 \times 10^{-4}$ | 0.5 | 0.5 | 0.5 |
| $h$ | | | 4 | 3112 | 4 | 4 | 4 |
| $\varepsilon$ | | | 0.84 | 0.57 | 0.09 | 0.01 | 0,02 |
| $a_0$ | kpc | | 0.005 | 1 | 10 | 8 | 4 |
| $\mathcal{M}$ | $10^9\,\mathcal{M}_\odot$ | 201.8 | 0.52 | 85.5 | 111.5 | 5.73 | − 1.43 |
| $L$ | $10^9\,L_\odot$ | 13.13 | 0.003 | 4.95 | 6.46 | 2.29 | − 0.57 |
| $f$ | | 15.4 | 173 | 17.3 | 17.3 | 2.5 | 2.5 |
| $\bar{\varrho} = \mathcal{M}/\frac{4}{3}\pi\varepsilon a^3{}_0$ | $\mathcal{M}_\odot/\mathrm{pc}^3$ | | $1.2 \times 10^6$ | 35.8 | 0.296 | 0.267 | − 0.267 |

The parameters of the model are subjected to the condition that a ring-like mass distribution and everywhere non-negative total density of the component could be provided.

The mass of the nucleus has been determined by means of the virial theorem. In an earlier paper (Einasto, 1969b) the mass has been found from the luminosity of the nucleus and its accepted mass-to-light ratio (Spinrad, 1966).

The calculated descriptive functions are presented graphically in Figures 1–6. In general, the results are similar with our previous ones (Einasto, 1969b; Einasto and Rümmel, 1969). Near the centre of the galaxy the functions are being improved to allow for the virial theorem, not used in our earlier papers.

## 4. Discussion

### A. MASS DISTRIBUTION

Our model differs in two points from the models obtained by earlier authors (Schmidt, 1957; Brandt and Scheer, 1965; Roberts, 1966; Gottesman *et al.*, 1966): the central concentration of mass is much higher, and the total mass smaller (Einasto, 1969b,

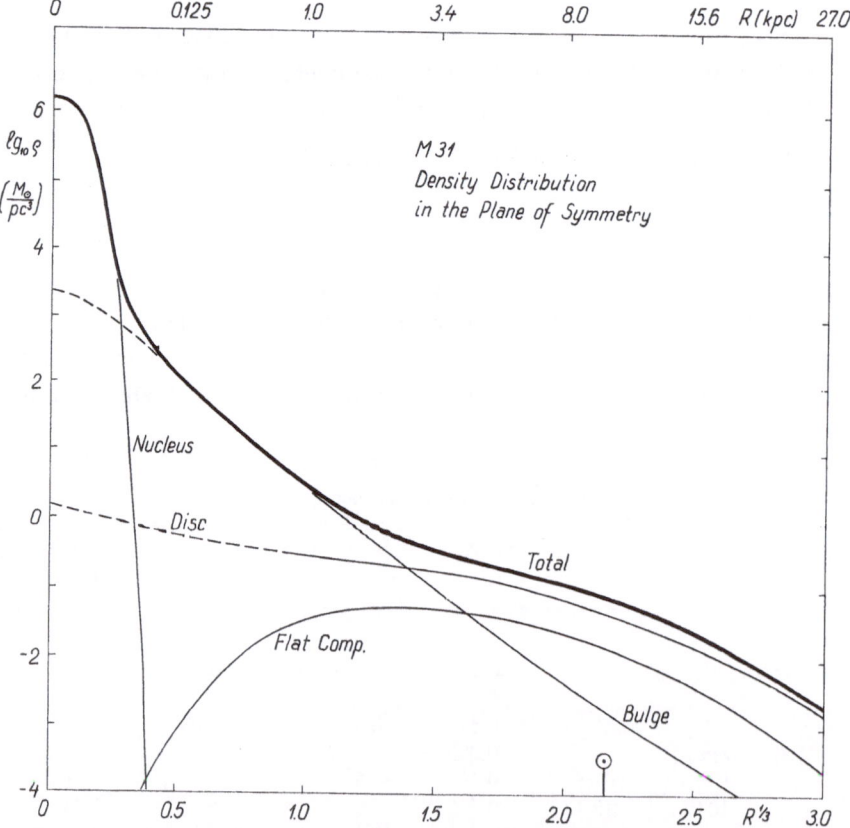

Fig. 1.   The density distribution of the components of M 31 in the plane of symmetry. Dashed parts of the curves are interpolated.

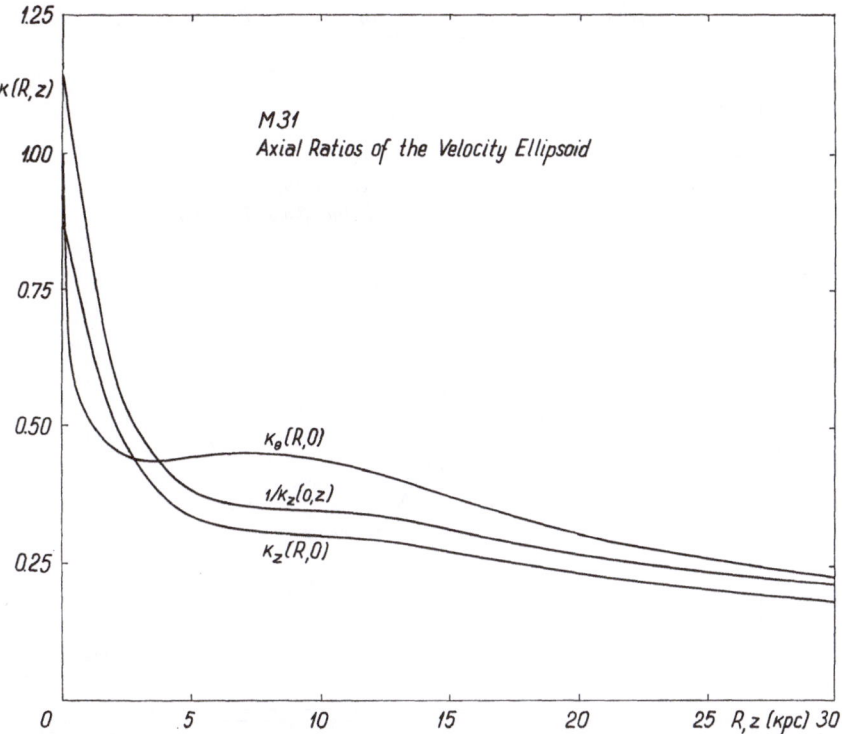

Fig. 2.   The axial ratios of the velocity ellipsoid in the plane of symmetry and in the axis of M 31.

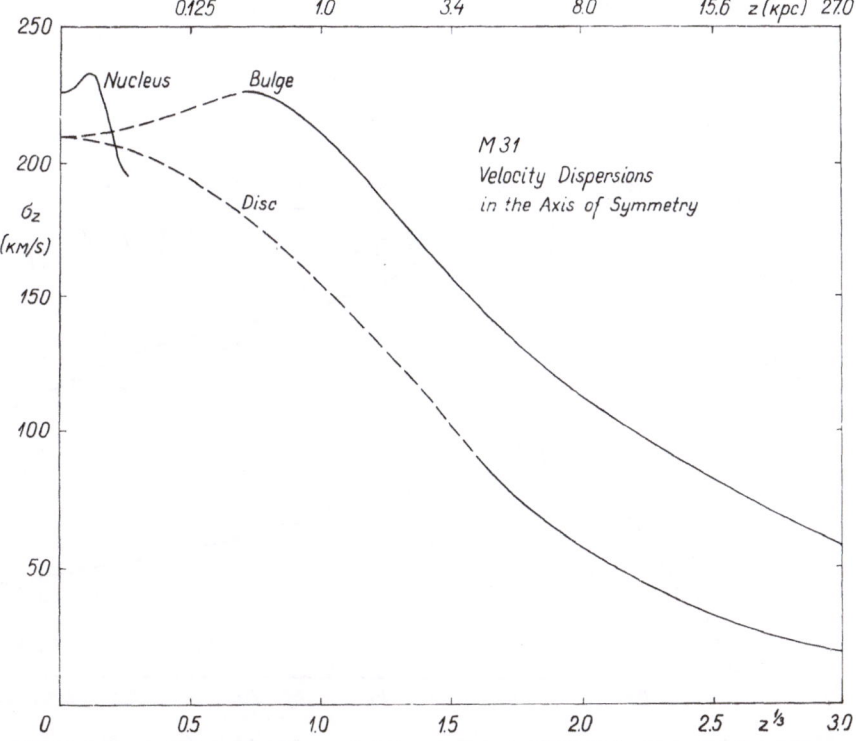

Fig. 3.   The velocity dispersions of the components of M 31 in the axis of symmetry.

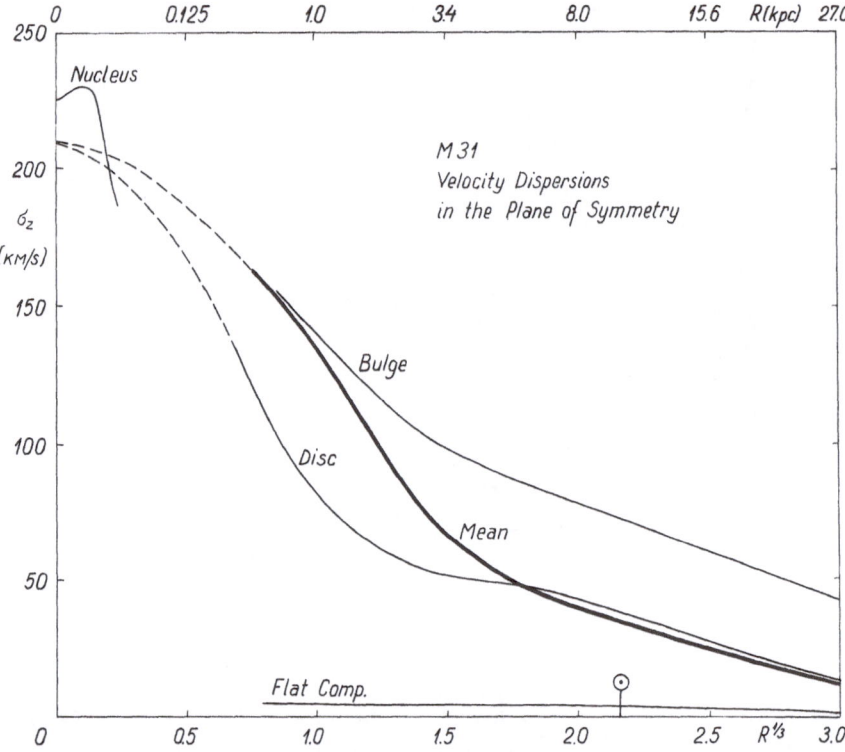

Fig. 4.   The velocity dispersions of the components of M 31 in the plane of symmetry.

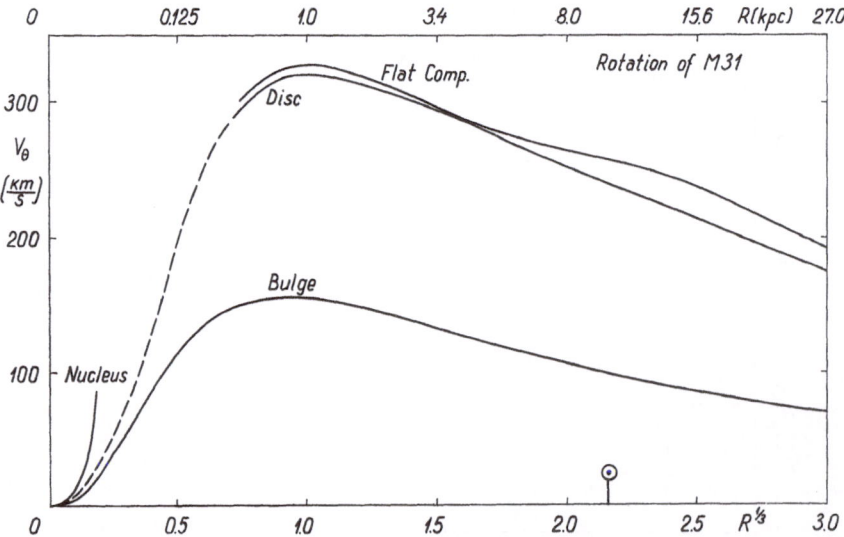

Fig. 5.   The rotation velocities of the components of M 31 in the plane of symmetry.

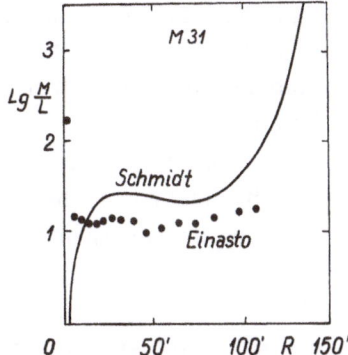

Fig. 6.   The mass-to-luminosity ratio in M 31 according to Schmidt's (1957) model and to our present model.

Table III). In both cases the differences can be explained by various circular velocity curves adopted.

In the central region the velocities found earlier from the 21-cm radio-line measurements are underestimated due to the unsufficient correction for the antenna smearing effect. The rotation velocities, derived optically for the stellar component of the galaxy, cannot be identified with the circular velocities, as the pressure term (velocity dispersion) in hydrodynamical equations is predominating.

The great masses are found in most cases as a result of approximating the observed rotation velocities with a generalized Bottlinger law

$$V_\theta = \frac{V_0 R}{[1 + (R/R_0)^n]^{3/2n}},$$          (22)

($V_0$, $R_0$ – constants) and of identifying the rotation velocities with the circular velocity.

We have shown (Einasto, 1969a) that the generalized Bottlinger law cannot be applied to the circular velocity, as in this case great masses at very large distances from the centre of the galaxy occur. This is impossible due to the tidal effect of nearby galaxies (King, 1962).

Small radial gradient of the rotation velocity, observed in the periphery of some galaxies, in particular, in the Andromeda galaxy, is probably to be explained in another way, for instance as the appearance of systematic streaming motion in the galaxy.

### B. MASS-TO-LIGHT RATIO

The mean mass-to-light ratio found, $f = 15.4$ is normal for a Sb galaxy. The flat population and the disc have also acceptable values, $f = 2.5$ and $f = 17.3$ respectively. The value for the bulge, $f = 17.3$ is a preliminary one, and a further study is needed.

The mass-to-light ratio for the nucleus, $f = 17.3$, seems at first glance to be too large. To explain this value we must suppose that the nucleus (a) consists of very old physically evolved stars, and (b) is dynamically not evolved.

The mean relaxation time of the nucleus is of the same order ($10^{10}$ yr) as the age of the whole galaxy. Therefore the nucleus is dynamically indeed little evolved and has lost only a small fraction of his low-mass stars. As the nucleus has had too little time to form dynamically by star-star encounters, it must be formed in the proto-galaxy stage of the galaxy evolution.

The metal-content of stars in the nucleus is normal (Spinrad, 1966). Therefore, if the high mass-to-light ratio and the great age of the nucleus will be confirmed, we must conclude that in the nucleus the metal-enrichment has taken place in a very early stage of the galaxy evolution.

**Note added in proof.** The mass $\mathscr{M} = 5.2 \times 10^8 \, \mathscr{M}_\odot$, obtained from the virial theorem for the nucleus, and the corresponding mass-to-light ratio $f = 170$ does not agree with the value $f = 17$ derived spectroscopically (Spinrad, 1966). This discrepancy may be removed, supposing that the nucleus contains besides stars an invisible central body – a dead quasar (Lynden-Bell, in press) or an object of unknown nature (Ambartsumian, 1958). In this case the virial theorem must be modified, and we get (Einasto, 1970, in preparation) for the point mass $\mathscr{M} = 1.4 \times 10^8 \mathscr{M}_\odot$, supposing $\mathscr{M} = 0.5 \times 10^8 \mathscr{M}_\odot$ for the mass of the stellar component of the nucleus (Einasto, 1969b).

## Acknowledgements

It is a pleasure to thank Dr. G. Kuzmin for helpful discussion, and Mrs. Liia Einasto for the programming of computations.

## References

Arp, H.: 1964, *Astrophys. J.* **139**, 1045.
Ambartsumian, V. A.: 1958, in *La structure et l'évolution de l'univers*, Stoops, Bruxelles, p. 241.
Baade, W. and Arp, H.: 1964, *Astrophys. J.* **139**, 1027.
Baade, W. and Swope, H. H.: 1963, *Astron. J.* **68**, 435.
Brandt, J. C. and Scheer, L. S.: 1965, *Astron. J.* **70**, 471.
Einasto, J.: 1968a, *Publ. Tartu Astron. Obs.* **36**, 357.
Einasto, J.: 1968b, *Publ. Tartu Astron. Obs.* **36**, 442.
Einasto, J.: 1969a, *Astron. Nachr.* **291**, 97.
Einasto, J.: 1969b, *Astrofiz.* **5**, 137.
Einasto, J.: 1969c, *Astrofiz.*, in press.
Einasto, J.: 1970, in preparation.
Einasto, J. and Rümmel, U.: 1969, *Astrofiz.*, in press.
Gottesman, S. T., Davies, R. D., and Reddish, V. C.: 1966, *Monthly Notices Roy. Astron. Soc.* **133**, 359.
King, I.: 1962, *Astron. J.* **67**, 471.
Kuzmin, G.: 1953, *Publ. Tartu Astron. Obs.* **32**, 332.
Kuzmin, G.: 1961, *Publ. Tartu Astron. Obs.* **33**, 351.
Kuzmin, G.: 1964, *Publ. Tartu Astron. Obs.* **34**, 10.
Roberts, M. S.: 1966, *Astrophys. J.* **144**, 639.
Schmidt, M.: 1957, *Bull. Astron. Inst. Netherl.* **14**, 17.
Spinrad, H.: 1966, *Publ. Astron. Soc. Pacific* **78**, 367.

# 10. A COMPARISON OF DYNAMICAL MODELS OF THE ANDROMEDA NEBULA AND THE GALAXY

V. C. RUBIN and W. K. FORD, JR.

*Department of Terrestrial Magnetism,*
*Carnegie Institution of Washington, Washington, D.C., U.S.A.*

**Abstract.** (1) From new radial velocities of 67 H II regions in M 31, rotational velocities and a mass model of M 31 are derived, and compared with the rotation curve and Schmidt mass model of our galaxy. (2) It is shown that in M 31 the distribution of H II regions as identified by Baade agrees with the distribution of neutral hydrogen determined from 21-cm observations. Also, the rotation curve derived from the H II velocities outside of the nucleus is similar to the rotation curve derived from 21-cm H I observations.

Because individual H II regions can be identified and studied in M 31, it is possible to determine the rotation as a function of distance from the center and to derive a mass model for that galaxy in more detail than is possible for almost any other spiral galaxy except our own. Early in the study of galaxies, it was assumed that M 31 and our Galaxy were very similar; it has been popular in recent years to emphasize their differences. Perhaps surprisingly, the dynamical model for M 31 which results from our recent spectroscopic survey of emission regions in M 31 (Rubin and Ford, 1970) resembles the current model of our Galaxy.

## 1. Mass Models for M 31 and for the Galaxy

Spectra of 67 H II regions from 3 to 24 kpc from the nucleus of M 31 have been obtained with the DTM image tube spectrograph attached to the Lowell and Kitt Peak telescopes. Radial velocities, principally from Hα at a dispersion of 135 Å/mm, have been determined with an accuracy of 10 km s$^{-1}$ for most regions. Rotational velocities have been calculated assuming circular motions only.

For the region interior to 3 kpc where no emission regions have been identified, a narrow [N II] $\lambda6583$ emission line is observed. Along the major axis, velocities from this line indicate a rapid rotation in the nucleus, rising to a maximum circular velocity of $V=225$ km s$^{-1}$ at $R=400$ pc, and falling to a deep minimum near $R=2$ kpc. This is shown in Fig. 1. Details of this work are published elsewhere (Rubin and Ford, 1970).

Along the minor axis of M 31, the [N II] $\lambda6583$ emission line velocities exhibit a series of maxima and minima, with typical velocities of almost $\pm100$ km s$^{-1}$ over regions of 500 pc in the galaxy, with respect to the central velocity. The most notable departure from the systemic velocity is near $R=1$ kpc on the SE (far) side, where velocities of $\pm100$ km s$^{-1}$ radial from the center are observed. It was Münch's (1960) observations of similar velocities from the [O II] $\lambda3727$ doublet that led him to infer that gas is streaming from the nucleus of M 31. The [N II] observations confirm this.

*Becker and Contopoulos (eds.), The Spiral Structure of Our Galaxy, 61–68. All Rights Reserved.*
*Copyright © 1970 by the I.A.U.*

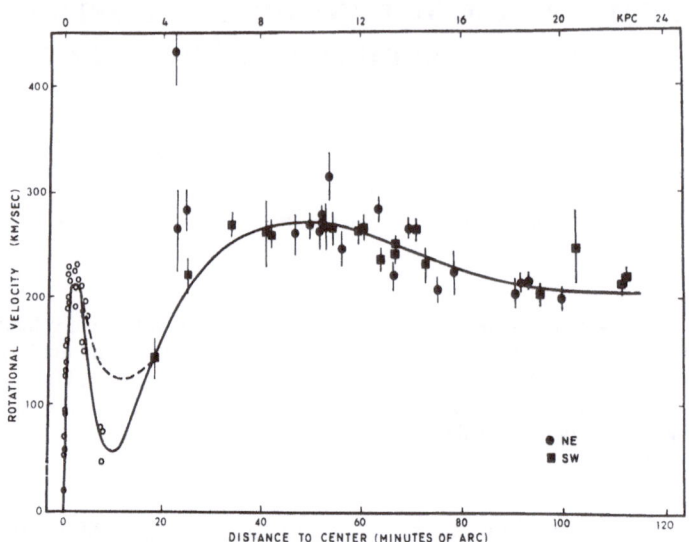

Fig. 1.   Rotational velocities in M 31. Open circles are velocities from [N II] $\lambda$ 6583 emission line near nucleus; filled circles and squares are from H II regions (individual regions or means of several regions in an OB association). Error bars indicate average error of rotational velocity, calculated from the range of individual velocities of emission regions in the association, or from the observational errors, whichever is larger. Solid curve is one representation of rotation curve, which is least squares fit of 4th order polynomial for $R > 12'$. Dashed curve near $R = 10'$ is a second rotation curve with higher inner minimum. © 1970 University of Chicago.

If we make the assumption that the gas and the stars in the nucleus are moving together in the gravitational field of the galaxy, then the following disk model of M 31 results. There is a dense, rapidly rotating nucleus, of total mass $6 \pm 1 \times 10^9 M_\odot$. Outside of the nucleus near $R = 2$ kpc the density is very low and the rotational velocities are very small.

In some regions from 500 pc to 1.4 kpc (most noticeably on the SE side) ionized gas is observed moving out from the nucleus with a velocity which decreases with increasing distance from the center. The very low rotational velocities here indicate that the gas must be ejected from the nucleus with almost no angular momentum. Our observations are not yet extensive enough to decide if the radial motions are widespread at this distance from the nucleus.

Beyond $R = 4$ kpc, the mass of the galaxy increases approximately linearly to about $R = 14$ kpc, and more slowly thereafter. The total mass to $R = 24$ kpc is $1.85 \pm 0.1 \times 10^{11} M_\odot$; one-half of this mass is located in the disk interior to $R = 9$ kpc.

For M 31, the mass of neutral hydrogen determined from 21-cm observations (Burke *et al.*, 1964) is between 6.7 and $14 \times 10^9 M_\odot$; half of it is contained in the disk interior to $R = 13$ kpc. The ratio of H I to total mass in M 31 out to $R = 24$ kpc is thus between 4 and 8%.

We now wish to compare the parameters derived for M 31 with those of our Galaxy. There is a built-in bias, however, for many of the dynamical parameters for our Galaxy have previously been adopted to make the general features of our Galaxy conform with those of M 31. We first consider the rotation curves; we use the mass model of Schmidt (1965). Because we are interested in the large scale features of our Galaxy, this model is more suited to a comparison than the detailed rotation curve, $4 < R < 10$ kpc, of Shane and Bieger-Smith (1966). Schmidt notes that circular velocities for our Galaxy are known only for 7 values of $R$, $R \leqslant 10$ kpc; beyond $R = 10$ kpc, the rotation curve is an extrapolation, chosen so that the density decreases as $R^{-4}$. These known values for $V_{rot}$ are plotted in Figure 2 (Rubin and Ford, 1970), along with the rotation curve adopted by Schmidt (1965). The two innermost points come from the observations of Rougoor and Oort (1960). We have also plotted in Figure 3 the rotation curve for M 31.

The rotation curve adopted by Schmidt, but modified to include the two innermost points, is rather similar to the rotation curve of M 31, although the very deep minimum near $R = 2$ kpc is not shown for our Galaxy. There is evidence, however, that the

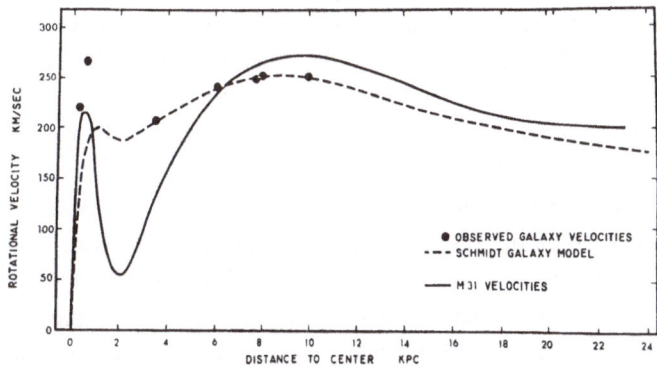

Fig. 2. Comparison of rotation curves of M 31 and the Galaxy, as a function of distance from the center. Filled circles are observed rotational velocities for the Galaxy (Rougoor and Oort, 1960; Schmidt, 1965). © 1970 University of Chicago.

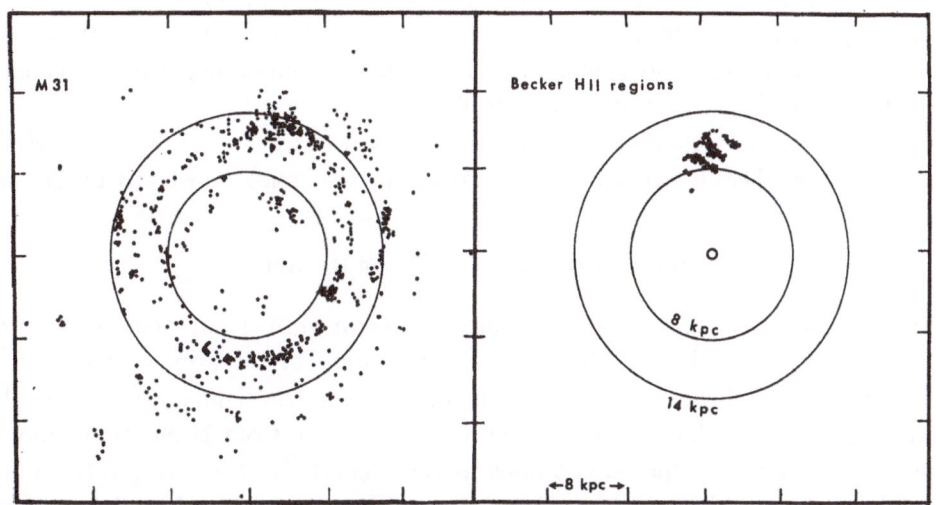

Fig. 3. (Left): Projection on plane of M 31 of 688 H II regions identified by Baade. (Right): Projection on plane of our Galaxy of H II regions located by Becker (1964). For both figures, circles mark distances 8 and 14 kpc from the center.

rotation velocities in our Galaxy are very small just outside of the rapidly rotating nucleus. This evidence comes from the 21-cm observations toward the nucleus. Both Rougoor (1964) and Burke and Tuve (1964) noted that their observations implied very low circular velocities in this region, although the final rotation model adopted by Rougoor (1964) was a compromise, and a less deep minimum was used. Burke and Tuve concluded that either $\omega$ or $d\omega$ ($\omega = V/R$) must increase with increasing $R$, or the arms in our Galaxy must be leading, to agree with the observed 21-cm velocities near $l^{II} = 0°$.

From the rotation curve for the Galaxy, Schmidt determined a total mass of $M = 1.8 \times 10^{11} M_\odot$; of which $1.5 \times 10^{11} M_\odot$ is contained within $R = 24$ kpc. Thus with the current distance scale for M 31, M 31 is about 20% more massive than the Galaxy. The maximum circular velocity for M 31 is $270 \pm 10$ km s$^{-1}$ while it is 250 km s$^{-1}$ for our Galaxy. Hence, merely decreasing the distance of M 31 would not improve the coincidence of the two rotation curves, although it would decrease the mass of M 31.

For the disk of our Galaxy, the model which emerges from the 21-cm observations is as follows (Oort, 1968; Kerr and Westerhout, 1965; Woltjer, 1965). The central Sagittarius A source is surrounded by a rapidly rotating disk; rotational velocities are about 200 km s$^{-1}$ at $R = 100$ pc, and are then approximately constant near $V = 250$ km s$^{-1}$ to $R = 1$ kpc. Between 1 and 2 or 3 kpc, the gas density is low, and there is a large systematic radial component in the velocities, decreasing from $V = 200$ km s$^{-1}$ to $V = 50$ km s$^{-1}$. Rotational velocities in this region are sometimes very low. Oort (1968) has suggested as possible causes of the expansion motions the action of asymmetrical gravitational fields in the Galaxy, the pressure of magnetic fields, or eruptive activity near the nucleus, with the last possibility being the least likely. Beyond $R = 4$ kpc, the rotational velocities increase to a maximum $V = 250$ km s$^{-1}$ near $R = 10$ kpc, and decrease slowly thereafter. The ratio of H I mass to total mass for our Galaxy is 4–7% (Kerr and Westerhout, 1965; Westerhout, 1968).

Although there exists in M 31 no source like the Sagittarius A source in the center of our Galaxy (Pooley and Kenderdine, 1967) the remaining model of the Galaxy described above applies almost equally well to M 31.

In Table I, we have collected some parameters for M 31 and our Galaxy. In M 31, at $R = 10.8$ kpc, there is a region which dynamically resembles the vicinity of the sun.

## 2. Spiral Structure in M 31 and the Galaxy

*Distribution of* H II *regions*: Baade identified 7 arms in M 31, from a combination of features: dust, gas, OB stars. The average separation of the arms is 3 or 4 kpc. It is generally noted that these spiral arms in M 31 are less tightly wound than the spiral arms in our Galaxy as outlined by H II regions and from 21-cm observations. However, if we restrict the observations to H II regions alone, the ring 8 to 14 kpc from the nucleus in M 31 exhibits a distribution of H II regions (Baade and Arp, 1964) which in some sectors is not unlike the distribution of H II region near the sun as determined by Becker (1964).

TABLE I

Comparison of models of our Galaxy and M 31

| | Our Galaxy | M 31 |
|---|---|---|
| Nucleus | $R < 200$ pc, rapidly rotating, $V_{max} \sim 200$ km s$^{-1}$ $R = 1$ kpc, gas leaving nucleus, $V \sim 100$ km s$^{-1}$ $1 < R < 4$ kpc, rotational velocity low, small mass density | Similar to our Galaxy |
| $M_{nucleus}$ $M_R < 24$ kpc | $10^{10}\,M_\odot$, $R < 900$ pc (Rougoor, 1964) $1.5 \times 10^{11}\,M_\odot$ 50% within $R = 8.5$ kpc | $6 \pm 1 \times 10^9\,M_\odot$, $R \sim 1000$ pc $1.8 \pm 0.1 \times 10^{11}\,M_\odot$ 50% within $R = 9$ kpc |
| $V_{max}$ | $250$ km s$^{-1}$ at $R = 9 \pm 1$ kpc | $270 \pm 10$ km s$^{-1}$ at $R = 10 \pm 1$ kpc |
| At $R = 10$ kpc | $V/R = 25$ km s$^{-1}$ kpc$^{-1}$ $A = +15$ km s$^{-1}$ kpc$^{-1}$ $B = -10$ km s$^{-1}$ kpc$^{-1}$ Mass surface density $= 114\,M_\odot$ pc$^{-2}$ Log density grad. $= -1.8$ | At $R = 10.8$ kpc: $V/R = 25$ km s$^{-1}$ kpc$^{-1}$ $A = +14.6$ km s$^{-1}$ kpc$^{-1}$ $B = -10.3$ km s$^{-1}$ kpc$^{-1}$ Mass surface density $180 \pm 30\,M_\odot$ pc$^{-2}$ Log density grad. $= -1.2$ |
| To $R = 24$ kpc | $M_{HI}/M = 4 - 7\%$ | $M_{HI}/M = 4 - 8\%$ $13$ km s$^{-1}$ in rotational component |
| Vel. dispersion about rotation curve | Orion, Sag. HII regions $< 10$ km s$^{-1}$ in radial velocity (Miller, 1968) | |

In Figure 3, the H II regions identified by Baade are plotted on the plane of M 31; a distance $D = 690$ kpc and an inclination $\xi = 79°$ have been adopted. The ring 8 kpc $< R <$ 14 kpc lies between the two circles. Also shown to the same scale is the distribution of H II regions near the sun (Becker, 1964). In this region in M 31, little or no clear separation into arms is apparent. It is uncertain just what an astronomer living at $R = 11$ kpc in M 31 would deduce as the positions of the spiral arms in his galaxy; it is unlikely that he would discern 7 distinct arms. Thus, in M 31 from the distribution of H II regions alone, at a distance analogous to that of the sun, there is little evidence concerning the spacing of the spiral arms.

H II *regions and neutral hydrogen in M 31*: Roberts (1966, 1968) has shown that in M 31 the maximum in the H I distribution occurs near $R = 50'$ (10 kpc), while for the Sc galaxies studied, the maximum in the H I distribution lies well outside the spiral arms. Further, Hodge (1969) has found no correlation between the H I intensity and the distribution of H II regions in 25 spiral galaxies. For M 31, a count of identified H II regions within $\pm 5.3$ of the major axis, as a function of distance from the center, correlates well with the H I density deduced by Burke *et al.* (1964) from 21-cm observations with a 10' beam along the major axis. This is shown in Figure 4, where the NE and the SW sides are plotted separately. Along the NE major axis, the number of identified H II regions is generally a maximum where the H I density is a maximum; along the SW major axis the distribution of H II regions is more nearly constant, as is the H I density.

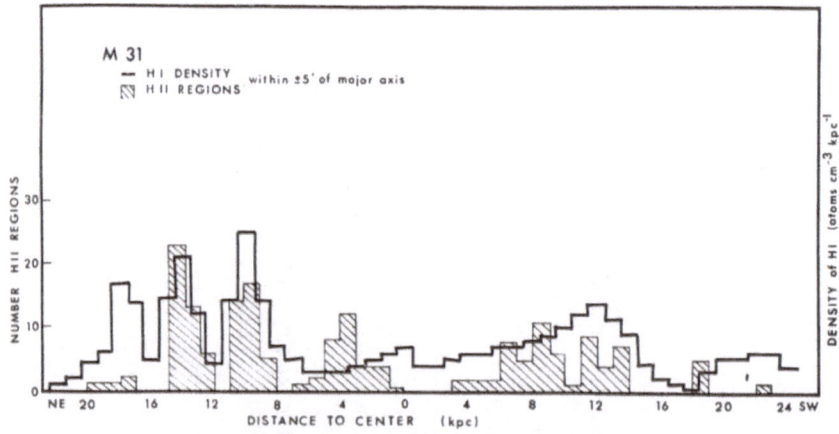

Fig. 4.    Number of H II regions in M 31 within $\pm 5'$ of the major axis, as a function of distance from the center, compared with the density of H I atoms within $\pm 5'$ of the major axis.

H II *velocities and* H I *velocities in M 31*: The circular velocities of the neutral hydrogen and the H II regions are also similar, for regions outside of the nucleus. In Figure 5 (Rubin and Ford, 1970) the 21-cm rotation curve from Burke *et al.* (1964) is plotted, together with the velocities of the observed H II regions within $\pm 5'$ of the major axis. The agreement is excellent, although the asymmetry observed in the 21-cm velocities is not apparent in the optical velocities.

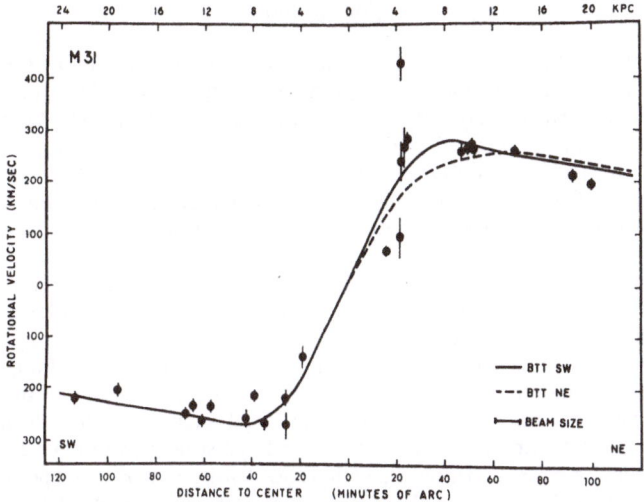

Fig. 5. Rotational velocities for 25 emission regions in M 31 within ±5′ of the major axis, as a function of distance to the nucleus. Error bars are the average error of the rotational velocity. Solid line is rotation curve of Burke *et al.* (1964) from 21-cm observations of SW major axis; dashed curve is rotational curve from observations of NE major axis. Note that the asymmetry observed in the 21-cm velocities is not apparent in the optical velocities. © 1970 University of Chicago.

In conclusion, it appears that to the accuracy of the velocities from the optical spectra, there are no major differences between the large scale features of M 31 and our Galaxy; nor between the H I velocities and H II velocities within M 31.

## References

Baade, W. and Arp, H. C.: 1964, *Astrophys. J.* **139**, 1027.

Becker, W.: 1964, *Z. Astrophys.* **58**, 202.

Burke, B. F. and Tuve, M. A.: 1964, IAU-URSI Symposium No. 20, p. 183.

Burke, B. F., Turner, K. C., and Tuve, M. A.: 1964, *Carnegie Inst. Wash. Year Book 63*, p. 341.

Hodge, P. W.: 1969, *Astrophys. J.* **155**, 417.

Kerr, F. J. and Westerhout, G.: 1965, *Stars and Stellar Systems* **5**, 167.

Miller, J. S.: 1968, *Astrophys. J.* **151**, 473.

Münch, G.: 1960, *Astrophys. J.* **131**, 250.

Oort, J. H.: 1968, in *Galaxies and the Universe* (ed. by L. Woltjer), Columbia University Press, New York, p. 1.

Pooley, G. G. and Kenderdine, S.: 1967, *Nature* **214**, 1190.

Roberts, M.: 1966, *Astrophys. J.* **144**, 639.

Roberts, M.: 1968, in *Interstellar Ionized Hydrogen* (ed. by Y. Terzian), Benjamin, New York, p. 617.

Rougoor, G. W.: 1964, *Bull. Astron. Inst. Netherl.* **17**, 381.

Rougoor, G. W. and Oort, J. H.: 1960, *Proc. Nat. Acad. Sci. U.S.A.* **46**, 1.

Rubin, V. C. and Ford, W. K., Jr.: 1970, *Astrophys. J.* **159**.

Schmidt, M.: 1965, *Stars and Stellar Systems* **5**, 513.

Shane, W. W. and Bieger-Smith, G. P.: 1966, *Bull. Astron. Inst. Neth.* **18**, 263.

Westerhout, G.: 1968, in *Interstellar Ionized Hydrogen* (ed. by Y. Terzian), W. A. Benjamin, Inc., New York, p. 638.

Woltjer, L.: 1965, *Stars and Stellar Systems* **5**, 531.

## Discussion

*J. H. Oort:* Referring to the model of the mass density distribution you propose I find it very difficult to accept that the mass density would show a pronounced dip in the region around 1 kpc from the center, for two reasons. The mass density in the central part must be largely due to old population II objects, which should be well mixed, and should have a smooth distribution. In the second place the rotation curve which Münch (cf. *Carnegie Inst. Yearbook* **63**, Rept. Mt. Wilson and Palomar Obs., p. 29) found for the *stellar* component in the central region shows no sign of the dip you found for the gas. His rotation curve, if correct, shows again that there is a smooth decrease of mass density and no dip.

I believe, therefore, that the dip you have discovered in the rotation curve for the gas indicates that in the region concerned there is a concentration of gas of very low angular momentum, which might well have come from the nucleus.

*V. Rubin:* We agree that it is reasonable to assume that the old population II objects in the central region of M 31 have a smooth distribution. However, detailed photometry of the major axis does not exist, so this must remain a conjecture. For the initial analysis of our velocity data, we have assumed that the gas is moving in equilibrium in the large scale gravitational field of the galaxy. From this assumption, a mass model results with a low density near $r = 1.6$ kpc. In support of this assumption we note that (1) Babcock's observations (*Lick Observ. Bull.* No. 498, 1939) of the H and K lines and the G-band from the stellar component also indicate a minimum in the velocities at $r = 1.6$ kpc from the nucleus; (2) measures on our plates of the stellar absorption feature at 5270 Å (FeI, the Fraunhofer E line) indicate a dip in the velocities, in agreement with the gas velocities. However, velocity measurements from broad absorption lines are of lower accuracy than velocities measured from emission lines. Additional observations at higher dispersion, both of emission and absorption lines, will help to settle this question.

Finally, even if the similarity of the motion of the stars and the gas is confirmed, there still remains the possibility that we are not observing equilibrium motions in the large scale gravitational field of the galaxy, and that the observed velocities in the inner regions cannot be used to determine the mass model there.

# 11. OB STARS ON THE OUTERMOST BORDERS OF M 31

F. BÖRNGEN, G. FRIEDRICH, G. LENK, L. RICHTER, and N. RICHTER

*Karl-Schwarzschild Observatorium Tautenburg, Jena, DDR*

Van den Bergh (1964) has selected and mapped OB associations in M 31 on the basis of plates taken with the Tautenburg 52 inch Schmidt camera. The selection was done by the blinking method. The task of the present investigation is to find if there exist OB stars (single or in association) on the outermost borders of this stellar system. For this purpose we measured the brightness of all stellar objects in UBV down to the magnitude 20ᵐ0 (B) on Tautenburg plates in a special test field (Figure 1).

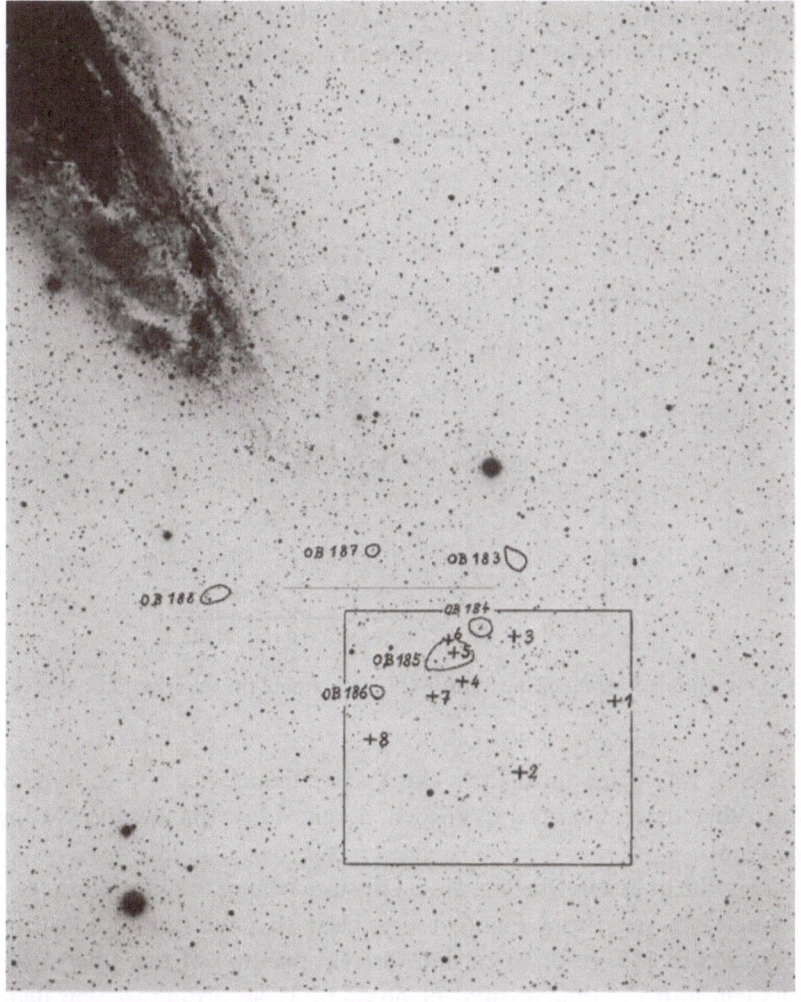

Fig. 1.   The test field and the associations OB 183 to OB 188 in the SW-section of M 31. The single OB stars discovered in the field are crosses.

*Becker and Contopoulos (eds.), The Spiral Structure of Our Galaxy, 69–71. All Rights Reserved.*

It includes 0.26 square degrees and its centre has a distance of $104' = 22$ kpc from the centre of M 31. This test field includes too Baade's field IV with the photoelectric standards in UBV observed by Arp (Baade and Swope, 1963) and the OB associations OB 184, OB 185 and OB 186 of Van den Bergh. The total number of stars brighter than $20^m0$ (B) was 996. The stars within the associations OB 184 and OB 185 are not included in these statistics. From this number, 704 objects could be measured in all 3 colours. We found among them 23 blue objects with $U-B \leqslant -0^m25$. The mean error of the brightness on the basis of 4 plates in each colour does not exceed $\pm 0^m07$ for stars of $19^m0$ (B).

By plotting these 23 blue objects in a two-colour diagram it was found, that at least 8 of them must be OB stars belonging to M 31. Figure 2 shows the position of these stars in a two-colour diagram. In Table I the magnitude (V) and colour indices of these objects are compiled. Their absolute magnitudes were determined assuming an absorption free distance modulus of $24^m30$ for M 31 and a general interstellar absorption of $0^m45$ in the line of sight according to a reddening of $E(B-V) = 0^m15 \pm 0^m06$.

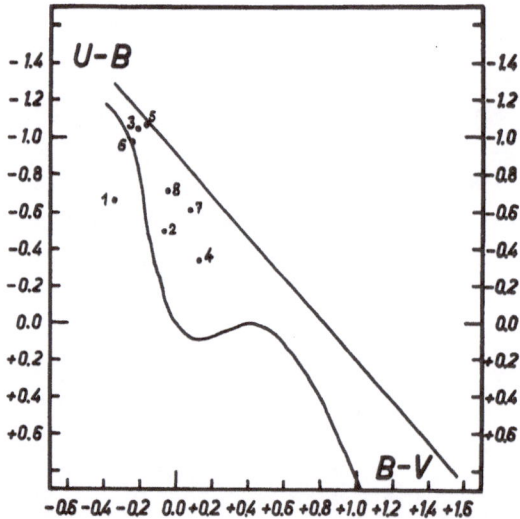

Fig. 2.   The two-colour diagram of the 8 single OB stars, found on the outermost borders of M 31.

This reddening was determined from the 8 stars themselves and is in good agreement with other determinations (Van den Bergh, 1964; Baade and Swope, 1963; Börngen, 1966).

There arises the question whether these OB stars belong to associations or whether they are single objects. From Figure 1 it is clear that the stars 5 and 6 belong to the association OB 185. In order to find the real position of the other 6 objects we projected them into the fundamental plane of M 31 according to Van den Bergh's Figure 5 assuming also a tilt angle of $12°3$. In our Figure 3 these objects are marked by crosses. There is no indication that these stars are members of any association.

TABLE I

Data for 8 stars attributed to M 31

| Star No. | V | U–B | B–V | M$_V$ |
|---|---|---|---|---|
| 1 | 19$^m$.01 | − 0$^m$.65 | − 0$^m$.33 | − 5$^m$.74 |
| 2 | 18 .34 | − 0 .50 | − 0 .07 | − 6 .41 |
| 3 | 19 .14 | − 1 .04 | − 0 .21 | − 5 .61 |
| 4 | 19 .18 | − 0 .34 | + 0 .13 | − 5 .57 |
| 5 | 19 .31 | − 1 .07 | − 0 .17 | − 5 .44 |
| 6 | 19 .65 | − 0 .98 | − 0 .25 | − 5 .10 |
| 7 | 19 .78 | − 0 .62 | + 0 .08 | − 4 .97 |
| 8 | 19 .51 | − 0 .71 | − 0 .04 | − 5 .24 |

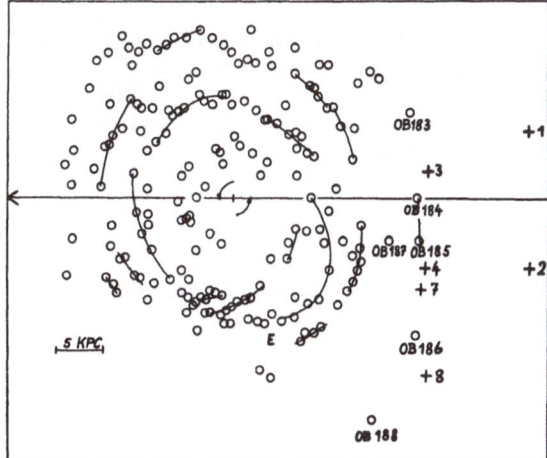

Fig. 3. Distribution of Van den Bergh's associations and of the 6 single OB stars of Table I (crosses) in the fundamental plane of M 31. In the figure the major axis of M 31 is horizontal. The tilt angle was assumed 12°.0.

They seem to be isolated and the question of their origin might be important. Two of them (1 and 2) have distances of 30 kpc from the centre of M 31. These are the largest distances found up to day for starlike objects belonging physically to M 31.

Finally it may be mentioned that in a recent investigation of another field of 0.045 square degrees by Börngen (1966), which is located at a distance of about 45–50 kpc from the nucleus of M 31, no OB stars belonging to M 31 were found. The total number of measured stars in this field down to 20$^m$.5 (B) was 301.

A more detailed information about the results presented here will be published in *Astronomische Nachrichten*.

### References

Baade, W. and Swope, H. H.: 1963, *Astron. J.* **68**, 435.
Börngen, F.: 1966, *Mitt. Karl-Schwarzschild Obs. Tautenburg* Nr. 28.
Van den Bergh, S.: 1964, *Astrophys. J. Suppl. Ser.* **9**, 65.

# 12. OBSERVATIONS OF RADIO CONTINUUM EMISSION FROM M 31

G. G. POOLEY

*Mullard Radio Astronomy Observatory, Cambridge, England*

**Abstract.** Observations of continuum emission from M 31 have been made with the Cambridge One-mile radio telescope (Pooley, 1969). Two observing frequencies were used; maps of the whole of the visible nebula were obtained at 408 MHz, and spectral data for the central region at 1407 MHz. The results show that the radiation from the disc is confined to the nucleus and to the population I spiral arms.

The nuclear region may be described in terms of two spherical radio components, with diameters of 200 pc and 1 kpc. Any more compact source at the nucleus has a luminosity less than $\frac{1}{20}$ of that of the source Sgr A in our Galaxy.

The intensity of the radio emission from the spiral arms is closely correlated with the number of H II regions visible. The main spiral arms, corresponding to Baade's arms 4 and 5, cross the major axis at about 8 and 12.5 kpc from the nucleus. The spectrum of the radiation shows that it is non-thermal in origin; the spectral index is 0.8. The intensity is less than that which would be observed from our own Galaxy at the same distance.

## Reference

Pooley, G. G.: 1969, *Monthly Notices Roy. Astron. Soc.* **144**, 101.

# 13. KINEMATICS OF M 33, M 51 AND THE L. M. C.

## G. MONNET

*Observatoire de Marseille, Marseille, France*

## 1. M 33

First interferometric results on M 33 in the Hα line at a dispersion of 20 Å/mm, involving 1048 radial velocities, have already been published (Carranza *et al.*, 1968). A new analysis with 3000 measures more will be published soon.

The main results are (see Figure 1):

(a) The arms and the disk of ionized hydrogen show pure circular motions, i.e. no systematic expansion greater than 5 km s$^{-1}$.

(b) The rotational velocity $\Theta$ in the arms at a distance $\varpi$ from the nucleus is given by $\Theta$ km s$^{-1}=38\,\varpi^{1/2}$ kpc ($\varpi<3.5$ kpc). The distance assumed is 690 kpc.

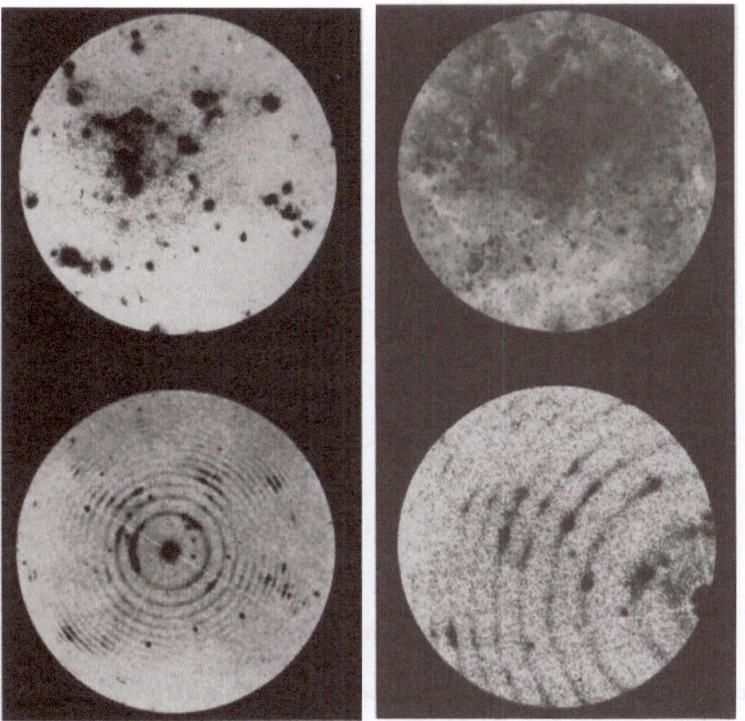

Fig. 1. This plate shows a field of 6′ diameter centered on M 33. *Upper right:* taken with a blue filter at the 77 inch Telescope of the Haute Provence Observatory. *Upper left:* taken with an interference filter 4 Å wide, in the Hα line with the 77 inch Telescope. *Lower right:* Pérot Fabry rings projected on M 33. Dispersion 20 Å/mm. Spatial resolution 2.5 second of arc with the 77 inch Telescope. *Lower left:* Pérot-Fabry rings projected on M 33. Dispersion 20 Å/mm. Spatial resolution: 1 second of arc. Taken with the Palomar 200 inch Telescope.

*Becker and Contopoulos (eds.), The Spiral Structure of Our Galaxy, 73–78. All Rights Reserved.*
*Copyright © 1970 by the I.A.U.*

(c) The mass of luminosity ratio increases from ~1 in the center to 10–15 at 7 kpc from the nucleus – in good agreement with previous results of Brandt (1965) and Gordon (1969). This suggests that even in an early type galaxy like M 33, the bulk of the mass distribution comes from population II stars.

(d) On the major axis, the rotational velocity in the concavity of the arms is smaller – by about 15 km s$^{-1}$ – than in the arms. This remarkable property seems to be closely related to the gravitational waves postulated by Lin and Shu (1964) to explain the persistence of spiral patterns in the disk of galaxies. However Lin's theory foresaw that a gravitational potential created by a density concentration in the arms gives a velocity perturbation where the disk material rotates faster in the concave side of the arms, in contradiction with the experimental result. (See Figure 2).

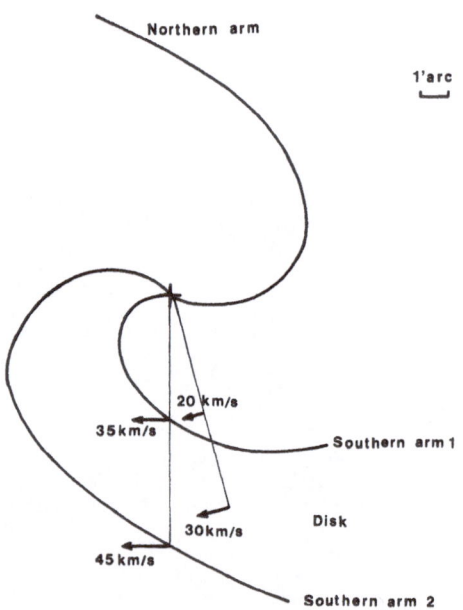

Fig. 2.   Observed velocity perturbation in M 33.

## 2. M 51

A detailed kinematic study of M 51 has been given by Carranza *et al.* (1969).

The two photographs show M 51 (Figures 3 and 4) in blue light (stars continuum) and in Hα light. One can see that the well-known spurs are no longer visible in Hα. Very often pieces of spiral arms in our Galaxy are called spurs, for instance in Orion or in Vela: they have probably nothing to do with spurs, as they are rich in bright HII regions. On the other hand, they may be like broken fragments of spiral arms as one can see in M 51 itself and on many other galaxies.

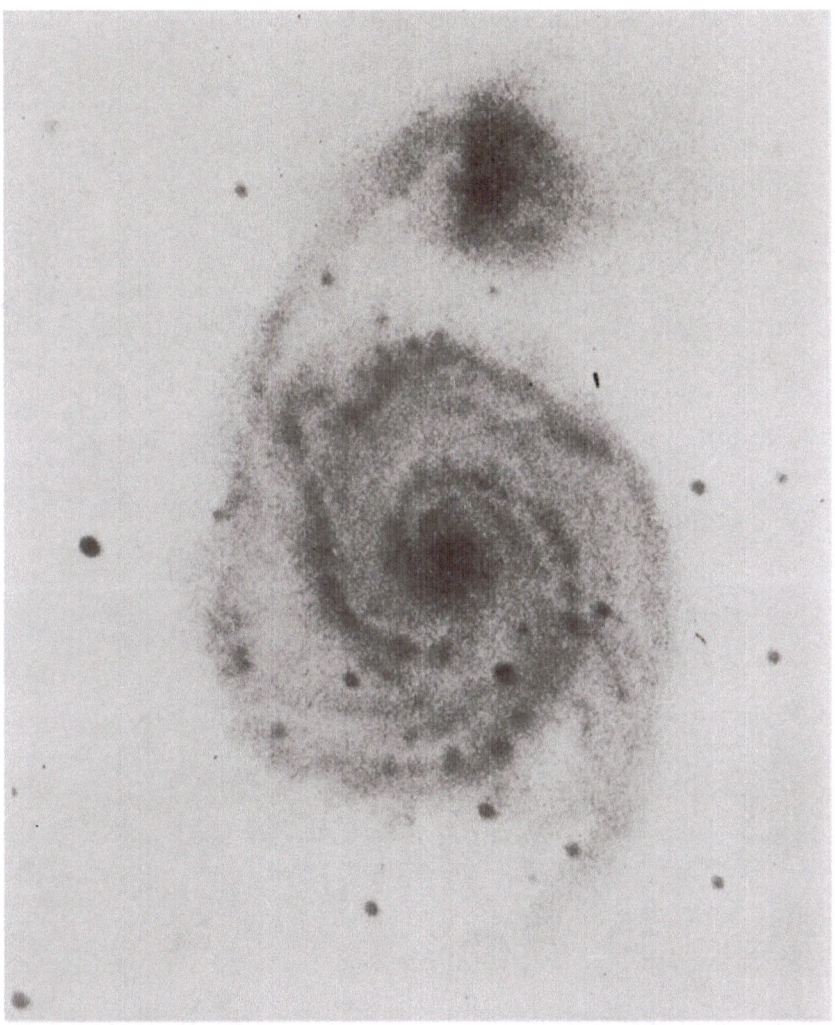

Fig. 3. Photograph of M 51 taken through a color filter centered at 4600 Å. One can see numerous inter-arm links between the main arms.

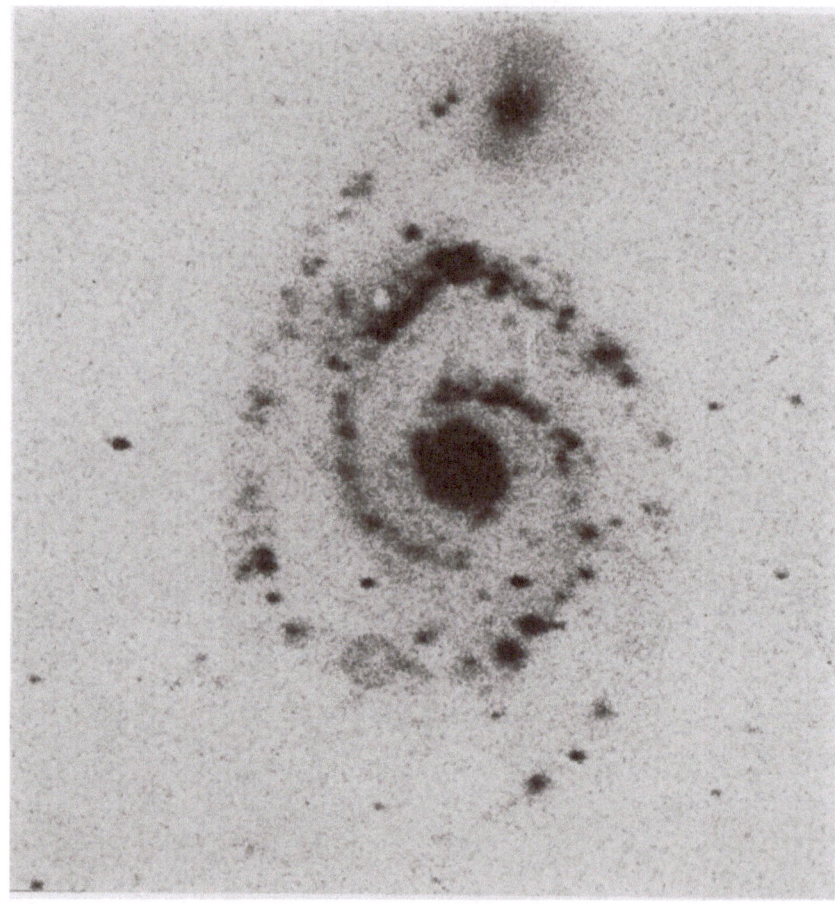

Fig. 4.   Photograph of M 51 in Hα light (interference filter 8 Å wide). The arms are specially well defined from HII regions.

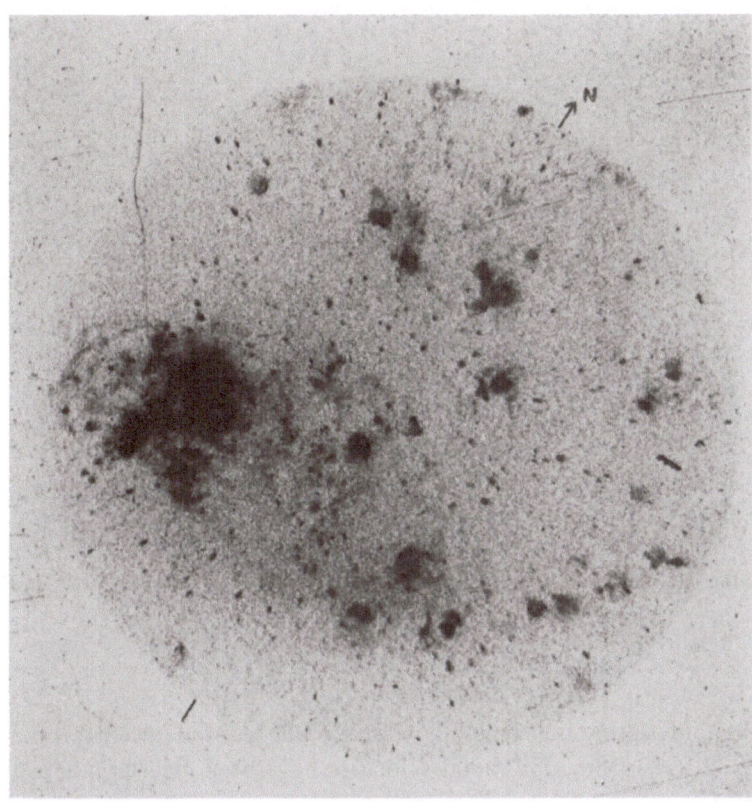

Fig. 5.   Large Magellanic cloud. Photograph in Hα light with an interference filter of 10 Å wide. Field 4°.5 in diameter.

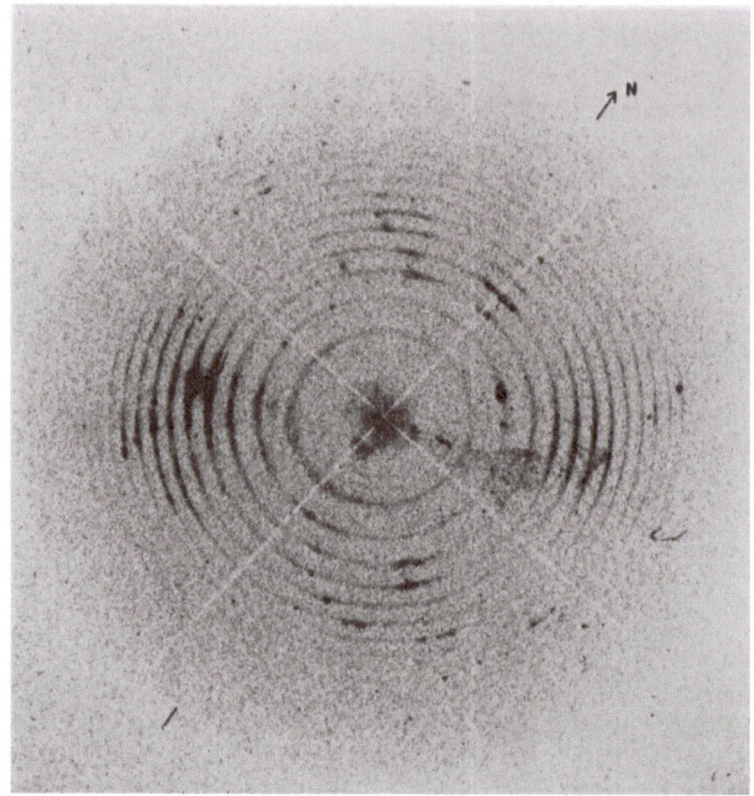

Fig. 6.    Large Magellanic cloud. The same field as in Figure 5. Rings of Pérot-Fabry in Hα ($p = 1060$).

## 3. L. M. C.

Pérot-Fabry interferograms (Figures 5 and 6) have been recently obtained in the L.M.C. by Y. Georgelin. 350 radial velocities have been measured at a dispersion of 20 Å/mm.

The first results are:

– Systematic velocity (sun):                253   km s$^{-1}$
– Maximum gradient of radial velocity:       12   km s$^{-1}$ degree$^{-1}$
to be compared with the Feast (1964) value:  14   km s$^{-1}$ degree$^{-1}$
– Mean velocity of the Doradus Nebula:      261   km s$^{-1}$
                      (Feast, 1964:  259.7 km s$^{-1}$)

### References

Brandt, J. C.: 1965, *Monthly Notices Roy. Astron. Soc.* **129**, 309.
Carranza, G., Grillon, R., and Monnet, G.: 1969, *Astron. Astrophys.* **1**, 479.
Carranza, G., Courtès, G., Georgelin, Y., Monnet, G., and Pourcelot, A.: 1968, *Ann. Astrophys.* **31**, 63.
Feast, M.: 1964, *Monthly Notices Roy. Astron. Soc.* **127**, 195.
Gordon, J.: 1969, Ph.D. Thesis.
Lin, C. C. and Shu, F. H.: 1964, *Astrophys. J.* **140**, 646.

# 14. THE STELLAR VELOCITY FIELD IN M 51

S. M. SIMKIN

*Columbia University, New York, N.Y., U.S.A., and*
*Kitt Peak National Observatory\*, Tucson, Ariz., U.S.A.*

**Abstract.** Radial velocities have been measured from the absorption lines on two image tube spectra of M 51. These velocities show large deviations from the 'smoothed' rotation curve for that object. The measurements seem to indicate that both the stars and the gas move in the same way.

At present, almost all of our information about the velocity fields in external spiral galaxies comes from observations of their gaseous component. It is now possible, with the help of image tubes, to observe the stellar motions in the disk regions of these objects as well. This note describes the initial results of such observations on M 51.

In March of 1968 two spectra were obtained of the integrated light from the disk stars in M 51. These were taken with the cassegrain spectrograph, at 100 Å/mm dispersion, and a Carnegie image tube at the 84″ telescope of Kitt Peak Observatory. The spectrograph slit for each of these plates was in an East-West position, perpendicular to the line of nodes for this object as determined by the Burbidges (1964) and Carranza *et al.* (1969). Figure 1 shows the slit positions. These were chosen to intercept the 'red' spiral arms, found in composite photographs of this object by Zwicky (1955) and Sharpless and Franz (1963). The slits also cross the well defined 'blue' arms. Thus, the composite spectra are formed primarily from the stars that participate in the spiral structure.

Radial velocities at various positions along the slit were determined by measuring redshifts for the absorption lines Hδ, Hζ, Hη, and Hθ. The emission lines Hβ, Hγ, and Hδ were also measured where present and reduced separately. The measurements were made on a Mann two-coordinate measuring machine and reduced using a two-dimensional dispersion curve, obtained from full slit comparison spectra, which gives corrections for changes in dispersion and scale as a function of both wavelength and slit position. The principal errors in these measurements arise from the difficulty of centering on the lines, which are intrinsically broad and slightly distorted by the large grain size of the image-tube phosphor. The probable errors calculated for velocities obtained at the same position from different lines and those calculated for velocities obtained from several independent measurements of the same line are of the same magnitude and amount to about 60 km s$^{-1}$ for absorption lines and 30 km s$^{-1}$ for emission lines.

The results of these measurements are plotted in Figures 2 and 3. The velocities given are heliocentric but no other corrections have been applied. In the same figures are indicated the expected radial velocities at these positions if the only motions present are circular ones following the average rotation curve for M 51 found by the

\* Operated by The Association of Universities for Research in Astronomy, Inc., under contract with the National Science Foundation.

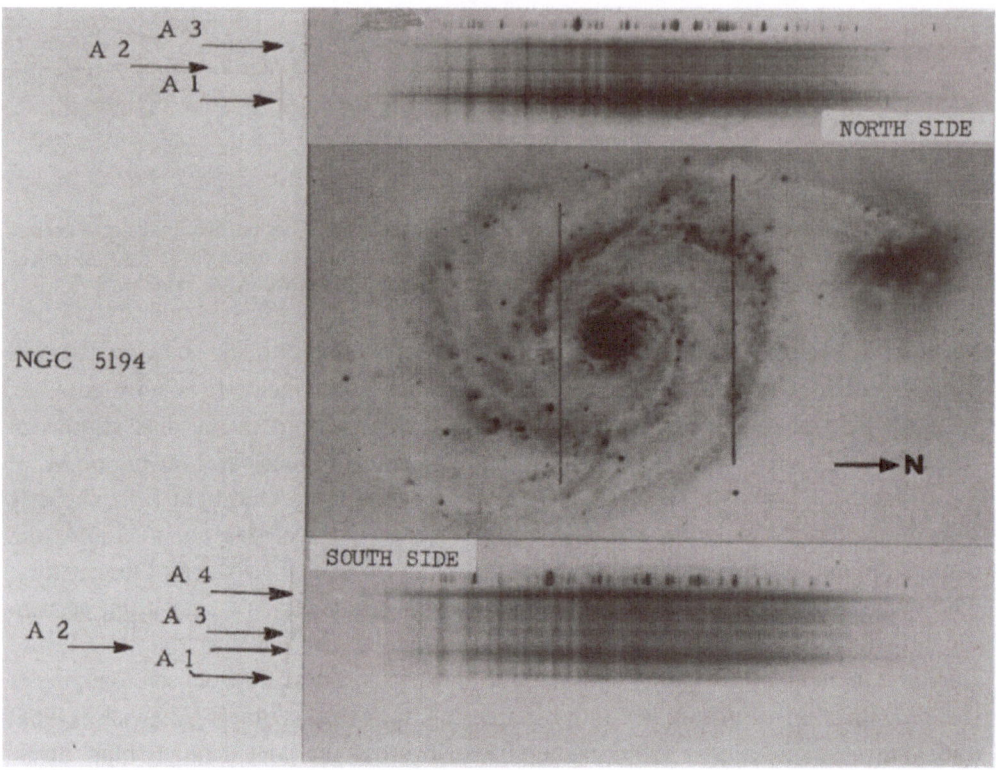

Fig. 1.    Spectra and slit positions. The photograph is from a plate taken at Kitt Peak Observatory,
April 1968. (103aO + GG 13).

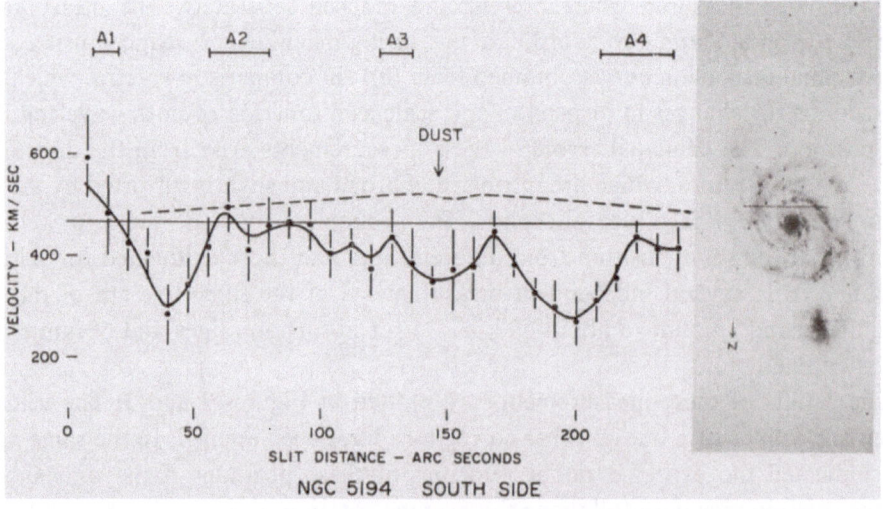

Fig. 2.    Absorption line measurements with error bars indicating twice the calculated probable error.
The dashed line is the expected velocities, the solid straight line the recessional velocity for the
system [Burbidge (1964); Carranza *et al.* (1969)].

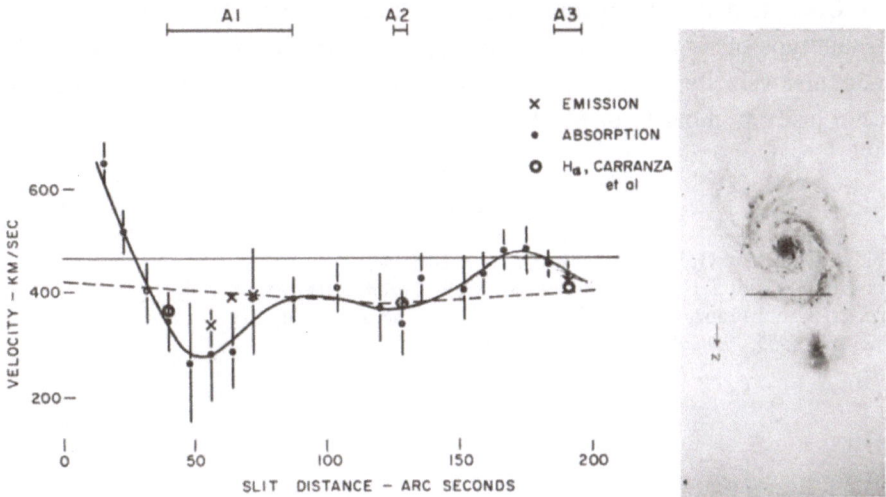

Fig. 3.   The symbols as in Figure 2.

Burbidges (1964) and Carranza *et al.* (1969). In addition, Figure 3 contains some points derived from emission line measurements on the northern spectrum and some points determined by Carranza *et al.* (1969) using Hα interferometry. In spite of the considerable deviation of the absorption line measurements from the projected 'average circular velocities' at these points, the velocities obtained from the ionized gas evidently agree with those from the stars. The agreement between the present measurements and the observations of Carranza *et al.* indicates that the absorption line data do not suffer from any large systematic error in the velocity zero point.

Inspection of Figures 2 and 3 suggests two possible conclusions. The most obvious inference is that the motions of the stars as well as the gas in the disk of M 51 show significant deviations from 'average projected circular velocity'. These deviations are not, however, as significant at points on or near the line of nodes. Since the motions of the stars and the gas are similar, they cannot be explained as the result of hydro-dynamical effects alone but must be caused by some process that effects both the stellar and the gaseous components of the disk. The lack of notable deviations at the points near the line of nodes may indicate that the non-circular velocities are primarily in the $z$ direction, as found by the Burbidges for the gas near the nucleus. This possibility can only be decided by further observations along the line of the nodes.

The second impression gained from Figures 2 and 3 is that the absorption line velocities tend to be closest to the 'average projected circular velocities' near the outer parts of the blue spiral arms. Clearly this is only a tentative suggestion and also requires further observations.

It is possible that the velocity deviations reported here are the result of the per-turbing influence of NGC 5195. There is, however, no direct evidence for this in the present observations. It is, moreover, the author's impression, gained from inspection of lower dispersion spectra of NGC 628 and 1232 and from conversations with

A. D. Code, that the absorption lines in spectra of other face on, Sc objects exhibit the same type of non-circular velocities as found for M 51. If this is so, then such non-circular velocities may be a general feature of well defined, late spiral systems and not just a peculiarity of M 51.

## References

Burbidge, E. M. and G. R.: 1964, *Astrophys. J.* **140**, 1445.
Carranza, G., Grillon, R., and Monnet, G.: 1969, *Astron. Astrophys.* **1**, 479.
Sharpless, S. and Franz, O. G.: 1963, *Publ. Astron. Soc. Pacific* **75**, 219.
Zwicky, F.: 1955, *Publ. Astron. Soc. Pacific* **67**, 232.

## 15. Hɪɪ REGIONS IN NGC 628, NGC 4254 AND NGC 5194

K. CHUVAEV and I. PRONIK

*Crimean Astrophysical Observatory, Nauchny, Crimea, U.S.S.R.*

Multicolour observations of galaxies are being carried out at the prime focus of the 2.6 m Schajn telescope using an image converter and 6–9 colour filters. The effective wavelengths are approximately 3600, 3730, 4400, 4680, 5090, 5280, 6090, 6600 and 7400 Å. The filters for 3730, 5090 and 6600 Å are centred on emission lines. For absolute calibration extrafocal star images are used.

The observations of the Sc galaxies NGC 628, NGC 4254 and NGC 5194 were carried out in 1965–69. Two of the galaxies are single, but NGC 5194 is double.

The energy distributions for the central region and for dozens of bright patches in each galaxy have been determined. The mean colour characteristics of the central regions and bright patches are given in Table I. Here $n$ is the total number of bright patches measured, while $n_1$ of them are Hɪɪ regions; $J_\lambda$ are the corresponding brightnesses in the wavelength $\lambda$.

TABLE I

Ratios of brightnesses in three colours

| NGC | $n$ | $n_1$ | Central region | | Bright patches | |
|---|---|---|---|---|---|---|
| | | | $\dfrac{J_{3600}}{J_{4350}}$ | $\dfrac{J_{4350}}{J_{5550}}$ | $\dfrac{J_{3600}}{J_{4350}}$ | $\dfrac{J_{4350}}{J_{5550}}$ |
| 628 | 60 | 42 | 0.4 | 0.54 | 0.8 | 1.7 |
| 4254 | 53 | 38 | 0.58 | 0.81 | 1.1 | 1.7 |
| 5194 | 85 | 58 | 0.46 | 0.70 | 1.2 | 1.4 |

The central regions (with diameters ∼400 pc) of NGC 4254 and NGC 5194 are bluer than that of NGC 628. Moreover there are huge Hɪɪ regions in the centres of NGC 4254 and NGC 5194, with sizes of the order of the corresponding central discs. The brightnesses of the central Hɪɪ regions near Hα are by 10–20% higher than the continuous spectrum. No emission has been found near Hα in the central disc of NGC 628. Compact groups of stars have been observed in the central discs too.

Most patches in spiral arms have more or less smooth spectral energy distribution (except in the Hα region) due to stellar radiation. As it is seen from Table I the star patches are bluer than the central regions of the galaxies. Moreover, the colour characteristics of the blue spectral region of the patches of NGC 4254 and NGC 5194 are like those of B 3 stars, and of the green like A–F stars. The spiral arm patches of NGC 628 are redder than those of NGC 4254 and NGC 5194. All bright star patches under consideration are not younger than A–F stars.

*Becker and Contopoulos (eds.), The Spiral Structure of Our Galaxy, 83–86. All Rights Reserved.*
*Copyright © 1970 by the I.A.U.*

In most patches the spectral continuum breaks at the Hα region where a rise of radiation is observed (Figure 1). In some of H II regions one can suspect the presence of nebular lines too. The upper limit of masses of H II regions in spiral arms of all galaxies is equal to $10^5 \, M_\odot$, but perhaps it must be lowered by one order of magnitude, because the filamentary structure of gas nebulae has not been taken into account.

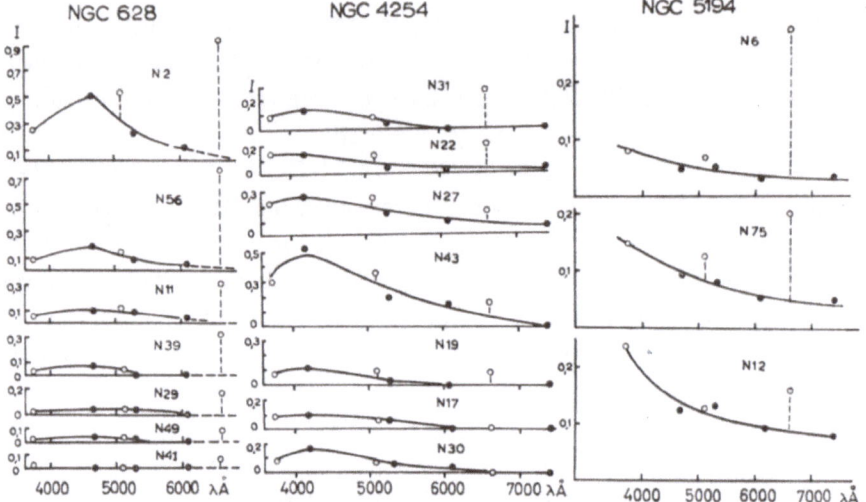

Fig. 1.    Energy distribution in the spectra of some typical spiral arm patches, shown in Figures 2, 3, 4. Circles indicate the positions of possible emission lines.

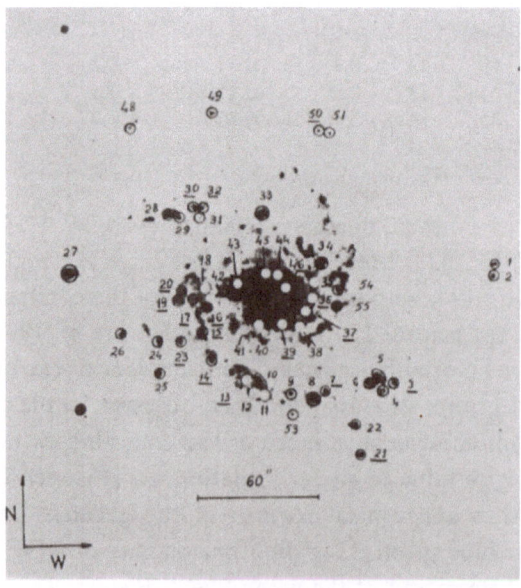

Fig. 2.    Spiral arm patches considered in NGC 4254. Underlined numbers are for the patches without Hα emission.

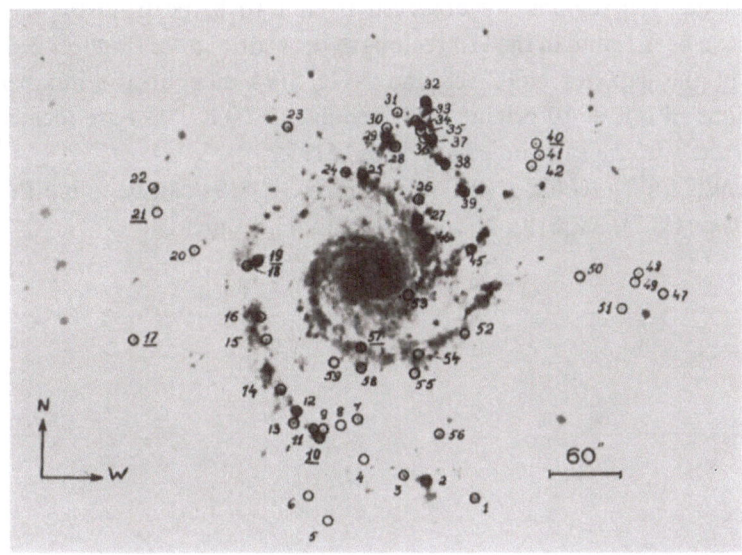

Fig. 3.   The same as in Figure 2 for NGC 628.

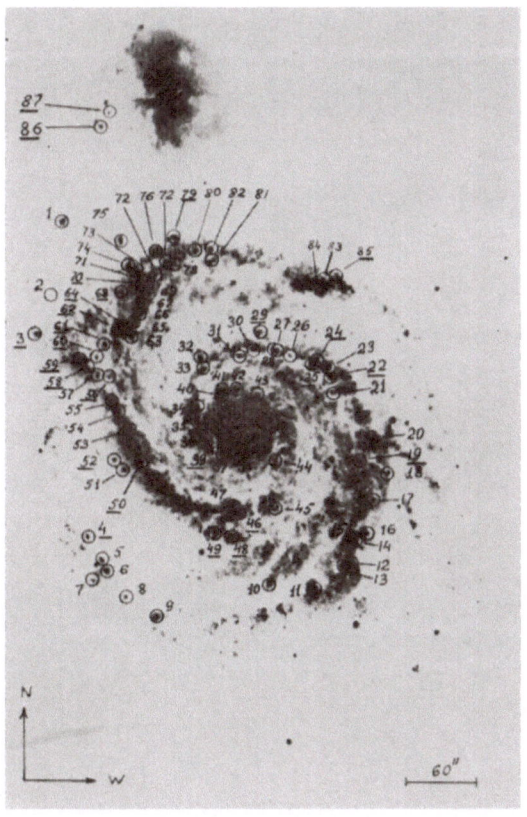

Fig. 4.   The same as in Figure 2 for NGC 5194.

In all three galaxies H II regions are observed from 1 to 8 kpc from the centre, the nearest ones have been found in the H II regions of the central discs (Figures 2, 3, 4). The spiral arms of the galaxies NGC 628 and NGC 5194 differ in the number of brightest H II regions: 8 out of 10 brightest gas nebulae of NGC 5194 are located in the spiral arm directed to the satellite galaxy NGC 5195.

The detailed results for NGC 628, NGC 4254 and NGC 5194 can be found in *Izv. Krym. Astrofiz. Obs.* (1967), **38**, 219; **40**, in press; and **43**, in press.

# 16. SPIRAL STRUCTURE AND DISTRIBUTION
# OF STELLAR ASSOCIATIONS IN NGC 6946

E. YE. KHACHIKIAN and K. A. SAHAKIAN

*Byurakan Astrophysical Observatory, Erevan, Armenia, U.S.S.R.*

**Abstract.** The associations of NGC 6946 outline its spiral arms. There is no relation between the colour or magnitude of the associations and their distance from the centre. Their mean absolute magnitude is $-11^{m}.1$ and their mean colour index near zero.

## 1. Introduction

Some photometric results concerning a number of stellar associations in the spiral galaxy NGC 6946 are discussed.

The observational material was obtained with the 2-m Schmidt camera of Tautenburg Observatory (DDR) by one of us (Khachikian). The results of a detailed two-colour surface photometry investigation of this galaxy have been published earlier (Börngen *et al.*, 1966). The method of observations is described there as well.

NGC 6946 is a Sc type galaxy. There are a great number of stellar associations in it. The galaxy has a small nucleus and two strong spiral arms with many branches. The remarkable feature of NGC 6946 is the abundance of supernova explosions during the last 60 years. The last explosion took place in 1968 (Wild, 1968). This fact allows to consider NGC 6946 as a giant galaxy (Kukarkin, 1965; Ambartsumian, 1965).

We have adopted the following data for this galaxy: distance modulus $(m-M)_0 = 28.3$, absorption $A_{pg} = 2^{m}.3$, absolute magnitude $M = -20^{m}.6$. (A more detailed discussion of these data is presented by Khachikian and Sahakian (1970).)

## 2. Distribution of Associations

The chart of distribution for all the 148 measured associations is given in Figure 1. The associations are shown projected on the spiral structure. It seems that in general associations are located in spiral arms. Two spiral arms are coming out directly from the nucleus. These arms split into three branches at a distance of 1' from the centre of the galaxy. The two innermost branches are the richest in associations, containing 35 of them each. Outside the arms the number of associations decreases abruptly.

## 3. Magnitudes of Associations

The absolute magnitude of associations varies between $-9^{m}.6$ and $-12^{m}.8$, the mean being $-11^{m}.1$. These data are in good agreement with Markarian's (1959) and Bok and Bok's (1962) observations. According to them the mean absolute magnitudes of the associations of M 101, M 51 and LMC are $-11^{m}.1$, $-11^{m}.2$ and $-10^{m}.6$ respectively.

*Becker and Contopoulos (eds.), The Spiral Structure of Our Galaxy, 87–90. All Rights Reserved.*
*Copyright © 1970 by the I.A.U.*

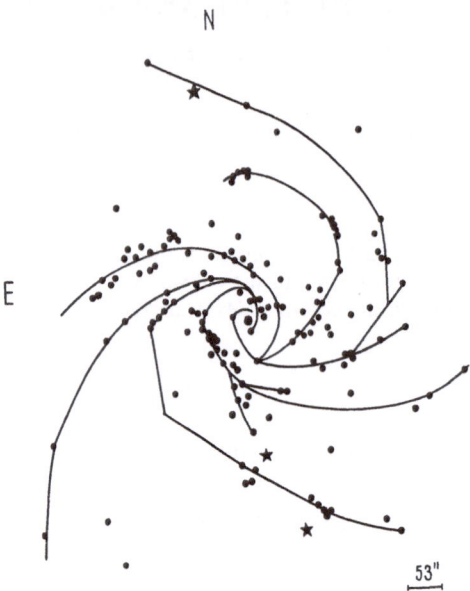

Fig. 1.   The distribution of the associations in NGC 6946 together with the galaxy's spiral structure.

Let us examine the distribution of associations according to their magnitude. For this purpose the associations have been divided into four groups in the following way:

$$
\begin{aligned}
&\text{I group}      && B > 20\overset{m}{.}0 \\
&\text{II group}  && 19\overset{m}{.}5 < B \leqslant 20\overset{m}{.}0 \\
&\text{III group} && 19\overset{m}{.}0 < B \leqslant 19\overset{m}{.}5 \\
&\text{IV group}  && B \leqslant 19\overset{m}{.}0 .
\end{aligned}
$$

In Figure 2 the distribution of each of these groups is given. Among the groups there is no clear-cut distribution depending on the distance from the centre. It is necessary to note that the brighter the associations, the more they form concentrations. The associations with $B > 19\overset{m}{.}5$ do not show such a feature, while about half of the associations with $B < 19\overset{m}{.}5$ form concentrations of 2, 3 and 4. Some of these are located in those parts of spiral arms where branches are formed. It is interesting to note that the majority of associations with $19\overset{m}{.}0 < B < 19\overset{m}{.}5$ are located in the two inner spiral arms. Among the faint associations with $B > 20\overset{m}{.}0$, four are forming a chain which is located near the nucleus between the two spiral arms.

## 4. Colour Indices of Associations

The observed colour indices of associations vary from $-0\overset{m}{.}1$ to $+1\overset{m}{.}3$. With the adopted value of colour excess $0\overset{m}{.}6$, the real indices fall in the interval $-0\overset{m}{.}7$, $+0\overset{m}{.}6$. The mean colour index for all associations is $(B-V)_0 \simeq 0\overset{m}{.}0$. The majority of associations have colour indices between $+0\overset{m}{.}3$ and $+0\overset{m}{.}8$. In Figure 3 the distribution of associations all over the galaxy according to their colour is given. As it is seen from Figure 3

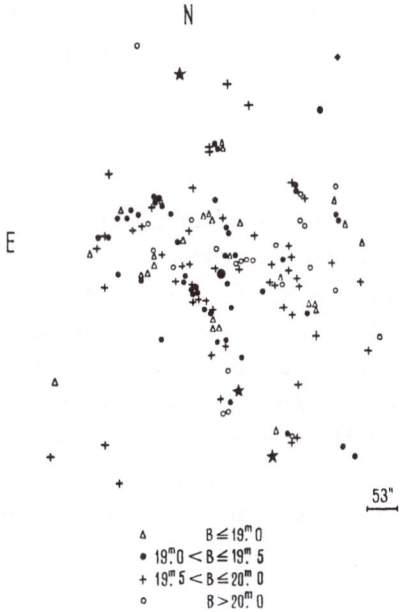

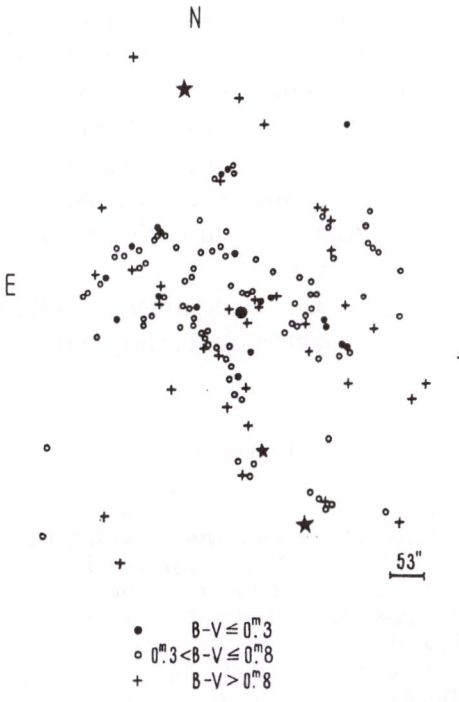

Fig. 2.   The distribution of the different groups of associations along the galaxy: ◯) $B > 20^m.0$; +) $19^m.5 < B \leqslant 20^m.0$; ●) $19^m.0 < B \leqslant 19^m.5$; △) $B \leqslant 19.0$.

Fig. 3.   The distribution of the associations depending on their colour: ●) $B-V \leqslant +0^m.3$; ◯) $+0^m.3 < B-V \leqslant +0^m.8$; +) $B-V > +0^m.8$.

there is no correlation between the colour of associations and their distance from the nucleus. Associations with different colours are found in all spiral arms and at any distance.

## 5. H II Regions in NGC 6946

It is known that stellar associations often contain emission nebulae. As the O-associations include stars of earlier spectral classes, they are responsible for regions of ionized hydrogen around them. When the quantity of hydrogen is high enough, these regions can be observed as emission nebulae. Therefore most of the associations can be observed as H II regions and vice-versa. According to Hodge (1967) NGC 6946 contains 40 H II regions. We can expect that most of them coincide with associations investigated by us. In fact 26 H II regions have been identified with our associations. Among them there are objects with colours varying from $-0^m45$ to $+0^m35$. However most of them have negative colour indices. The mean value of $(B-V)_0$ for the H II regions identified as associations is $-0^m07$.

## 6. Conclusions

(1) The galaxy NGC 6946 is rich in stellar associations, that outline its whole spiral structure. The brighter associations have a tendency to form concentrations. This is observed especially near the points where spiral arms split into branches. The majority of associations are found in the two inner spiral arms of the galaxy.

(2) The associations do not show any relation between their colour or magnitude and their distance from the centre of the galaxy. The same phenomenon has been also noticed by Markarian (1959) in the associations of M 51 and M 101. The associations apparently have similar characteristics, not depending on their position in the galaxy. Therefore the process of stellar formation occurs simultaneously all along the spiral arms. Hence, the sources that lead to the formation of associations are distributed more or less uniformly along the arms.

(3) The mean absolute magnitude of associations is $-11^m1$ and their mean colour index near zero. NGC 6946 is probably a giant galaxy with absolute magnitude about $-20^m6$.

## References

Ambartsumian, V.: 1965, *Astrofiz.* **1**, 473.
Bok, B. and Bok, P.: 1962, *Monthly Notices Roy. Astron. Soc.* **124**, 435.
Börngen, F., Kalloglian, A., Khachikian, E. Ye., and Eynatian, J.: 1966, *Astrofiz.* **2**, 431.
Hodge, P.: 1967, *An Atlas and Catalogue of H II Regions in Galaxies*, Seattle.
Khachikian, E. Ye. and Sahakian, K. A.: 1970, *Astrofiz.*, in press.
Kukarkin, B.: 1965, *Astrofiz.* **1**, 465.
Markarian, B.: 1959, *Soobshch. Byurak. Obs.* **26**, 3.
Wild, P.: 1968, IAU Circ. No. 2057.

PART II

# OBSERVATIONS OF SPIRAL STRUCTURE IN OUR GALAXY

# A. RADIO OBSERVATIONS

# 17. SPIRAL STRUCTURE OF NEUTRAL HYDROGEN IN OUR GALAXY

F. J. KERR

*University of Maryland, College Park, Md., U.S.A.*

**Abstract.** This paper discusses the evidence on the hydrogen spiral structure which is available from 21-cm observations. The problems involved in deriving a hydrogen map are discussed, and one interpretation of the spiral pattern is presented. Some of the major characteristics of the H I in spiral arms are discussed, and a comparison is made between the kinematics of H I and H II.

## 1. Introduction

I wish to discuss some of the observational data about hydrogen in spiral arms, partly as an introductory review, and partly to present some new results. We know quite clearly that H I concentrates to elongated features, which are presumably spiral arms and their minor components, but the detailed pattern is far from clear. The difficulties come from:

(1) The enormous amount of fine structure detail, which makes it hard to recognize the main features, or even to know whether there is a regular pattern.

(2) As is well known, distances are kinematic, and other methods of deriving distance can only help to a small extent.

## 2. Velocity-Distance Problem

In getting kinematic distances, we assume that differential rotation is always the most important motion, and usually consider that rotation is circular and axisymmetric, or at least follows a regular pattern. A great deal of attention has been given to derivation of the best possible rotation curve. The only method available for the main part of the Galaxy is from the 21-cm observations themselves, using the well-known method in which the extreme velocity on a line profile is considered to represent the circular velocity at the tangential point, the position on the line of sight which is closest to the galactic center. Extrapolation to the region outside the sun is commonly done through a model such as the Schmidt model, which is designed to fit the inner region results, with some assistance from optical data near the sun.

Figure 1 shows again a pair of such curves from 21-cm observations at Parkes, giving the variation of the (assumed circular) rotational velocity in the galactic plane with distance from the center, for the first and fourth quadrants of longitude (the so-called 'northern' and 'southern' sides of the Galaxy). We see the well-known north-south asymmetry, the approximate similarity of the two curves, the large-scale oscillation which is presumably related to end-on spiral arms, and the small-scale fluctuations.

Rotation curves derived by other workers from 21-cm observations (e.g., Shane

*Becker and Contopoulos (eds.), The Spiral Structure of Our Galaxy, 95–106. All Rights Reserved.*
Copyright © 1970 by the I.A.U.

and Bieger Smith (1966) in the north and Bajaja *et al.* (1967) in the south) agree quite well with the Parkes curves. The detailed differences that are found are probably due to the use of different beamwidths and bandwidths. There is also an interesting field of study in the differences between rotation curves at different *z*-levels with respect to the galactic plane, or the curve obtained always from the latitude where the hydrogen density is a maximum, and not restricted to the galactic plane. In the present context, however, these minor differences do not seriously affect the spiral mapping problem, where the greatest difficulty is the restriction to kinematic distances.

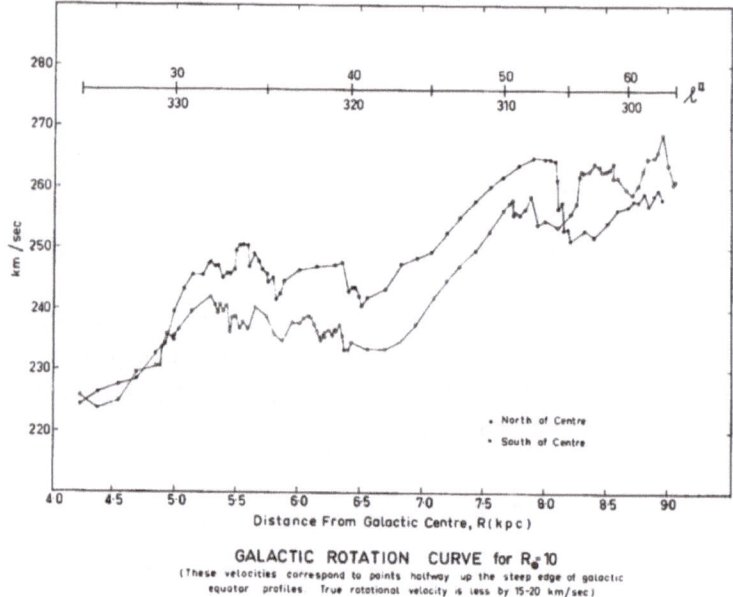

GALACTIC ROTATION CURVE for $R_e$ 10
(These velocities correspond to points halfway up the steep edge of galactic equator profiles. True rotational velocity is less by 15-20 km/sec)

Fig. 1.   Galactic rotation curves for the northern and southern sides of the center, derived from tangential point observations, assuming circular rotation (Kerr, 1964).

The north-south asymmetry shows immediately that the motion is not entirely circular; perhaps there is an oval distortion or some other large-scale effect. The velocity fluctuations are related to the fine structural details, and these interfere seriously with the possibility of drawing a precise map of the spiral structure. For any small feature, there is no way for us to distinguish between its peculiar velocity and the rotational component, and distances must be uncertain by an amount related to the peculiar or non-circular component. Such errors are proportionately greatest at short distances, where the hydrogen picture is consequently weakest, but they must be present all over the Galaxy. A further and probably related complication is that the very presence of a spiral feature distorts the velocity pattern in its vicinity, as will be discussed by some of the theoretical workers. Clearly the structural and kinematic problems must be solved together.

We cannot expect, as used to be thought, that we can ever produce a fully detailed

map of the hydrogen structure on a kinematic basis alone. A better understanding
of the large-scale velocity field should give us a reasonable picture of the major spiral
arms, but accurate positions for the smallest hydrogen features will always elude us.

## 3. Mapping

Figure 2 shows one version of the hydrogen spiral pattern, projected on to the galactic
plane; this is based on observations by Kerr and Hindman (Parkes) and Henderson
(Green Bank). In the outer region, the Schmidt model and circular motion have been
assumed. Within these limitations, a clear trailing pattern can be seen. Near the sun,
a Carina-Cygnus arm is shown, but it is not clear whether this is actually continuous
through the sun. As stated above, the possibility of peculiar motions makes this part
of the diagram quite uncertain in detail.

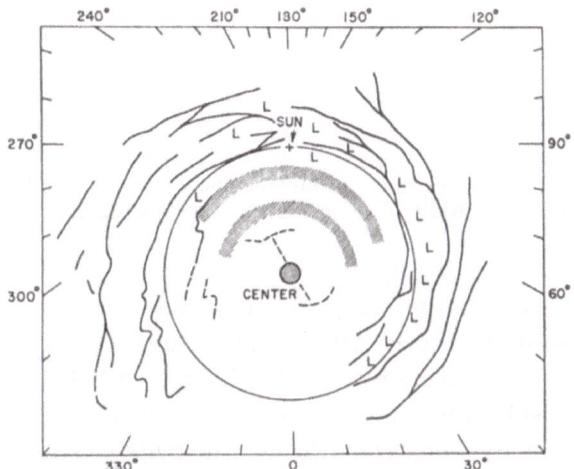

Fig. 2.   Sketch of the main features of the neutral hydrogen spiral structure, from observations of
Kerr (1969a), Hindman (1969) and Henderson (1967). Structural details are not shown in the inner
region, owing to the large uncertainty in the distance. Regions of low hydrogen density
are indicated by L.

Inside the sun's position the two major arms, the Sagittarius arm and the Norma-
Scutum arm, are only roughly sketched, and no attempt has been made to represent
the detail in this region. Further in are the 'expanding' 3-kpc arm and a possible
rudimentary bar of high-velocity gas.

The great complexity of the distribution in the inner region is clearly seen in
Figure 3, due to G. Knapp, in which the identifiable features are drawn in a velocity-
longitude plot, with the best estimate of their continuity. The solid-line features are
at positive latitudes, and the dashed-lines are negative; the width of the lines gives
an indication of the importance of each feature. The latitude characteristics of the
features will be discussed in more detail later.

The Sagittarius arm, Norma-Scutum arm, and the 3-kpc arm can be clearly seen in the diagram, as well as another feature near the center at $-70$ to $-50$ km s$^{-1}$. Each major arm shows considerable branching and splitting, and there are suggestions of cross-linkages. It must be remembered, however, that features which are adjacent or in contact on this diagram are not necessarily near to each other in space, as their latitudes may be different and also different peculiar velocity components may give them different distances.

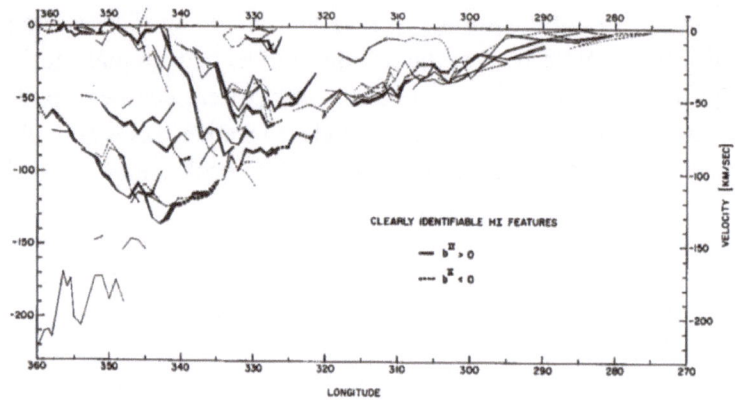

Fig. 3. Velocity-longitude plot of clearly-identifiable spiral features, with the best estimate of their continuity. The width of a line indicates the importance of the feature.

To go from this stage to a real map, i.e., to convert velocity into distance, requires a full understanding of the velocity field, and also a resolution of the velocity ambiguity for each component, to decide between the 'near' and 'far' locations. The smaller features at least must have non-circular components of velocity, and are consequently difficult to locate in position. The velocity ambiguity can sometimes be resolved, particularly for the major features, but not sufficiently completely for a good map to be drawn. There is a strong suggestion that the two major arms are curving away in a trailing sense. This can be seen in Figure 4 for the Sagittarius arm, where the wide feature at $-55$ km s$^{-1}$ and the narrower one at $-22$ km s$^{-1}$ probably belong to a single trailing arm, with a pitch angle of 8°. The arm trails in a similar way in the first quadrant, but the feature is not regular enough to be drawn in over its entire length.

Partial evidence on the location of the inner arms can be obtained from considerations of tangency. We saw earlier that the major oscillations in the rotation curves indicate the tangential directions of the major arms. Also, in a plot of the integrated hydrogen (Kaplan, unpublished), the Norma-Scutum arm shows up strongly in cross section on both sides, centered at about $l^{II} = 330°$ and 31°. On the southern side, the arm appears to be centered a few tenths of a degree above the equator, while on the northern side it is centered right on the equator. The Sagittarius arm can also be seen in cross-section at $l^{II} = 303°-305°$ and at 55°, but less strongly than is the case for the Norma-Scutum arm.

These tangential directions also show up clearly as steps in the continuum back-

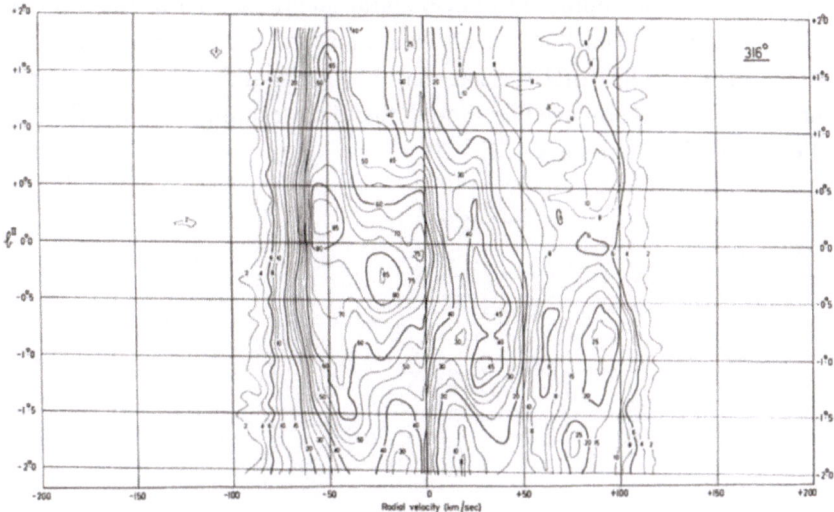

Fig. 4. Contour diagram in velocity-latitude plane for $l^{II} = 316°$ (Kerr, 1969a).

ground, as was first pointed out by Mills (1958), and in the clustering of discrete sources, mainly H II regions. The total flux in low-latitude sources is plotted as a function of longitude in Figure 5, from the observations of Beard *et al.* (1969).

In addition to the Norma-Scutum and Sagittarius arms, the Carina and Cygnus directions stand out as important concentrations of various constituents, both radio and optical. These results suggest that a Carina-Cygnus arm passes through the sun, but the continuity is difficult to establish in this nearby region. There appear to be fairly clear gaps in the pattern around longitudes 295° and 60°.

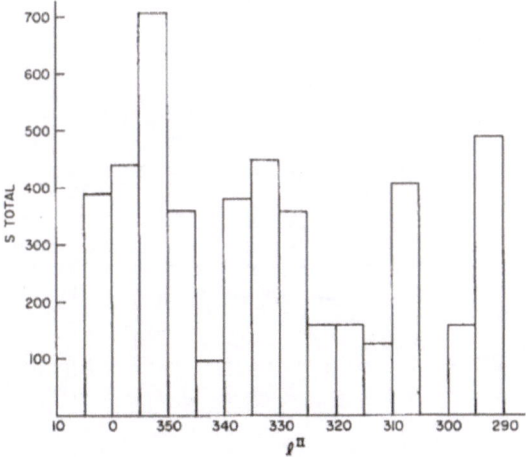

Fig. 5. Histogram of the total flux in discrete sources for $|b| < 2°$, as a function of longitude. (From observations of Beard *et al.*, 1969.)

It is well known that radio and optical spiral diagrams tend to look different, and it has often been suggested that gas and stars may be subject to different dynamical influences – in particular, that the gas may be significantly affected by magnetic or other nongravitational forces. We should therefore examine the relative kinematic properties of H I, H II and stars.

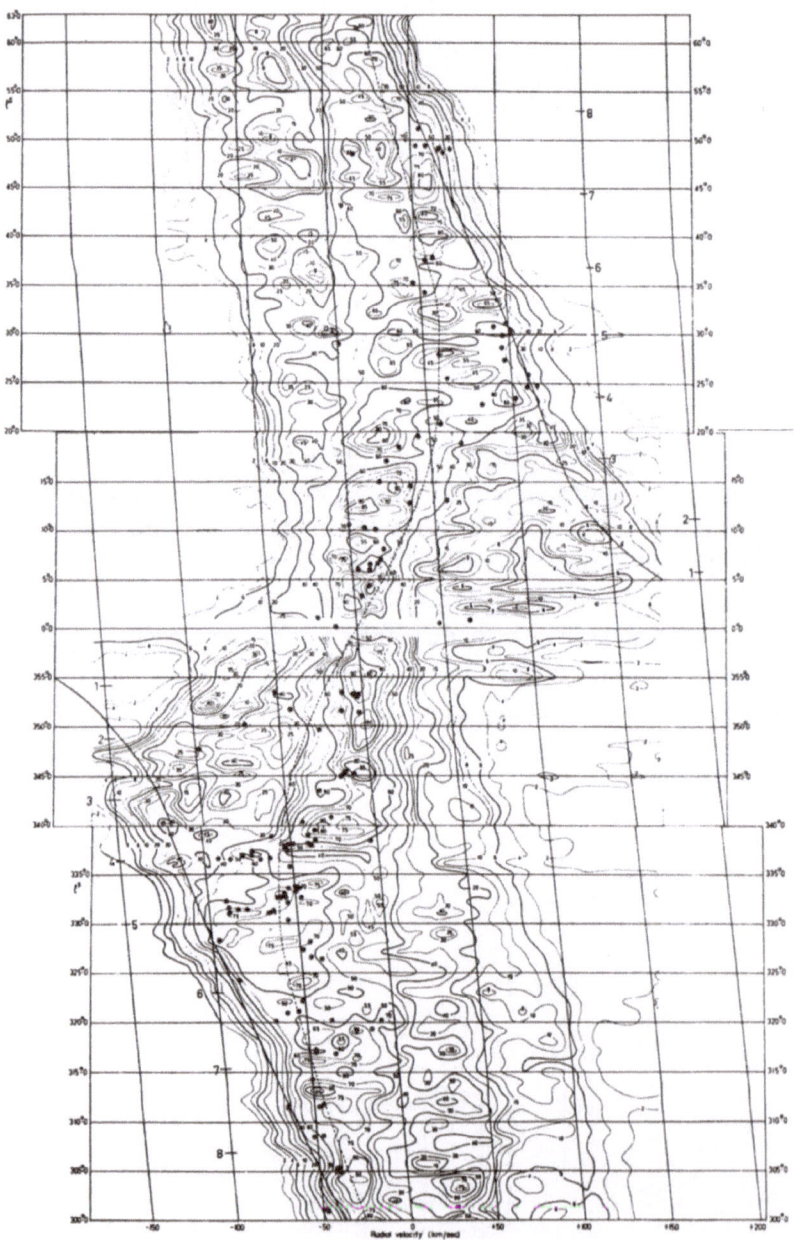

Fig. 6.   Comparison between recombination-line velocities (black dots), measured by Mezger *et al.* (1970), and H I velocity-longitude contour plot (Kerr, 1969a).

### 4. H I and H II

A relatively new radio approach to the spiral structure problem is through recombination line observations, which give evidence on the distribution and motions of H II regions. Dr. Mezger has given a detailed account of the radio work on H II regions, but I will present some collaborative work on comparisons between the kinematical properties of H I and H II.

Figure 6 shows such a velocity-longitude plot, to be published by Mezger *et al.* (1970), in which they superpose their H II observations from a joint NRAO-CSIRO survey at Parkes, Australia, on an H I contour map from Parkes (Kerr, 1969a). There is a line of H II sources following the main ridge of local low-velocity H I. The higher-velocity H II regions avoid the nuclear section, but otherwise they lie close to the H I ridges that are associated with the main spiral arms. Individual H II regions show their own peculiar velocities, but there is no sign of a systematic difference between H I and H II motions exceeding a few kilometers per sec.

An approximate rotation curve can also be derived for H II regions by fitting an upper envelope to the ensemble of the recombination line measures. This has been done by Mezger *et al.*, who obtained a curve that is close to the H I rotation curve (Figure 7).

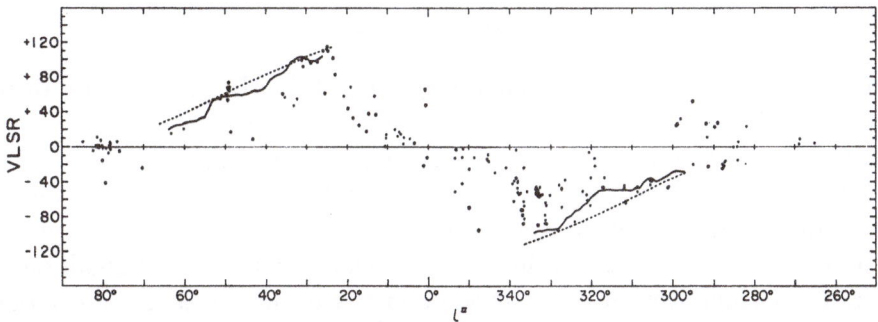

Fig. 7.  Rotation curve derived from the upper envelope of the recombinationline velocities of Mezger *et al.* (1970).

Another type of H I-H II comparison is obtained from a study of 21-cm absorption, i.e., the absorption of continuum emission from H II regions in cooler H I clouds in the foreground. From such absorption observations, it is possible in many cases to gain information about the distances of H II regions which are too far away to be measurable optically. We cannot get precise absolute distances in this way, but we can place the H II regions in relation to the H I emission features. Kerr and Knapp (1969) have shown that the H II regions can reasonably be located in this way in the main H I spiral arms.

Additional evidence comes from the velocities of absorption features in the 18-cm lines of OH and the 6-cm line of $CH_2O$. These measures are not numerous enough to give an independent spiral picture, but the various types of velocity measures for gas in spiral arms are in approximate agreement. One direction in which we can expect good progress in the near future is in the detailed comparison of all the various types of measurement in as many directions as possible.

## 5. Gas and Stars

Several recent solar-motion solutions have been obtained with respect to the local neutral hydrogen, and all gave results that are close to the 'standard' solar motion, which is derived from the mean motion of the common stars. (For a detailed review, see Kerr, 1969b.) Solar-motion solutions with respect to the young Population I stars give results that differ a little from the standard solar motion, but such solutions depend greatly on selection effects and must be largely influenced by the irregularities of motion that occur over the several kiloparsecs in which the young stars can be seen.

Any systematic difference between the motions of gas and young stars cannot exceed 3–4 km s$^{-1}$, which is the uncertainty produced by the irregularities in the regional motions in our vicinity. In regions of extreme irregularity, such as the portion of the Perseus arm studied by Rickard (1968) and Miller (1968), there are large departures from circular motion, but the significant result for the present discussion is that the H I, H II, Ca and stars all appear to be moving essentially together.

## 6. Some Properties of H I in Spiral Arms

Although we cannot yet set up a definitive picture of the H I spiral pattern, there are a number of things we can say about the H I in spiral arms.

Firstly, the H I is clearly concentrated to spiral-like features. A major pattern can be seen, but there appears to be a high degree of splitting and interconnection of the main features. Even the major arms themselves show a longitudinal substructure in the form of variations of density and peculiar velocity, on a scale of about one kiloparsec. The regional velocity variations are in fact related to the density variations along an arm, as if each of the semi-discrete patches has its own peculiar velocity component. The Gould belt, the anomalous-velocity region of the Perseus arm, and the concentration of gas in the vicinity of W 51 may be considered as examples of three types of localized deviations from a smooth pattern.

One of the most important things that the observationalists can say to the theoreticians is that the pattern is highly irregular and fragmented, though certainly not as much so as in some of the pathological cases we can find in external systems. We should try to set up some measure of the 'degree of regularity' that we think we can see.

The density contrast between arms and interarm regions is an important quantity, but this is difficult to specify exactly in most of the Galaxy, because adjacent features tend to overlap each other on our contour maps. This arises from the internal velocity and latitude spread of the features themselves, and is not due to instrumental effects. In the few places where arms do stand out clearly, such as the regions outside the sun's distance from the center at longitudes around 40° and 240°, there is a clear arm-interarm contrast of density of over 10 to 1, but we do not know whether this degree of contrast exists in other parts of the Galaxy, especially in the inner part. One type of observation suggesting a low degree of contrast was first pointed out by Burke some years ago. In some parts of the Galaxy, the spiral arm peaks stand on top of a much broader background, as if there is a distributed component in addition to the arms. On the other hand, we often see matter apparently connected with the spiral arms, but extending a long way from the galactic plane. See for example the results of Goniadski in Argentina, quoted by Kerr (1969b).

The neutral hydrogen layer is very flat over most of the Galaxy, but measurable deviations are found for many of the spiral features. These are of the order of tens of parsecs, and the interesting thing is that a typical arm stays consistently above or below the plane for a considerable part of its length. This property assists in establishing the continuity of a particular feature (Henderson, 1967). Henderson, and also Varsavsky, have suggested that successive arms may be alternately above and below the plane, but I think the overall arrangement is probably not as regular as this.

Early 21-cm results showed that the hydrogen layer is twisted in the outer parts in a systematic way, downwards on one side, and upwards on the other. We do not yet know for certain whether this effect is internally or externally produced, and thus whether it is relevant to the spiral structure problem. It is clear, however, that spiral structure is strongly established in these twisted outer regions. The normal relief map, which is based on the hydrogen centroid, suggests a gradual bending of a smooth layer, but this is an oversimplification. The hydrogen is strongly concentrated into arms, and in some places one sees two clearly distinct arms at about the same velocity, but at different latitudes. (Note that equality of velocity does not necessarily mean equality of distance.) An example is shown in Figure 8 (Hindman, 1969).

The hydrogen layer also broadens out considerably in the outer regions, as shown by Van Woerden and others. The thickness between half-density points increases from

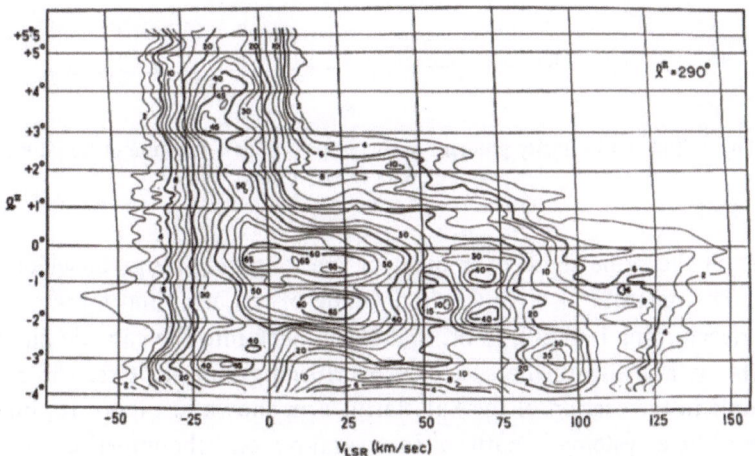

Fig. 8.   Contour diagram in velocity-latitude plane for $l^{II} = 290°$ (Hindman, 1969), showing the complexity of the spiral structure in the turned-down outer part of the Galaxy.

80–100 pc near the center and 200 pc in the solar region to 500 or 1000 pc further out.

One of the interesting effects frequently found inside a spiral arm is an 'overturning motion'. For example, in the 3-kpc arm, the upper portion is approaching us more rapidly than the lower part, as first shown by Rougoor (1964). More recent data from the Parkes survey are given in Figure 9, which shows that the overturning motion

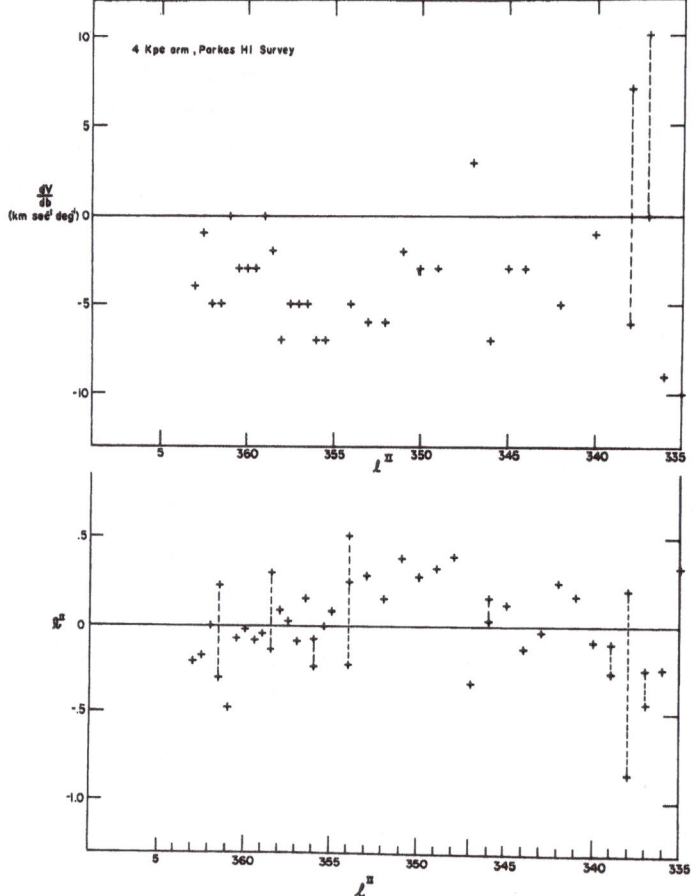

Fig. 9.   The latitude and velocity gradient in the 3-kpc arm as a function of longitude, from Parkes survey data (Kerr, 1969a).

continues in the same sense over most of the longitude range in which we can see the arm. The diagram also shows the latitude of the arm, and there is a suggestion that the overturning is acting to bring the arm back into the galactic plane. Unfortunately, this is not always the case. An overturning motion is seen in many places, for example in the section of the Sagittarius arm shown in Figure 10, but there does not seem to be a systematic pattern. In several regions, the upper part is moving out from the center more quickly, but the reverse is also found. These effects may well be related to helical magnetic field patterns in spiral arms.

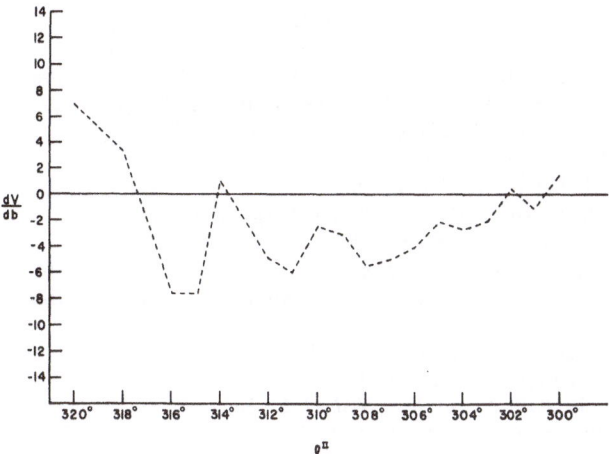

Fig. 10.   Velocity gradient in part of the Sagittarius arm, from Parkes survey data.

## 7. Symmetry

I have placed great stress on irregularity. Another boundary condition to be placed on galactic models can come from considerations of symmetry. The velocity asymmetry has been discussed already. We have also studied the integrated hydrogen in various regions of the Galaxy and find some fairly clear results (Kaplan, unpublished).

(1) Inside the circle through the sun, the amount of hydrogen in the first and fourth quadrants is closely the same, except for a slight excess near the center in the fourth quadrant ('southern' side).

(2) The thickness of the hydrogen layer is noticeably larger in the fourth quadrant than the first in the general vicinity of the tangential point regions.

(3) Outside the solar circle, there is a distinct excess of hydrogen in the fourth quadrant over that in the first, and the effect persists at all latitudes in the disk layer.

These conclusions can be influenced by the velocity field and by optical depth considerations, but we have found no reasonable model to explain the observed results other than a real excess of material on one side over the other.

## 8. Conclusion

The 21-cm observations show clear evidence of a spiral pattern in the Galaxy, probably somewhere in the Sb-Sc range. The detailed pattern is not yet clear, but a great deal of information is available on the characteristics and location of the spiral features.

## Acknowledgement

This work has been supported by the U.S. National Science Foundation.

# References

Bajaja, E., Garzoli, S. L., Strauss, F., and Varsavsky, C. M.: 1967, IAU Symposium No. 31, p. 181.
Beard, M., Day, G. A., and Thomas, B. M.: 1969, *Australian. J. Phys.*, in press.
Henderson, A. P.: 1967, Ph.D. Thesis, University of Maryland.
Hindman, J. V.: 1969, in preparation.
Kerr, F. J.: 1964, IAU Symposium No. 20, p. 81.
Kerr, F. J.: 1969a, *Australian. J. Phys. Astrophys. Suppl.,* No. 9, 1.
Kerr, F. J.: 1969b, *Ann. Rev. Astron. Astrophys.* **7**, 39.
Kerr, F. J. and Knapp, G.: 1969, *Australian. J. Phys. Astrophys. Suppl.*, in press.
Mezger, P. G., Wilson, T. L., Gardner, F. F., and Milne, D. K.: 1970, *Astron. Astrophys.* **4**, 96.
Miller, J. S.: 1968, *Astrophys. J.* **151**, 473.
Mills, B. Y.: 1958, IAU Symposium No. 9, p. 431.
Rickard, J. J.: 1968, *Astrophys. J.* **152**, 1019.
Rougoor, G. W.: 1964, *Bull. Astron. Inst. Netherl.* **17**, 381.
Shane, W. W. and Bieger Smith, G. P.: 1966, *Bull. Astron. Inst. Netherl.* **18**, 263.

# 18. THE DISTRIBUTION OF HII REGIONS

P. G. MEZGER

*National Radio Astronomy Observatory\*, Green Bank, W. Va., U.S.A., and
Max-Planck-Institut für Radioastronomie, Bonn, Germany*

**Abstract.** The distribution of optically observed HII regions and OB stars with galactic longitude indicates that it is primarily determined by extinction by interstellar dust. Thus optical observations can, at the best, reveal the local structure in the vicinity of the sun. Radio observations, on the other hand, are not affected by dust. Thus the distribution of galactic radio sources, which peaks in the northern part at about $l^{II} = 17°.5$, must be related to the large-scale structure of our Galaxy. Two radio recombination line surveys of the northern and southern sky yield kinematic distances. If only the 'giant HII regions' are retained, the following distribution is obtained: (1) Only 5 giant HII regions are found within the 4 kpc arm. (2) The bulk of the giant HII regions is concentrated in a ring between 4 and 6 kpc from the galactic center. (3) There are other concentrations of giant HII regions indicating the existence of the Sagittarius and Perseus arm. (4) The three features revealed by optical observations of HII regions in the vicinity of the sun cannot be matched with the large-scale distribution outlined by giant HII regions. This is particularly true for the so-called Orion arm. (5) At distances beyond 13 kpc from the galactic center virtually no giant HII regions are found. (6) The surface density of giant HII regions attains its maximum between 4 and 8 kpc; the surface density of neutral hydrogen (HI) attains its maximum between 11 and 15 kpc, but the actual space density of HI in the region 4 to 8 kpc may still be rather high.

## 1. Spiral Arms and Star Formation

Spiral arms in our galaxy appear to be defined primarily by a density of the interstellar matter higher than that in the interarm region and by the existence of hot, massive early-type stars which ionize the surrounding gas. According to a theory advanced by Lin and Shu (1964, 1966), the spiral pattern is maintained by a density wave which rotates between 4 and 12 kpc from the galactic center with an angular velocity lower than that of the general galactic rotation of the Population I material. As the interstellar material passes through the density wave potential minimum, it is compressed and star formation on a large scale – as observed in external galaxies – may thus be initiated. It was first suggested by Clark (1965) and subsequently confirmed by various authors (for a comprehensive review see Mebold, 1969) that the interstellar matter in spiral arms consists of dense, cool clouds embedded in (and possibly in pressure equilibrium with) an intercloud gas with temperatures of several thousand degrees K. It may well be that O-star clusters and associations form out of these dense and cool clouds.

## 2. Star Formation and HII Regions

The radio flux from an ionization bounded HII region is directly related to the flux of Lyman continuum photons from the exciting star(s). It is found that most of the

\* The NRAO is operated by Associated Universities, Inc., under contract with the National Science Foundation.

*Becker and Contopoulos (eds.), The Spiral Structure of Our Galaxy, 107–121. All Rights Reserved.
Copyright © 1970 by the I.A.U.*

H II regions observed as thermal radio sources require one or more early-type O-stars for their ionization, i.e. stars with main sequence lifetimes of a few $10^6$ years. My associates and I have investigated in some detail the obvious relation between H II regions and regions of star formation in a number of papers, which are summarized in a recent review paper (Mezger, 1969) or which are in preparation (Felli and Churchwell, 1969), respectively. In the context of the present review paper on galactic structure the following results are relevant:

(a) Most H II regions which are strong thermal radio sources consist of one or more compact components of high electron density but small linear dimensions, which are embedded in an extended low-density H II region. It appears that the compact H II regions represent very early evolutionary stages of subgroups, which Blaauw (1964) observes in O-star associations (Schraml and Mezger, 1969).

(b) These compact components, even in nearby extended H II regions, are often heavily obscured by dust. As a rule, their exciting stars cannot be seen (Schraml and Mezger, 1969).

(c) On the other hand, most of the optically observed H II regions (e.g. those catalogued by Sharpless, 1959) are only weak thermal radio sources. It appears that they represent later evolutionary stages, where the ionized gas has expanded and the circumstellar dust clouds of the exciting stars have been dispersed or destroyed (Felli and Churchwell, 1969).

It is well known that, owing to the extinction by interstellar dust, optical observations of H II regions and their exciting stars are generally limited to distances $\lesssim 3$ kpc from the sun while radio waves are hardly affected by the interstellar medium. However, in comparing the spiral structure of our galaxy as outlined by optical and radio observations of H II regions the additional selection effect that radio and optical observations pertain to different evolutionary stages of H II regions and O-star clusters should be considered, too.

## 3. Optical Observations of H II Regions

Rather than reviewing all the earlier optical work pertaining to H II regions and spiral structure I will refer in this section to what appears to be the most complete set of observations of OB-stars and H II regions.

Miss Sim (1968), based on the Hamburg-Warner and Swasey survey, investigated the distribution of $OB^+$, OB and $OB^-$-stars as a function of both galactic longitude (1968) and latitude (private communication). The former results are reproduced in Figure 1a–c. Using the catalog of H II regions by Sharpless (1959) I have prepared the corresponding diagram for the longitude distribution of optically visible H II regions* in Figure 1d. The correlation between H II regions and early-type stars is best for the $OB^+$-stars. In latitude, the half power width of the Sharpless H II regions is about $3°$, that of the OB-stars is between $4°$ and $5°$ (Sim, private communication).

---

* The percentage of H II regions in Figure 1d refers to the total number of 224 H II regions in the longitude range $0° \leqslant l^{II} \leqslant 180°$.

The distribution of both OB-stars and H II regions exhibits a minimum between $20° < l^{II} < 90°$, with the exception of the Cygnus X region ($l^{II} \simeq 76°$). As will be shown later, it is, however, that very longitude range which coincides with the most active regions of star formation in our galaxy. Obviously, obscuration by dust clouds is the predominant factor that determines the distributions of OB-stars and H II regions in Figure 1a–d.

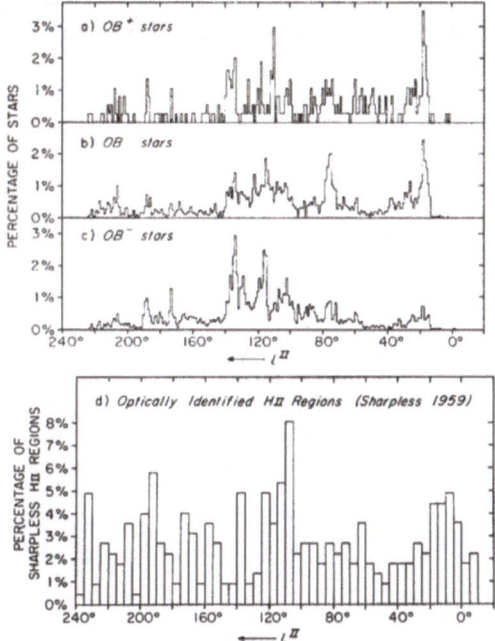

Fig. 1   (a)–(c). Distribution of OB-stars in galactic longitude. – (d) Distribution of optically visible H II regions in galactic longitude.

Courtès *et al.* (1968) investigated the local distribution of H II regions, using both photometric distances of exciting stars and kinematic distances derived from optically determined radial velocities. Figure 2, reproduced from their paper, shows the distribution of H II regions based on photometric distances. There appear reasonably well outlined parts of the Perseus, Orion, and Sagittarius arm. At a longitude of 330° three distant H II regions are seen which are probably members of the Norma-Scutum arm. The pitch angle of the three arms is about 20°.

## 4. Radio Continuum Surveys

The era of high resolution surveys of the galactic continuum radiation in the GHz range was opened with a present-day classic paper by Westerhout (1958). In this frequency range the thermal radiation from ionized hydrogen begins to predominate over the diffuse non-thermal background radiation. Westerhout discovered three basic characteristics of the galactic continuum radiation:

(1) The bulk of the galactic continuum radiation comes from a narrow range centered about the galactic plane (referred to as disk component). It consists of a thermal component with a very narrow distribution in latitude (HPW $\simeq 1°6$) and a non-thermal disk component with a considerably wider HPW of $\simeq 4°2$.

(2) As a function of galactic longitude the thermal component attains a maximum at $l^{II} \simeq 26°$, whereas the non-thermal component increases steadily towards the galactic center. From this latter result Westerhout concluded that most of the ionized hydrogen must be concentrated in a ring just outside the 4 kpc arm.

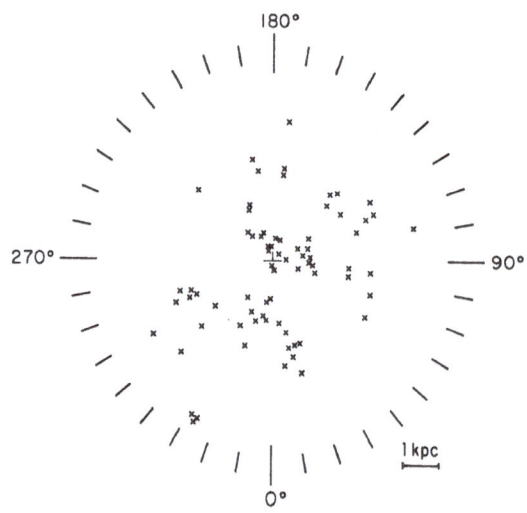

Fig. 2.   Projected positions of optically visible HII regions, based on photometric distances of their exciting stars.

(3) Superimposed on the disk components are a number of discrete sources compiled in the Westerhout catalog; these are presently referred to by their W-numbers.

Subsequent surveys have refined Westerhout's observations (e.g. Altenhoff *et al.*, 1960) or expanded them to the southern part of our galaxy (e.g. Mathewson *et al.*, 1962) but did not essentially change Westerhout's basic conclusions.

It took exactly 10 years before another big step forward was made in the radio continuum surveys of our galaxy. Stimulated by the detection of radio recombination lines (following section) and a limited survey at 5 GHz with the NRAO 140-ft telescope (Mezger and Henderson, 1967) Altenhoff, using the NRAO 140-ft telescope at 2.7 GHz carried out a complete survey of the galactic plane visible from Green Bank within the latitude limits $b^{II} = \pm 2°$. A first report on his results was published in 1968. Owing to both the higher angular resolution of the 140-ft telescope and the higher sensitivity of modern broadband radiometers equipped with low-noise preamplifiers, Altenhoff has detected about ten times as many sources as given in Westerhout's source catalog. It turned out that in Altenhoff's survey Westerhout's thermal disk component is nearly completely resolved into individual thermal sources *. In Table I recent high resolution continuum surveys of the northern and southern part of the Galaxy are listed.

In the remainder of this section I will refer mainly to three northern surveys by Altenhoff *et al.* (1969). These surveys were made with telescopes of different sizes; owing to an appropriate frequency selection (Table I) between 1.4 and 5 GHz, the angular resolution of the three surveys was nearly identical, i.e. $\simeq 10'$. The same reduction method was applied, including the separation between sources and background.

This procedure yielded another new and important characteristic of galactic sources. Their spectra could now be determined from the peak flux densities (or main beam brightness temperatures, respectively) at the three frequencies without any additional assumptions on the source size which make earlier work on galactic spectra so highly unreliable. Altenhoff *et al.* (1969) determined spectra of 206 galactic sources out of which 141 have spectral indices $(S \propto v^{\alpha}) \, \alpha \gtrsim -0.3$ and therefore could be thermal. An unexpectedly large fraction (31%) of the investigated sources have spectral indices $\alpha \lesssim -0.3$ and thus are obviously non-thermal. Altenhoff (1968) could not find a strong correlation between the distribution of these non-thermal sources and the diffuse non-thermal background radiation. However, at least in some cases non-thermal sources appear to be closely associated with HII regions (e.g. W 28, W 49, W 51).

Figure 3 shows the longitudinal distribution of sources. All three surveys have in common the longitude range $12° \leqslant l^{\mathrm{II}} \leqslant 55°$. The two surveys at 2.7 and 5 GHz cover the additional range $345° \rightarrow 0° \rightarrow 12°$. Only the 2.7 GHz survey extends beyond 55°; that part of the reduction is preliminary and the results are therefore indicated by dashed lines. (The hatched areas pertain to the 5 GHz H 109$\alpha$ and continuum survey of northern sources which will be discussed in the following section.)

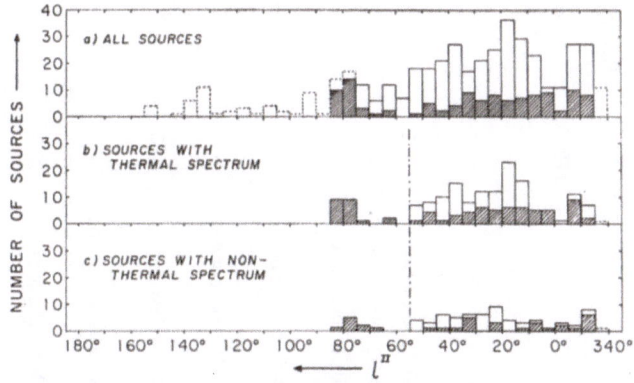

Fig. 3   (a)–(c). Distribution of galactic radio sources in galactic longitude, as observed at $v = 2.7$ GHz. The shaded areas pertain to the H 109$\alpha$ observations at $v = 5$ GHz.

---

* From this result it cannot be implied, however, that most of the ionized hydrogen in our galaxy is concentrated in completely ionized HII regions since radio radiation depends on the emission measure, i.e. the square of the electron density integrated along the line of sight. Thus radio surveys are heavily weighted towards regions of hot, dense plasma.

## TABLE I

### Recent high resolution galactic continuum surveys

| ν/MHz | Telescope | HPBW | Detection limit (peak flux density) | $l^{II}$ | $b^{II}$ | Completeness | References |
|---|---|---|---|---|---|---|---|
| 408 | Mills cross | 2'.8 in EW | | 210°–0° 0°–50° | | 72 selected regions | (1) |
| 1420 | NRAO 300-ft. telescope | 10'.2 | 0.5 f.u. | 12°–55° | ±5° | | (2) |
| 1420 | Parkes 210-ft. telescope | 14'. | | 280°–355° | ±2° | | (3) |
| 2700 | NRAO 140-ft. telescope | 11'.0 | 2.0 f.u. | 345°–0° 0°–155° | ±2° | | (2) |
| 2700 | Parkes 210-ft. telescope | 7'.3 | 1.0 f.u. | 288°–0° 0°–38° | ±2° | | (4) |
| 5000 | NRAO 140-ft. telescope | 6'.5 | | 347°–0° 0°–210° | | 120 individual sources | (5) |
| 5000 | Parkes 210-ft. telescope | 4'.0 | | 210°–0° 0°–50° | | 63 selected regions | (6) |
| 5000 | Fort Davis 85-ft. telescope | 10'.8 | 3.0 f.u. | 340°–0° 0°–55° | ±3° | | (2) |
| 15375 | NRAO 140-ft. telescope | 2'.0 | | 350°–0° 0°–210° | | 22 selected regions | (7) |

(1) Shaver and Goss (1969); (2) Altenhoff et al. (1969); (3) Hill (1968); (4) Staff of CSIRO, Div. of Radiophys. (1969); (5) Reifenstein et al. (1970); (6) Goss and Shaver (1969); (7) Schraml and Mezger (1969).

Part (a) of Figure 3 shows the distribution of all sources observed at 2.7 GHz. Parts (b) and (c) are the corresponding distributions of thermal and non-thermal sources. Note that the sum of thermal and non-thermal sources for a given longitude interval is usually smaller than the total number of all sources observed at 2.7 GHz. The larger fraction of sources are H II regions and it appears to be permissible, therefore, to compare the general source distribution in Figure 3a with that of optically identified H II regions in Figure 1d. The difference in the two distributions is evident and there appears to be – in the large-scale distribution at least – an anticorrelation rather than a correlation. The distribution of radio sources exhibits a minimum about the galactic center, increases rapidly to a maximum about 17.5° and gradually tapers off towards the galactic anticenter which is an obvious 'zone of avoidance' of radio sources. There are secondary maxima in the source distribution about 37.5°, (62.5°), 77.5°, (92.5°), 132.5° and 152.5°. The secondary maxima whose longitudes are given in brackets hinge on an increased source number in one longitude interval only and statistical fluctuations therefore cannot be excluded completely. It is of interest to note, however, that only these somewhat uncertain maxima in the source distribution are matched by corresponding maxima in the optical distribution (Figure 1d).

This obvious anticorrelation in the distribution of optically identified H II regions and radio sources appears to be the joint result of two selection effects discussed in Section 2 (radio observations favor O-star associations and clusters in very early evolutionary stages) and Section 3 (obscuration by dust especially in regions of active star formation). It is thus clear that only radio observations can further the investigation of the large-scale spiral structure of our galaxy.

The latitude distribution of radio sources and optically identified H II regions given in Figure 4a–d tends to confirm this conclusion. The distribution of radio sources is obviously the result of the superposition of a narrow distribution representing the intrinsically intense, distant sources and a much wider distribution representing nearby and intrinsically weak sources. The optically identified H II regions show only the wide distribution representing the local objects.

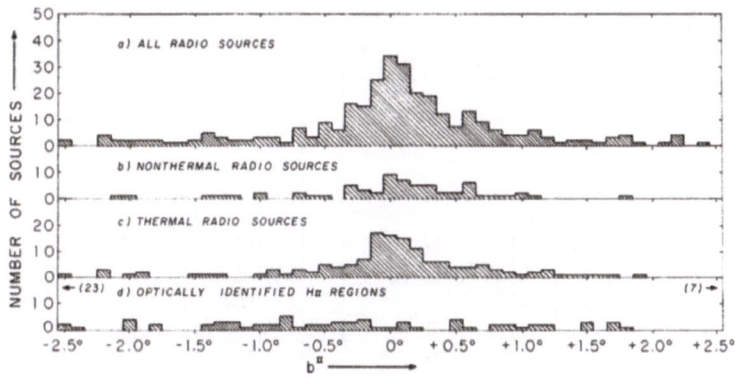

Fig. 4    (a)–(c). Distribution of galactic radio sources in galactic latitude as observed at $\nu = 2.7$ GHz. – (d) Distribution of optically visible H II regions in galactic latitude.

It is of interest to note that the non-thermal sources have a smoother distribution in longitude (Figure 3c) and a wider distribution in latitude (Figure 4b) as compared to the thermal radio sources. They may represent a distribution of objects somewhere between the very distant 'giant H II regions' and the intrinsically weak H II regions.

The large fraction of non-thermal sources with relatively high surface brightness and small angular dimensions is still a puzzle and can hardly be explained by an average birth rate of one supernova per hundred years.

## 5. Radio Recombination Line Surveys

It was a tantalizing fact for observers of the galactic continuum emission to know that they could observe H II regions at the opposite side of our galaxy but at the same time to know that there was no way to determine their distances. This was one of the main incentives for us to search for radio recombination lines which could provide us with radial velocities and hence kinematic distances of H II regions. Three limited recombination line surveys (Mezger and Höglund, 1967; Dieter, 1967; McGee and Gardner, 1968) followed the first unambiguous and quantitative observation of the H 109α-line by Höglund and Mezger (1965). Subsequently, these surveys were rendered obsolete by two rather complete H 109α line surveys of the northern and southern galaxy. Data pertinent to these two surveys are given in Table II.

Owing to some overlap in the two surveys the total number of radial velocities of individual H II regions is 201 rather than the sum of the two numbers in the sixth column, 213. Selection criterion for sources to be included in the two surveys was a value of the peak antenna temperature of $T_c \geqslant 1$ K. We feel, however, that some sources in the range $1.3 \text{K} \geqslant T_c \geqslant 1 \text{K}$ may have been missed. The distribution of all sources included in the two surveys is shown in Figure 5a. Owing to the different characteristics

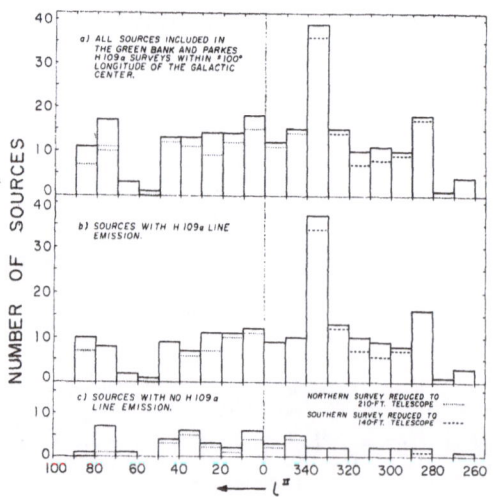

Fig. 5   (a)–(c). Distribution of radio sources included in northern and southern H 109α surveys. Dashed and dotted line refers to distributions reduced for instrumental selection effects (see text).

## TABLE II

The NRAO-MIT and NRAO-CSIRO-MIT H 109α surveys of northern and southern galactic sources

| Telescope | HPBW | $\eta_B$ | $T_e$ (min) | $S_5$ (min) | Total number of sources | Sources with H 109α emission Number | % | Sources without H 109α emission Number | % | References |
|---|---|---|---|---|---|---|---|---|---|---|
| NRAO 140-ft. | 6'.5 | 0.7 | 1 0 | 3.8 | 120 | 82 | 68 | 38 | 32 | (1) |
| Parkes 210-ft. | 4'.0 | 0.45 | 1.0 | 2.6 | 151 | 131 | 87 | 20 | 13 | (2) |

(1) Reifenstein et al. (1970); (2) Wilson et al. (1970).

of the two telescopes (see first three columns of Table II) the Parkes 210-ft telescope is more sensitive to point sources but less sensitive to extended sources than the NRAO 140-ft telescope. The dashed and dotted lines in Figure 5 represent the southern survey if observed with the 140-ft telescope and the northern survey if observed with the 210-ft telescope, respectively. However, the thus corrected source distributions have to be viewed with caution, since we can only estimate those sources which would have been rejected but not those which would have been added to either survey.

Figures 5a, b show the distribution of observed sources with and without H 109α emission, respectively. Sources with line emission are considered to be thermal, those without line emission, non-thermal. The source distributions Figures 5a–c in the longitude interval 345° through 0° to 85° are given in Figure 3 as shaded areas. It is of interest to note that for the northern part of our galaxy the percentage of sources without H 109α emission (32%) is nearly the same as the percentage of non-thermal sources (31%) identified on grounds of their continuum spectral indices (Section 4), thus confirming the unexpectedly high fraction of non-thermal sources. For the southern part the fraction of sources with no H 109α emission (13%) is considerably lower. This surprising result has yet to be confirmed by an investigation of the continuum spectra of southern sources.

Radial velocities of H II regions were obtained from their recombination line emission with a typical accuracy of $\pm 1$ km s$^{-1}$. Kinematic distances were derived using the Schmidt (1965) model of galactic rotation. This procedure implies:

(1) That the kinematics of ionized and neutral hydrogen are identical.

(2) That the distance ambiguity of H II regions inside the solar circle can be resolved.

The first question has been investigated in three papers (Dieter, 1967; Kerr et al., 1968; Mezger et al., 1970) and an affirmative answer was obtained. The second problem, the distance ambiguity, can in principle be resolved from observations of the radio absorption spectrum of the interstellar matter between the source and the sun. To date, most of the relevant absorption measurements were made at 21 cm. It appears, however, that some of the molecular lines, e.g. those emitted by the OH- and H$_2$CO-molecules, may be better suited for this purpose since they are generally seen only in absorption.

Figure 6 shows the projected positions of all H II regions whose kinematic distances were obtained. Both 'near' and 'far' distances are given for those sources whose distance ambiguity could not be resolved. The resulting distribution is typical for similar investigations of the galactic structure: The source distribution exhibits a more or less radial structure, with the source density decreasing with increasing distance from the sun. Such a result should be anticipated, since we obviously observe the superposition of a local distribution of intrinsically weak H II regions and another distribution much more narrowly confined to the galactic plane which represents the distant 'giant' H II regions (see Section 4). It is this latter group of giant H II regions, however, which appears to outline the spiral structure in external galaxies (see, e.g., Hodge, 1969a). For an elimination of the local weak H II regions, we define a giant

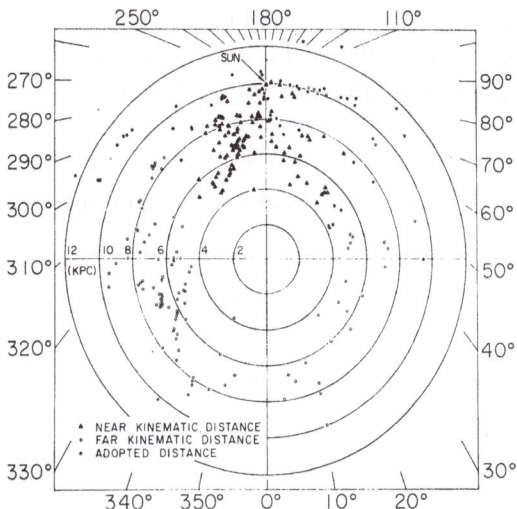

Fig. 6.   Projected positions of Hɪɪ regions whose H 109α emission was detected.

Hɪɪ region somewhat arbitrarily by the condition

$$\left[\frac{S_5}{\text{f.u.}}\right]\left[\frac{D_{\text{near}}}{\text{kpc}}\right]^2 \geqslant 400$$

with $S_5$ the continuum flux density at 5 GHz and $D_{\text{near}}$ the 'near' kinematic distance of the source. In this way we eliminate all Hɪɪ regions with flux densities less than four times the flux density of the Orion Nebula. The remaining distribution of giant Hɪɪ regions is shown in Figure 7. The clustering of sources around the sun has disappeared but the general source density is still considerably higher on our side of the galactic center. This result is not unexpected since both H 109α surveys are complete for giant Hɪɪ regions only out to distances of about 10 kpc from the sun. The minimum peak flux density of sources in any future recombination line survey has to be decreased by at least a factor of 4 if one wants to include all giant Hɪɪ regions within the solar circle. It appears doubtful if this can be achieved with any radio telescope presently in operation.

### 6. Spiral Structure and Radial Distribution of Hɪɪ Regions and Neutral Gas

We expect the spiral structure of our galaxy to be outlined by Hɪɪ regions, especially by giant Hɪɪ regions. The distribution of giant Hɪɪ regions as derived from kinematic distances (Figure 7), however, does not reveal a clear-cut spiral structure. This may be partly due to instrumental selection effects. But one should be aware of the fact that giant Hɪɪ regions in external galaxies in most cases do not outline a very clear-cut spiral pattern either (e.g. Hodge, 1969a). It appears that in most cases where photographs of external galaxies show a well-defined spiral pattern, this is the combined result of the presence of Hɪɪ regions and the presence of luminous early-type stars (Hodge, private communication). Within the spiral arms, the giant H II regions very often exhibit a rather patchy distribution. We know that H II regions, especially those with a high surface brightness, must have rather short lifetimes, even if compared with the main sequence lifetime of OB-stars. It appears that giant H II regions are indicators of presently active regions of star formation, whereas OB-star associations and clusters represent those regions where star formation on a large scale happened during the past $10^6$ to $10^7$ years.

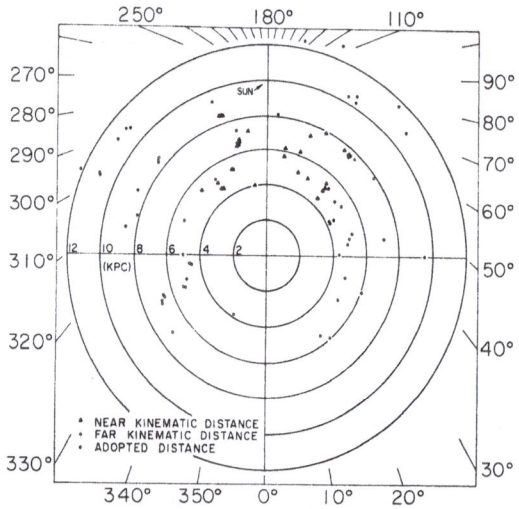

Fig. 7.   Projected positions of giant H II regions only.

In Figure 7 one recognizes a ring of giant H II regions between roughly 4 and 6 kpc. There are similar but not as conspicuous concentrations of giant H II regions farther out from the galactic center which may outline parts of spiral arms, but there is no unambiguous way to connect these features in the northern and southern part of the galaxy. This is certainly in part a consequence of the fact that kinematic distances become highly unreliable at low galactic longitudes. Therefore, we may expect to get a more clear-cut picture by combining the local distribution of H II regions based on photometric distances (Courtès et al., 1968, and Figure 2) with the distribution of giant H II regions (Figure 7). The result is shown in Figure 8. It is obvious that the high pitch angles obtained for the three pieces of spiral arms in the vicinity of the sun are incompatible with the large-scale spiral pattern as outlined by giant H II regions. However, I want to reiterate that we are dealing here with two different classes of objects, viz. optically observed, intrinsically weak H II regions on the one side and giant H II regions on the other, which indicate the birth of O-star associations but are primarily observable by their radio radiation.

What do we know about the radial distribution of H II regions in our galaxy? Hodge (1969b) has determined the radial distribution of giant H II regions* in a

---

* At present it is not possible to define the H II regions observed by Hodge in a quantitative way as we do for the radio 'giant' H II regions. We can only guess that this must be similar objects.

number of external galaxies. Roberts (1968) found that the distribution of HII regions and neutral hydrogen in some external galaxies, which he investigated, are markedly different. The HII regions generally appear to be concentrated in an inner ring, the HI in an outer ring with the two rings scarcely overlapping. Westerhout, as early as 1958, suggested such a distribution of ionized and neutral gas in our own galaxy, but his interpretation was later questioned by Mathewson *et al.* (1962). With the present data we can investigate the radial distribution of HII regions in our own galaxy in a quantitative way. Figure 9a shows the number of giant HII regions in rings 1 kpc wide each. Only the five giant HII regions possibly located inside the 4 kpc arm are uniformly spread out in this graph between 0 and 4 kpc. The distance ambi-

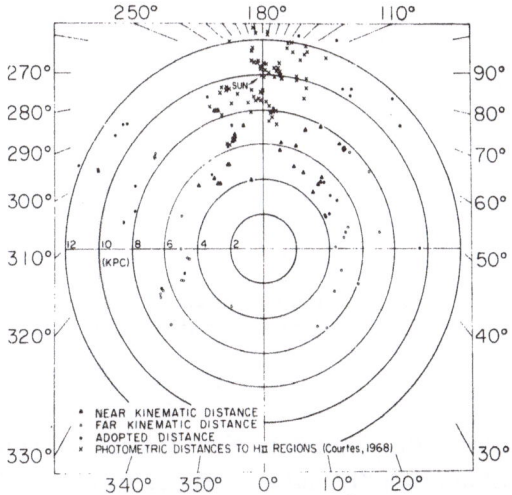

Fig. 8.   Combination of projected positions of optically visible local HII regions (Figure 2) and giant HII regions (Figure 7).

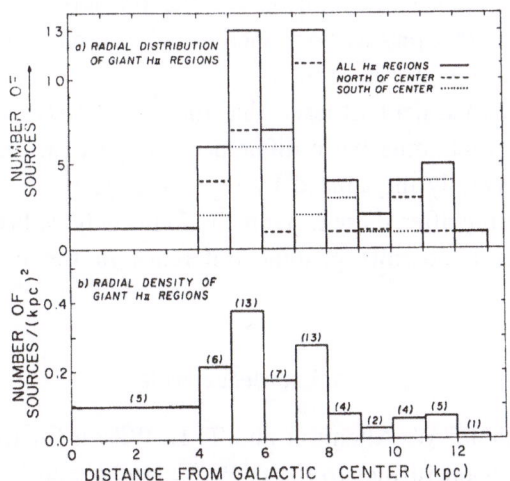

Fig. 9   (a)–(b). Radial distribution of giant HII regions.

guity does not affect the radial position but only the azimuthal position of an H II region. As mentioned earlier, our radio survey of giant H II regions is complete out to a distance of 10 kpc only. Therefore, a more complete survey will certainly increase the number of giant H II regions and will also slightly alter the shape of the distribution function by giving more weight to the outer rings. In Figure 9b the corresponding source density is shown. There is a maximum between 5 and 6 kpc and a secondary maximum between 7 and 8 kpc. The most obvious feature in the distribution of giant H II regions is the broad maximum between 4 and 8 kpc and the virtual non-existence of giant H II regions outside 12 kpc.

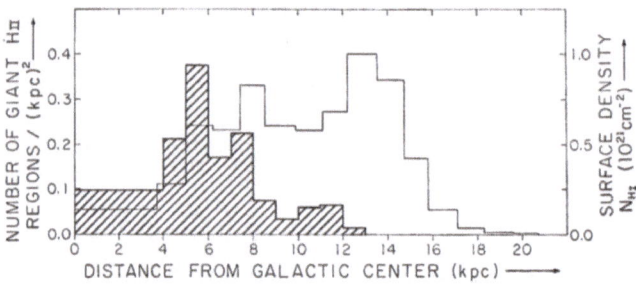

Fig. 10.   Comparison of density of giant H II regions and surface density of neutral hydrogen as a function of the distance from the galactic center.

In Figure 10 the density distribution of giant H II regions (Figure 9b) is compared with the radial distribution of the surface density of neutral hydrogen as recently derived by Van Woerden (Oort, 1965). The latter distribution attains its maximum value at about a distance of 13 kpc from the galactic center, where no giant H II regions are observed. It thus appears that a high density of neutral hydrogen is a necessary but not a sufficient condition for the formation of O-star clusters and associations and that other physical parameters – such as e.g. Lin's density wave – play an important role in this process, too.

It should be mentioned here, at least, that the two H 109α surveys of the northern and southern sky reveal some basic differences. We are still investigating selection effects which are certainly introduced by the two different radio telescopes used in these surveys. It is only after these instrumental effects have been sorted out that we can try to interpret a possible genuine difference in the northern and southern distributions.

### Acknowledgements

It is a pleasure to thank W. Altenhoff and T. L. Wilson for their help in preparing this review paper. I want further to thank P. W. Hodge and Miss E. Sim for helpful comments.

# References

Altenhoff, W.: 1968, in *Interstellar Ionized Hydrogen* (ed. by Y. Terzian), Benjamin, New York, p. 519.
Altenhoff, W., Mezger, P. G., Wendker, H., and Westerhout, G.: 1960, *Veröff. Univ. Sternwarte Bonn* **59**, 48.
Altenhoff, W., Downes, D., Goad, L., Maxwell, A., and Rinehart, R.: 1969, in press.
Blaauw, A.: 1964, *Ann. Rev. Astron. Astrophys.* **2**, 213.
Churchwell, E. and Felli, M.: 1969, several papers, in preparation.
Clark, B. G.: 1965, *Astrophys. J.* **142**, 1398.
Courtès, G., Georgelin, Y., Monnet, G., and Pourcelot, A.: 1968, in *Interstellar Ionized Hydrogen* (ed. by Y. Terzian), Benjamin, New York, p. 571.
CSIRO: 1969, in preparation.
Dieter, N. H.: 1967, *Astrophys. J.* **150**, 435.
Felli, M. and Churchwell, E.: 1969, in preparation.
Goss, W. M. and Shaver, P. A.: 1969, in preparation.
Hill, E. R.: 1968, *Australian. J. Phys.* **21**, 735.
Hodge, P. W.: 1969a, *Astrophys. J. Suppl. Ser.* **18**, 73.
Hodge, P. W.: 1969b, *Astrophys. J.* **155**, 417.
Höglund, B. and Mezger, P. G.: 1965, *Science* **150**, 339.
Kerr, F. J., Burke, B. F., Reifenstein, E. C., Wilson, T. L., and Mezger, P. G.: 1968, *Nature* **220**, 1210.
Lin, C. C. and Shu, F. H.: 1964, *Astrophys. J.* **140**, 646.
Lin, C. C. and Shu, F. H.: 1966, *Proc. Nat. Acad. Sci. Am.* **55**, 229.
Mathewson, D. S., Healey, J. R., and Rome, J. M.: 1962, *Australian. J. Phys.* **15**, 354, 369.
McGee, R. X. and Gardner, F. F.: 1968, *Australian. J. Phys.* **21**, 149.
Mebold, U.: 1969 *Beiträge zur Radioastronomie* **1**, 97.
Mezger, P. G.: 1969, *Colloque Internat. Liège No. 16*, in press.
Mezger, P. G. and Henderson, A. P.: 1967, *Astrophys. J.* **147**, 471.
Mezger, P. G. and Höglund, B.: 1967, *Astrophys. J.* **147**, 490.
Mezger, P. G., Wilson, T. L., Gardner, F. F., and Milne, D. K.: 1970, *Astron. Astrophys.* **4**, 96.
Oort, J. H.: 1965, *Trans. IAU* **12A**, 789.
Reifenstein, E. C., Wilson, T. L., Burke, B. F., Mezger, P. G., and Altenhoff, W. J.: 1970, *Astron. Astrophys.* **4**, 357.
Roberts, M. S.: 1968, in *Interstellar Ionized Hydrogen* (ed. by Y. Terzian), Benjamin, New York, p. 617.
Schmidt, M.: 1965, *Stars and Stellar Systems* **5**, 513.
Schraml, J. and Mezger, P. G.: 1969, *Astrophys. J.* **156**, 269.
Sharpless, S.: 1959, *Astrophys. J. Suppl. Ser.* **4**, 257.
Shaver, P. A. and Goss, W. M.: 1969, in preparation.
Sim, M. E.: 1968, *The Royal Observatory, Edinburgh, Publ.* **6**, No. 5.
Westerhout, G.: 1958, *Bull. Astron. Inst. Netherl.* **14**, 215.
Wilson, T. L., Mezger, P. G., Gardner, F. F., and Milne, D. K.: 1970, in preparation.

# Discussion

*Van Woerden:* The neutral-hydrogen surface densities quoted by Mezger have been derived by me from Westerhout's (1957, *Bull. Astron. Inst. Netherl.* **13**, 201) 'cross-sections' at constant $l^{II}$, by integrating hydrogen densities along lines perpendicular to the plane and then averaging the resulting surface densities in concentric rings around the center. In comparing these H I surface densities with the numbers of H II regions, one should bear in mind that the effective layer thickness of H I increases outward; the radial distribution of H I volume density differs less from that of H II regions than does the distribution of surface density.

# 19. A MOTION PICTURE FILM OF GALACTIC 21-CM LINE EMISSION

G. WESTERHOUT

*University of Maryland, College Park, Md., U.S.A.*

Since 1964 we have been observing 21-cm line profiles in a new survey of the neutral hydrogen distribution in the neighborhood of the galactic plane with the 300-foot radio telescope of the National Radio Astronomy Observatory* in Green Bank, W.Va. This is the largest telescope available for 21-cm line work; it has a beamwidth of 10 min of arc and is equipped with an excellent line receiver. Since it seems unlikely that an extensive hydrogen-line survey will be made with any larger telescope, we felt that for reference purposes a concerted effort should be made to obtain as many 21-cm data as possible pertaining to the structure of the Galaxy with this telescope. The data have been presented in the form of contour maps giving the intensity of the 21-cm line radiation as a function of right-ascension and velocity at constant declination. A series of contour maps was distributed to the astronomical community in 1966 as the first edition of the *Maryland–Green Bank Galactic 21-cm Line Survey*. The second edition, containing 1200 pages and approximately 1800 maps, was distributed in the summer and fall of 1969. It is expected that additional contour maps, completing the survey as originally planned, covering a latitude range from $b^{II} = +1°$ to $-1°$, $l^{II} = 11°$ to 235° ($b^{II} = +3°$ to $-3°$ between $l^{II} = 100°$ and 145°), will be finished by the summer of 1970. Scans were made across the galactic equator with a stationary telescope, so that the declination is constant through each scan; the declination intervals varied from 4 to 6 min of arc. Eventually, we plan to cover a strip from $b^{II} = +5°$ to $-5°$ between $l^{II} = 11°$ and 235°, containing 225000 independent points at intervals of 6 min of arc, with an effective beamwidth of 12.5 min of arc, a velocity resolution of 2 km s$^{-1}$, and a total of $1.2 \times 10^8$ individual intensities.

In our search for a way to examine this mass of data in a convenient form, we have found the contour maps excellent for the study of small-scale features. But by presenting them in the form of motion picture film, one will obtain an overall view of a large section of sky and its variations in a reasonably short amount of time. A section of such a film, covering a region $30° \times 6°$, was shown at the Symposium. In order to produce the film, we converted the survey data into contour maps in galactic latitude and velocity, so that each frame of the film gives intensity as a function of $b$ and $v$ for constant $l$. The contour maps from which the film was constructed were produced by computer and photographed from the face of a cathode-ray tube. Each frame was photographed 7 times, so that when the film is shown at a speed of 24 frames per sec, one beamwidth in longitude passes by in approximately 2.5 sec. Three frames of the film are reproduced in Figure 1. The successful production of this film is mainly due

* The National Radio Astronomy Observatory is operated by Associated Universities, Inc., under contract with the National Science Foundation.

*Becker and Contopoulos (eds.), The Spiral Structure of Our Galaxy, 122–125. All Rights Reserved.*
*Copyright © 1970 by the I.A.U.*

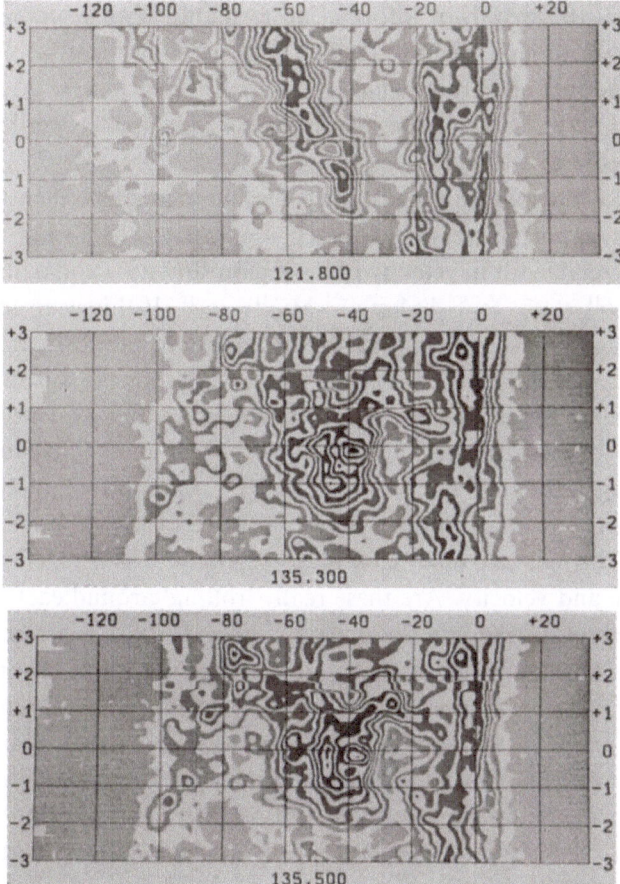

Fig. 1.  A sample of three frames from the motion picture film for galactic longitudes 121.8, 135.3 and 135.5. Coordinates are radial velocity with respect to the local standard of rest, and galactic latitude. The contour intervals are approximately 14 K in brightness temperature. Contour lines with higher temperatures are darker.

to the extremely competent computer programming, data analysis and organization of the reductions by Mr. H. U. Wendlandt.

The film displays a part of the sky in which our coverage in latitude is already considerably larger than $b^{II} = +1°$ to $-1°$, namely the region from $l^{II} = 108°$ to $138°$, $b^{II} = +3°$ to $-3°$. The contour maps shown in the film are corrected for the effect of the far-out sidelobes of the antenna, caused by irregularities in the telescope surface (the 'error beam'; a considerable fraction of the radiation comes from a region about $6°$ in diameter centered on the main beam of the antenna). This correction results mainly in an enhancement of small details with respect to the overall background, but does not produce major changes in the overall pattern. We found that for studies of large-scale galactic structure, the uncorrected maps in the published survey are more than adequate, as long as one is not interested specifically in detailed intensity comparisons.

The film shows the wealth of information available in this survey. In particular, we found that even though we have studied this region in detail using the published survey maps, a number of features which show up very clearly were not seen previously on the published maps, showing the importance of having an overall rapid view over a relatively large piece of sky.

Some of the features which stand out in the film should be mentioned here. The Perseus arm, the important feature between $v = -40$ and $-60$ km s$^{-1}$, appears to have a very appreciable tilt in velocity in the region studied: its maximum varies from $-60$ km s$^{-1}$ at $b^{II} = +3°$ to $-40$ km s$^{-1}$ at $b^{II} = -3°$. If this were to be interpreted as a distance effect due to differential galactic rotation, the top part of this section of the Perseus arm would be 2.5 kpc further away from the sun than the bottom part. As Kerr mentions in his review, there is substantial evidence for a 'rolling' motion in the spiral arms; in this interpretation, the 'circular velocity of rolling' would be somewhat larger than 10 km s$^{-1}$ at $b^{II} = \pm 3°$ (or approximately $z = \pm 200$ pc). But is this the right interpretation? At several points the arm appears to break up into three very distinct maxima, each approximately $1°5$ in extent, and each at a different latitude and velocity. Are these really 'rolling' around each other?

Both from a study of the film and from a study of the regular contour maps, the Perseus arm has the distinction of being a very massive feature over at least 90° in galactocentric longitude. A first look at the film already makes obvious a statement which has been made in the past: a spiral arm has regions of relative 'quiescence', interspersed by regions where the velocity profile widens considerably, has many maxima at the same or different latitude and an integrated intensity which is much higher than that of the adjacent regions. If such a high-density, high-velocity-width region is to be associated with a region of star formation, the question needs to be answered why such regions occur over only limited portions of spiral arms, typically of the order of 1 kpc in length.

From a study of both the contour maps and the film, it is clear that the intensities of inter-arm regions can be extremely low, often more than 10 times lower than the peak intensities in the arms. But a very interesting phenomenon is the formation of connections between arms in the 30° strip covered by the film. Two such connections form, and remain over longitude ranges of approximately 5°, between the Orion arm and the Perseus arm, and at least one such connection forms between the Perseus arm and the so-called Outer arm. At one point, all three arms appear to be joined by a collection of inter-connected intensity peaks. The frequency of these inter-connections and their variation with longitude give the impression that we might well be dealing here with the 'feathers' observed relatively frequently in the photographs of other galaxies.

Finally, a few remarks about the work in progress at Maryland connected with this survey. We plan to convert the entire survey into a collection of contour maps, each at constant longitude, and corrected for the effect of the far-out (error-beam) sidelobes. These maps will then likewise be put in the form of a motion picture film. A statistical study is underway of small-scale features in several different regions of the Galaxy. Preliminary results indicate that there is a preponderance of small peaks with sizes close to the resolution limit of the telescope, more or less independent of the distance to the sun. This raises the question whether or not the entire interstellar medium is broken up into individual small 'clumps', together blending into the spiral arms. Or do we see here the cool condensations in a hot medium?

Mr. R. H. Harten is studying the large-scale streaming characteristics of the gas in the arms. Examination of the existing data gives evidence for several distinct types of motion: a 5 km s$^{-1}$ streaming motion in the direction of galactic rotation, a helical magnetic field and its implied motion, and a general 'tumbling' of the spiral arms. All of these put together might indicate a helical streaming of the gas with a velocity of the order of 10 km s$^{-1}$ and a skew angle of about 30°, in the direction of galactic rotation. Smaller features seem to indicate the presence of rope-like structures wound around the main body of the arms and streaming some 5 km s$^{-1}$ faster again than these arms.

## Acknowledgement

These investigations are supported by the U.S. National Science Foundation.

# 20. SPIRAL STRUCTURE OF THE GALAXY DERIVED FROM THE HAT CREEK SURVEY OF NEUTRAL HYDROGEN

H. WEAVER

*Radio Astronomy Laboratory, University of California, Berkeley, Calif., U.S.A.*

**Abstract.** The extensive Hat Creek survey of neutral hydrogen combined with southern observations provides the basis for a new discussion of the spiral structure of the galaxy. The purpose of this investigation is to provide a general picture of the galaxy. It is found that the pitch of the spiral arms is approximately 12°.5 and that there are many spurs and interarm features as we observe in external galaxies.

The sun is not located in a major spiral arm, but rather in a spur or offshoot originating near or at the Sagittarius arm, which is a major structure in the galaxy. The young stars in the general vicinity of the sun delineate this spur, not a major arm structure. The stars and the gas are in agreement in indicating a large pitch angle (20°–25°) for this local structure, which differs from the smaller pitch angle for the arms which form the system as a whole.

In the presentation a computer-produced movie of the galaxy based on Hat Creek hydrogen contour maps similar to those in Figure 1 was shown. It was used to illustrate generally the complexity of the gas structure and, in particular, to show (i) observational aspects of the spur in which the sun is located and (ii) the point of origin of the so-called Perseus arm.

## 1. The Hat Creek Survey of Neutral Hydrogen

The recently automated 85-foot telescope at the Hat Creek Observatory has been used in conjunction with the 100-channel receiver to complete a neutral-hydrogen survey in the region of the galactic plane (Weaver and Williams, 1969). At each half-degree interval throughout the longitude range 10° to 250° a uniformly-spaced sequence of 81 neutral-hydrogen profiles was observed over the latitude range −10° to +10°. The observations comprising the survey are thus spaced by 0°.25 (one-half beam width) in latitude and by 0°.5 (one beam width) in longitude. The data from the 501 observed latitude cuts are displayed in the form of computer-produced contour maps of which samples are shown in Figure 1. The instrumental and statistical data that characterize the observations are brought together in Table I.

### TABLE I

Observational data relating to Hat Creek survey of neutral hydrogen

| | |
|---|---|
| Sky coverage: | Every 0°.25 in galactic latitude between limits −10° to +10°; every 0°.5 in galactic longitude over range 10° to 250° (19521 profiles) |
| Frequency coverage: | Each profile covers 250 km s$^{-1}$; 200 information points per profile |
| Frequency resolution: | 10 kHz (2.11 km s$^{-1}$) spaced every 5 kHz |
| Data display: | Contour maps as shown in Figure 1. Contour intervals $T_A = 1°, 2°, 4°, 8°, 15°, 20°, 30°,\ldots$ |
| Integration time: | 72 sec per profile |
| RMS fluctuation: | 0.18 K (system noise ⩽ 150 K) |
| Angular resolution: | 35'.5 HPBW |

*Becker and Contopoulos (eds.), The Spiral Structure of Our Galaxy, 126–139. All Rights Reserved.*

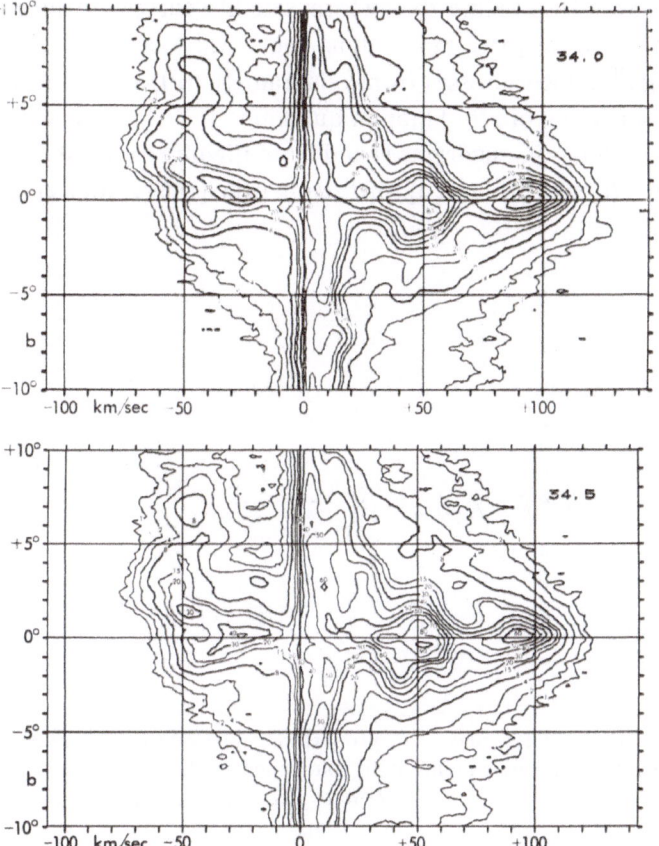

Fig. 1.   Sample of a page from the Hat Creek survey of neutral hydrogen.

Completion of this extensive survey makes possible a new discussion of the spiral structure in the galaxy.

## 2. A Method of Deriving Spiral Structure from Observations of Neutral Hydrogen

To illustrate the method of analysis to be used in determining spiral structure, we take as a mathematical model of the galaxy the smooth, continuous spiral shown in Figure 2a. We assume that over the $R$-range of interest (3 to, say, 15 kpc) the rotation curve of the model galaxy is a continuous monotonic function $\omega(R)$ of $R$ alone. Under these conditions the spiral in Figure 2a, if observed from the sun shown in Figure 2a at galactocentric distance $R_0$, would produce in the observational longitude, radial velocity ($l^{II}$, $v$) coordinate system the characteristic pattern of loops and sine-like curves shown in Figure 2b. Points on the inner part of the spiral, for which $R < R_0$, transform to points in quadrants II and IV of Figure 2b and characteristically lie along loops. Points on the outer parts of the spiral ($R > R_0$) transform to points in quadrants I and III of Figure 2b and form sine-like curves. At those points at which the spiral crosses $R = R_0$ (we move along the spiral locus so that $R$ increases) the corresponding $l^{II}$, $v$ locus crosses from quadrant II to III or IV to I, and thus may prevent the completion of a full loop in quadrants II or IV. It may readily be shown that all complete loops in quadrants II and IV in Figure 2b are tangent to the envelope curve $v = R_0 [\omega (R_0 \sin l^{II}) - \omega (R_0)] \sin l^{II}$, from which the rotation curve $\omega (R)$ is found observationally.

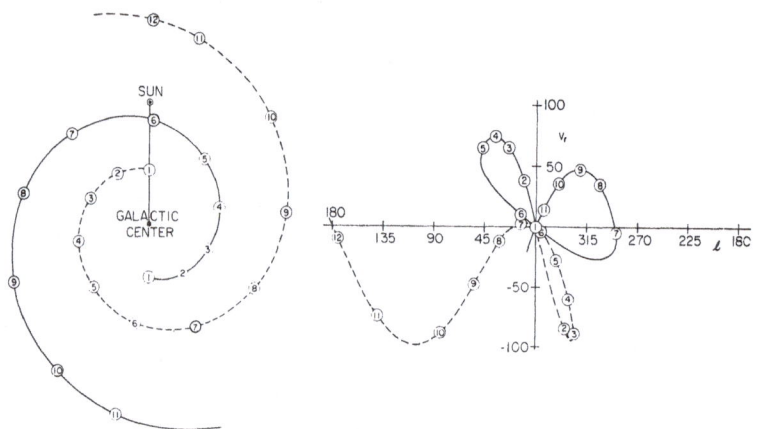

Fig. 2.    (a) (left) Model spiral galaxy. – (b) (right) longitude-velocity ($l^{II}$, $v$) diagram for the spiral shown on the left. Corresponding points in the spiral and its ($l^{II}$, $v$) diagram may readily be identified ($l$ means $l^{II}$).

    The pattern shown in Figure 2b is typical; the precise form and phase of any $l^{II}$, $v$ locus is determined, of course, by the form and phase of the generating spiral, $R(\theta)$, and by the rotation curve, $\omega (R)$.

    Real galaxies (see the *Hubble Atlas of Galaxies*, Sandage, 1961) are more complex than a pair of smooth spirals differing in phase by 180°. Real galaxies consist, characteristically, of spiral arms (often broken into segments which may differ in pitch and phase) plus interarm links, spurs, and other less regular features. If we investigate a slightly more realistic galaxy with such features as are illustrated in Figure 3a, we still obtain in the observational $l^{II}$, $b^{II}$ coordinate system a diagram typically like that of Figure 2b, except that it contains, additionally, bits and pieces of curves which are the $l^{II}$, $v$ transformations of the interarm links, spurs, segments and so forth, pictured in Figure 3a.

    The general approach in this discussion of spiral structure will be to construct the observationally determinable $l^{II}$, $v$ diagram and from that to infer the nature and form of the spiral galaxy.

### 3. Construction of the $l^{II}$, $v$ Diagram

The concentrations of hydrogen which form the gas arms in the galaxy are of finite cross section. Investigations which we cannot describe in detail here show that within

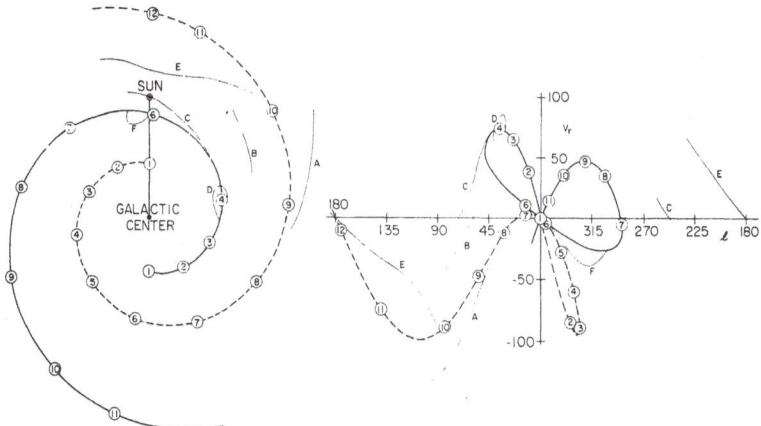

Fig. 3.  (a) (left) Model spiral galaxy with sample spurs, interarm, and other features; – (b) (right) longitude-velocity ($l^{II}$, $v$) diagram for the structures shown on the left.

a gas arm there are many subunits or subconcentrations which show continuity over significant ranges of longitude. As a working model, we imagine the arm to be made up of subunits or subconcentrations drawn out along the arm. Characteristically, then, the cross section of the arm at a given $\theta$ value would show gas concentrations at slightly different $R$ and $z$ (height above the plane) values. Observing from the sun, we would see these subconcentrations within the arm at slightly different latitudes, $b^{II}$, and velocities, $v$, at some given longitude, $l^{II}$. On each observed contour map, which, specifically is a $b^{II}$, $v$ map for a given $l^{II}$, we measure the velocity of each of the sub-concentrations within an arm or arm-like structure. In the present investigation in which the goal is to produce an overall picture of the galaxy, we ignore minor latitude differences within an arm structure and treat all subconcentrations within an arm structure similarly, projecting them onto the galactic plane.

To construct the $l^{II}$, $v$ diagram we require for our analysis, we plot each measured velocity as a function of longitude; to show relatedness of subconcentrations we connect with a vertical line all those points representing subconcentrations within an arm or arm-like structure.

In the observed $l^{II}$, $v$ diagram, Figure 4, the points in the longitude range 10° to 230° are from the Hat Creek Survey; those in the longitude range 300 to 360 are from the observations by Kerr (1969); those indicated by open triangles in the longitude range 290 to 310 are from Varsavsky (1969), and the points shown as filled triangles in the longitude range 280 to 300 are H II velocities from Wilson (1969). These latter have been added to strengthen the diagram in an important longitude range in which adequate H I observations are currently unavailable.

## 4. The Pitch of the Spiral Arms

If a galaxy has generally smooth continuous arms as shown in Figure 5a, we can provide a general characterization of the spiral structure by specifying the pitch angle

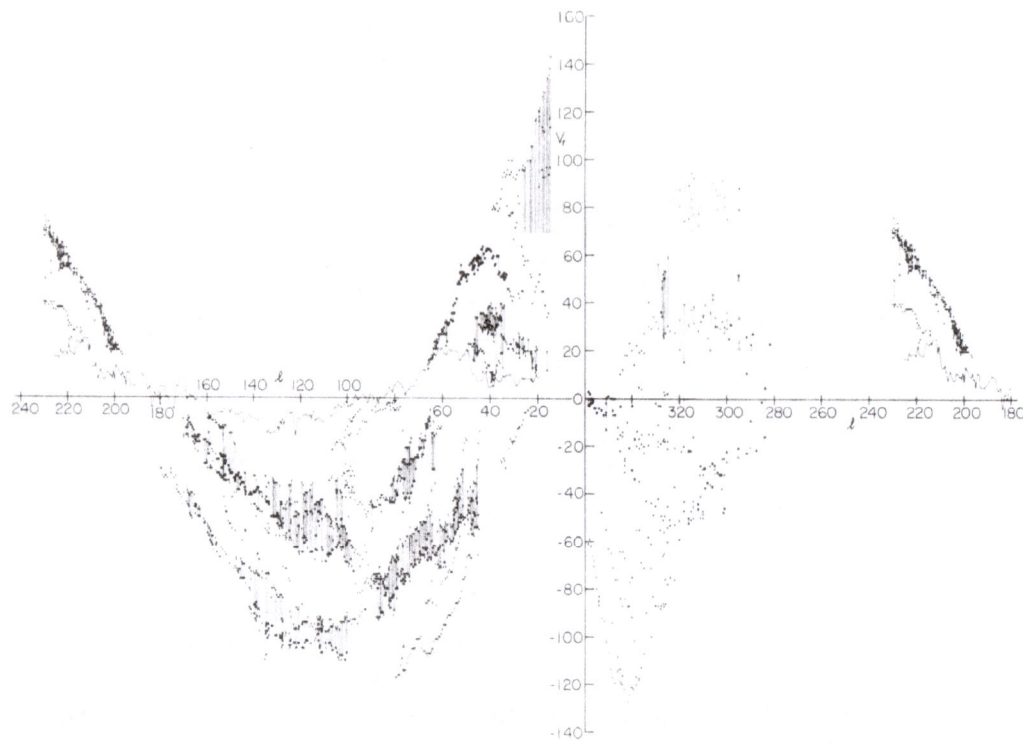

Fig. 4.   Observed $l^{II}$, $v$ diagram for the galaxy. Intensity of features is indicated by size of symbol. See Section 4 for detailed expansion of data sources. The open squares represent points on the expanding 3 kpc arm.

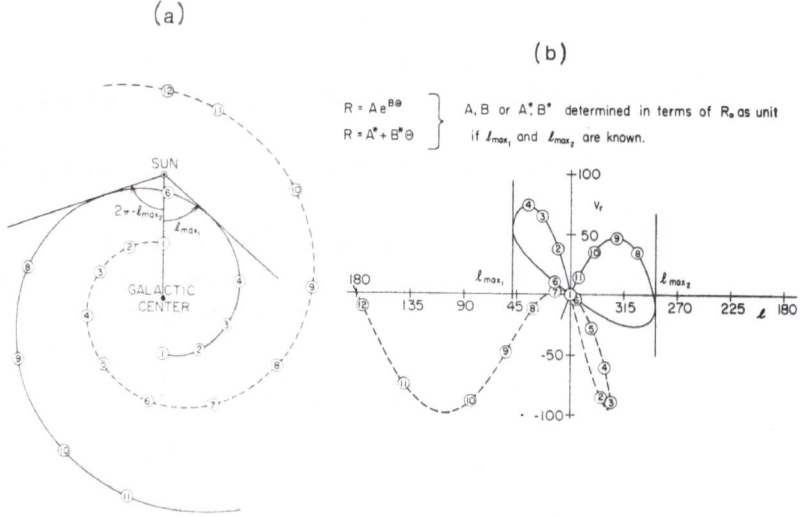

Fig. 5.   Diagram illustrating the tangents $l_{max1}$ and $l_{max2}$ (see text).

of the spiral. For a galaxy such as that shown in Figure 5a there exist well-defined tangents to the arms. Figure 5b displays the observationally determinable $l^{II}$, $v$ diagram corresponding to the spiral in Figure 5a. The longitudes $l_{max1}$ and $l_{max2}$ at which tangency occurs can be found from observation as illustrated in Figure 5b. In turn, knowing $l_{max1}$ and $l_{max2}$, one may employ numerical methods to find values of the parameters $A$, $B$ or $A^*$, $B^*$ (defined in Figure 5) which characterize the arm of the galaxy contained within the longitude range between the tangents. For the galaxy we

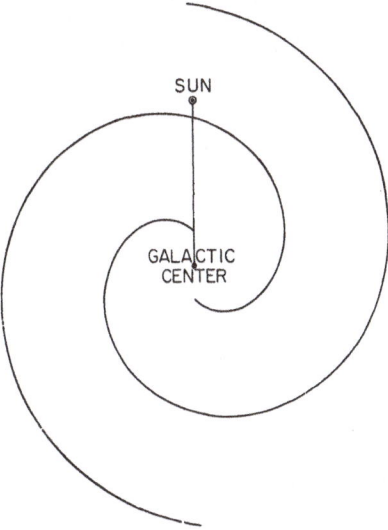

Fig. 6.  A smoothed galactic model inferred from the ($A^*$, $B^*$) values derived from observed $l_{max1}$ and $l_{max2}$ values.

find $l_{max1} = 50°.5$, $l_{max2} = 284°.0$ (details of the determination are given elsewhere by Weaver, 1969), and from these values. $A = 0.410$, $B = 0.245$; $A^* = 0.202$, $B^* = 0.220$. The pitch angle of the arm is $12°.5$ in the solar vicinity. The overall form of the galaxy deduced from the ($A^*$, $B^*$) values is shown in Figure 6.

## 5. A More Detailed Picture of the Galaxy

The galaxy is more complex than indicated by Figure 6; there are more features in the galactic $l^{II}$, $v$ diagram than can be accounted for by the spirals in Figure 6. To obtain a more detailed picture of the spiral structure in the system, we represent the arm and arm-like features of the $l^{II}$, $v$ diagram by segments of curves as illustrated in Figure 7. With the Schmidt (1965) rotation curve, we transform these segments in the $l^{II}$, $v$ plane to the $R$, $\theta$ plane, and obtain the picture shown in Figure 8.

The two sides of the galaxy appear rather different in character in Figure 8. The difference is not physically real; it is caused by the scarcity of observations from the southern sky. The southern data cover only the latitude range $+2°$ to $-2°$ at intervals of $1°$ in longitude. While it is evident from these observations that there are many

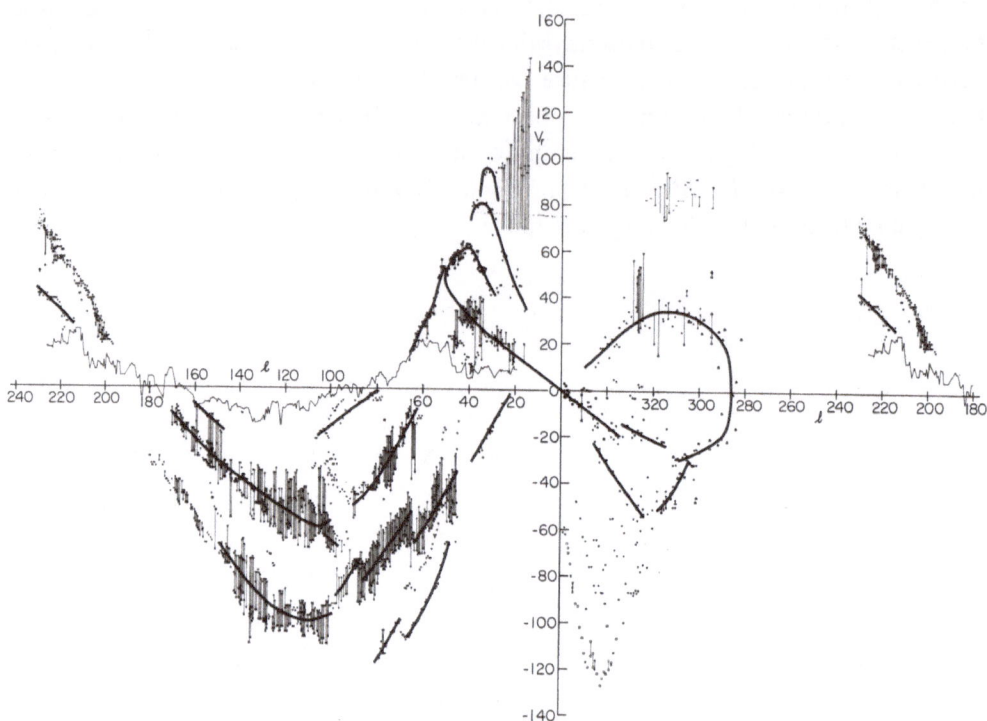

Fig. 7.    Representation of ($l^{II}$, $v$) features.

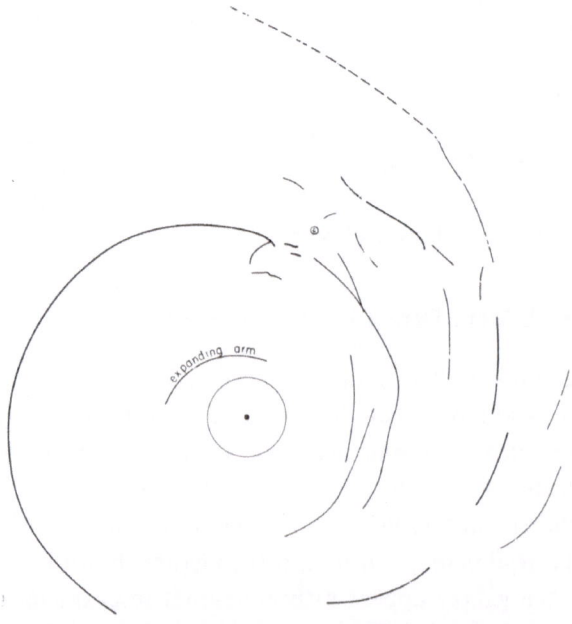

Fig. 8.    Observational picture of the galaxy derived by transforming the loci in Figure 7 to $R$, $\theta$ through the rotation curve. The lack of detail on the left-hand side of the diagram is observational only and reflects no real aspect of the galaxy (see text).

details in the southern structure, those details often cannot be traced with complete certainty; they have therefore not been drawn in Figure 7 or Figure 8. For the northern-sky section of Figure 8 the data used in the analysis cover the latitude range $-10°$ to $+10°$ at every $0°.5$ in longitude. Such extensive sky coverage makes the tracing of spiral structure much more certain.

In Figure 8 the structure on the right-hand side of the spiral should be taken as more characteristic of the galaxy. On the left-hand side of the spiral only gross features have been depicted; many other spiral features are present in this part of the diagram, but they have not been entered in the picture because of lack of precise information.

The spiral shown in Figure 8 resembles a late Sb or an Sc galaxy that might well be broadly characterized by the smooth spiral shown in Figure 6. It contains many spurs and inter-arm fragments of the type visible on pictures of galaxies.

## 6. The Local Hydrogen Structure: Model Calculations

That the sun is located in an arm-like structure is clearly indicated by many observations; hydrogen is seen everywhere around us. However, the sun is not in a major arm, but rather in an offshoot or spur of the Sagittarius arm.

A variety of radio observations indicate the orientation of the feature in which the sun is located. The 21-cm continuum map by Westerhout (1958) shows the Cygnus complex at $l^{II} \sim 80°$ as a clearly separate entity. The southern 21-cm continuum map by Mathewson *et al.* (1961) shows the Vela complex at $l^{II} \sim 264°$ as a separate unit similar in angular size to Cygnus. These two complexes, separated by very nearly 180° in longitude, are the result of our looking fore and aft along the arm-like structure in which the sun is located.

How such a structure, seen from within, will reveal itself in the hydrogen line radiation is indicated in Figures 9, 10, and 11 on the basis of a model (Figure 9) in which we show the sun inside a spiral structure. At such a location, gas surrounds

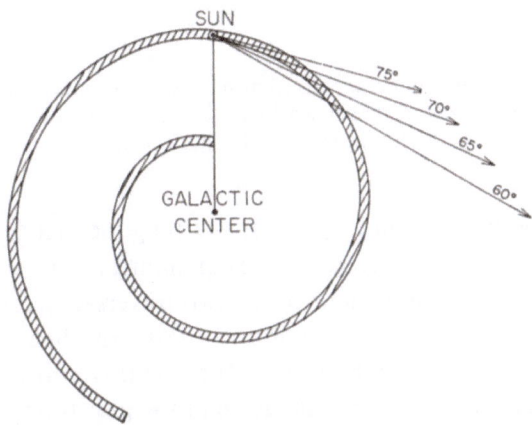

Fig. 9. A model spiral gas structure containing the sun. Lines of sight at $l^{II} = 60°, 65°, \ldots$ are illustrated.

the sun in every direction. Lines of sight are shown at $l^{II}=60°$, $65°$, .... The line of sight at $60°$ passes through the gas immediately surrounding the sun; it traverses the gas structure a second time at a distance of several kpc from the sun. The same situation is found for the line of sight at $65°$ except that the second traversal is less distant from the sun. Finally, however, at some longitude $l_{c1}$ the line of sight no longer traverses the structure twice. At $l_{c1}$, and for $l^{II}>l_{c1}$, there is only a single traversal of the structure. A corresponding critical line of sight at longitude $l_{c2}$ exists at approximate longitude $360°-l_{c1}$, at which there is again a change from single to double traversal. The exact numerical value of $l_{c2}$ depends upon the form (particularly upon the curvature and width) of the gas structure.

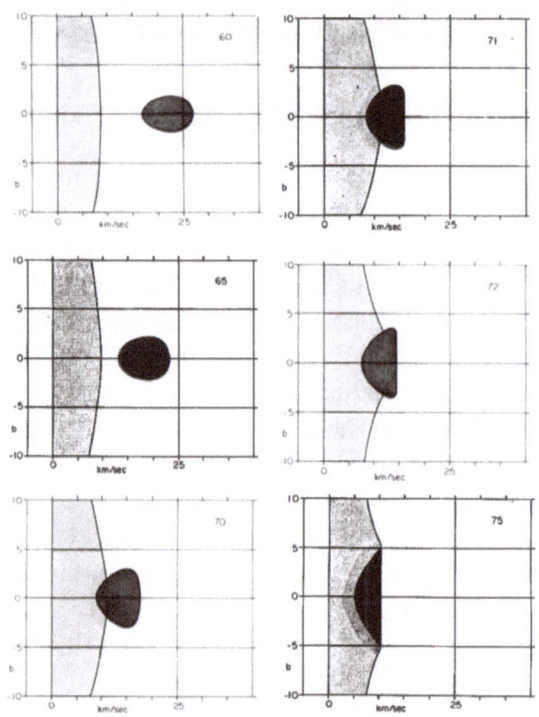

Fig. 10.   Theoretical $b^{II}$, $v$ contour diagrams (shown in outline form) computed for the model shown in Figure 9. The series is especially designed to show the merging of the second-traversal feature and the surrounding gas (see text).

The critical longitudes $l_{c1}$ and $l_{c2}$ divide the longitude circle into two sectors. In the inner sector extending from $l_{c2}$ to $l_{c1}$ and containing $l^{II}=0°$, a line of sight traverses the structure twice. (This will be termed the two-traversal sector.) In the outer sector extending from $l_{c1}$ to $l_{c2}$ and containing $l^{II}=180°$, the line of sight traverses the structure only once. (This will be termed the one-traversal sector.) We omit consideration of special cases in which curvature of the gas structure is such that one or both critical $l$-values cease to exist or become multiple.

To delineate in detail the form of the gas structure in which the sun is located, we must specify three distances $d_1$, $d_2$, $d_3$ for each longitude in the two-traversal sector, and one distance $d_1$ at each longitude in the one-traversal sector. Specifically,

$d_1$ = the distance from the sun to the point at which the line of sight leaves the gas structure;

$d_2$ = the distance from the sun to the point at which the line of sight re-enters the gas structure;

$d_3$ = the distance from the sun to the point at which the line of sight leaves the gas structure a second time.

In the one-traversal sector the quantities $d_2$ and $d_3$ do not exist. In the two-traversal sector as $l^{II} \to l_{c1}$, $d_1 \to d_2$. As the line of sight moves from the two-traversal to the one-traversal sector, $d_1$ has a discontinuity at $l_{c1}$, at which longitude $d_1 \to d_3$ in numerical value.

Observationally, we determine the Doppler counterparts of $d_1$, $d_2$, $d_3$, not the distances themselves. Moreover, by the methods of measurement described in Section 3, we would not measure Doppler counterparts of $d_2$ and $d_3$ separately, but the Doppler counterpart of some average $\langle d_2, d_3 \rangle$, the precise character of which would depend mainly upon the density distribution within the gas structure. For the purposes of this investigation, the Doppler counterpart of the average $\langle d_2, d_3 \rangle$ is adequate; $\langle d_2, d_3 \rangle$ defines the run of the gas structure, it only fails to provide information on the width of the gas structure, which we here do not specifically investigate.

In the case of $d_1$, however, an effort has been made to define the 'edge' of the gas structure, that is, an effort has been made to determine the Doppler counterpart of $d_1$ in the belief that $d_1$ will be physically more meaningful than $\langle 0, d_1 \rangle$. Details of the observational procedure of making the determination are given elsewhere (Weaver, 1969).

In Figure 10 we show in outline theoretical $b^{II}$, $v$ contour diagrams for the $l^{II}$-values considered in the model in Figure 9. The 'surrounding' gas is of wide latitude distribution since it is nearby; it starts at $v=0$ and extends to a $v$-value corresponding to $d_1$ for $l^{II} \leqslant l_{c1}$. For $l^{II} \leqslant l_{c1}$ the distribution extends to $v_{max}(l^{II})$, which is unrelated to $d_1$. The second feature, shown darker in the $b^{II}$, $v$ diagrams, is of smaller latitude extent than the surrounding gas; it arises from the line of sight's second traversal of the gas structure. For $l^{II}=60°$, the feature representing the second traversal in the model appears at $\langle v \rangle = 22$ km s$^{-1}$. As $l^{II} \to l_{c1}$ the feature representing the second traversal (i) moves in toward the 'surrounding' gas, finally overlapping it, and (ii) increases in extent in latitude since the average distance $\langle d_2, d_3 \rangle$ decreases. Finally, at longitude $l_{c1}$ the separate feature arising from the second traversal merges; there is a single distribution starting at $v=0$ and extending to $v_{max}(l^{II})$. There may be folding in the latter distribution, however, in that gas from two points on the line of sight may have the same velocity. Such double points, like the separate feature, are shown dark in Figure 10.

Figure 11 displays the theoretical $l^{II}$, $v$ diagram corresponding to the model in Figure 9 and the theoretical $b^{II}$, $v$ maps in Figure 10. In accord with the convention of Section 3, the distant, second-traversal feature is represented by mean-value points (the Doppler counterparts of $\langle d_2, d_3 \rangle$). The 'surrounding' gas is shown as a continuous line which represents the Doppler velocity equivalent to $d_1$ for $l^{II} \leq l_{c1}$. The discontinuity at $l_{c1}$ occurs because a line of sight for which $l^{II} > l_{c1}$ enters the single-traversal sector and there intersects the circle of maximum velocity.

The mergence of the second-traversal feature into the surrounding gas is clearly shown in Figures 10 and 11.

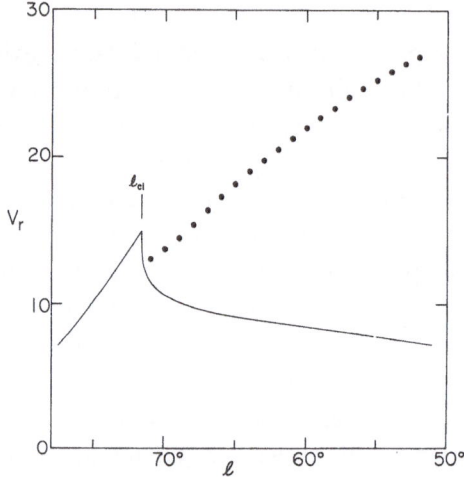

Fig. 11.    Theoretical ($l^{II}$, $v$) diagram for the model galaxy in Figure 9 derived from $b^{II}$, $v$ contour diagrams like those shown in Figure 10 (see text).

In the earlier galactic model displayed in Figure 3a there is a spur (Feature C) going through the sun in the same general manner as we have discussed above. The velocity-longitude counterpart of Feature C is shown in Figure 3b. The spur in Figure 3a has no thickness and thus differs from that discussed in Figures 9, 10, and 11. As we look fore and aft along the zero-thickness spur in Figure 3a, we find that in the $l^{II}$, $v$ diagram Feature C extends to $v = 0$, where it shows a discontinuity. As we look fore and aft along the spur of finite thickness, we see that there is also a discontinuity in the features arising from the second traversal by the line of sight. The second-traversal features merge into the surrounding gas distribution in the $l^{II}$, $v$ diagram, however, not the value $v = 0$. The 'surrounding' or local gas is the connection between the separately-seen fore and aft sections of a spur of finite thickness that contains the sun.

## 7. The Local Hydrogen Structure: Observations

In the observed $l^{II}$, $v$ diagram, Figure 4, there are strong features in the vicinity of $l^{II} = 60°$ and $l^{II} = 220°$ that are analogous to Feature C or to the second-traversal feature in Figure 11. We note particularly that they merge into the distribution representing the surrounding gas in precisely the manner predicted by the model study and shown in Figure 11.

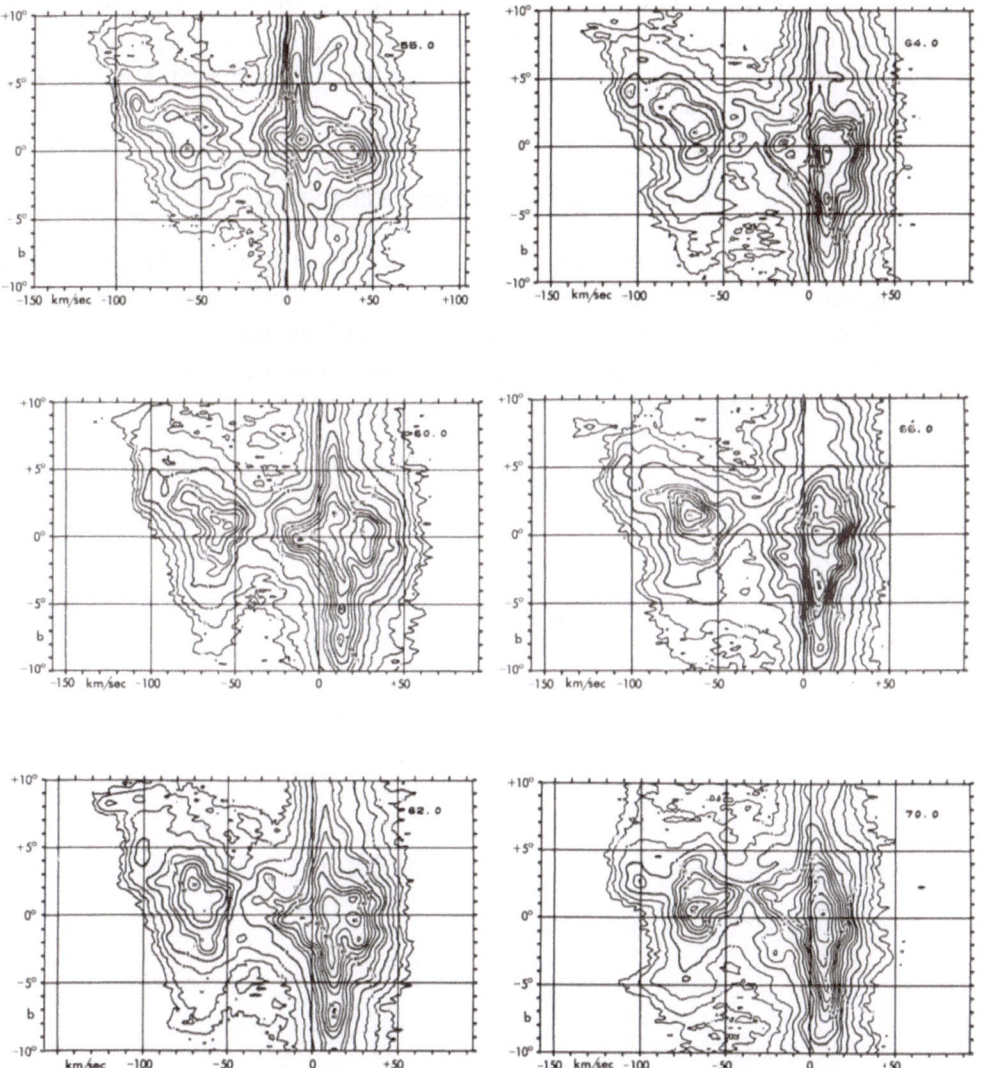

Fig. 12. Observed $b^{II}$, $v$ contour diagrams illustrating the behavior of the spur containing the sun. The marked similarity to the predicted (theoretical) $b^{II}$, $v$ diagrams is striking. Note in particular how the separate feature merges into the surrounding gas.

In Figure 12 we display the observational counterpart of Figure 10. The similarity of Figures 10 and 12 is striking. In the observational diagram the feature representing the second traversal by the line of sight is at first ($l^{II} = 55$) clearly separated from the surrounding gas. As $l^{II}$ increases, the separate feature increases in latitude coverage and merges with the surrounding gas precisely as predicted for a gas structure containing the sun.

In Figure 8 the structure derived by transformation of the features at $l^{II} \sim 60°$, $l^{II} \sim 220°$ in Figure 7 is shown as a spur originating at or in the vicinity of the

Sagittarius arm. It is not shown as continuous through the sun since only the features represented by lines in Figure 7 were transformed in Figure 8. Clearly, however, the connection between the parts of the spur through the sun is provided by the surrounding gas as the model study predicts.

## 8. Local Hydrogen Structure and Stars

In Figure 13 we superimpose on Figure 8 the picture of local galactic structure defined by young stars (Schmidt-Kaler, 1964). There is no disagreement between the young stars and the gas in defining the local structure. The long-standing apparent disagreement between the pitch angle of the arms derived from the stars and that found from the gas arose because of an inappropriate comparison.

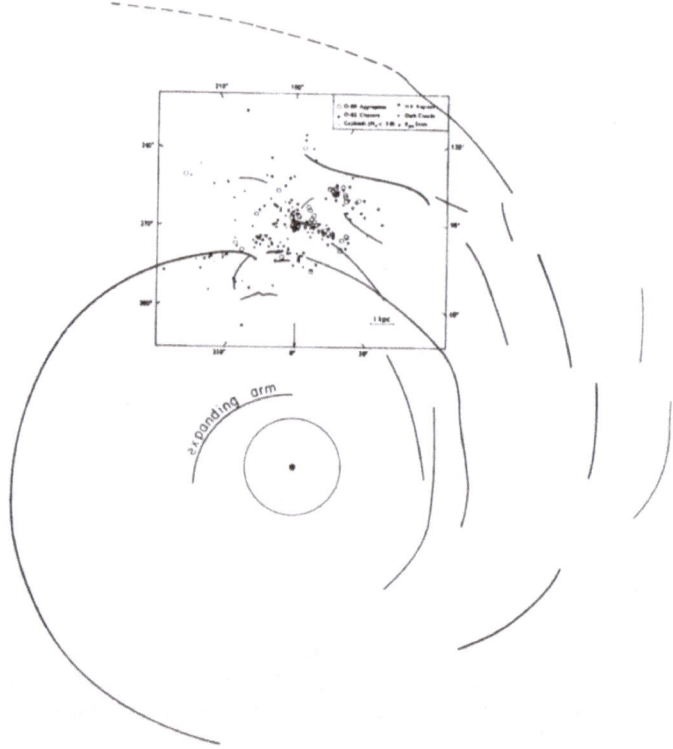

Fig. 13. The observed local distribution of young stars superimposed on the observationally determined distribution of neutral hydrogen. Each distribution was derived completely independently of the other. The agreement is remarkably good. The diagram provides an explanation of the causes of the long-standing apparent disagreement between the pitch angle derived from the (local) stars and the (galactic-scale) neutral hydrogen structure.

The observed stars are located primarily in the local spur that contains the sun. The local spur does indeed have a large pitch angle, 20° to 25°, when defined either by stars or by gas. The pitch angle of the large-scale arm structure derived from hydrogen observations is much smaller, of the order of 12°.5. The discrepancy arose when the large star-derived pitch angle for the *local* structure was compared with the much smaller gas-derived pitch angle of the *large-scale* structure. There is no disagreement of pitch angle when the comparison is confined to the local structure in which the observed stars are located. There are no appropriate stellar observations to test agreement of stars and gas in defining the pitch angle on the galactic scale.

## References

Kerr, F. J.: 1969, *Australian. J. Phys. Astrophys. Suppl.* **9**, 1.
Mathewson, D. S., Healey, J. R., and Rome, J. M.: 1961, *Australian. J. Phys.* **15**, 354.
Sandage, A.: 1961, *The Hubble Atlas of Galaxies*, Carnegie Institution of Washington, Washington.
Schmidt, M.: 1965, *Stars and Stellar Systems* **5**, 528.
Schmidt-Kaler, T.: 1964, *Trans.IAU* **12B**, 416.
Varsavsky, C.: 1969, private communication.
Weaver, H.: 1969, in preparation.
Weaver, H. and Williams, D. R. W.: 1969, in preparation.
Westerhout, G.: 1958, *Bull. Astron. Inst. Netherl.* **14**, 215.
Wilson, T.: 1969, Ph.D. Thesis, M.I.T.

# 21. GALACTIC CONTINUUM SOURCE COUNTS

## T. L. WILSON

*National Radio Astronomy Observatory, Charlottesville, Va., U.S.A.*

The CSIRO 11 cm continuum survey of the galactic plane from $l^{II} = 280°$ to 37° has recently been finished. I have used this survey to compare the number of sources in various 5° intervals north and south of the galactic center. The NRAO-CSIRO-MIT H 109α line survey (Wilson *et al.*, 1970) found 39 sources lying between 330° and 340°, while the NRAO-MIT (Reifenstein *et al.*, 1970) survey found only 14 between 20° and 30°. If the galaxy were symmetric, one would expect roughly equal numbers. One suspects a selection effect is present, and a rough check is very worthwhile. The CSIRO survey is particularly valuable since it is the highest resolution survey covering large portions of the galaxy on both sides of the center.

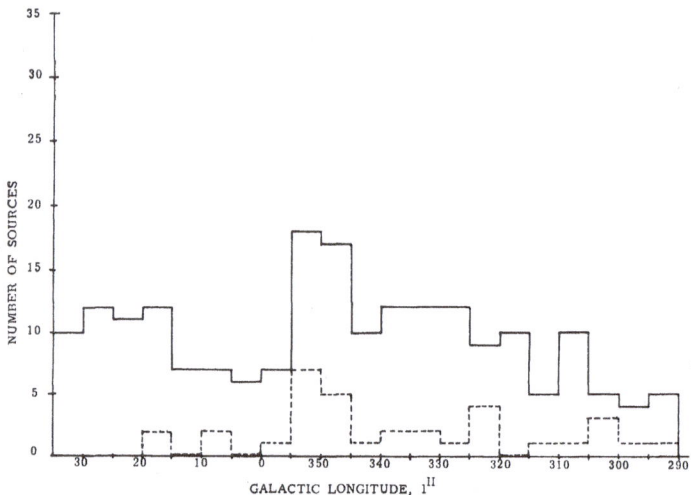

Fig. 1.   Plot of number of sources per 5° interval, as a function of $l^{II}$. Solid line – source number. Dotted line – number of optically identified sources.

We have included only those sources which would be present in the H 109α line surveys, that is those with peak brightness temperatures above 2 K. These, averaged over 5° of longitude, form the solid curve, while the dotted curve is the number of optical identifications per 5° interval. The optical sources are taken from the RCW catalogue. The number of sources in the two zones 20°–30° and 330°–340°, are roughly the same, so that one concludes that the larger numbers found in the south are due to some selection effect, probably the higher angular resolution of the 210′ radio telescope.

An interesting point is the large peak between 345° and 355°. This is probably mostly due to a nearby cluster of sources, although the H 109α line shows that 5 sources possess rather high velocities. This hypothesis is supported by the somewhat higher number of optical identifications and the low radial velocity found from the recombination line surveys. The velocities are shown in Figure 2.

The average velocities in the 345°–355° zone are about $-30$ km s$^{-1}$, far below the maximum Schmidt model velocity $-50$ km s$^{-1}$ for sources outside the 4 kpc arm at this longitude. The velocities in the 330°–340° and 20°–30° zones are much higher,

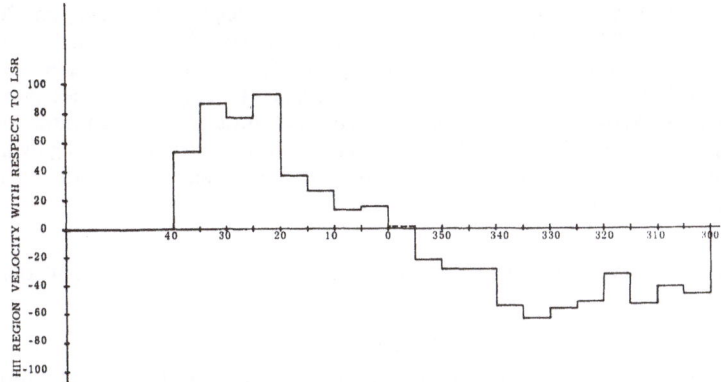

Fig. 2.    H II region VLSR's averaged over 5° intervals, plotted against $l^{II}$.

although not equal. These last two zones, between 4 and 6 kpc from the center, contain most of the giant H II regions in the galaxy. They do not show up well in the 11 cm survey because they are quite distant and their individual sources cannot be resolved. The nearby sources bias us very strongly and only a knowledge of their distance, through their radial velocities, can tell us which are the nearby and which are the distant, significant features.

### Acknowledgements

I am indebted to Dr. F. J. Kerr for valuable discussions on this subject. The NRAO is operated by Associated Universities, Inc., under contract with the National Science Foundation. The CSIRO Survey was performed by M. Beard, G. A. Day, W. M. Goss, F. J. Kerr, and B. M. Thomas. I am indebted to Mr. Day for a preliminary version of the map of the region from 6° to 26°.

### References

Reifenstein, E. C. III, Wilson, T. L., Burke, B. F., Mezger, P. G., and Altenhoff, W. J.: 1970, *Astron. Astrophys.* **4**, 357.
Wilson, T. L., Mezger, P. G., Gardner, F. F., and Milne, D. K.: 1970, in preparation.

## 22. MATTER FAR FROM THE GALACTIC PLANE
## ASSOCIATED WITH SPIRAL ARMS

J. H. OORT

*Leiden Observatory, Leiden, The Netherlands*

**Abstract.** Neutral hydrogen concentrations were studied in the region between 48° and 200° longitude (new) and $+6°$ and $+20°$ latitude. Gas associated with individual arms beyond the sun has been found up to distances from 1 to 2 kpc from the galactic plane with an average density between 1 and 2% of that in the arm centres. The gas with the highest negative velocities shows a different behaviour. This has no counterpart in the plane, and it shows little or no concentration towards lower latitudes. Contrary to the hydrogen associated with the arms its distribution is highly asymmetrical. It does not seem to exist at negative latitudes. It may be of a similar origin as the high-velocity gas in high galactic latitudes near the sun.

It is of interest to investigate how far spiral arms extend in a direction perpendicular to the galactic plane.

For the part inside the sun's orbit we have Schmidt's (1957) data; he finds that in this whole inner region there is still a density of 2.5% of that in the central disk at $z = 560$ pc. In an unpublished investigation based on observations with a much smaller beamwidth W. W. Shane finds for $R = 5$ kpc a density of 0.010 cm$^{-3}$ at $z = 520$ pc. For $R = 7$ kpc this density is reached at $z = 600$ pc. This should be compared with an average density in the plane (average of spiral arm and inter arm) of about 0.5 or 0.4, so Shane finds 2 to 2.5% density at $z = 520$ and 600 pc, in very good agreement with Schmidt. Simonson has found several discrete features at about 1000 pc from the galactic plane and probably connected with arms within 4 kpc from the centre. They have masses of the order of $10^5$ solar masses, and are similar to the clouds previously studied by Gail Smith and by Prata in the same general region.

In this inner region of the Galaxy it is difficult to decide whether such features are associated with specific arms (although this seems likely). The relation between halo features and spiral arms can, however, be observed quite well in the arms beyond the sun. An extensive investigation of the Perseus and outer arms has recently been made in Leiden by Miss Kepner. It covers the region between 48° and 200° longitude and $+6°$ and $+20°$ latitude.

Figure 1a shows three samples of the velocity distribution of hydrogen atoms at intermediate latitudes for the longitude range investigated, which includes the entire region where the Perseus arm is well defined, and a large part of the outer arm. The circles and points plotted give the velocities of well-defined humps, or, sometimes, shoulders in the line profiles. The vertical lines indicate the half widths of the components on the velocity scale. The hatched regions show the velocities of the principal arms in the galactic plane. One can see from this diagram that around the velocities corresponding to these various arms hydrogen is concentrated also at the higher latitudes. Although in the longitudes concerned the entire outer-arm structure is elevated above the average galactic plane, as part of the well-known bending of

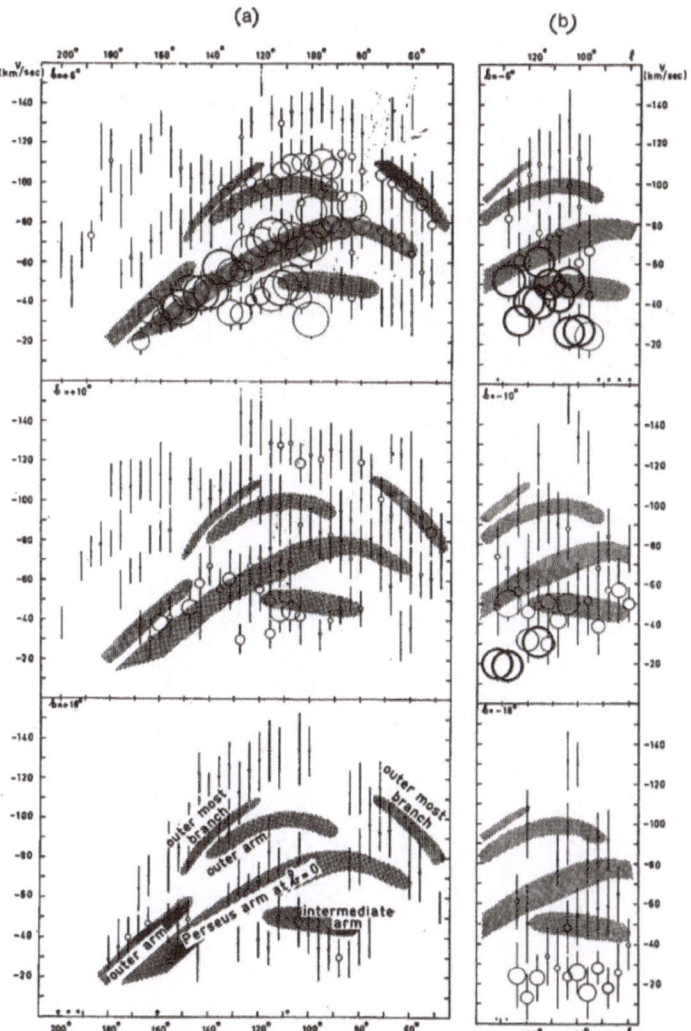

Fig. 1. (a) Three samples of the velocity distribution of hydrogen atoms at intermediate latitude. (b) The same for southern latitudes.

the outer parts of the galactic layer, this effect is small compared with the much higher latitudes considered in this study.

The effect of the clustering of the velocities around the same value at the various latitudes is clearly shown in Figure 2.

With plausible values for the distances to the various arms we find from Miss Kepner's data that in the Perseus arm the average density drops to 1% of the density in the centre of the arm only at $z$-distances varying between 700 and 2400 pc from the plane of symmetry of the arm, the median being about 1400 pc. In the majority of the longitudes there appears still to be at the latitudes concerned ($+10°$ to $+15°$) an observable concentration near the velocity of the arm.

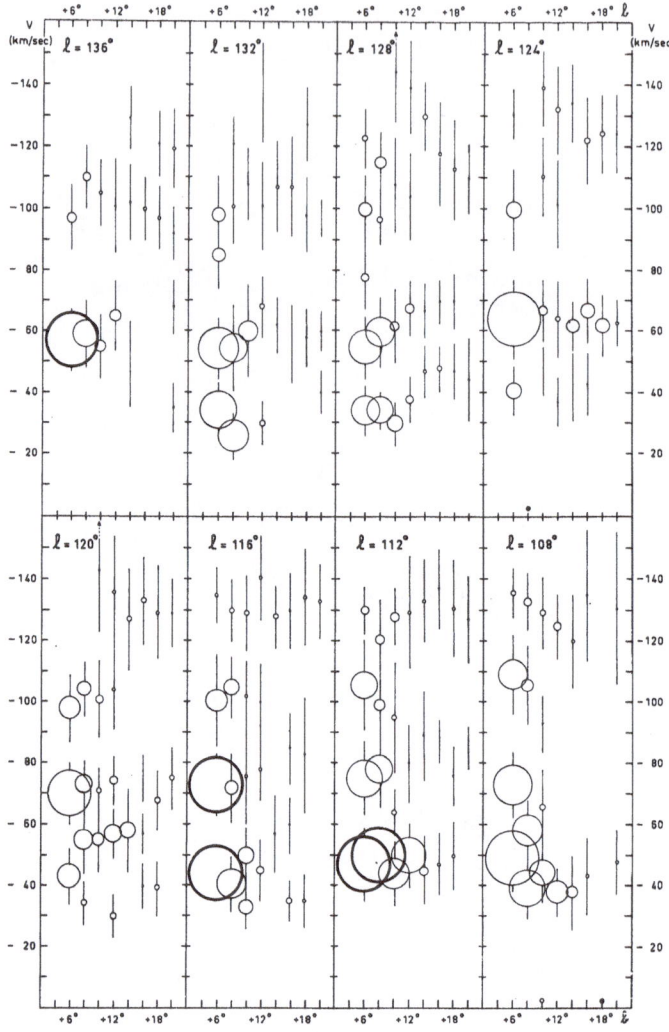

Fig. 2.   The effect of clustering of the velocities.

A similar extension to high $z$-values is found for the outermost branch of the outer arm. Here the density has dropped to about 2% at a distance between 1 and 2 kpc from the plane through the arm.

This is for Northern galactic latitudes. Observations for Southern latitudes are less complete, and have not yet been discussed in detail. A preliminary survey of the data by Mr. Hulsbosch indicates that the extension South of the plane is quite similar to that North of it (Figure 1b).

At the distance of the Perseus arm a star in the galactic plane would need a velocity component perpendicular to the plane of about 55 km s$^{-1}$ for reaching a maximum height of 1400 pc. In order that gaseous filaments could penetrate to such distances their velocity of expulsion from the disk must, of course, be very much higher, because of deceleration by the interstellar medium. In fact, they could probably escape from the disk only by an extremely violent event involving very large masses of gas.

One might wonder how under these circumstances the velocities of the gas observed at these heights could still show a distinct clustering around the velocity corresponding to that of the arm in the plane. This must be due to the fact that only the clouds expelled in directions making very large angles with the plane can manage to leave the central layer of gas.

At several positions in Miss Kepner's survey exceptionally high intensities are found, exceeding those in adjacent longitudes by factors from 4 to 10.

Table I shows the most striking examples found in this material.

TABLE I

Positions of high intensity

|  | $l^{II}$ | $b^{II}$ | $v$ (km s$^{-1}$) | $z$ (kpc) | mass H$_I$ ($10^6 M_\odot$) |
|---|---|---|---|---|---|
| I | 52° to 60° | 10° to 12° | $-78$ | 3.1 | 2.0 |
| II | 120 | 8      14 | $-62$ | 1.1 | 0.2 |
|  | 124 | 14     18 |  |  |  |
| III | 104 | 14     20 | $-50$ | 2.1 | 0.3 |

The brightness temperatures of these features are between 4° and 5°. Estimates of the H$_I$ mass are given in the last column, in units of $10^6$ solar masses. It will be seen that the masses are extremely high for matter pushed out to heights between 1 and 3 kpc from the galactic plane.

A similar phenomenon at negative latitudes was discovered long ago by Mrs. Hack and van Woerden, and has been discussed by Blaauw (1962). These authors found two clouds around $l^{II} = 135°$, $b^{II} = -11°$, with velocities of $-48$ and $-26$ km s$^{-1}$, respectively. They suggested that these masses may have been expelled from the h and $\chi$ Persei association, which lies at the same longitude. In that case the mean distance from the plane would be 0.4 kpc, and the total mass $0.18 \times 10^6 M_\odot$.

In the present material there is an indication of a similar mass concentration at a velocity of about $-35$ km s$^{-1}$ and at longitudes 132° and 128°, latitudes $+6°$ and $+8°$. However, more observations would be required to see whether or not this is connected with the phenomenon of Van Woerden, Hack and Blaauw.

It seems probable that all the gas at large distances from the plane will be swept down into the central layer in times of the order of a few times $10^7$ yr. It must, therefore, be replenished in a similar interval.

It follows that every few times $10^7$ yr the interstellar medium must be violently shaken up.

Before concluding this account of the observations in the lower intermediate Northern latitudes I should say a few words about the gas with velocity in excess of the velocities of known spiral arms. The primary incentive to the investigation was to search for such high-velocity gas at large distances from the sun.

It will be noted that over the greater part of the region surveyed there is hydrogen with negative velocity considerably exceeding that of the outermost spiral arms. This gas has a distribution in latitude which is strikingly different from that of the gas which is related to the arms: in the latitude interval studied it shows little if any tendency of concentration towards the lower latitudes. Moreover, this gas is *not* found in *negative* galactic latitudes.

This phenomenon was first discovered and extensively discussed by Habing (1966), who investigated a region from 42° to 142° longitude.

Especially noteworthy in the new survey are the high negative velocities (between $-100$ and $-120$ km s$^{-1}$) in the anticentre region. Can this be an outlying spiral arm whose motion has a radial component of 110 km s$^{-1}$ directed towards the galactic centre, – and which lies far from the galactic plane, on the opposite side of the outer arm studied by Lindblad (1967) which near $l^{II} = 180°$ has a latitude of $-9°$? Or is this high-velocity gas similar to that which has been observed in high galactic latitudes, but here at greater distances from the sun? If the latter interpretation is the correct one it probably applies to all of the hydrogen in the present survey observed at velocities exceeding that of the outermost branch of the outer arm. Much of this might possibly be gas falling from outside into the Perseus arm. The circumstance that the high-velocity gas discussed by Habing and by Miss Kepner shows the same asymmetry with respect to the galactic plane as the high-velocity clouds in high latitudes, both being predominantly observed at positive latitudes, is an indication that we are dealing with the same phenomenon in the two cases.

A full account of Miss Kepner's investigation will appear in *Astronomy and Astrophysics*.

## References

Blaauw, A.: 1962, in *The Distribution and Motion of Interstellar Matter in Galaxies* (ed. by L. Woltjer), Benjamin, New York, p. 63.
Habing, H. J.: 1966, *Bull. Astron. Inst. Netherl.* **18**, 323.
Lindblad, P. O.: 1967, *Bull. Astron. Inst. Netherl.* **19**, 34.
Schmidt, M.: 1957, *Bull. Astron. Inst. Netherl.* **13**, 247.

## Discussion

*Van Woerden:* Oort says that the gas at large distances from the plane will be swept back into the plane in about 30 million years and therefore has to be replaced. Although it is probable (but not at all certain) that in $10^8$ or $10^9$ years from now there will be similar amounts of gas far from the plane as today, I think there is no reason at all to view the replacement process along lines of detailed balancing. Our galaxy may look different then, at least in detail; we should not consider the gas distribution as an equilibrium configuration.

# 23. ON A POSSIBLE CORRUGATION
# OF THE GALACTIC PLANE

C. M. VARSAVSKY and R. J. QUIROGA

*Instituto Argentino de Radioastronomía, Buenos Aires, Argentine*

**Abstract.** We have studied the rotation curve of the Galaxy at different heights below and above the equator. In the course of this work we noticed that the maximum brightness temperature of hydrogen oscillates around the galactic plane following a fairly sinusoidal pattern. It is further noticed that the maximum temperature of hydrogen occurs right on the plane in the regions where the rotation curve has a form indicating solid body rotation. A rotation curve based on points of maximum hydrogen temperature does not differ appreciably from a rotation curve measured on the galactic plane.

When the 30-meter telescope of the I.A.R.-C.I.W. Radio Astronomy Station was put into operation in June, 1966, one of its first tasks was to check the rotation curve of the Galaxy that had been measured previously at Parkes (Bajaja *et al.*, 1967). Further work on the rotation curve involved measuring the curve at heights of 50 and 100 parsecs above, and 100 parsecs below, the galactic plane. The results, shown in Figure 1, indicate that at $z = 50$ pc the rotation curve obtained is practically identical with that at $z = 0$, while the curves at $z = \pm 100$ pc indicate slower rotation.

To carry out this work, observations were made at 1° intervals in longitude and, at each longitude, the latitude was chosen so that the tangential point would be at the proper height. Comparison of the observations at different latitudes gave indication that the maximum brightness temperature at the tangential point was not normally at $b^{II} = 0°$ but rather at some height below or above the equator.

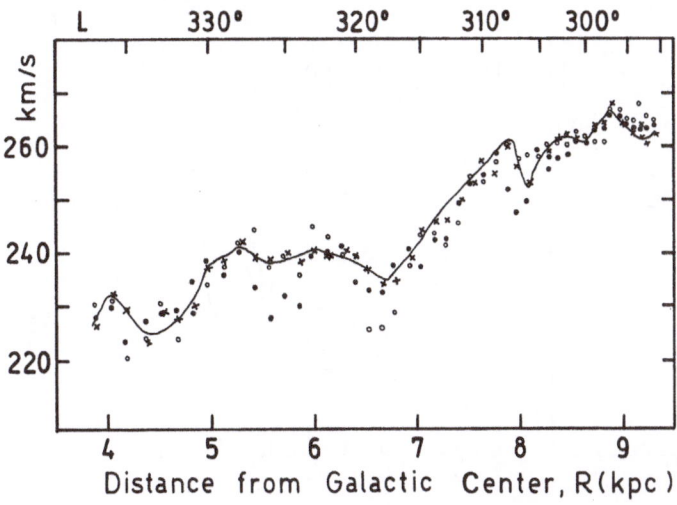

Fig. 1.   The rotation curve of the Galaxy at different heights. The full line shows the curve for $z = 0$. Open circles correspond to $z = 100$ pc; filled circles to $z = -100$ pc; crosses to $z = 50$ pc.

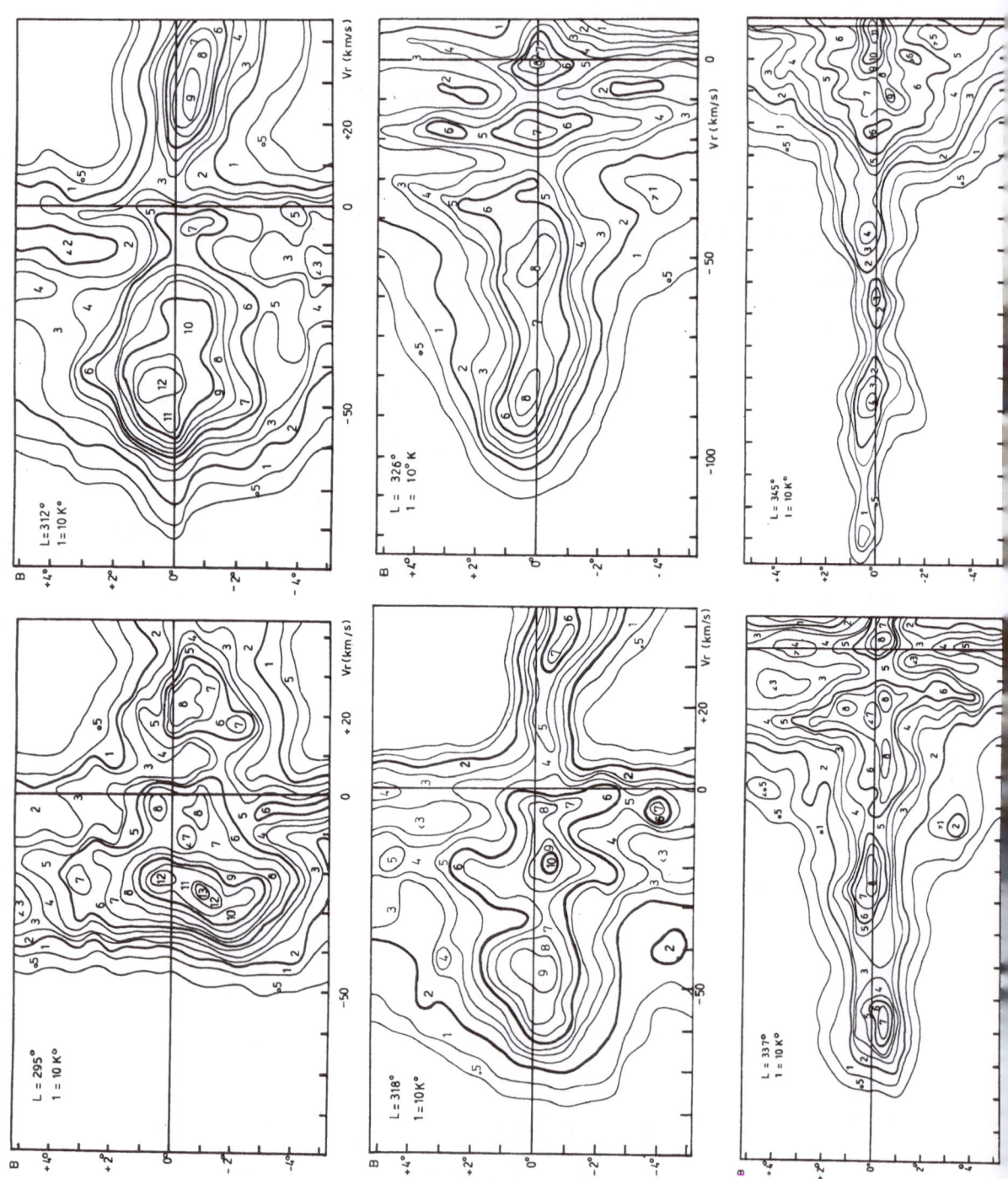

To study this effect in detail, as well as other properties of the hydrogen near the galactic plane that will be considered in a later publication, we carried out a survey in the region $281° \leqslant l^{II} \leqslant 345°$ and $-5° \leqslant b^{II} \leqslant 5°$, with spacings of $2°$ or $3°$ in $l^{II}$ and $0°.5$ in $b^{II}$. From the profiles we prepared contour diagrams of brightness temperature as a function of galactic latitude and radial velocity, for given longitudes. Samples of such contour diagrams are shown in Figure 2.

As can be seen easily from Figure 2, the points of highest brightness temperature with maximum radial velocity, that is, at the tangential point, occur at different heights below or above the equator. Figure 3 shows this effect in detail.

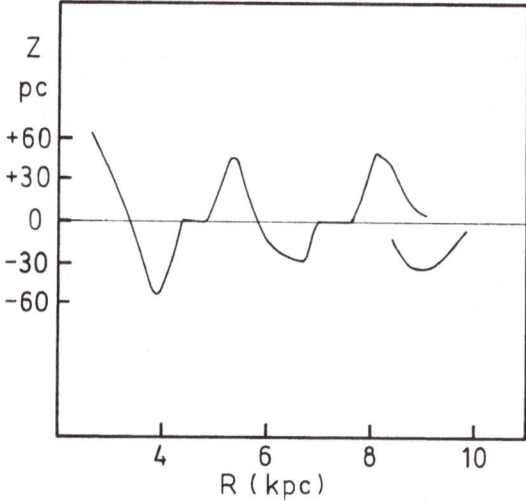

Fig. 3.   Height dependence of the points of maximum brightness temperature as a function of distance to the galactic center.

The curve in Figure 3 presents several interesting features. In the first place, the general shape of the curve is sinusoidal, and indicates that the galactic plane is 'corrugated', with a wavelength of about 3 kpc. Secondly, it is quite remarkable that the two regions, centered around 4.6 and 7.3 kpc from the center of the Galaxy, where the greatest density of hydrogen seems to be exactly on the equator, correspond to the portions of the rotation curve where the velocity varies linearly with distance, indicating solid body rotation. Thirdly, we see that for $8.4 \leqslant R \leqslant 9.0$ kpc there are two concentrations of hydrogen, one below and one above the galactic plane, with the same radial velocity.

Finally, Figure 4 shows a comparison of the rotation curve obtained at the plane with one using the points that have the highest brightness temperature. We see that both curves are essentially identical; in the second curve the dip at $R = 8.1$ kpc

Fig. 2a–f.   Contour diagrams of brightness temperature as a function of galactic latitude and radial velocity for different longitudes.

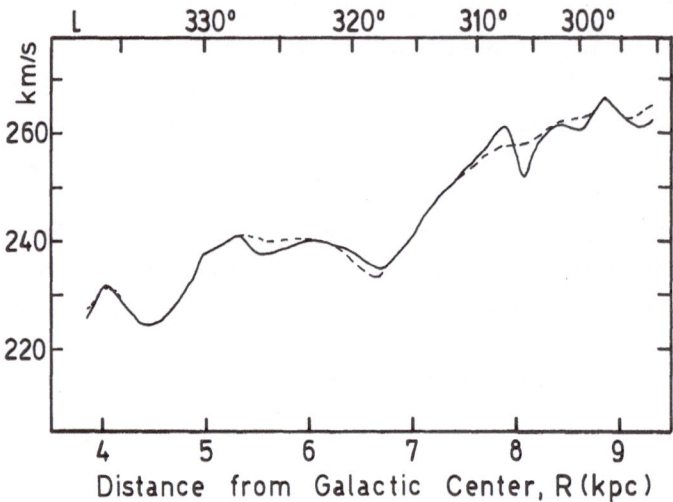

Fig. 4.    The rotation curve of the galaxy at $b^{II} = 0$ (full line) and at the heights of maximum brightness
temperature (dotted line).

disappears, and a few other points are shifted to slightly higher velocities, but not nearly enough to make the curve obtained for the fourth quadrant coincide with that obtained by Kerr or by Shane and Bieger-Smith (1966) for the first quadrant.

## Acknowledgement

One of us (R. J. Q.) wishes to thank the Consejo Nacional de Investigaciones Científicas y Técnicas of Argentina for a scholarship that made this work possible.

## References

Bajaja, E., Garzoli, S., Strauss, F., and Varsavsky, C. M.: 1967, IAU Symposium No. 31, p. 181.
Shane, N. W. and Bieger-Smith, G. P.: 1966, *Bull. Astron. Inst. Netherl.* **18**, 263.

# 24. A STUDY OF SPIRAL STRUCTURE FOR 270° ⩽ $l^{\mathrm{II}}$ ⩽ 310°

S. L. GARZOLI

*Instituto Argentino de Radioastronomía, Buenos Aires, Argentine*

A survey of the distribution of the intensity of the 21 cm-line of atomic hydrogen in the region with galactic coordinates $270° ⩽ l^{\mathrm{II}} ⩽ 310°$, $-3° ⩽ b^{\mathrm{II}} ⩽ 2°$, was made with the 100-foot radiotelescope and 56-channel receiver of the Radio Astronomy Station of the Instituto Argentino de Radioastronomía and the Department of Terrestrial Magnetism of the Carnegie Institution of Washington. The angular and velocity resolution of the system is 28′ of arc and 2 km s$^{-1}$.

The analysis of the data included the identification of hydrogen concentrations, and the study of their interconnection, which gives rise to the different structures that produce the spiral appearance of the Galaxy.

We found concentrations with velocities forbidden by Schmidt's mass model of the Galaxy. We also find that the sun seems to have a priviliged position when linear density variation is studied. This can be due to errors in distance determination, to a temperature effect (that was considered uniform throughout all concentrations) or to a velocity dispersion effect arising from the assumption of Gaussian components.

The above indicates that our results must be taken as a merely approximate picture of the actual situation.

We estimate the errors which may result from the mass model. For some longitudes, errors affecting the determination of distances can reach values up to 3.6 kpc for very small velocities (less than 10 km s$^{-1}$) while for greater velocities they can be of the order of 0.4 kpc.

From the general analysis of our observations it is obvious that the structures present a big amount of fine structure. Structures are made up by a sequence of concentrations with similar features (velocity, size, velocity dispersion, position relative to the galactic plane); sometimes very small clouds are resolved within these concentrations. This can be clearly seen in the case of local hydrogen.

We study the distribution of local hydrogen, and it is found that the sun is probably located in the inner edge of the local structure.

Small concentrations with narrow profiles and high temperature are easily found at zero velocity, especially at latitudes above and below the plane where the local structure and the local hydrogen can be separated.

In order to obtain an approximate picture of the Galaxy, a velocity-longitude diagram for the concentrations was drawn (Figure 1). This diagram should show continuities between concentrations, that is, spiral structures. However, we cannot predict, just from this diagram, the possible connections between concentrations, since they are at different distances from the galactic plane. Therefore their distribution in $z$ must also be considered.

Taking this into account and using criteria about the velocity dispersion and size

*Becker and Contopoulos (eds.), The Spiral Structure of Our Galaxy, 151–153. All Rights Reserved.*
*Copyright © 1970 by the I.A.U.*

of the concentrations we have drawn the general diagram presented in Figure 2.

The most important feature that appears from this graph is that, even when the structures approximately follow the general lines of spiral arms, they are broken into concentrations whose values of $N_H$, positions relative to the plane, size and dispersion in velocity are variable.

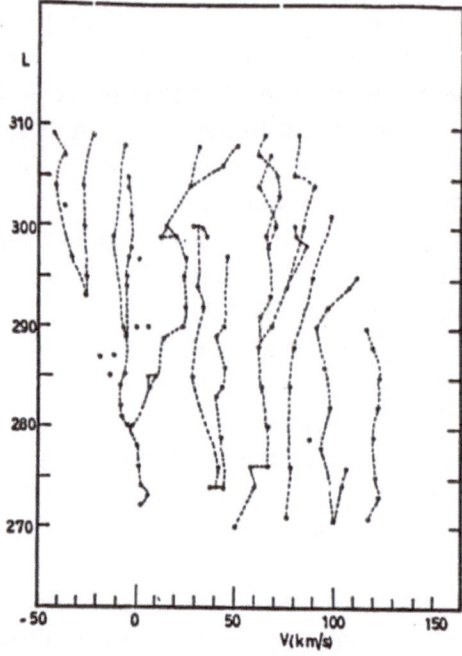

Fig. 1.   Velocity-longitude diagram for the hydrogen concentrations. The longitude and the velocity are those corresponding to the maximum temperature.

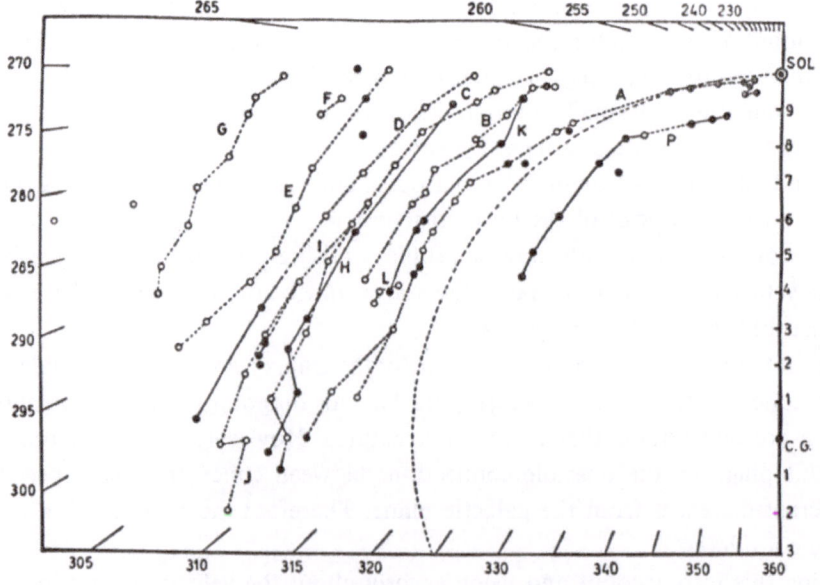

Fig. 2.   General diagram of Galactic structure.

A study of the distribution of hydrogen in the Carina and Sagittarius arms was made in order to determine their location and the possibility of fitting them to arms in the Eastern galactic hemisphere. Figures 1 and 2 let us conclude that the Carina structure does not go through the sun but tends towards Sagittarius.

The position of these structures does not agree with the position of the structures given by optical objects, but the region of low density of stars and gas near the Carina edge is real and not an effect caused by local obscuration; however, the stellar and gaseous 'holes' are shifted in longitude.

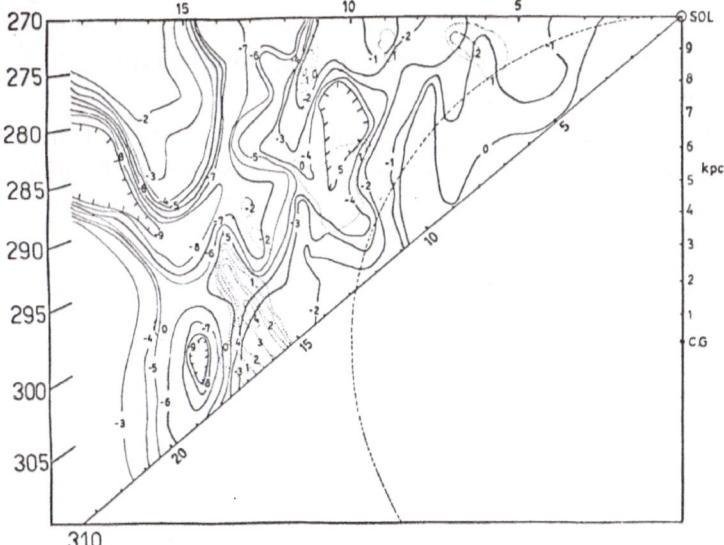

Fig. 3. The lines show the position relative to the plane of all the structures. Full lines denote negative $z$ values, dotted lines positive ones.

Concentrations with many different features were found. It is remarkable that many of them have high velocities and also that some of them are located at great negative distances from the plane.

The bending of the plane is obvious in the region studied, and it looks that the general tendency is in the direction of the Large Magellanic Cloud.

We also find concentrations with very low temperature and velocity dispersion, subtending small angles and located principally on and above the galactic plane in those regions where the main structures bend beneath it. We assume that they are the product of some secondary effect of the forces that caused the bending of the plane.

### Acknowledgement

This paper is a brief summary of my Ph.D. Thesis that will be published in full (in Spanish in the *Anales de la Sociedad Científica Argentina*). A more extended English version will be published elsewhere.

# 25. A PECULIAR NEUTRAL HYDROGEN CONCENTRATION AT $l^{II}=280°$, $b^{II}=-18°$

E. BAJAJA and F. R. COLOMB

*Instituto Argentino de Radioastronomía, Buenos Aires, Argentine*

The observations discussed in this paper were obtained with the 30-m telescope of the Instituto Argentino de Radioastronomía.

The feature described here was found during a general survey of a region bounded by $220° \leqslant l^{II} \leqslant 300°$, $-30° \leqslant b^{II} \leqslant -15°$ which will be published later. At $l^{II}=280°$, $b^{II}=-18°$ an emission appeared with a peak temperature of about 20 K, half width of 6 km s$^{-1}$ and $V_R = -30$ km s$^{-1}$.

This feature was rather unusual, so we began the observation of the surroundings of that point to the limits defined by the possibility of detection of the signals. A total of 200 points were observed distributed according to Figure 1, which represents approximately the extension of the concentration.

Every profile has been studied by fitting gaussian curves after subtracting local

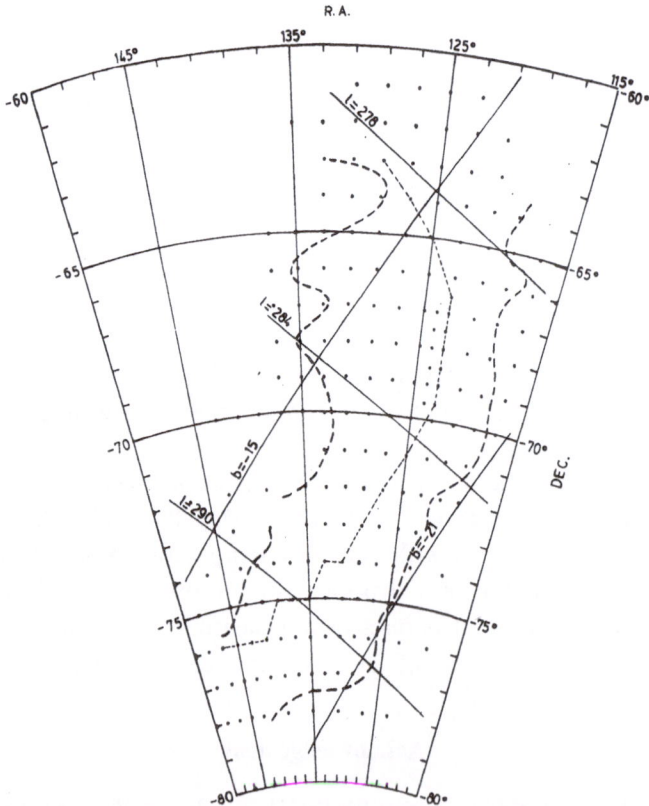

Fig. 1.   Grid of observed points. Heavy dashed line indicates the limits of the clouds. Light dashed line shows the maximum peak antenna temperature.

*Becker and Contopoulos (eds.), The Spiral Structure of Our Galaxy* 154–156. *All Rights Reserved.*

hydrogen contributions. In this way we obtained the peak temperature, dispersions and velocities which could give us some idea of the different physical parameters of the cloud. In many cases several peaks were seen simultaneously with velocities distributed between $-10$ and $-50$ km s$^{-1}$.

The strongest feature appears at $V$ between $-26$ and $-42$ km s$^{-1}$ and shows an elongated shape along $\alpha = 123°$ from $\delta = -66°$ to $\delta = -70°$ where the axis of the concentration turns and becomes parallel to the galactic plane.

The general appearance of the cloud resembles the one described by Verschuur (1969) and Wesselius (1969).

This cloud, located at $l^{II} = 103°$, $b^{II} = +69°$, presents also a turning point, but the direction of the arm parallel to the galactic plane, is in the opposite sense.

In the regions of the cloud where the intensities are strongest, there is a relative absence of local hydrogen. A similar characteristic has been pointed out by Cugnon (1968) for another feature at $l^{II} = 349°$, $b^{II} = +3°$.

The dimensions and mass of the cloud depend on the assumed distance. Table I shows, for several assumed distances: the distance to the galactic plane $z$, the distance to the galactic center $R$, dimensions $L_1 \times L_2$ and mass $M$. The mass has been obtained from $N_H$ contour maps for velocities ranging from $-19$ to $-50$ km s$^{-1}$.

TABLE I

Data for the observed cloud

| $r$ (kpc) | $z$ (kpc) | $R$ (kpc) | $L_1 \times L_2$ (kpc $\times$ kpc) | $M$ ($M_\odot$) |
|---|---|---|---|---|
| 0.1 | 0.03 | 10 | $0.009 \times 0.026$ | $2.6 \times 10^2$ |
| 1 | 0.3 | 9.8 | $0.087 \times 0.262$ | $2.6 \times 10^4$ |
| 10 | 3.1 | 12.4 | $0.87 \times 2.62$ | $2.6 \times 10^6$ |
| 50 | 15.4 | 49 | $4.35 \times 13.1$ | $6.5 \times 10^7$ |
| 100 | 30.9 | 93.7 | $8.7 \times 26.2$ | $2.6 \times 10^8$ |

The analysis of the velocity distribution along the cloud's axis reveals a continuous variation from $-20$ to $-40$ km s$^{-1}$.

Two interpretations are possible for this feature. If the velocity variation is due to the effect of the line of sight angle with respect to the true center mass velocity, that angle would be 70° and the absolute value of the velocity 90 km s$^{-1}$ pointing to increasing declinations and approaching us.

Another possible interpretation is that the cloud is rotating. The implications of this hypothesis will not be considered here because they depend strongly on the assumed distance. No peculiar optical feature was found which could be associated with this cloud. No special absorption has been seen in the area and there are no stars available with measured interstellar lines. Thus, any distance assumed is merely hypothetical. More objects of this kind are being looked for.

A more detailed study of this concentration will be published elsewhere.

# References

Cugnon, P.: 1968, *Bull. Astron. Inst. Netherl.* **19**, 363.
Verschuur, G. L.: 1969, *Astron. Astrophys.* **1**, 473.
Wesselius, P. R.: 1969, *Astron. Astrophys.* **1**, 476.

# 26. CONTRIBUTION TO THE STUDY OF THE DISTRIBUTION OF NEUTRAL HYDROGEN IN THE REGION $230° \leqslant l^{II} \leqslant 280°$

D. GONIADZKI and A. JECH

*Instituto Argentino de Radioastronomía, Buenos Aires, Argentine*

**Abstract.** A sky survey of the 21 cm hydrogen line has been made with the 100-foot Radiotelescope of the I.A.R.-C.I.W. Radio Astronomy Station in the region $230° \leqslant l^{II} \leqslant 280°$, $-15° \leqslant b^{II} \leqslant -3°$.

We study the distribution of the local hydrogen and that in the Orion, Intermediate and Perseus arms. We find a new structure that starts at $l^{II} = 265°$. We also study the concentrations which lie far below the plane; some of them seem to be related to Lindblad's G arm.

## 1. Introduction

A sky survey has been made in the 21 cm emission line of neutral hydrogen covering the region with the galactic coordinates (new) $230° \leqslant l^{II} \leqslant 280°$ and $-3° \geqslant b^{II} \geqslant -15°$. The spacing was of 5° in longitude and of 1° in latitude. A total of 143 points in the sky were observed, each of them three times at different moments.

These data have been analyzed by means of contour diagrams which give the antenna temperature as a function of latitude and radial velocity for a given galactic longitude. Some of the graphs obtained are shown in Figures 1a, b, c.

In order to study continuous structures we made schematic plots using as radial velocity for each concentration the one corresponding to maximum temperature; their connection has been studied following the work of Lindblad (1967).

For some concentrations it was possible to separate the corresponding component by Gaussian analysis and in such cases $N_H$ was evaluated from the formula given by Van de Hulst *et al.* (1954). A value of 135 K has been taken for the kinetic temperature, since this was the highest value found in our scale. Velocity dispersion was taken as the width at $T_b = 0.5 \ T_{b\,max}$. Distance estimates were based on Schmidt's (1965) model of the Galaxy.

## 2. The Observed Features

We have adopted the notation used previously by Höglund (1963) and Lindblad (1967). For $l^{II} > 180°$ Lindblad found mainly four features.

(a) *Structure A* or local hydrogen whose typical characteristics are a very narrow velocity dispersion and great spread in latitude. The concentrations have velocities near zero. Figure 2 shows the mean value of those found in each latitude studied. The velocity of the local hydrogen appears to be negative at $l^{II} = 280°$, near the plane. This is due to the effect of an overlapping of the local hydrogen, the Orion arm and a contribution of the Carina arm (Garzoli, 1969). Beneath the plane the concentrations have velocities more nearly zero, as it is shown by the contour diagrams corresponding to that longitude.

At low latitudes we are in a region of great hydrogen concentration, due to the effect

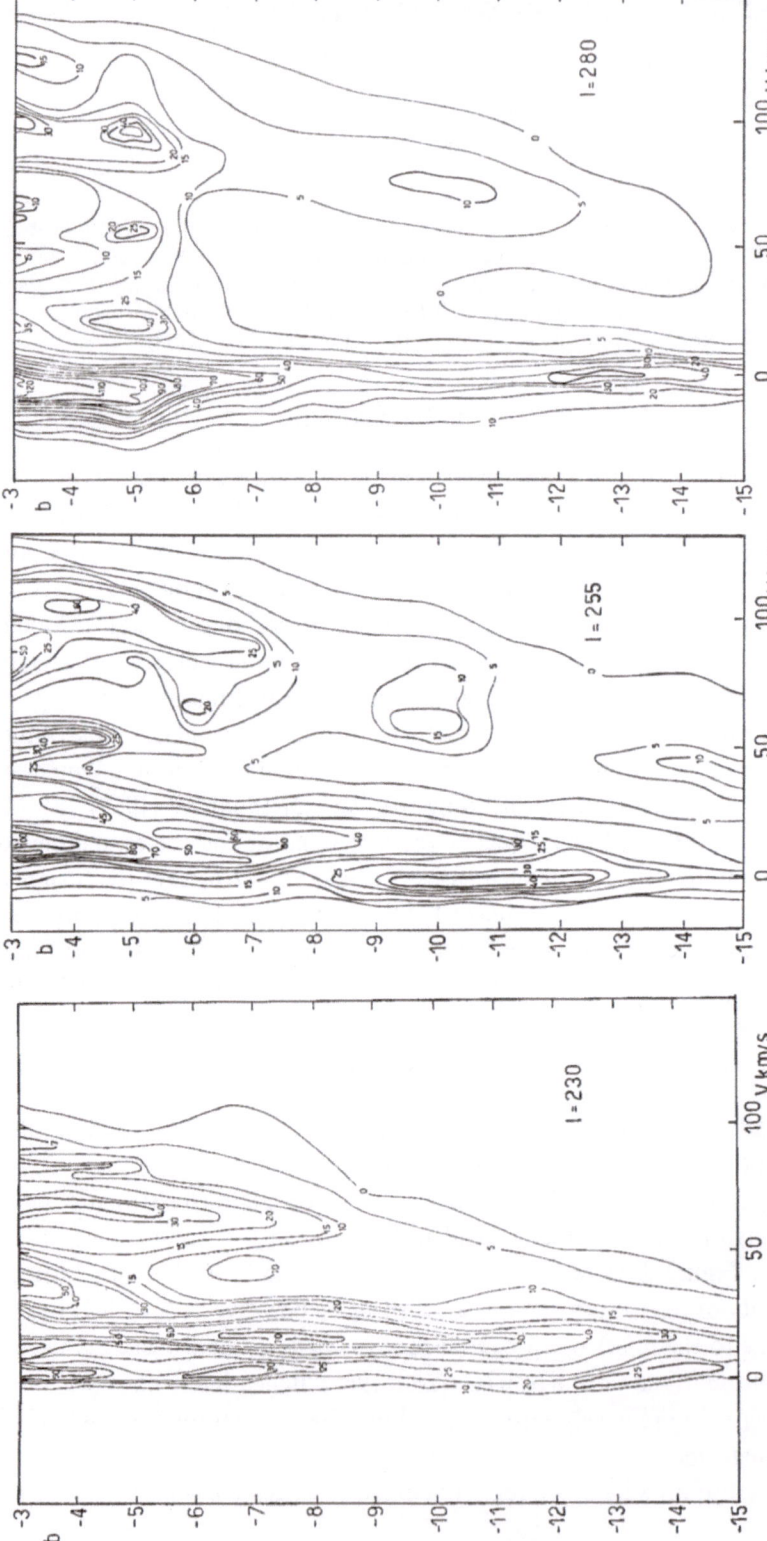

Fig. 1.   Contour maps of antenna temperature as function of galactic latitude and radial velocity at longitudes, $l^{III} = 230°$, $255°$, $280°$.

of the local structure or spiral arm which goes through the sun. For this reason local concentrations are the ones identified at most negative latitudes. On the basis of the circular galactic rotation model local hydrogen should become negative at $l^{II} = 270°$; since it is positive for $270° \leqslant l^{II} \leqslant 280°$, these local concentrations are probably affected by peculiar motions.

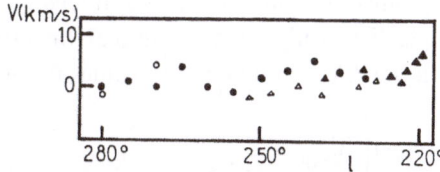

Fig. 2. Radial velocity – longitude diagram for the maximum temperature of the local hydrogen. Open circles represent Garzoli's data, open triangles are from the Kootwijk survey and full triangles are from the Dwingeloo survey. Full circles correspond to the present work.

Lindblad has suggested a theoretical model of a shell in expansion. The predictions of this model agree with our observations.

A full analysis into Gaussian components has not been made and consequently the mean values for velocity dispersion and the average number of atoms observed along the line of sight are not given. Typical average values are:

$$\sigma \approx 2\text{--}5 \text{ km s}^{-1}, \quad N_H \approx 6 \times 10^{20} \text{ at cm}^{-2},$$

which agree quite well with Lindblad's results.

(b) *Structure H or Orion arm.* Figure 3 shows radial velocities found for each longitude for structure H. The contour diagrams show that at $l^{II} = 230°$ the maximum intensity falls below the plane, around $b^{II} = -7°$, and approaches it sharply as longitude increases. At $l^{II} = 260°$ it is entirely on the plane and then it seems to dip down again.

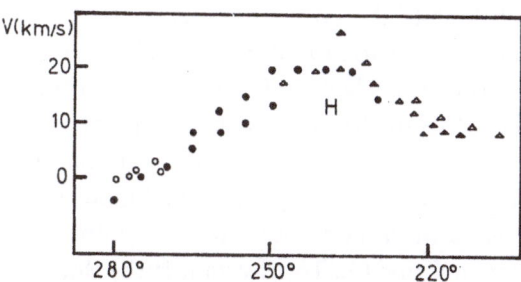

Fig. 3. Radial velocity – longitude diagram for the local arm (feature H); the symbols are the same as in Figure 2.

At $l^{II} = 230°$ this structure is quite separated from the local hydrogen, but near $l^{II} = 265°$ it overlaps with the local gas. Typical values for $\sigma$ and $N_H$ are 10 km s$^{-1}$ and $4 \times 10^{21}$ at cm$^{-2}$ respectively; both are considerably greater than the corresponding values found by Lindblad because the analysis into Gaussian curves has not been done.

As we can see, if circular motion is admitted, the most intense concentrations seem to be located behind the sun, with respect to the galactic center, and go away from it following an approximately ring-shaped path as $l^{II}$ increases. This is shown in Figure 5.

(c) *Structure I or intermediate arm.* It is a very continuous structure located on the plane, with little spread in latitude. For this reason it has not been found by Lindblad at longitudes near ours, since he only observed intermediate latitudes with the Dwingeloo 25 m telescope. The only data available for $210° \leqslant l^{II} \leqslant 235°$ are those obtained at Kootwijk with the 7.5 m antenna. The values found are shown in Figure 4,

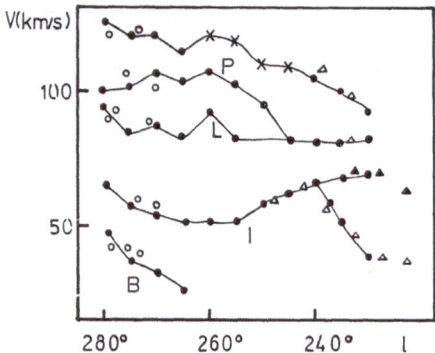

Fig. 4.   Radial velocity – longitude diagram for the structures B, I, L and P. The symbols are the same as in Figure 2.

which also shows the agreement of our observations with those made by Lindblad and Garzoli. The latter refers to it as structure C. Typical values for $\sigma$ and $N_H$ are 13 km s$^{-1}$ and $2 \times 10^{21}$ at cm$^{-2}$ respectively. For $203° \leqslant l^{II} \leqslant 212°$ Lindblad finds a value of $13 \times 10^{21}$ at cm$^{-2}$ for $N_H$. He also finds that the data for structure I are more similar to those found in the other hemisphere for the external arm, but he thinks it is doubtful that both belong to the same structure.

(d) *Structures P and L or Perseus arm.* Raimond's (1966) opinion is that the Perseus arm consists of both structures. He calls L the structure with greatest intensity and lowest velocity and he refers to the minor components as P. Structure L is the one with greatest spread in b; for this reason it is the only one measured by Lindblad with the 25 m telescope in a region overlapping ours. Figure 4 shows the data obtained and gives a comparison between them with those of Lindblad and Garzoli (the latter refers to these structures as F and G). To ascertain if the different concentrations are interconnected it is necessary to make a finer mesh in longitude and observe closer to the plane. The values of $N_H$ and $\sigma$ are likely to be highly variable and are affected by big errors due to the presence of many mingled structures.

(e) *Structure B.* For longitudes between 265° and 280° we observe a structure which does not appear at lower longitudes and consequently has not been studied by Lindblad. We call it B following Garzoli. It is located on the plane and we cannot say if it is related or not to the structure studied before. Its behaviour is shown in the radial velocity vs. longitude diagram of Figure 4.

Figure 5 shows the general diagram of galactic structure for all observed features on the basis of Schmidt's rotation model.

### 3. Concentrations at Intermediate Latitude

All concentrations found are shown on Figure 6. The crosses are concentrations with little spread in latitude near $b^{II} = -5°$. It is impossible to determine $\sigma$ without analysis into Gaussian components since they are mixed with other arms.

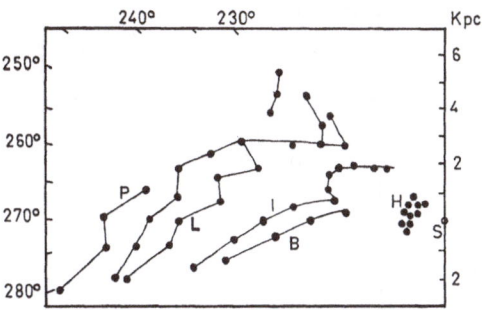

Fig. 5. General diagram of galactic structure (distances are based on Schmidt's model of the Galaxy).

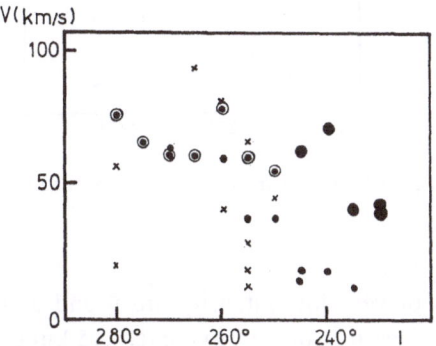

Fig. 6. Radial velocity – longitude diagram for the concentrations at intermediate latitudes.

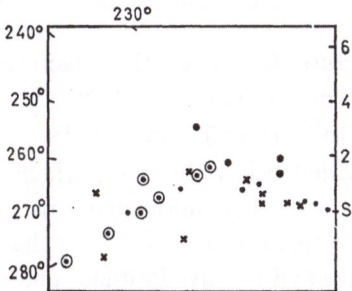

Fig. 7. Location of the concentrations at intermediate latitudes projected on the galactic plane.

At latitudes between $-7°$ and $-8°$ and longitudes between $230°$ and $245°$ there are concentrations denoted by big dots, with $\sigma = 14$ km s$^{-1}$. Velocities correspond to the intermediate arm. These concentrations seem to be continued by the ones that start at $l^{II} = 250°$ which have a rather higher velocity dispersion and are located at $-10° \geqslant b^{II} \geqslant -11°$. The velocities are similar to those corresponding to structure L and this structure may affect the width. These concentrations are denoted by circled dots. This whole feature is very likely a continuation of feature G of Lindblad.

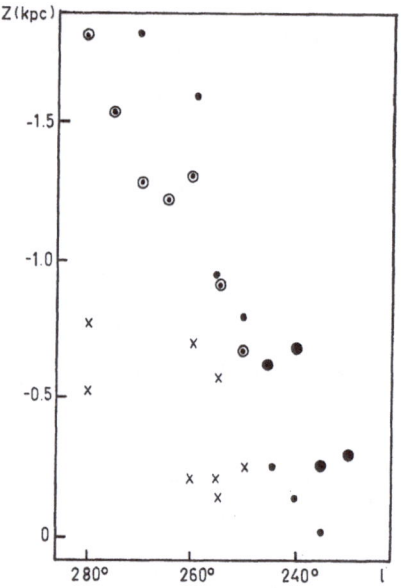

Fig. 8.  Vertical distribution of the concentrations at intermediate latitudes.

Other concentrations of very low intensity are found at more negative latitudes, between $-13°$ and $-15°$ with $\sigma$ of approximately 15 km s$^{-1}$; they are denoted by small dots.

## 4. Conclusions

The distribution of the structures found near the galactic plane shows good agreement with that found by Lindblad at $230° \leqslant l^{II} \leqslant 240°$ and that found by Garzoli at $270° \leqslant l^{II} \leqslant 280°$. They tend to be extended in latitude.

Those concentrations located below the plane which seem related to Lindblad's arm G, appear as extensions of the Intermediate arm away from the plane. The angular extent in latitude of this arm is lower than that of the Perseus arm. More observations at both negative and positive latitudes and at every degree in longitude are needed to test these conclusions.

## Acknowledgements

We wish to express our sincere thanks to Dr. Carlos M. Varsavsky for his generous and constant encouragement and guiding throughout this work.

One of us (D. G.) wishes to thank the Consejo Nacional de Investigaciones Científicas y Técnicas for a scholarship that made this work possible.

## References

Garzoli, S.: 1969, *Anales de la Sociedad Científica Argentina*, in press.
Höglund, B.: 1963, *Ark. Astron.* **3**, No. 19.
Lindblad, P. O.: 1967, *Bull. Astron. Inst. Netherl.* **19**, 34.
Raimond, E.: 1966, *Bull. Astron. Inst. Netherl.* **18**, 191.
Schmidt, M.: 1965, *Stars and Stellar Systems* **5**, 513.
Van de Hulst, H. C., Müller, C. A., and Oort, J. H.: 1954, *Bull. Astron. Inst. Netherl.* **12**, 117.

# 27. RADIAL VELOCITIES OF NEUTRAL HYDROGEN IN THE ANTICENTER REGION OF THE GALAXY

L. VELDEN

*Astronomische Institute der Universität Bonn, Bonn, Germany*

**Abstract.** An observational material of 21-cm H I emission-line profiles is investigated by a statistical method to derive the kinematical properties of the interstellar gas in the region of the galactic anticenter. A description of the method used as well as the results obtained, concerning deviations from a circular rotation, are given.

This report on radial velocities of the neutral hydrogen in the anticenter region is based on 1253 observations of 21 cm-emission-line profiles. The observations were done at the 25-m telescope of the University of Bonn (Velden, 1970).

Figure 1 shows a schematic diagram of the observational program in galactic coordinates. The observations are arranged as 11 cross sections of the galactic plane, indicated by vertical lines in the longitude range $l^{II} = 120°$ to $240°$ with a spacing of $10°$. Each cross section consists of line profiles observed at fixed positions in the latitude range from $-30°$ to $30°$ with a spacing of $\frac{1}{2}°$ in latitude. No observations were done at longitudes $160°$ and $210°$. The dashed horizontal lines separate the observed positions into three latitude ranges later used.

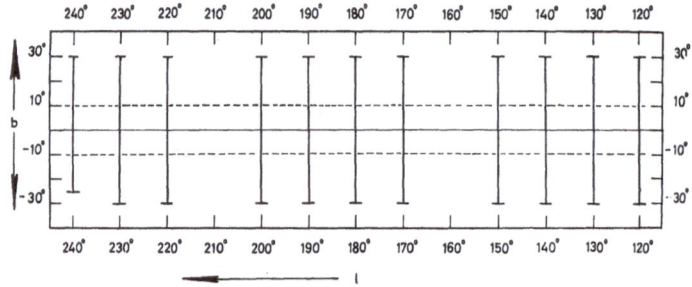

Fig. 1. Schematic diagram in galactic coordinates of the region observed. The vertical lines indicate the cross sections covered by fixed position 21-cm observations.

The observations were done to win a broad base for a statistical investigation of the neutral hydrogen properties as seen under the influence of varying differential galactic rotation. A further intention was to derive criteria for the known spiral features outside the anticenter and try to follow them through the anticenter region. The investigations done for this purpose showed, though not quite unexpectedly, a kinematical effect, the description of which was made the object of this paper.

The reduction procedure (Schwartz, 1967; Grahl *et al.*, 1968) yielded the line profiles in the form of brightness temperature vs. radial velocity with respect to the local standard of rest. For the velocity reduction the standard solar motion was used.

The kinematical investigation was done in a strictly statistical sense. The line profiles of a latitude range at a fixed longitude were used for the calculation of a mean velocity. This was achieved by weighting the radial velocities of the line profiles by their respective brightness temperatures. For a latitude range the mean velocity thus derived may be called 'profile area weighted mean velocity'. To smooth out the random motions of individual objects and derive the large scale motions mean radial velocities were determined for latitude ranges of at least 20°.

The method of area-weighting used is handicapped against calculating mass-weighted mean velocities, by giving less weight to the same physical objects at greater distances. On the other hand the calculation of mass-weighted mean velocities would be based on, especially in the anticenter region, highly doubtful individual distances derived from a model of galactic rotation.

An advantage of the method used is the inclusion of the material at velocities forbidden in the sense of a circular rotational model.

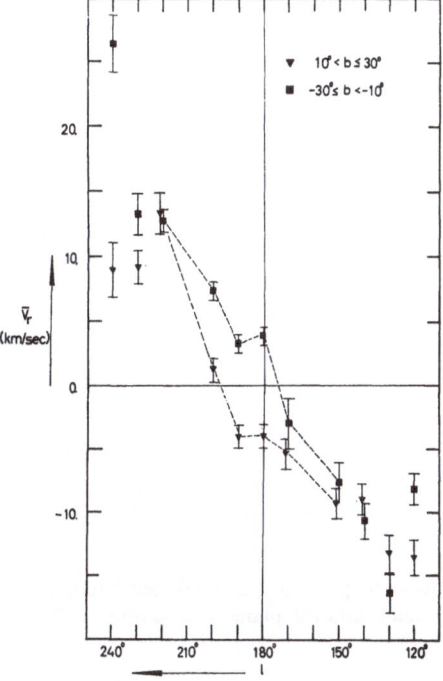

Fig. 2.   Velocity-longitude diagram of the observations at intermediate latitudes above ($10 < b^{\mathrm{II}} \leqslant 30$, triangles) and below ($-30 \leqslant b^{\mathrm{II}} < -10$, squares) the galactic plane. The error bars indicate the computed rms-errors.

Figure 2 shows a velocity-longitude diagram for the mean velocity points of the latitude ranges $-30°$ to $-10°$ represented by squares and $10°$ to $30°$ plotted as triangles. The error bars indicate the computed rms errors. The velocities above and below the galactic plane show a splitting around the anticenter of $\pm 4.5$ km s$^{-1}$. The extension of the velocity splitting to higher and lower galactic longitudes remains uncertain since the line of sight does there obviously cut into more distant material because of the known bending of the galactic plane. The velocity splitting at the anticenter can be removed by a correction for a relative velocity of gas and LSR of $-12.5$ km s$^{-1}$ in the direction of the galactic north pole.

Figure 3 demonstrates the mean velocity-longitude relation of the same intermediate latitude observations combined now into one class as open circles. A formal correction of 12.5 km s$^{-1}$ of the LSR velocity in the galactic pole direction was applied. A rough fit to the mean velocity points is achieved by the dashed curve which was calculated from Schmidt's model of galactic rotation for material assumed at a distance of 0.8 kpc from the sun in the galactic plane.

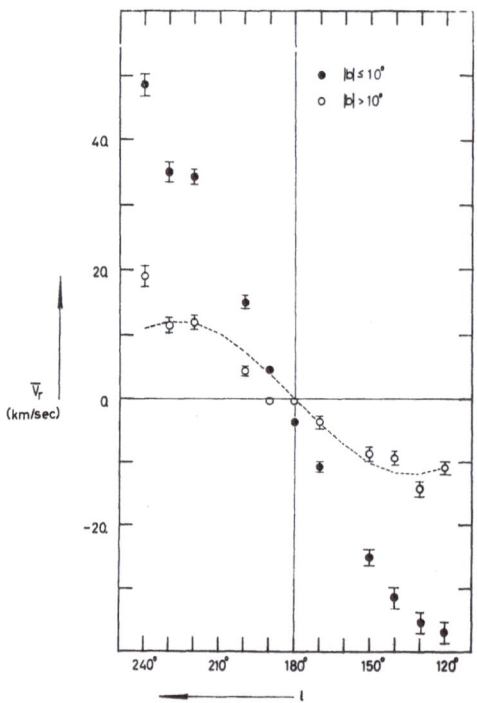

Fig. 3.    Velocity-longitude diagram of the combined intermediate latitude observations $(10 < |b^{\mathrm{II}}| \leqslant 30$, open circles) and the plane observations $(0 \leqslant 10|b^{\mathrm{II}}| \leqslant 10$, filled circles).

The filled circles represent the mean velocities of the low latitude observations between $-10°$ and $10°$. Though a rather smooth curve can be drawn for their velocity-longitude relation, a systematic deviation to negative velocities of about 3.5 km s$^{-1}$ is apparent near the anticenter.

Figure 4 shows the influence of this on a distance determination in a polar diagram of distance from the sun vs. galactic longitude. The mean velocities of the low latitude class are transformed to mean distances by Schmidt's model of circular galactic rotation and plotted as open circles. A significant distance jump does appear at the anticenter direction. Recalculating the mean distances, under the idea of an

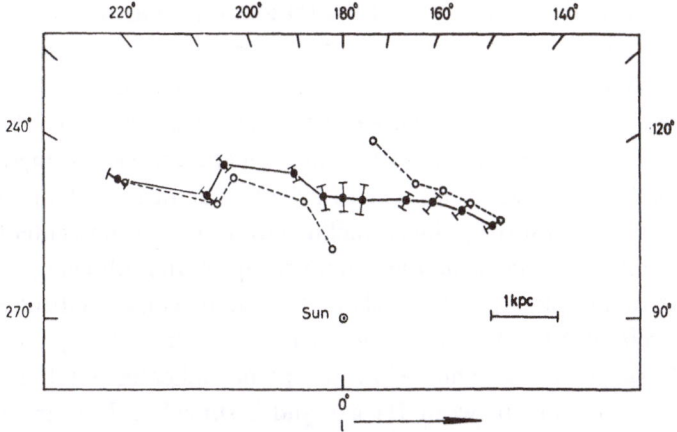

Fig. 4.   Polar diagram of distance from the sun vs. galactic longitude. The open circles represent the mean velocity points of Figure 3 of the plane material transformed by Schmidt's (1965) rotational model to mean distance points. The filled circles show the mean distances after a formal correction for an outward motion of the LSR of 3.3 km s$^{-1}$.

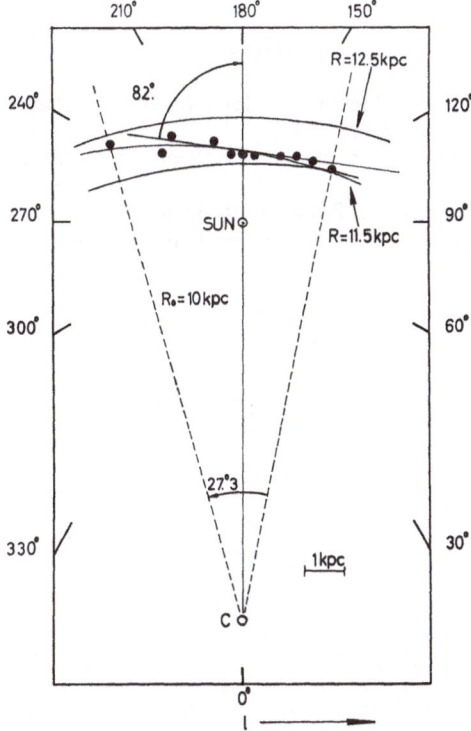

Fig. 5.   Polar diagram of mean distances from the galactic center vs. longitude of the plane observations.

in general smooth spatial distribution of the H I gas, for an assumed outward motion of the LSR of 3.3 km s$^{-1}$ yields the line of filled circles. The distance for the anticenter point is derived from the slope of the mean velocity curve.

Figure 5 does need a warning because of its rather suggestive character. The mean distance points are plotted with respect to the galactic center. An increase of mean distances of 0.8 kpc is found for a range of 27° of galactocentric longitude. A calculated mean distance line for the points is inclined to the center-anticenter line with 82°.

The observational facts presented are summed up by the following.

(1) The mean radial velocities observed at intermediate galactic latitudes are roughly described by a circular rotation law. The splitting of the velocity-longitude curves of the latitude ranges above and below the plane indicates a relative velocity of 12.5 km s$^{-1}$ in z-direction between H I gas and assumed LSR near the anticenter.

(2) A motion of approach of 3.3 km s$^{-1}$ between H I gas of low latitude and LSR is observed in the anticenter region, which does not seem present in the gas at intermediate latitudes.

(3) The mean distance of the plane material increases in general with increasing galactic longitude in the region observed.

(4) The slight bulge of the mean distance points towards lower distances from the sun near the anticenter direction indicates an only low value for the self-absorption of the 21-cm emission line.

## References

Grahl, B. H., Hachenberg, O., and Mebold, U.: 1968, *Beiträge zur Radioastronomie* **1**, 3.

Schmidt, M.: 1965, *Stars and Stellar Systems* **5**, 513.

Schwartz, R.: 1967, *Forschungsberichte des Landes Nordrhein-Westfalen*, Nr. 1844, Westdeutscher Verlag Köln-Opladen.

Velden, L.: 1970, 'Atlas of 21-cm Line Profiles of the Outer Part of the Galaxy in the Region of the Galactic Anticenter', *Beiträge zur Radioastronomie*, in preparation.

## 28. RADIO OBSERVATIONS OF SOME DETAILS IN THE HI LOCAL SPIRAL ARM

N. V. BYSTROVA, J. V. GOSSACHINSKY, T. M. EGOROVA,
V. M. ROZANOV*, and N. F. RYZHKOV

*Pulkovo Observatory, Leningrad, U.S.S.R.*

The observations were made with the large Pulkovo radiotelescope (beamwidth 7′, bandwidth 20 and 10 kHz). Figure 1 contains our drift curves across the cluster NGC 2264. The details on them may be identified with Raimond's (1966) clouds, but the proofs for the reality of the connection between the cluster and the clouds are not very reliable. The cloud '*b*' is believed to be connected with NGC 2264 because of their close neighborhood. But on the other side of the galactic plane in this region we have found a bright and narrow (∼1°) detail, whose middle falls exactly on the

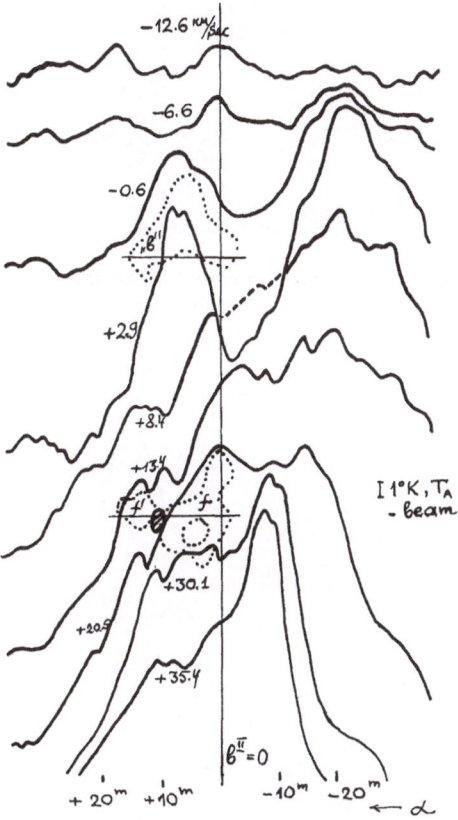

Fig. 1. Drift curves across the cluster NGC 2264 at different velocities relative to the LSR. The coordinates are RA and antenna temperature. Dots represent the contours of the clouds found by Raimond, a line across them represents our scans.

* V. M. Rozanov died on August 12, 1968, at the age of 29.

*Becker and Contopoulos (eds.), The Spiral Structure of Our Galaxy, 169–172. All Rights Reserved.*
*Copyright © 1970 by the I.A.U.*

western border of the H II region around the star $\lambda$ Orionis (Figure 2). Figure 3 contains the computed drift curves at different velocities for $\Delta f = 20$ kHz according to the model of the neutral hydrogen expanding envelope proposed by Wade (1957) around the H II region $\lambda$ Ori. Figure 4 represents our observations. Although some details, in particular the narrow detail at the velocity $-1.4$ km s$^{-1}$, are very similar to the calculated ones, our observations differ from Wade's model in that they do not have

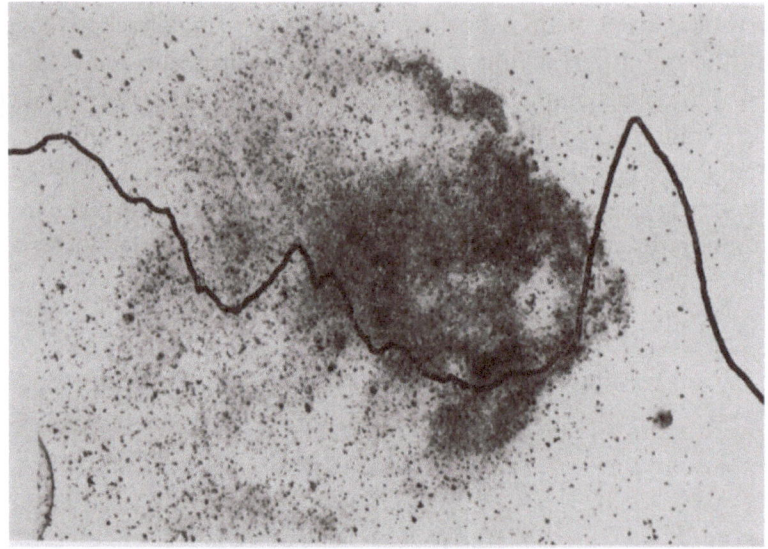

Fig. 2.    Part of the drift curve at velocity $-1.4$ km s$^{-1}$ across the star $\lambda$ Ori, put on the photograph of this H II region.

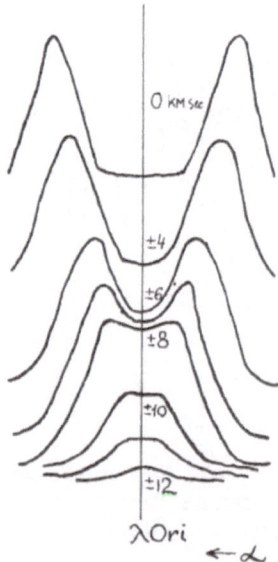

Fig. 3.    Expected drift curves across $\lambda$ Ori according to the model by Wade (1957) for a 20 kHz band.

any velocity symmetry relative to zero; in fact the details are situated only at the positive side. This fact may eliminate one half of the proposed envelope. There is no symmetry relative to the central star and no coincidence of the velocity of the narrowest detail with the velocities of the central star and nebula. Figure 5 demonstrates the drift curve obtained across both objects, NGC 2264 and λ Ori nebula. Maybe the old Menon's idea about the connection between the branching of the gaseous arm and the formation of the clusters is correct. The details in the HI distribution look like the split branches from the main body of the arm (Figure 5).

On Figure 6 are our drift curves across the region of the magnetic field reversal near $l^{II} = 180°$, $b^{II} = -30°$ (Gardner *et al.*, 1967). Taking into account the model on Figure 2 we may conclude that there is a hole with diameter approximately 50 pc in

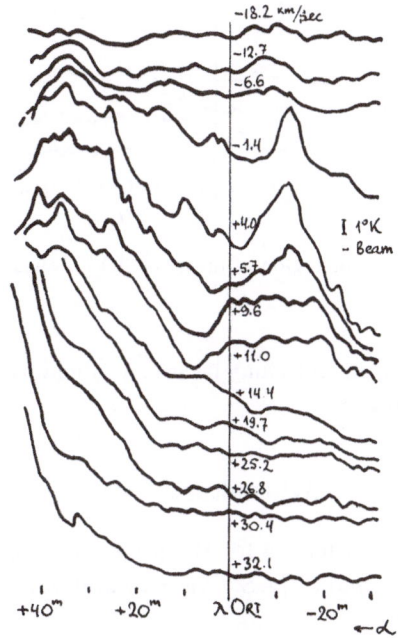

Fig. 4. Observed drift curves at different velocities across λ Ori.

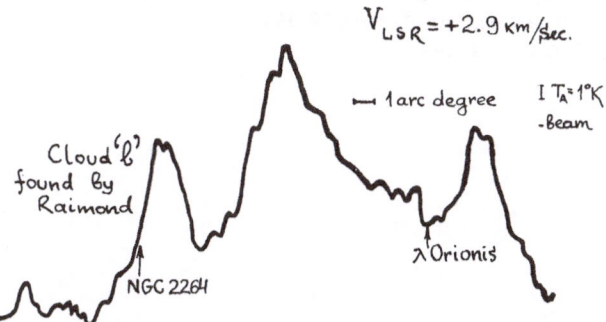

Fig. 5. Drift curve across NGC 2264 and across the HII region λ Orionis. $V = +2.9$ km s⁻¹ relative to LSR, $\Delta f = 10$ kHz.

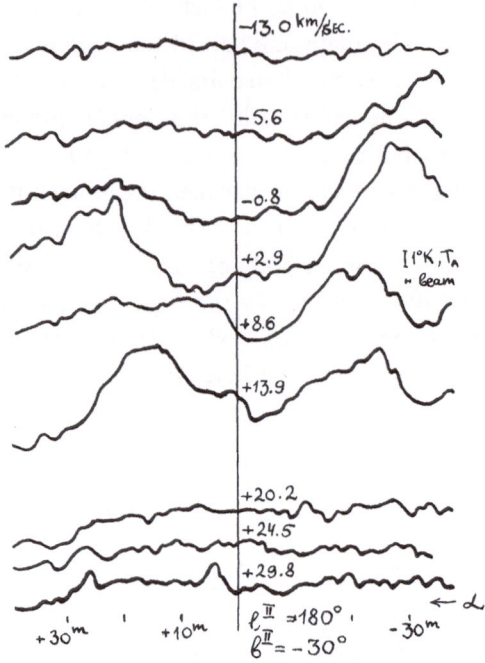

Fig. 6.   Drift curves across the region of magnetic field reversal at $l = 180°$, $b = -30°$.

the gaseous tongue in question. Here may be something like a half-envelope expanding with the velocity 12–13 km s$^{-1}$.

## Acknowledgements

Our thanks are due to Z. A. Alferova for the programming of observations and for further computer reductions and to N. S. Evgrafov and V. A. Jakovlev for their taking part in the observations.

## References

Gardner, F. F., Whiteoak, J. B., and Morris, D.: 1967, *Nature* **214**, 371.
Raimond, E.: 1966, *Bull. Astron. Inst. Netherl.* **18**, 191.
Wade, C. M.: 1957, *Astron. J.* **62**, 148.

# 29. HYDROGEN IN THE REGIONS OF FOUR STELLAR CLUSTERS

H. M. TOVMASSIAN

*Byurakan Astrophysical Observatory, Erevan, Armenia, U.S.S.R.*

Young clusters of O type and the corresponding stellar associations, together with the associated hydrogen complexes, make up the spiral arms of our Galaxy. Thus the investigation of the distribution of hydrogen and its relation to the clusters is of definite interest.

The present report gives preliminary results of observations of the hydrogen line in the regions of four clusters, aiming to reveal the gas clouds associated with these clusters.

The observations were made in 1965 with the 210-feet radio telescope of the Australian National Radio Astronomical Observatory at Parkes. Most of the observations were made with the 48-channel receiver with bandwidths of the channels of 33 kHz, and some of them with the 15-channel narrow-band receiver with bandwidths of the channels of 9 kHz. The noise level has been kept almost the same during both observations by increasing the integration time with the narrow-band receiver. Peak-to-peak fluctuations with both receivers were about 1.5 K.

The observed area in each case exceeds that of the corresponding clusters by several hundred times. Most of the scans have been made along the galactic plane through the positions of the clusters or shifted by 6', 12', 18' and 30' from them. Two scans have been made across the plane.

The results on the following four clusters – NGC 2264, NGC 2353, NGC 2362, and NGC 3293 – out of 15 observed are presented below.

*NGC 2264:* The neutral hydrogen in the region of NGC 2264 and of the stellar association Mon OB1, which contains this cluster, has been investigated by Menon (1956) and later by Raimond (1966) with better frequency and angular resolutions. In this region Raimond has detected three large H I clouds, two of which, he suggested, are associated with NGC 2264. But the positions of these clouds and their sizes raise doubt as to their immediate connection with the cluster. Most likely these clouds are connected with the association Mon OB1 in general.

The observations at Parkes, made with better angular resolution than that of Raimond, have detected only a small neutral hydrogen cloud in the direction of the cluster (Figure 1). Its angular dimensions are about 30' and the brightness temperature is almost 20 K. The radial velocity of this cloud (reduced to LSR) is $+10$ km s$^{-1}$ and agrees rather well with that of the cluster, which is equal to $+7.5$ km s$^{-1}$. Adopting the distance of NGC 2264 to be 760 pc (Raimond, 1966) the linear diameter of the cloud is about 8 pc, the density about 10 atoms per cm$^3$ and the mass about 70 $M_\odot$. Thus the sum of masses of neutral and ionized hydrogen (Tovmasjan, 1967), associated with NGC 2264 is about 80 solar masses.

*NGC 2353:* The line profiles taken through the position of this cluster have revealed no neutral hydrogen cloud structure. All profiles at velocities near the radial velocity of the cluster are sufficiently smooth. Thus there is no noticeable amount of H I connected with NGC 2353. The attempts to detect H II in the vicinity of this cluster were also negative (Tovmasjan, 1967).

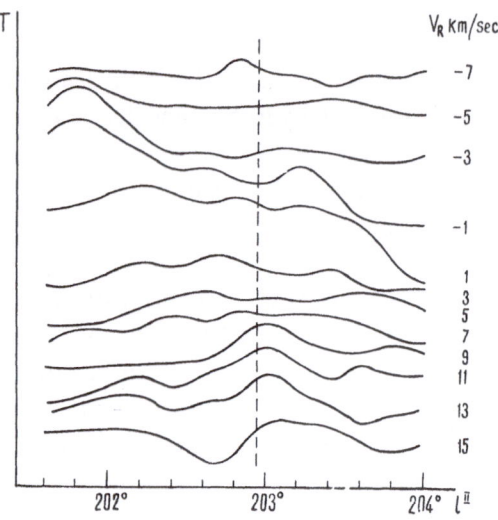

Fig. 1.  Line profiles taken through the position of NGC 2264 with the narrow-band receiver.

*NGC 2362:* This is a small O-type cluster with angular dimensions equal to about 7'. Line observations have revealed a neutral hydrogen cloud of rather symmetrical form at a distance of about 30' to the southwest from its centre. The diameter of the cloud is about 25'. The cloud is seen at radial velocities in the range of 30 km s$^{-1}$ while the radial velocity of the cluster is 15 km s$^{-1}$. Both the position and the difference in radial velocities suggest that the cloud is not connected with NGC 2362. It is rather associated with a radio source of comparable size at an adjacent position detected by continuum observations (Tovmasjan, 1967). The position of NGC 2362 and both radio sources (in line emission and continuum) are shown in Figure 2.

Thus no hydrogen is associated with this cluster as well.

*NGC 3293:* The region of this cluster is very rich with large H II complexes (Menon, 1956). The hydrogen line observations have also revealed here a complex structure.

The line profiles show the existence of a cloud which is most likely associated with NGC 3293. The radial velocity of the cloud is about −14 km s$^{-1}$, which is just that of the cluster. The centre of this cloud is projected at the southern edge (in galactic coordinates) of the cluster. The diameter of the cloud is about 20 pc and the brightness temperature is ∼24 K. The density in this cloud is ∼1.5 atoms per cm$^3$ and its mass is about 150 $M_\odot$. These quantities are obtained assuming that the cloud is at the distance of NGC 3293, which is 2.6 kpc. If only this neutral hydrogen cloud is associated with NGC 3293 then its total gas content is about 300 $M_\odot$ since the mass of H II there is 120 $M_\odot$ (Tovmasjan, 1967).

Another feature is probably associated with NGC 3293. This feature may be treated assuming a model of an expanding shell (Figure 3). The radial velocity of the shell is about $+2$ km s$^{-1}$ and differs from that of the cluster by 16 km s$^{-1}$, since the latter is $-14$ km s$^{-1}$. At the distance of the cluster the outer and the inner radii of the shell are $\sim 50$ pc and $\sim 35$ pc respectively. With the peak brightness temperature of the shell equal to $\sim 30$ K the mean density of hydrogen atoms in it is about 2 atoms per cm$^3$. The total mass of the shell is about 15000 $M_\odot$. However, because of the large difference in radial velocities of the cluster and of the shell we can not be sure that the shell is really connected with NGC 3293.

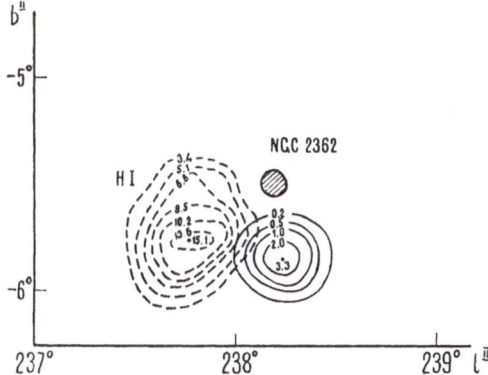

Fig. 2. Map of the region of NGC 2362 with the sources of continuum and line emission. The contours of equal temperature for the neutral hydrogen source are shown for radial velocities of about $+30$ km s$^{-1}$.

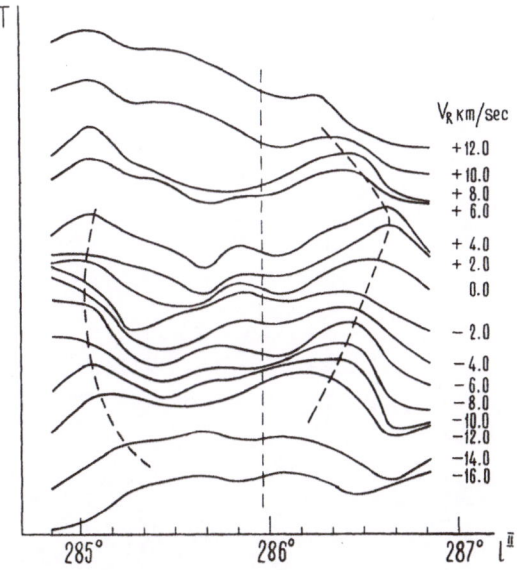

Fig. 3. Line profiles through the position of NGC 3293 obtained with the narrow-band receiver.

The radial velocity of the small cloud in NGC 2264 is $\sim 7$ km s$^{-1}$. The radial velocities for the NGC 3293 are reduced to the Sun.

## Acknowledgements

I am grateful to Drs. E. G. Bowen and J. G. Bolton for permission to use the facilities of the Australian National Radio Astronomical Observatory. My thanks are due to Mrs Liesel Scholem, who has reduced the profiles on the Sydney CDC-3200 computer of the SCIRO Computing Research Section, and Messrs M. Shahoyan and E. Shahbazian, who have helped me in the reduction and analysis of the observations.

## References

Menon, T. K.: 1956, *Astron. J.* **61**, 9.
Raimond, E.: 1966, *Bull. Astron. Inst. Netherl.* **18**, 191.
Tovmasjan, H. M.: 1967, IAU Symposium No. 31, p. 37.

# 30. THE DISTRIBUTION IN GALACTIC LONGITUDE OF OBSERVABLE PULSARS

G. LYNGÅ

*Lund Observatory, Lund, Sweden*

I have studied the galactic longitude distribution of low latitude pulsars on the assumption that pulsars are evenly distributed in the galactic disk but that their observability is severely impaired by spiral features, rich in H II regions, in which the high electron density causes a high dispersion measure for a comparatively short optical path. The assumption of even distribution in the galactic disk seems natural as no correlation between positions of pulsars and spiral tracers is known, and since supernovae in external galaxies are not known to confine to spiral arms.

There are 14 pulsars in the consistently surveyed interval $l^{II} = 250°-70°$, $b^{II} = -5°-+5°$. In the intervals $l^{II} = 285°-295°$ and $l^{II} = 305°-45°$ the observations are made against the background of the Carina arm and the Sagittarius arm, respectively. Intervals $l^{II} = 250°-285°$ and $l^{II} = 45°-70°$ are considered to be inter-arm regions. The interval $l^{II} = 295°-305°$ contains a relatively weak link of H II regions, and is for the present discussion also considered as an inter-arm region. With these definitions the pulsar statistics are presented in Table I.

### TABLE I
Statistics for low latitude pulsars

| Description of regions | Total interval in $l^{II}$ | Number of pulsars | Mean dispersion measure |
|---|---|---|---|
| Towards spiral features | 110° | 4 | 50 pc cm$^{-3}$ |
| Inter-arm regions | 70° | 10 | 107 pc cm$^{-3}$ |

The data of Table I clearly show that, on the whole, pulsars are seen at larger distances in the inter-arm regions than towards spiral features. If it is assumed that the pulsars viewed towards spiral features evenly fill out the volume considered, then a mean electron density $\bar{n}_e = 0.04$ cm$^{-3}$ can be derived and this value in turn corresponds to an effective horizon at 3 kpc for the inter-arm directions. This calculation assumes a constant luminosity function.

The mean distance from the apparent galactic plane for the pulsars studied would, if pulsars are distributed symmetrically in relation to the actual galactic plane, imply a position of the sun 40 pc above this plane.

## Acknowledgement

I thank Professor B. Y. Mills for helpful comments and for the communication of unpublished data.

# 31. PULSAR DISTANCES, SPIRAL STRUCTURE AND
# THE INTERSTELLAR MEDIUM

B. Y. MILLS

*Cornell-Sydney University Astronomy Center, University of Sydney, Australia*

**Abstract.** The distances of all pulsars are calculated on the assumption that they are immersed in a uniform medium of average electron density 0.06 cm$^{-3}$. It then appears that the pulsars are concentrated towards the local and Sagittarius spiral features and that their mean height above the plane is consistent with that of known supernova remnants. The mean distances appear to be approximately correct, but individual distances are uncertain by about a factor of two. Evidence from radio continuum results supports this model of the ionized interstellar medium.

Several attempts have been made to estimate the distances of the pulsars using theoretically based models for the free electron distribution in the interstellar medium. A different approach is to make a crude first order approximation that the pulsars are immersed in a uniform dispersing medium, calculate the distances and directions of all pulsars on this assumption and compare the results with the properties of other galactic components, particularly supernova remnants, which would appear to have a common origin. Discrepancies revealed in this comparison may then be used to refine the assumed model of the interstellar medium and improve the distance scale.

For reasons discussed later I have taken the free electron density of the uniform medium to be $n_e = 0.06$ cm$^{-3}$. In Figure 1 the distribution of the 40 pulsars of known dispersion measure has been plotted as projected on to the galactic plane, on the assumption that their distances are directly proportional to their dispersion measures ($\int n_e \, dl$), using the value of $n_e$ above. The pulsar parameters are given in Table I.

If the pulsars and the dispersing medium are uniformly distributed throughout the galactic disc, it would be expected that the distribution in Figure 1 would show circular symmetry. It does not. We may disregard the absence of high dispersion pulsars in the smaller sector between $l^{II} = 55°$ and $l^{II} = 195°$ because this area is outside the field of view of the Molonglo Radio Telescope. This instrument is responsible for the discovery of all but one of the high dispersion pulsars in the remaining sector. However, in this larger sector the distribution also does not appear symmetric and displays features which appear to be closely related to the well known HI spiral pattern.

Practically all the pulsars may be associated with the two groupings shown on the diagram which reproduce quite well in direction and distance the local and Sagittarius spiral features. The high dispersion pulsars marking extensions of these features appear to be particularly significant. This grouping appears too striking to be ascribed to chance, and the crude first order approximation to the free electron distribution would seem to be more accurate than one might reasonably expect.

With this suggestion in mind let us look briefly at some of the relevant observed properties of the interstellar medium. Firstly, it has been shown by Gould (1969) that the free-free emission and absorption processes in the ionized medium agree with

*Becker and Contopoulos (eds.), The Spiral Structure of Our Galaxy, 178–181. All Rights Reserved.*
*Copyright © 1970 by the I.A.U.*

## TABLE I

The galactic coordinates and derived distances of 40 pulsars

| Designation | $l$ | $b$ | $\int n_e dl$ | $z^a$ | $d^a$ |
|---|---|---|---|---|---|
| | o | o | cm$^{-3}$ pc | pc | pc |
| MP 0031 | 111 | − 69 | 12 | −186 | 70 |
| MP 0254 | 271 | − 55 | 10 | −137 | 96 |
| CP 0328 | 145 | 0 | 27 | 0 | 450 |
| MP 0450 | 217 | − 34 | 25 | −253 | 346 |
| NP 0527 | 184 | − 7 | 50 | −102 | 830 |
| NP 0532 | 185 | − 6 | 56 | − 98 | 930 |
| PSR 0628 − 28 | 237.0 | − 16.7 | 10 | − 48 | 16C |
| MP 0736 | 254 | − 9 | 100 | −260 | 1640 |
| CP 0808 | 140 | + 34 | 6 | + 56 | 83 |
| AP 0823 + 26 | 197 | + 32 | 19 | +168 | 268 |
| PSR 0833 − 45 | 264 | − 3 | 63 | − 54 | 1050 |
| CP 0834 | 219.7 | + 26.3 | 13 | + 95 | 194 |
| MP 0835 | 260 | 0 | 120 | 0 | 2000 |
| MP 0940 | 278 | − 3 | 145 | −127 | 2420 |
| PP 0943 | 230 | + 45 | 9.5 | +112 | 112 |
| CP 0950 | 228.9 | + 43.7 | 3 | + 34 | 34 |
| MP 0959 | 281 | − 1 | 90 | − 26 | 1500 |
| CP 1133 | 241.9 | + 69.2 | 5 | + 78 | 30 |
| MP 1154 | 297 | − 0.1 | 270 | − 8 | 4500 |
| AP 1237 + 25 | 250 | + 86 | 8.5 | +142 | 10 |
| MP 1240 | 302.0 | − 1.0 | 220 | − 64 | 3660 |
| MP 1426 | 313 | − 6 | 60 | −104 | 1000 |
| MP 1449 | 315.3 | − 5.3 | 90 | −138 | 1500 |
| PSR 1451 − 68 | 313.9 | − 8.6 | 12 | − 23 | 198 |
| HP 1507 | 90 | + 53 | 15.5. | +206 | 234 |
| MP 1530 | 326 | + 2 | 20 | + 12 | 332 |
| AP 1541 + 9 | 18 | + 46 | 35 | +420 | 405 |
| MP 1642 | 15 | + 26 | 40 | +282 | 600 |
| MP 1706 | 7 | + 15 | 10 | + 43 | 161 |
| MP 1727 | 341 | − 9 | 140 | −365 | 2420 |
| MP 1747 | 344 | −11 | 40 | −128 | 655 |
| PSR 1749 − 28 | 1.6 | − 1.0 | 51 | − 15 | 850 |
| MP 1818 | 25 | + 5 | 70 | +103 | 1160 |
| MP 1911 | 31 | + 7 | 75 | +152 | 1240 |
| CP 1919 | 55.8 | + 3.5 | 12.6 | + 13 | 210 |
| PSR 1929 + 10 | 48 | − 4 | 8 | − 9 | 134 |
| JP 1933 + 16 | 52 | − 2 | 143 | − 87 | 2500 |
| AP 2015 + 28 | 68 | − 4 | 14 | − 16 | 233 |
| PSR 2045 − 16 | 30.5 | − 33.1 | 11 | −105 | 162 |
| PSR 2218 + 47 | 98 | − 8 | 44 | −102 | 726 |

[a] In the final columns are listed $z = \dfrac{\int n_e dl}{0.06} \sin b$; and $d = \dfrac{\int n_e dl}{0.06} \cos b$

radio observations of emission at high radio frequencies and absorption at low frequencies if an electron temperature of about 6000 K is assumed. There is no evidence for a significant component of low temperature electrons produced by cosmic ray ionization, except possibly in some directions of high H I concentration.

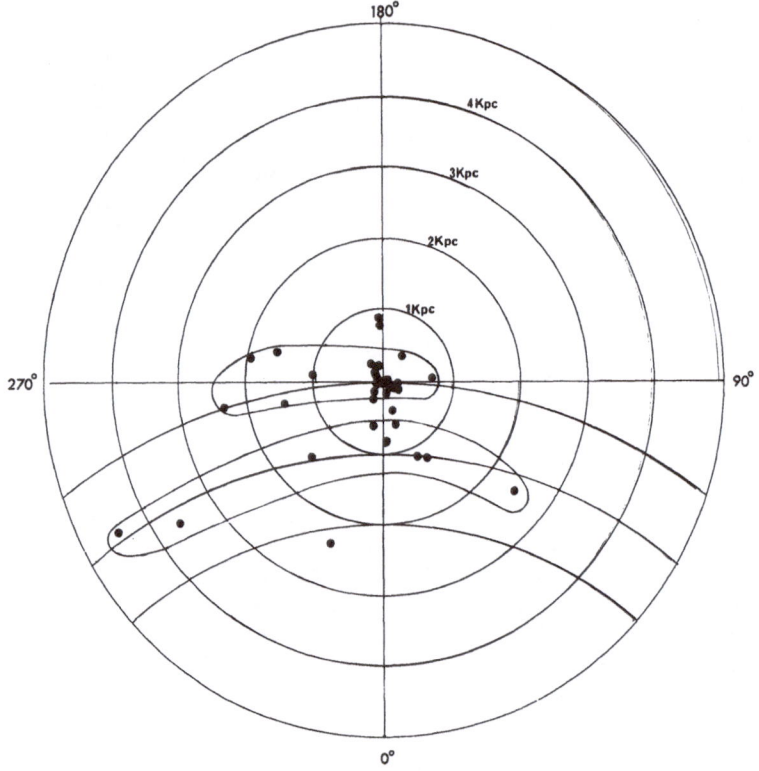

Fig. 1.   The distribution of pulsars in the galactic plane on the assumption of a uniform dispersing medium of density $\langle n_e \rangle = 0.06$ cm$^{-3}$.

Gould also finds that the local value of the mean square electron density $\langle n_e^2 \rangle \simeq 0.06$ cm$^{-6}$. However, we have seen that a plausible distribution of pulsars is yielded by $\langle n_e \rangle = 0.06$ cm$^{-3}$ or $\langle n_e \rangle^2 = 0.0036$ cm$^{-6}$. Distances derived from this value of $n_e$ are also consistent with a common origin for pulsars and supernova remnants. The former have a mean height above the plane, $\langle |z| \rangle \simeq 110$ pc, while for the supernova remnants in the same part of the Galaxy $\langle |z| \rangle \simeq 85$ pc according to a recent tabulation of Milne (1969). In view of the uncertainty in both distance scales the agreement is good and the distributions in $|z|$ otherwise appear very similar. The identifications of the Vela (PSR 0835-45) and Crab (NP 0532) pulsars with the corresponding supernova remnants yield $\langle n_e \rangle \simeq 0.1$ and $\langle n_e \rangle \simeq 0.03$ respectively. Accordingly, there appears little doubt that, in the interstellar medium, $\langle n_e^2 \rangle \gg \langle n_e \rangle^2$. This inequality implies that the distribution of the ionized medium is very irregular, but we have seen that the dispersion measures appear to give a good indication of pulsar distances. We must suppose that the free-free processes ($\propto \langle n_e^2 \rangle$) are dominated by dense clouds and the pulsar dispersions ($\propto \langle n_e \rangle$) by a much more uniform tenuous intercloud medium. The temperature of the ionized intercloud medium cannot be substantially less than that of the clouds or its effects would be more apparent in the low frequency absorption measurements. Thus ionization by dilute UV radiation seems very likely.

Further evidence for a hot intercloud medium comes from some recent H-line observations of Radhakrishnan and Murray (1969), who find that the intercloud hydrogen atoms are at a very much higher spin temperature than those in the H I clouds responsible for the greater part of the observed galactic emission and absorption of the H-line. In one direction at least the spin temperature exceeds 1000 K.

A plasma with $\langle n_e \rangle \simeq 0.06 \text{ cm}^{-3}$ and $T \simeq 6000 \text{ K}$ is unlikely to be closely confined to the galactic plane, certainly not if there is any tendency towards equipartition of turbulent energies. Ultraviolet ionization would be possible to great heights and, in general, one might expect such a plasma to exist at heights at least as great as those to which the magnetic fields extend. A measure of this extent is given by the galactic distribution of synchrotron radiation which is much wider than that of the H-line emission, for which the mean height $\langle |z| \rangle$ is similar to that of the pulsars. Accordingly, the assumption that all pulsars lie within the dispersing layer appears to be very reasonable. If this were not so it would be difficult to explain the resulting very high $z$ distances of some of the pulsars.

A better fit to the Sagittarius arm distance and the Vela and Crab supernovae is obtained if it is assumed that, within the spiral arms, the mean electron density is greater than $0.06 \text{ cm}^{-3}$ and, between the arms, it is less. However, the small amount of available data does not appear to justify the assignment of actual numbers at this stage. Even if improved statistical distances can be obtained there appears little prospect of appreciably reducing the uncertainty in the distances of individual pulsars, indeed the only refinement which appears justifiable is an allowance for the dispersion measure of known H II regions when they lie in front of a pulsar. Only two of the pulsars, PSR 0833-45 and MP 0736, are obviously affected but the H II region has low surface brightness and is believed to be very close so that corrections should not be large. Quantitative information is not at present available.

To conclude, it appears that the simplest possible model for the dispersing medium yields distances which are probably individually accurate to a factor of two and yields the pulsar statistics with much greater accuracy. There is a striking association of a group of ten pulsars with the accepted location of the Sagittarius spiral arm. The great majority of the remainder appears to be associated with the local spiral feature. This interpretation of the data follows from the assumption of a fairly uniform distribution of free electrons extending to much greater heights above the plane than the dense H I clouds.

## References

Gould, R. J.: 1969, *Australian J. Phys.* **22**, 189.
Milne, D. K.: 1969, 'Non-thermal Galactic Radio Sources', submitted to the *Australian J. Phys.*
Radhakrishnan, V. and Murray, J. D.: 1969, *Proc. Astron. Soc. Australia* **1**, 215.

# 32. ON THE PHYSICAL CONDITIONS IN INTERSTELLAR Hɪ GAS

M. GREWING, U. MEBOLD and K. ROHLFS

*Max Planck-Institut für Radioastronomie und Institut für Astrophysik, Bonn, Germany*

**Abstract.** Weighted average values of the ionization ratio $n_e/n_H$ for the interstellar gas can be obtained from a comparison of pulsar data and 21-cm emission measurements. For high latitude pulsars this procedure is straight forward and reasonably free from assumptions. The resulting ionization ratios are high ($n_e/n_H = 0.39$). If this is compared with values as given by the theory of ionization and heating of interstellar gas by subcosmic rays, temperatures for the 'neutral' gas above $10^4$ K are obtained. For low latitude pulsars this procedure depends to a much larger extend on the assumed distance of the objects and on the ionization theory. Here the values found are $n_e/n_H = 0.07$ and $T_e = 2000$ K.

Through the discovery of pulsating radio sources (Hewish *et al.*, 1968) a new parameter of the interstellar medium has become available to observation. From pulse delay measurements at different frequencies we immediately obtain

$$\text{DM} = \int_0^d n_e \, ds \quad [\text{cm}^{-3} \, \text{pc}].$$

(1)

Due to the relative cosmic abundance of the chemical elements in interstellar space the source for the electrons in (1) must mainly be the hydrogen gas. Therefore it is reasonable to compare (1) with an expression that is formed similar, but with the electron density replaced by the density of neutral hydrogen

$$\text{HM} = \int_0^d n_H \, ds \quad [\text{cm}^{-3} \, \text{pc}],$$

(2)

the integral in (2) also extends from the observer up to the pulsar. Dividing (1) and (2) we have for an arbitrary distribution of $n_e$ and $n_H$

$$\frac{\text{DM}}{\text{HM}} = \frac{\int_0^d n_e \, ds}{\int_0^d n_H \, ds} = \frac{\int_0^d \left(\frac{n_e}{n_H}\right) \cdot n_H \, ds}{\int_0^d n_H \, ds} = \overline{\left(\frac{n_e}{n_H}\right)}.$$

(3)

DM/HM is a weighted average of the ionization ratio $n_e/n_H$, the weighting factor being $n_H$, i.e. the volume density of *neutral hydrogen*. Assuming approximate pressure equilibrium in interstellar space, we have

$$n_H T_k \approx \text{const}$$

(4)

*Becker and Contopoulos (eds.), The Spiral Structure of Our Galaxy, 182–185. All Rights Reserved.*

so that

$$\frac{\mathrm{DM}}{\mathrm{HM}} = \frac{\displaystyle\int_0^d \frac{n_e}{n_H} \cdot \frac{\mathrm{d}s}{T_k}}{\displaystyle\int_0^d \frac{\mathrm{d}s}{T_k}} = \overline{\left(\frac{n_e}{n_H}\right)}. \tag{5}$$

Hence DM/HM is the 'harmonic mean value' of the ionization ratio $n_e/n_H$.

Unfortunately HM as given in Equation (2) cannot be observed directly. While 21-cm absorption measurements do extend up to the unknown distance, $d$, of the pulsar, they do not give $\int n_H \, \mathrm{d}s$ but $\int (n_H/T_s) \, \mathrm{d}s$. $T_s$ however, can be determined unambiguously only in exceptional cases.

If the radiation can be assumed optically thin the emission spectrum of the 21-cm line emission does give $\int_0^\infty n_H \, \mathrm{d}s$. Since the evidence available today points towards the pulsars as members of the disk population of our galaxy, we obviously must have

$$\mathrm{HM} \leqslant \int_0^\infty n_H \, \mathrm{d}s = c \cdot \int_{-\infty}^\infty T_b(v) \, \mathrm{d}v = \mathrm{HM}_{\mathrm{obs}} \tag{6}$$

and thus

$$\mathrm{DM/HM}_{\mathrm{obs}} \leqslant \mathrm{DM/HM} = \overline{(n_e/n_H)}. \tag{7}$$

Therefore the observed DM/HM is a *lower* limit to the harmonic mean ionization ratio.

Gordon *et al.* (1969) showed that the maximum optical depth of the 21-cm line emission for the direction of the three pulsars AP 0823, CP 0950 and CP 1133 is $\tau \ll 1$, so that (6) can be applied. For the other high latitude pulsars of Table I this cannot be proved, but as stated by Clark (1965), strong absorption is extremely unlikely to occur at high galactic latitude. Column 6 of this Table, therefore gives lower limits to the harmonic mean ionization ratio of the interstellar gas between pulsar and observer.

One possible explanation for the high ionization ratio might be given by the hypothesis, that the path intersects an HII region. An inspection of the distribution of O and B stars yielded however only very few objects, where this could be accepted (see Column 8).

The only other possibility is a high intrinsic ionization ratio of the interstellar medium. This is provided by the ionization theory of interstellar gas by subcosmic ray particles (Pikel'ner, 1968; Hjellming *et al.*, 1969; Field *et al.*, 1969; Spitzer and Tomasko, 1968) possibly supplemented by ionization by soft, diffuse X-rays (Silk and Werner, 1969). In this theory a unique relation exists between the ionization ratio and the electron temperature. If the theory of Hjellming *et al.* (1969) is used, the mean temperatures given in Column 7 of Table I are obtained.

These temperatures are considerably higher than those obtained by other means for the HI medium. They depend, however, only on the observed ratio DM/HM and

TABLE I

Observational data and electron temperatures for the gas in front of high latitude pulsars

| (1) | (2) | | (3) | (4) | (5) | (6) | (7) | (8) |
|-----|-----|-----|-----|-----|-----|-----|-----|-----|
| Object | $l^{II}$ | $b^{II}$ | DM | HM | $\tau_{max}$ | DM/HM | $T_{min}$ | H II reg |
| AP 0823 | 197 | $+32$ | 19.2 | 92 | $<0.2$ | 0.21 | $>10000$ | – |
| CP 0950 | 230 | $+44$ | 3.0 | 86 | $<0.1$ | 0.035 | 450 | – |
| CP 1133 | 240 | $+70$ | 5.0 | 97 | $<0.1$ | 0.052 | 1000 | – |
| MP 0450 | 217 | $-34$ | 25 | 100 | – | 0.25 | $>10000$ | – |
| MP 0736 | 254 | $-9$ | 100 | 950 | – | 0.11 | 6500: | $\zeta$ Pup, $\gamma^2$ Vel |
| CP 0808 | 140 | $+34$ | 5.8 | 80 | – | 0.07 | 2500 | – |
| CP 0834 | 220 | $+26$ | 12.8 | 130 | – | 0.10 | 5400 | – |
| JP 0943 | 228 | $+42$ | 20 | 98 | – | 0.20 | $>10000$ | – |
| AP 1237 | 254 | $+86$ | 8.6 | 25 | – | 0.34 | $>10000$ | – |
| HP 1506 | 90 | $+53$ | 19.6 | 50 | – | 0.39 | $>10000$ | – |
| AP 1541 | 18 | $+46$ | 35 | 135 | – | 0.26 | $>10000$ | – |
| PSR 1642 | 14 | $+26$ | 33 | 350 | – | 0.095 | 5000: | $\zeta$ Oph |
| MP 1727 | 341 | $-9$ | 140 | 600 | – | 0.23 | $>10000$ | I Sco |
| MP 1747 | 344 | $-11$ | 40 | 540 | – | 0.074 | 2600 | I Sco |
| PSR 2045 | 30 | $-33$ | 11.4 | 250 | – | 0.046 | 720 | – |
| PSR 2218 | 98 | $-8$ | 43.8 | 690 | – | 0.063 | 1600 | – |

on the ionization theory. It is quite well conceivable, that this theory still needs improvement, but it is difficult to imagine a theory, that gives ionization ratios of the interstellar gas as high as 0.4 together with temperatures much below a few $10^3$ K.

If this interpretation is right, a considerable part of the observed line width of the 21-cm emission would be purely thermal in origin, so that the importance of internal turbulence in cosmic clouds could be smaller, than hitherto assumed. The general gas pressure of an intercloud medium with a temperature of several $10^3$ K could be a great help to assist the stability of an average interstellar cloud.

The neutral hydrogen near the galactic plane cannot be considered to be optically thin, and therefore absorption effects in the H I gas must be taken into account when estimating eq. (2) for low latitude pulsars. It has been possible to measure the 21-cm absorption for 3 pulsars. Since the distance, $d$, to PSR 1749 probably is much smaller than the extend of the galactic disk in the same direction, the estimate for HM obtained from the 21-cm profiles is so much larger than the relevant value, that the resulting inequality is of little significance. There thus remain two objects with known absorption and spectra (CP 0328 and NP 0532). As shown by Mebold (1969) the absorption is mainly produced by a cold gas component. We therefore considered a two-component model where the cold gas produces the absorption, while the hot gas is responsible for most of the emission. For the cold gas an electron temperature of $T_e = 55$–$60$ K was inferred from the spectra. The amount of hot gas necessary to produce the remainder of the observed emission profile could then be determined for two model gas distributions, viz. model I, where the cold clouds lie in front of all the hot gas, and model II, where half of the hot gas is in front and half of it behind the cold clouds. The latter model is probably closer to reality.

Due to the uncertain value of the distance, $d$, of the two pulsars it is doubtful how much of the hot gas visible in the 21-cm emission spectrum is situated in front of the pulsar and thus has to be included in (7). That part of the observed DM which is produced by the cold gas can be determined from the measured amount of cold hydrogen and any one of the ionization theories. It is for both pulsars very small and all uncertainties in both the temperature and the ionization theory are of no consequence. The remaining part of the DM then must be caused by the 'hot gas component'. Depending on the distance assumed and which model distribution for the hot gas is accepted values of $0.04 < DM/HM < 0.14$ are found, the most probable value being 0.07, corresponding to a $T_e = 2000\,\mathrm{K}$ according to the theory of Hjellming et al. (1969). For a more extensive discussion of these low latitude pulsars, see Rohlfs et al. (1969).

There thus seems to be a marked difference between the mean $n_e/n_H$ values obtained for high and low latitude objects indicating a temperature gradient within the hot H<small>I</small> gas component in the sense, that the temperature increases with increasing distance from the galactic plane.

## References

Clark, B. G.: 1965, *Astrophys. J.* **142**, 1398.
Field, G. C., Goldsmith, D. W., and Habing, H. J.: 1969, *Astrophys. J.* **155**, L149.
Gordon, C. P., Gordon, K. J., and Shalloway, A. M.: 1969, *Nature* **222**, 129
Hewish, A., Bell, S. J., Pilkington, J. D. H., Scott, P. F., and Collins, R. A.: 1968, *Nature* **217**, 709.
Hjellming, R. M., Gordon, C. P., and Gordon, K. J.: 1969, *Astron. Astrophys.* **2**, 202.
Mebold, U.: 1969, *Beiträge zur Radioastronomie* **1**, 97.
Pikel'ner, S. B.: 1968, *Soviet Astron.* **11**, 737.
Rohlfs, K., Mebold, U., and Grewing, M.: 1969, *Astron. Astrophys.* **3**, 347.
Silk, J. and Werner, M. W.: 1969, *Astrophys. J.* **158**, 185.
Spitzer, L. and Tomasko, M. G.: 1968, *Astrophys. J.* **152**, 971.

# B. OPTICAL OBSERVATIONS

# 33. LOCAL STELLAR DISTRIBUTION AND GALACTIC SPIRAL STRUCTURE

S. W. McCUSKEY

*Warner and Swasey Observatory, Cleveland, Ohio, U.S.A.*

**Abstract.** Aside from the well-known spiral arm tracers such as the OB associations, young galactic clusters, WR stars and possibly the long-period classical cepheids, the more common stars in the neighborhood of the sun within 2 kpc show little or no relationship to the local spiral structure of the galaxy.

## 1. Introduction

The role played by primary spiral arm tracers in delineating the galactic structure in our region of the galaxy is well known. Intrinsically bright young stars such as the OB stars and early Be stars, the Wolf-Rayet stars and possibly the brightest Cepheids seem to be distributed in localized sections of spiral arms. Concentrations of these objects in the O-associations, the galactic clusters having early O and B stars, and in association with ionized hydrogen regions strengthen the probability that in our neighborhood we see the following well established features as summarized by Bok (1967):

| Region | Galactic Longitude $l^{II}$ | Distance |
|---|---|---|
| Perseus | 90–140° | 2–3 kpc |
| Local | 60–210° | Sun on inside edge |
| Sagittarius | 330–0–30° | 2 kpc |
| Carina (section) | 280–310° | 1.4–5.5 kpc |
| Centaurus (link) | 310–330° | 2 kpc (weak) |

To these the radio astronomers have added the Norma-Scutum spiral arm extending broadly from $l^{II}=30°$ to $l^{II}=327°$ at about 4 kpc from the sun at its nearest point.

More recently Westerlund (1969) has investigated the OB associations in the filamentary nebulous regions of RCW 86 (Rodgers *et al.*, 1960) and RCW 103 at $l^{II}=315°$ and 332° respectively. He finds distances consistent with segments of the Sagittarius spiral arm extending from $l^{II}=328–337°$, the distance from the sun at these galactic longitudes being about 2.5 kpc and 1.4 kpc, respectively. The Norma arm is identified by the Ara grouping at $l^{II}=337°$, distance 3.5 kpc and the RCW 103 association and Norma grouping at $l^{II}=332°$, distance about 3.9 kpc.

Bok and associates are investigating photometrically in depth several other areas in the southern sky. These will be reported upon by him and will not be considered in detail here.

In the present paper my principal purpose will be to examine briefly the current status of any relationship which may be present between the general stellar population and the spiral structure in our neighborhood.

*Becker and Contopoulos (eds.), The Spiral Structure of Our Galaxy, 189–198. All Rights Reserved.*
Copyright © 1970 by the I.A.U.

## 2. High Luminosity Stars

First we shall summarize a few of the facts concerning the well-known spiral arm tracers. The distributions of two principal stellar contributors, the *OB stars* and the *Wolf-Rayet stars*, have already been exhibited in a diagram by Klare and Neckel (1967) and one by Smith (1968). The first shows the space distribution of $OB^+$ and $OB^°$ stars as a function of position in the galactic plane. The second indicates the positions of known Wolf-Rayet stars.

Although there is considerable scatter in the distribution of the WR stars, two concentrations of these stars coincide remarkably well with the local and the Sagittarius spiral arms as defined by the OB stars. The WR stars toward the galactic center from the sun are systematically about 0.6 kpc more distant than the greatest concentration of OB stars in this direction. At the same time a similar trend is evident in the grouping in the local spiral arm toward Cygnus. Here the difference is less clearly defined. If we were to assume that the OB concentration and that of the WR stars in the Sagittarius direction were equidistant, then an adjustment of 0.9 mag. in the mean absolute magnitude, $M_v$, in one group, or the other, or both in combination would be indicated.

While both the OB stars and the WR stars associated with the local spiral arm indicate a very low inclination of this structure to the line at $l^{II} = 90°$, those outlining the Sagittarius arm convey the picture of a much greater inclination, something like 25°. The scatter in the distribution of the WR stars makes very uncertain their assignment to any spiral pattern at large distances from the sun.

It has been shown by Klare and Neckel (1967) that the $OB^-$ stars (primarily main sequence B1–B3 objects) are not in general clearly associated with spiral structure. There is an elongated narrow concentration of these at $l^{II} = 130°$ extending from 1.5 to 3 kpc from the sun. This is similar but much more elongated than a similar concentration for the $OB^+$ and $OB^°$ stars. The reality of such an extended group and its existence for any length of time in a differentially rotating galaxy may be questioned.

This broad brush treatment of the OB stars must, of course, be ultimately supplemented by studies of their distribution from more precise photometric and spectrographic data. Space density functions have been calculated from more accurate data for several galactic longitudes in the northern hemisphere in an effort to see whether the Perseus spiral arm is evident at the galactic anti center and whether another spiral arm exists farther from the sun as suggested by the radio astronomers. These analyses give the space density, *stars per* $10^6$ pc$^3$, as a function of distance for five northern galactic longitudes (Figure 1). At $l^{II} = 129°$ the maximum density associated with the Perseus spiral arm is clearly apparent but no clear-cut maxima in space density occur beyond 3 kpc from the sun. Nor is there any perceptible indication of an extension of the Perseus spiral arm into the galactic anticenter region.

*Emission B-stars* of early type have been shown by Schmidt-Kaler (1964, 1966) to be distributed near the galactic plane in regions resembling parts of spiral arms. Behr (1965, 1966) has questioned this, pointing out that the dispersion in absolute magnitude of these stars is much larger than that found by Schmidt-Kaler. As a result many of the structural details shown by the latter are smeared out. Recent work by Wray (1966), however, concerning the distribution of Be stars in the southern Milky Way simply indicates a clumpiness in their distribution with an indication of possibly an inclined structure to the galactic plane.

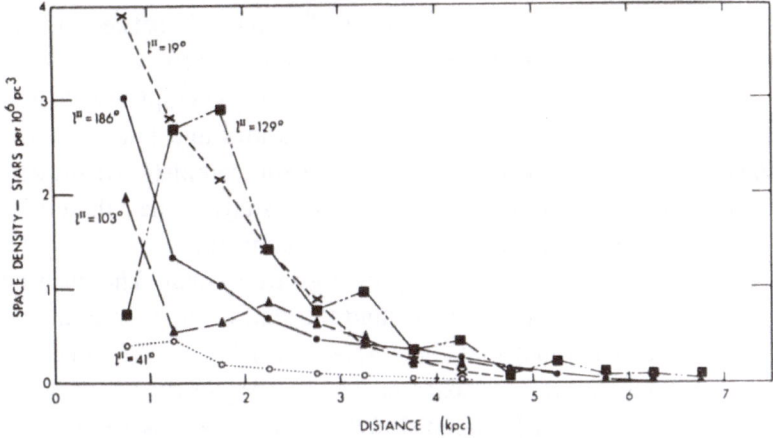

Fig. 1. Space density distribution of OB stars as a function of distance at five galactic longitudes.

The paper by Schmidt-Kaler (1964) contains a figure which illustrates well the indication of spiral structure among the Be stars. The mean absolute magnitudes for these stars are, according to Schmidt-Kaler, $M_v \sim -4.2$ for classes Bpe, B0(III–V) e to B 2 (III–V) e. A recent study by Crampton (1968) of the kinematics of Be stars essentially confirms these intrinsic luminosities.

Cepheid variables with $M_v < -4.3$ show some indication of concentrations along the currently recognized spiral structure. However, the recently published data by Fernie and Hube (1968) leave considerable doubt concerning this point. A plot of the positions on the galactic plane for those classical cepheids with periods longer than 10 days ($M_v < -4.3$) differs considerably from that shown by Kraft (1965), in which an association between the bright cepheids and the recognized spiral structural details of the galaxy was more marked.

Fernie (1968) has pointed out that with a foreknowledge of the spiral structure one can see that the classical cepheids do show a weak association with the structure indicated by other types of objects. The sun appears to be on the *outer edge* of a local spiral arm, however, rather than on the inner edge as indicated by other spiral arm tracers. Incompleteness in the data may have a strong influence on the apparent distribution of these objects.

## 3. The Common Stars

In this grouping we include the vast majority of stars, constituting the main sequence from B 5 to F 5 and the normal giant stars from F 8 to M. Because progress on detailed spatial analyses of these groups is slow and since a recent summary (McCuskey, 1965) has not been seriously changed by added material, we summarize briefly the present state of affairs.

Space density contours on the galactic plane indicate large concentrations of B 5 stars at $l^{II} = 165°$ between 500 and 1000 pc from the sun; of B 8–A 0 stars at $l^{II} = 100°$ and 135° within 500 pc of the sun; of a high concentration of B 8–A 0 stars near $l^{II} = 300°$ at a distance of 1 kpc and another at $l^{II} = 200–220°$ between 1.5 and 2.0 kpc. This clumpy distribution of stars of moderate age ($1–5 \times 10^8$ yr) shows little correlation with the large-scale spiral structure. It should be emphasized, however, that these detailed analyses based on spectral classifications and photometry carried out at the Warner and Swasey Observatory are far from complete. At present five areas of the Milky Way in the southern sky are under study. When these are finished a better picture of the local stellar system can be obtained.

Among the older stars similar local groupings are evident. The most conspicuous of these is the group of F 0–F 5 stars around the sun. It has been pointed out before (McCuskey and Rubin, 1966) that these appear to form a local concentration near the sun. An average decrease in space density of 65% in the first 600 pc from the sun occurs. A recent study by Rydgren (1969) indicates that this is not due to a statistical fluctuation of randomly distributed stars. Rydgren's analysis of the space motions of 96 nearby F-stars places a maximum age of $2 \times 10^7$ years for the local concentration. He suggests that epicyclic resonance may be responsible for the present grouping, and that the grouping is a transient phenomenon.

The G 8 III–K 3 III stars appear to be rather uniformly distributed in the region of the sun. One concentration at $130° < l^{II} < 150°$ in the range $r = 100–500$ pc coincides approximately with the B 8–A 0 and A 2–A 5 groupings here.

The local concentrations and the rather high population densities near the sun shown by these studies are very much influenced by uncertainties in the evaluation of the interstellar absorption. On the face of it, however, the localization of intermediate age stars within 250–500 pc of the sun may indicate their preference for the inner part of the local spiral arm. But any striking evidence of association between them and the spiral structure is not present.

Analyses of the stellar distribution toward the galactic center (unpublished) and toward the galactic anticenter (McCuskey, 1967) have permitted the construction of space density profiles showing the variation of the number of stars per 1000 pc$^3$ over a range of 4 kpc centered on the sun. These have been constructed for several spectral groups: B 5, B 8–A 0, A 1–A 5, A 7–F 5, G 8 III–K 3 III and F 8 III–M 5 III. The profiles (Figures 2 and 3) show:

(a) A steeper decline in space density of B 8–A 0 stars toward the galactic center than toward the anticenter – another indication of a possible concentration of these stars near the inner edge of the local spiral arm.

(b) The concentration of late A and the F stars near the sun, mentioned above.

(c) A marked steady decrease in space density of giant G 8–K 3 stars from the galactic center direction toward the anticenter.

(d) A remarkably constant space density of F 8 III – M 5 III stars over the entire span of distance included.

No indication of major connections between these groups and the local spiral structure is apparent.

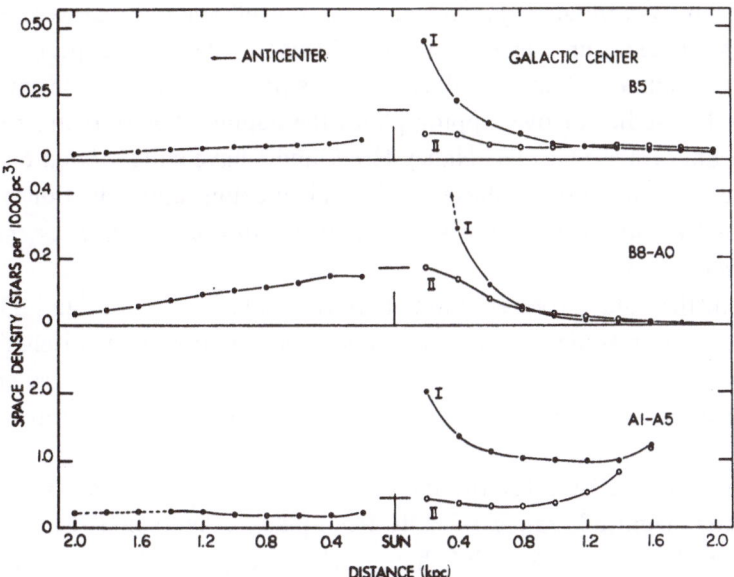

Fig. 2.   Space density profile for early-type stars from galactic anticenter to center directions through the solar neighborhood. Curves I and II refer to calculations using high and low estimates of interstellar absorption, respectively. Bars at the solar position refer to space densities for the region within 100 parsecs of the sun.

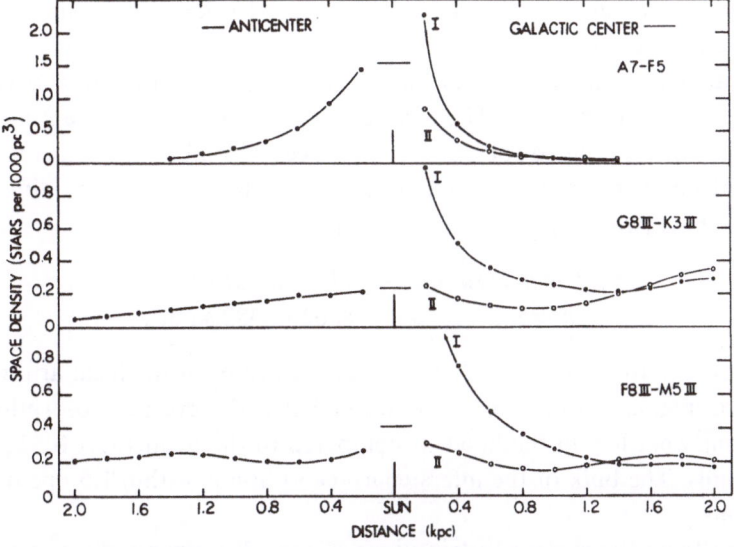

Fig. 3.   Space density profiles for main sequence F and late-type giant stars. See caption to Figure 2 for details.

## 4. New Survey of B 8–A 3 Stars

In a further attempt to associate in some detail the early A-stars with the local spiral structure a comprehensive survey of the entire galactic belt visible from Cleveland was begun three years ago. The use of a thin $1°8$ ultraviolet transmitting prism (dispersion 1100 Å/mm at H$\gamma$) attached to the Burrell-Schmidt telescope and Kodak IIa O plates with exposure times of 20 min yield clearly identifiable spectra for B 8–A 3 stars brighter than $V \sim 13$ mag. Galactic clusters provide a magnitude calibration for the survey. By studies of overlapping plates the counts of A-stars are reduced to a reasonably uniform system. Dr. Nancy Houk and I have collaborated on the identification and counting of these objects. Independent counts and reductions for overlap, etc. were finally combined into the accepted *number of* B 8–A 3 *stars per square degree brighter than V = 13*.

At present the data for galactic longitude ranges $95° < l^{II} < 150°$ and $55° < l^{II} < 75°$ are complete. In galactic latitude the range varies. Our present discussion will be limited to $-3° < b^{II} < +3°$. The data, however, extend somewhat farther from the galactic plane but not uniformly. One region embraces the local spiral arm and the other is an interarm region.

To supplement the data for the faint A-stars we have classified the stars in this group for the magnitude range $V = 7–10$ from plates taken some years ago with the $4°$ prism attached to the Burrell-Schmidt telescope (dispersion 280 Å/mm at H$\gamma$). Furthermore we have used the counts made years ago by Seydl (1929) from the Henry Draper Catalogue. Thus we have available for analysis values of log $N(7)$, log $N(10)$ and log $N(13)$, where $N(V)$ is the *number of A-stars per square degree brighter than magnitude V*, for each square degree, or combinations of square degrees, in the regions of the Milky Way outlined above.

The data on interstellar reddening published by Neckel (1967) and by FitzGerald (1968) have been used to obtain the interstellar absorption as a function of distance in small sub-areas of the regions. A ratio $A_v/E_{B-V} = 3.1$ was used to convert the color excesses into total absorption. The absorption data for the distance range 0–2.5 kpc from the sun was evaluated for areas of 1 sq deg or combinations thereof.

The total numbers of A-stars, adjusted for overlaps, magnitude differences, etc. counted in these two sections of the survey are:

> *Local spiral arm* – 34 731 in 581 sq deg.
> *Interarm*         –  8802 in 382 sq deg.

Thus we find 2.6 times as many stars per square degree in the local arm region as in the interarm region. At the same time we find that the average absorption, $\langle A_v \rangle$ at 1 kpc is somewhat less in the latter as compared to the former. At 0.5 kpc the same result prevails. The bulk of the interstellar absorption is within 1.5 kpc of the sun in both regions.

It is clear from the surface distributions of the 13th magnitude A-stars that large fluctuations occur from point-to-point along the galactic plane within distances of

1 to 1.5 kpc of the sun. A large part but not all of this is attributable to variations in the interstellar absorption. At $l^{II} = 130$–$140°$ above the galactic plane and at $l^{II} = 98$–$108°$ below the galactic plane, for example, excesses in the counts of A-stars are difficult to explain by a lack of obscuration there. On the other hand most of the variation in the numbers of these stars in the interarm region seems to be due to fluctuations in absorption.

The star-count and absorption data have been used in a standard way (McCuskey, 1966) to compute space densities for the A-stars. As parameters we have used $M_0 = +0.9$ and $\sigma_0 = 0.7$ mag., for mean absolute visual magnitude and dispersion respectively. These refer to unit volume of space. The calculations were performed on the Univac 1108 of the Jennings Computing Center at Case Western Reserve University. We are indebted to Mr. W. H. Wooden for carrying out these calculations. *There results the number of* B 8–A 3 *stars per* $10^3$ pc$^3$ as a function of galactic longitude at distances of 250, 500, 750, 1000, 1500 and 2000 pc from the sun for three zones of galactic latitude, $-3° < b^{II} < -1°$, $-1° < b^{II} < +1°$, and $+1° < b^{II} < +3°$.

From the space density analysis we draw the following conclusions:

(a) A definite density gradient in $-1° < b^{II} < +1°$ seems to exist across the local spiral arm, the number of A-stars at 250 pcs and at 500 parsecs at $l^{II} = 95°$ being about 5 times the number at $l^{II} = 150°$. A similar trend is evident for the zone $-3° < b^{II} < -1°$ but it is not as pronounced. For the region above the galactic plane this trend is not evident.

(b) This density gradient has largely vanished at $r = 1500$ and 2000 parsecs, and at $l^{II} = 145$–$150°$ there is some evidence for another high concentration of A-stars at these distances.

(c) At $l^{II} = 133°$ and around $l^{II} = 105°$ there are distinctly delineated maxima in the space density of A-stars, resembling large clusterings. These appear to be rather elongated regions, some 50 pc wide and 300 pc long. The reality of such 'cigar-shaped' high density regions may be questioned. But they do appear, not only in the distributions of A-stars but in those of other spectral classes. It has been suggested to us by Dr. V. C. Reddish that perhaps a cloud of relatively large particles, non-reddening, is embedded in the stellar distribution in such a way as to produce the elongated structures. This remains to be investigated.

(d) A large clustering of A-stars occurs at $l^{II} = 52$–$62°$ in the interarm region at $r = 250$ pc. It persists at $r = 500$ pc but has essentially disappeared beyond that distance. The general decrease of space density with distance is much more rapid in the interarm region than in the region of the local spiral arm.

(e) If we assume that the B 8–A 3 stars are associated, albeit loosely, with the local spiral structure then we might interpret the above results in the following way:

(i) If the sun is situated toward the inner edge of the local spiral arm (LSA), the heavy concentrations of A-stars at $r = 250$–$500$ pc near $l^{II} = 95$–$105°$ and $l^{II} = 132$–$135°$ imply their predominance also toward the inner edge of LSA.

(ii) The end of the space density decrease at $r \sim 800$–$1000$ pc around $l^{II} = 100°$ implies vaguely a boundary to LSA here.

(iii) The most rapid density decrease occurs at $l^{II} = 115$–$125°$, the radial extent of LSA here being 500–600 pc. This is about one-half of the width of the LSA as defined by other stellar populations.

(iv) If the sun is near the inner edge of LSA the distributions of the A-stars as noted above would indicate an angle of inclination to $l^{II} = 90°$ of some 25–30° for LSA. This is in conformity with the structure defined by other stellar objects but much larger than the inclination angle of the spiral structure indicated by the concentrations of H I.

It should be re-emphasized, however, that vagaries in the evaluation of the interstellar absorption enter the analyses strongly.

## 5. Giant M, S, C Stars

Blanco (1965) and Mavridis (1966) have summarized in considerable detail the information obtained from infrared surveys concerning the galactic distribution of these late-type giant stars. Briefly it may be said that:

(a) *The early-type M stars*, spectral classes earlier than M 5 III, are distributed somewhat irregularly along the galactic equator with concentrations at $l^{II} \sim 60°$ and $l^{II} \sim 240°$ suggestive of membership in the local spiral arm population.

(b) The giant stars of classes M 5–M 6.5 are more uniformly distributed between spiral arm and interarm regions but show an appreciable increase in numbers toward the galactic center.

(c) The very late M-stars, classes M 7–M 10, appear to be confined more or less uniformly to a thin disk centered on the galactic plane. The thickness of the disk is about 400 pc. These stars do not seem to be as concentrated toward the galactic center as do those of the M 5–M 6.5 group.

(d) The *S-type stars* are very rare in space. It has been shown by Westerlund (1964) that their distribution in the southern Milky Way is similar to that of the OB-associations. They are, therefore, probably spiral arm objects. On the other hand, there may be two types of S-stars (Keenan, 1954; Takayanagi, 1960) one of which is *not* associated with spiral structure. Much more needs to be done to determine distances for these objects before any definitive conclusions can be drawn.

(e) The *Carbon stars*, particularly those classified as N in the *Henry Draper Catalogue*, exhibit some tendency toward concentrations in spiral arm regions. Westerlund (1964) has examined in detail the surface distribution of these in the southern Milky Way. There is a preponderance of carbon stars toward Carina at $l^{II} = 290$–$300°$ and these appear to be distributed fairly uniformly along the spiral structure in that direction. Blanco (1965) has shown that these stars are more numerous toward the galactic anticenter than toward the galactic center. And Westerlund's (1964) survey of the southern sky does not indicate any penetration of the Sagittarius spiral arm in this direction. There is considerable evidence for clustering and pairing among the carbon stars.

(f) Space density analyses for the late-type giant stars are not available for a great many regions in the galactic plane. For the stars M 5 III and later, however, the space density, *number of stars per* $10^6$ pc$^3$, has been calculated as a function of distance for several areas (Figure 4). These data have been taken from the work by Westerlund (1959a, b, 1965), Albers (1962), McCuskey and Mehlhorn (1963), Blanco (1965), Hidajat and Blanco (1968), and McCuskey (1969). A relatively uniform space distribution of these stars (0.3–0.4 stars per $10^6$ pc$^3$) for regions away from the galactic center is evident from these surveys. There is, however, a considerable increase in space density toward the galactic center and a slow decrease toward the anticenter.

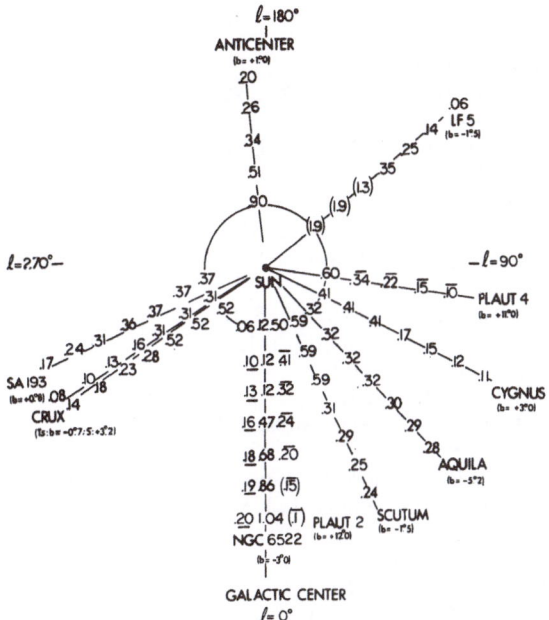

Fig. 4.   Space density of late-type giant M stars near the galactic plane. See text for details.

At comparable distances greater than 3 kpc from the sun there appear to be 2–4 times as many late M-stars toward the galactic center as in the opposite direction. There is little evidence that these stars are associated with the spiral structure of the galaxy.

## 6. Summary

We conclude from this survey that among the *stellar groups* thus far studied in sufficient detail

(a) The OB$^+$ and OB$^0$ stars, the WR stars, the Be stars of early type, the O-associations, some of the S-stars and the N-stars (carbon), appear to be linked to the spiral structure of the galaxy in our neighborhood.

(b) The B 5 and the B 8–A 3 main sequence stars, the classical cepheids, the early M-giant stars (including those that are variable) *possibly* are associated with the local spiral structure, although the evidence is not strong.

(c) The remaining upper main sequence stars, earlier than F 8, the yellow-red giants, the M-giants later than M 5, some of the S-stars and the R-carbon stars exhibit no discernable relationship to the spiral structure.

## References

Albers, H.: 1962, *Astron. J.* **67**, 24.
Behr, A.: 1965, *Z. Astrophys.* **61**, 182.
Behr, A.: 1966, *Z. Astrophys.* **63**, 133.
Blanco, V. M.: 1965, *Stars and Stellar Systems* **5**, 241.
Bok, B. J.: 1967, *Am. Scient.* **55**, 375.
Crampton, D.: 1968, *Astron. J.* **73**, 338.
Fernie, J. D.: 1968, *Astron. J.* **73**, 995.
Fernie, J. D. and Hube, J. O.: 1968, *Astron. J.* **73**, 492.
FitzGerald, M. P.: 1968, *Astron. J.* **73**, 983.
Hidajat, B. and Blanco, V. M.: 1968, *Astron. J.* **73**, 712.
Kraft, R. P.: 1965, *Stars and Stellar Systems* **5**, 157.
Keenan, P. C.: 1954, *Astrophys. J.* **120**, 484.
Klare, G. and Neckel, T.: 1967, *Z. Astrophys.* **66**, 45.
Mavridis, L. N.: 1966, in *Colloquium on Late-Type Stars* (ed. by M. Hack), Osservatorio Astronomico di Trieste, p. 420.
McCuskey, S. W.: 1965, *Stars and Stellar Systems* **5**, 1.
McCuskey, S. W.: 1966, *Vistas in Astronomy*. **7**, 141.
McCuskey, S. W.: 1967, *Astron. J.* **72**, 1199.
McCuskey, S. W.: 1969, *Astron. J.* **74**, 807.
McCuskey, S. W. and Mehlhorn, R.: 1963, *Astron. J.* **68**, 319.
McCuskey, S. W. and Rubin, R. M.: 1966, *Astron. J.* **71**, 517.
Neckel, T.: 1967, *Veröff. Landessternwarte Heidelberg-Königstuhl*, Band 19.
Rodgers, A. W., Campbell, C. T., and Whiteoak, J. B.: 1960, *Monthly Notices Roy. Astron. Soc.* **121**, 103.
Rydgren, A. E.: 1969, *Astron. J.*, in press.
Schmidt-Kaler, Th.: 1964, *Z. Astrophys.* **58**, 217.
Schmidt-Kaler, Th.: 1966, *Z. Astrophys.* **63**, 131.
Seydl, J.: 1929, *Publ. Prague*, No. 6.
Smith, L. F.: 1968, *Monthly Notices Roy. Astron. Soc.* **141**, 317.
Takayanagi, W.: 1960, *Publ. Astron. Soc. Japan* **12**, 314.
Westerlund, B.: 1959a, *Astrophys. J. Suppl. Ser.* **4**, 73.
Westerlund, B.: 1959b, *Astrophys. J.* **130**, 178.
Westerlund, B.: 1964, IAU-URSI Symposium No. 20, p. 160ff.
Westerlund, B.: 1965, *Monthly Notices Roy. Astron. Soc.* **130**, 45.
Westerlund, B.: 1969, *Preprints of the Steward Observatory* No. 21.
Wray, J. D.: 1966, A Study of Hα-Emission Objects in the Southern Milky Way, Ph.D. Thesis, Northwestern University.

# 34. REMARKS ON LOCAL STRUCTURE AND KINEMATICS

A. BLAAUW

*Kapteyn Laboratory, Groningen, The Netherlands*

**Abstract.** Attention is drawn to a few aspects of the state of motion of the local population which may become of importance for the study of local spiral structure. Uncertainties in the present knowledge of the local standard of rest are discussed, (a) with regard to the possible outward or inward motion with respect to the galactic centre, and (b) with regard to the component in the direction of circular motion. We furthermore draw attention to the quiet state of motion of the local interstellar gas and to the contrasting fairly high group velocities among the recently formed OB associations.

## 1. Introduction

Studies of the velocity distribution and space distribution of youngest stars and gas must now be regarded in the context of the very useful hypothesis of the density wave theory for spiral structure, which must be expected to lead to certain predictions with regard to the state of motion of these objects in and outside the spiral structure. For the understanding of the properties of the local population this theory should be expected to refer mostly to the Orion arm. The significance of the Orion arm in the context of the density wave theory is not quite clear, the indications are that it must be considered as a secondary feature. For such studies the following remarks may become useful.

## 2. The Local Deviation from Circular Motion

The radioastronomical as well as the optical radial velocity observations are always referred to the sun, and, normally, subsequently reduced to the 'local standard of rest' (LSR), i.e. the circular velocity around the galactic centre. This reduction is done by means of an assumed value of the velocity, $S$, of the sun with respect to the LSR. The velocity $S$ is not accurately known and its uncertainty in amount and direction are involved in such problems as: the possible common motion of gas and stars in the radial direction from the galactic centre outward or inward (general expansion or contraction), and the deviation from circular velocity for the velocity component in the direction of circular motion for these objects but especially for the gas (the possible lag of the gas). The following reviews the present uncertainties in these quantities.

Figure 1 represents the velocities projected on the galactic plane with respect to the sun: $U$ is the component toward $l^{II} = 180°$ (anticentre); $V$, toward $l^{II} = 90°$ (galactic rotation). The various points indicate the generally adopted values of $U$ and $V$ for the 'common' types of stars, giants and dwarfs of spectral types G, K and M taken from the compilation by Delhaye (1965), for the interstellar medium as a mean of various determinations (Takakubo, 1967; Venugopal and Shuter, 1967; Blaauw, 1952), and for the OB stars within 600 pc from Lesh (1968). The accuracy of these positions

is 1 to 2 km s⁻¹. For the spectral types A and F, the velocity pattern is rather
patchy and can best be described as a superposition of a number of discrete streamings.
The hatched sections of the diagram represent the approximate areas of these groupings
for the A-type stars according to Eggen (1963).

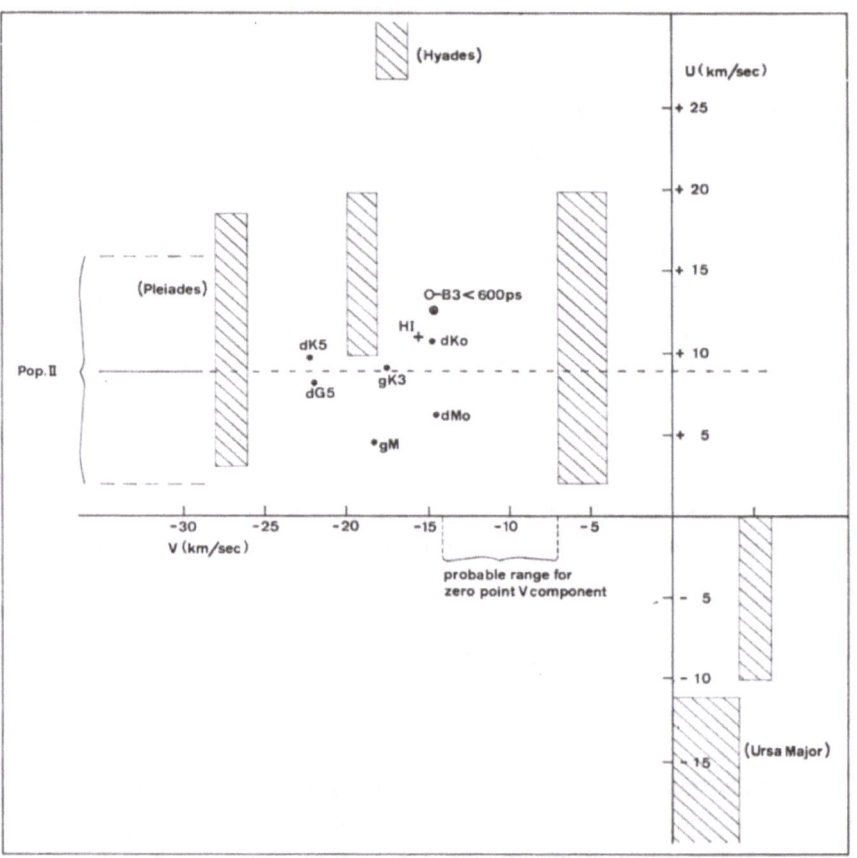

Fig. 1.   Velocity, with respect to solar motion, for the late-type 'common stars' and for the neutral
hydrogen and the O–B3 stars within 600 pc. Hatched areas indicate principal streamings in A-type
stars. Only the components $U$ (toward galactic anticentre) and $V$ (direction of
galactic rotation) are shown.

## A. VELOCITY WITH RESPECT TO GALACTIC CENTRE

Since no stellar, nor gaseous objects located at the galactic centre can be reliably used
(only radio measures are available and they concern objects in the interstellar medium
affected by strong turbulent motion), the most direct method consists of using those
stellar objects which may be expected to show no systematic expansion or contraction.
This is the case for objects belonging to Population II and to the old Disk Population.
We list below 4 categories for which a solution of the velocity component $U$ can be
derived from the literature.

RR Lyrae Variables, periods $>0.42\,d$ (Van Herk, 1965):

$$U = +23 \text{ km s}^{-1} \pm 9 \text{ (p.e.) from 76 stars;}$$

Subdwarfs (Deeming, 1961):

$$U = +2 \text{ km s}^{-1} \pm 18 \text{ (p.e.) from 103 stars;}$$

Planetary Nebulae (Minkowski, 1965):

$$U = +4 \text{ km s}^{-1} \pm 6 \text{ (p.e.) from 136 objects;}$$

Globular Clusters (Kinman, 1959):

$$U = +10 \text{ km s}^{-1} \pm 18 \text{ (p.e.) from 70 clusters.}$$

The differences between these determinations cannot be regarded as significant. A weighted mean value is:

$$\langle U \rangle = +9 \text{ km s}^{-1} \pm 5 \text{ (p.e.).}$$

This mean value, also represented in Figure 1, lies within the range of those for the various groups in that figure. There is, therefore, no evidence of expansion or contraction of these objects. However, the uncertainty is large and does not preclude a possible expansion or contraction of 10 km s$^{-1}$ or less. An expansion of $+7$ km s$^{-1}$ for matter in the solar neighbourhood, as proposed by Kerr (1962) in order to bring 21 cm observations in the first and fourth galactic quadrants on a common rotation curve, therefore cannot be ruled out or confirmed.

Obviously in view of the importance of this problem, attempts should be made to improve upon the determination of $\langle U \rangle$. At first thought, the high velocity objects in the disk population might seem to be an additional group for which the mean value of $U$ should be added to those listed above, since their velocity pattern might be expected to be smoothed out and symmetric in the $U$ component. This should be the case especially if the choice is limited to strong negative values of $V$, for instance $V < -50$ km s$^{-1}$ because then we are dealing with the more eccentric orbits. It appears that such a choice is not a suitable one because the distribution of the $U$ components is definitely asymmetric with an excess of positive values of the component $U$. This is shown, for instance, in the distribution of projected space velocities of the long period variables with periods shorter than 300 days, see Osvalds and Risley (1961), and also apparent if one plots the distribution of the $U$ components for the $K$ giants with $V < -50$ km s$^{-1}$ in the catalogue of Eggen (1966). Another demonstration of this asymmetry may be found in Miczaika's plot of the distribution of the apices of the high velocity stars as a function of galactic longitude of which a reproduction may be found in Oort's paper on stellar dynamics (Oort, 1965). This asymmetry is not yet well understood. It may well be related to the non-axisymmetrical gravitational field in the galactic plane. Further efforts to improve upon the determination of $\langle U \rangle$ therefore should be based preferably on old population stars with inclined orbits with respect to the galactic plane.

## B. POSSIBLE LAG WITH RESPECT TO CIRCULAR VELOCITY

Figure 1 shows that the $V$ components for the common stars are about $-18$ km s$^{-1}$. Since no direct way exists to measure the sun's velocity component $V$ with respect to the local standard of rest, an indirect estimate must be made. Axisymmetric dynamical theory for a well mixed steady state stellar system predicts a systematic lag with respect to the circular velocity for a given component of the stellar population as a consequence of the dispersion of the stellar velocities with respect to circular motion, and of the density gradient with increasing distance from the galactic centre. Denoting the logarithmic density gradient by $\delta \log v / \delta R$, the lag by $T$, the circular velocity by $\theta_c$, and the dispersion of stellar velocity components in the direction of $U$ by $\tau_U$, we have

$$\frac{\delta \log v}{\delta R} = T \frac{2\theta_c - T}{\tau_U^2} + \text{small known terms}.$$

The estimate of the density gradient is quite uncertain; on the one hand we know that the older population among the common stars probably has a gradient $\delta \log v / \delta R$ between $-0.15$ and $-0.30$ from studies at intermediate latitudes. But on the other hand the younger population may well show a density distribution related to the local spiral structure. If we use the value for the old component, we arrive at values of $T$ such that, if added to the $V$ components in the figure, they place the circular velocity point in the range between $V = -7$ and $-14$ km s$^{-1}$. This is indicated in the figure. If we are dealing with a different density gradient for the younger star population, the possible range may well shift several km farther towards more negative values of $V$.

It appears from the figure that a systematic lag for the neutral hydrogen in the solar neighbourhood and for the closely associated youngest stars (O–B 3) of up to 8 km s$^{-1}$ may well exist but it may also be negligible. The situation may be improved by systematic observations of the radial velocities of older population stars at intermediate galactic latitudes at longitudes 90° and 270°.

## 3. Local State of Motion of Gas and Youngest Stars

I should like to stress two points: the remarkably quiet state of motion of the local gas and the, in comparison to this, much more turbulent state of motion in the youngest stellar population.

## A. STATE OF MOTION OF LOCAL GAS

The most complete information on the motion of the local hydrogen comes from the 21 cm observations. At the lowest latitudes there is some difficulty in distinguishing nearer and more distant gas. It is therefore most instructive to consider the information derived from intermediate latitudes; reference is made in particular to the analysis of 21 cm profiles by Takakubo (1967). His Figures 1–7, dealing with the latitudes $+25°$, $+20°$, $+15°$, $+10°$, $-10°$, $-15°$ and $-20°$ clearly reveal the sinusoidal variation of the radial velocity of the main gaussian component notwithstanding the fact that the semi-amplitude on the average is only 3.8 km s$^{-1}$. There are no pronounced disturbances in the low velocity field. Fairly high negative velocities are observed at the galactic longitudes about 90° to 160° which partly must be ascribed to matter at large distances from the plane above the Perseus arm. Takakubo's figures may be considered to be representative for the volume up to about 400 pc from the sun. This includes most of the system of the Gould Belt of early type stars. Its existence also in the distribution of neutral hydrogen is well established, particularly by the excess densities at positive latitudes towards the galactic centre and negative latitudes toward the anticentre. In these directions outward motions in the neutral hydrogen, probably associated with the Gould Belt system, are observed, but the main feature is that of the differential galactic rotation.

## B. STATE OF MOTION OF YOUNGEST STARS

The youngest stars, spectral types O to B 3 are known to have smaller velocity dispersion than the later, older types. A.r.m.s. velocity in one component of about 10 km sec$^{-1}$ is usually found for the stars within, say, 600 pc. It would be wrong, however, to conclude from this that no pronounced streamings occur. This stellar population is known to be built up of a number of more or less isolated associations plus the somewhat older field stars. The nearest associations, Scorpio Centaurus, Per OB II, Ori OB I, Lac OB I, Cep OB II, Cep OB III, etc. contain most of the stars of types O to B 2 and about half of those of types B 3. Whereas within each of these associations the internal velocity dispersion is very small, a few km s$^{-1}$ at most, the associations as a whole have considerable relative velocities. Thus, the association Per OB II has a relative velocity, with respect to the average B star population, of about 15 km s$^{-1}$, and the average velocity of Scorpio Centaurus is about 8 km s$^{-1}$, and these two motions are approximately opposite. The velocity of the Orion association also is about 10 km s$^{-1}$. The differential galactic rotation, so clearly visible in the interstellar gas, is only a secondary phenomenon among the O to B 3 stars within 600 pc and the apparently irregular velocity pattern of the principal groupings prevails. One strongly gets the impression that in the process of star formation out of a quiet interstellar medium fairly large group motions are conveyed to the resulting associations. Conversely the state of motion of the youngest stars with this fairly high irregularity, is not necessarily an indication of large turbulence in the generating interstellar medium. For descriptions of the state of motion of the nearest O to B 3 stars, see Blaauw (1956), and Lesh (1968).

It may be noted that these large group velocities of the associations which imply considerable deviation from circular velocity, would, as seen from a distance, lead to quite erroneous location in the Galaxy if the standard relation between distance and radial velocity based on circular velocity is used. The Per OB II association, as seen from the Perseus arm would be misplaced by almost a kiloparsec. This implies that also distance determinations of distant associations by means of the standard galactic rotation curve may involve considerable accidental errors.

# References

Blaauw, A.: 1952, *Bull. Astron. Inst. Netherl.* **11**, 459.
Blaauw, A.: 1956, *Astrophys. J.* **123**, 408.
Deeming, T. J.: 1961, *Monthly Notices Roy. Astron. Soc.* **123**, 273.
Delhaye, J.: 1965, *Stars and Stellar Systems* **5**, 61.
Eggen, O. J.: 1963, *Astron. J.* **68**, 697.
Eggen, O. J.: 1966, *Roy. Obs. Bull.*, No. 125.
Kerr, F. J.: 1962, *Monthly Notices Roy. Astron. Soc.* **123**, 327.
Kinman, T. D.: 1959, *Monthly Notices Roy. Astron. Soc.* **119**, 559.
Lesh, J.: 1968, *Astrophys. J. Suppl. Ser.* **17**, 371.
Minkowski, R.: 1965, *Stars and Stellar Systems* **5**, 321.
Oort, J. H.: 1965, *Stars and Stellar Systems* **5**, 455.
Osvalds, V. and Risley, A. M.: 1961, *Publ. Leander McCormick Obs.* **11**, 147.
Takakubo, K.: 1967, *Bull. Astron. Inst. Netherl.* **19**, 125.
Van Herk, G.: 1965, *Bull. Astron. Inst. Netherl.* **18**, 71.
Venugopal, V. R. and Shuter, W. L. H.: 1967, *Astron. J.* **72**, 534.

# 35. GALACTIC CLUSTERS AND Hɪɪ REGIONS

W. BECKER and R. FENKART

*Astronomisches Institut der Universität Basel, Binningen, Switzerland*

The galactic clusters and the exciting stars of Hɪɪ regions are powerful indicators of the spiral structure since these objects lie on or very close to the zero age main sequence (ZAMS). For clusters this behaviour can be confirmed by direct photometric observation. The exciting stars of Hɪɪ regions are probably so young as not to have evolved from the ZAMS. The coincidence of these objects with the ZAMS indicates that there is little scattering in the absolute magnitudes and, as a consequence, little uncertainty in the determination of distance. This fact reveals itself very distinctly in the case of Hɪɪ regions, whose exciting stars define the spiral structure considerably better than the stars of the same early spectral types but not connected with Hɪɪ regions.

Within the last 3 years the contributions of various observatories have increased the number of galactic clusters with distances determined by three-colour photometric methods from about 150 to about 230.

Unfortunately the corresponding number of Hɪɪ regions with distances determined by the H$\beta$- and H$\gamma$- intensities of their exciting stars has not augmented in the same measure, namely from 55 to only 70. Some of these coincide with young clusters.

Even so, the total number of these objects has increased sufficiently to justify a new survey of their spatial distribution and, hopefully, stimulating new work in this field.

We have collected systematically the published results and combined them with our own. For the clusters it seems practical to determine the distances in a homogeneous way. Most authors use the B–V colour magnitude diagram (CMD) for the determination of age and distance modulus and the two-colour diagram (TCD) for the evaluation of the colour excess.

There are three reasons which lead us to the conviction that an other method is more practical, a method using both CMD's (B–V and U–B), but not the TCD. These reasons are:

(1) The apparent magnitude, necessary for the separation of physical and non-physical members of galactic clusters does not enter the TCD. The colour excess evidently should be determined only by the physical members. Since the B–V CMD alone in many cases allows no conclusive determination of the physical membership, both CMD's together complement each other and are the next best tool to proper motions.

(2) The colour excess can be more precisely determined by postulating the coincidence of the physical members of the cluster with the ZAMS of B–V *and* U–B CMD simultaneously and taking into account the fixed ratio of the excesses in U–B and B–V.

(3) The U–B CMD is often more reliable for the determination of the distance modulus, especially for young clusters, since the upper part of the ZAMS is much less steep here than in the B–V CMD, where it is almost vertical. In addition, the

determination of age, i.e. of the colour index of the whitest stars, is more precisely
done in the U–B CMD than in the B–V CMD, since an interval in U–B of 0.4 magn.
corresponds to an interval of only 0.1 magn. in B–V.

Using our method we had to modify the published distances in some cases by 10
to 30%, but mostly by less than 10%.

Figure 1 shows the distribution of the young galactic clusters (●) with earliest
type between O and B 2 to B 3. It does not diverge from the distribution of the H II
regions (○), also shown in this figure. Not considering a few 'drop-outs', the distribution defines three lanes suggesting parts of spiral arms. This suggestion is supported
by the fact that there exists a spiral galaxy, NGC 1232, whose spiral arms coincide
with these parts of *our* spiral arms regarding their width and their mutual distance,

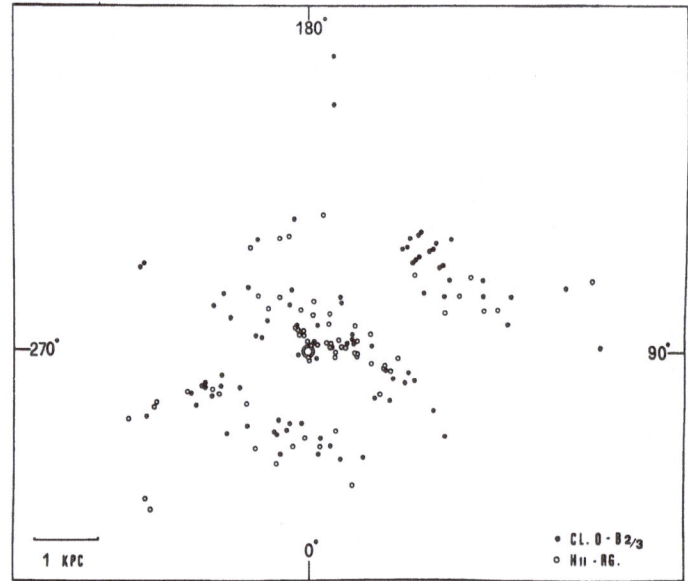

Fig. 1.    Space distribution of young galactic star clusters (●) and H II regions (○) in the galactic
plane. Parts of three spiral arms can be seen, the local arm (O), the Perseus arm (+I) and the
Carina-Sagittarius-Scutum arm (−I). The position of the sun is (⊙).

that is they agree even in pitch angle and distance from the respective nuclei (Figure 2).
The total diameter of NGC 1232 is according to its expansion velocity (Hubble
constant 100 km s$^{-1}$ Mpc$^{-1}$) practically the same as the one of the Milky Way
system.

Figure 3 shows the spatial distribution of the older clusters with earliest type later
than B 3, which seems quite random. However the number of such intermediate age
clusters is still too small to provide a basis for a successful investigation of the transition from spiral to random distribution.

The arrangement of the young clusters and the H II regions in the spiral arms is
due partly to selection effects caused by the observation programs and partly by
particularities of the spatial distribution.

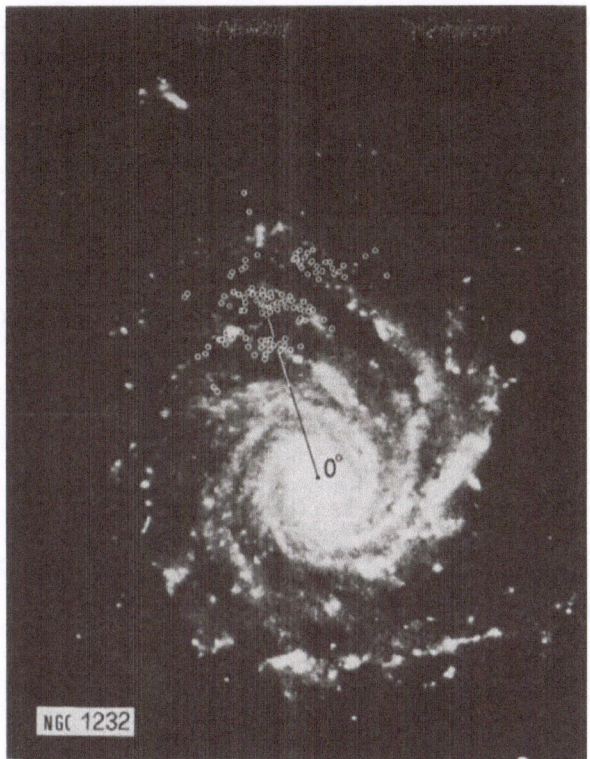

Fig. 2.   Young galactic star clusters and HII regions as projected in a corresponding scale to the spiral galaxy NGC 1232, which has the same linear diameter as our Galaxy.

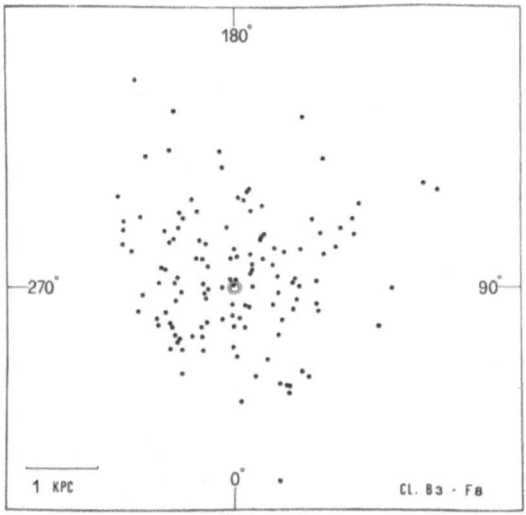

Fig. 3.   Space distribution of the old galactic star clusters.

If we permit ourselves to compare again the situation in NGC 1232 to our galaxy, we cannot expect to complete the picture in the directions $l^{II} = 200°$ to $250°$ substantially, because the spiral arms seem to dissolve in these directions. On the other hand, a completion of the spiral arm $+I$ in the direction $l^{II} < 95°$ meets observational difficulties, since its clusters will have distances over 3 kpc and will be influenced by heavy absorption of the Cygnus bifurcation. Between $l^{II} = 140°$ and $180°$, young objects seem to be missing completely in the arm $+I$.

Most interesting and relatively easy to complete is the picture in the region between $l^{II} = 300°$ and $330°$, which will be of great importance for a resolution of the discrepancy in the interpretation between Bok and ourselves. A better definition of the inter-arm regions in the directions around $l^{II} = 50°$ (Aquila-Vulpecula) and $l^{II} = 255°$ to $285°$ (Puppis) might prove to be of equal interest. We plan to concentrate our activities mainly on these problems.

### Acknowledgement

We thank the Schweizerischer Nationalfonds zur Förderung der Wissenschaftlichen Forschung for financial help.

# 36. A NEW INTERPRETATION OF THE GALACTIC STRUCTURE FROM HII REGIONS

G. COURTÈS, Y.P. and Y.M. GEORGELIN, and G. MONNET

*Observatoire de Marseille, Laboratoire d'Astronomie Spatiale, Marseille, France*

**Abstract.** From 6000 optical radial velocities of HII regions a new spiral structure (4 arms of pitch angle 20°) is found. The radial velocities of the observed HII regions are the same with the velocities of the HI regions. The kinematics of HII regions is similar to that of Cepheids and B stars.

A number of radial velocities of HII regions were first obtained by Courtès (1962, 1960); later Courtès *et. al.* (1968) collected 4000 optical radial velocities of 150 HII regions. Now, 6000 optical velocities have been measured and will be presented elsewhere (Georgelin, 1969). Figure 1 and Figures 2 and 4 of Courtès *et al.* (1969) show a comparison with recent surveys at 21 cm. There is a very close agreement between Hα velocities and the main maxima of the 21-cm line (as found by Courtès in 1959). The position of the main body of each spiral arm is then well defined from the distances of the HII regions. Those distances are, of course, those of the exciting stars which have been determined by spectrophotometric studies (Becker, 1963; Lyngå, 1964–65; Georgelin, 1969).

Radial velocities and spectrophotometric distances of HII regions give a model of the rotation curve of the Galaxy which is identical to the Schmidt curve (Courtès *et al.*, 1968). This agreement between the two rotation curves (Figure 5 of Courtès *et al.*, 1969) indicates that HII regions have the same rotational velocities as other population I components such as bright cepheids and neutral hydrogen. The constants of the solar motion $(U_0, V_0, W_0)$, of the galactic rotation $(A)$, and of the expansion $(c$ and $k)$ have been computed by the same method used by Kraft and Schmidt (1963) for the cepheids. The results of Georgelin (1969) are compared (Table 1 of Courtès *et al.*, 1969) with those of Kraft and Schmidt (1963) for the cepheids and those of Feast and Shuttleworth (1965) for the B stars.

The conclusion of a pitch angle of 20° (Courtès *et al.*, 1968) is of course in contradiction with the nearly circular arms given by the classical 21-cm model of Oort *et al.* (1958), but agrees with the details of the 21-cm data themselves, and we feel that this shortcoming of the earlier model is due to a misinterpretation of 21-cm data:

Kinematic distances inferred from 21-cm radial velocities are correctly defined only at angular displacements greater than 20° from the centre-anticentre direction. The existence of such an indeterminate sector leads to a fundamental ambiguity in the connection of the two fractions of each spiral arm across this sector. One can join, e.g., the Carina arm to the Cygnus arm (pitch angle 0° – 21 cm model) or to the Sagittarius arm (pitch angle 25° – optical model). In four sectors only, the distance of spiral arms is well known from radial velocities. But the pitch angle can only be marginally defined because this sector is too short in longitude (60°) and the precision of the kinematic distances at the extremity of each sector which defines the inclination.

*Becker and Contopoulos (eds.), The Spiral Structure of Our Galaxy, 209–212. All Rights Reserved.*
*Copyright © 1970 by the I.A.U.*

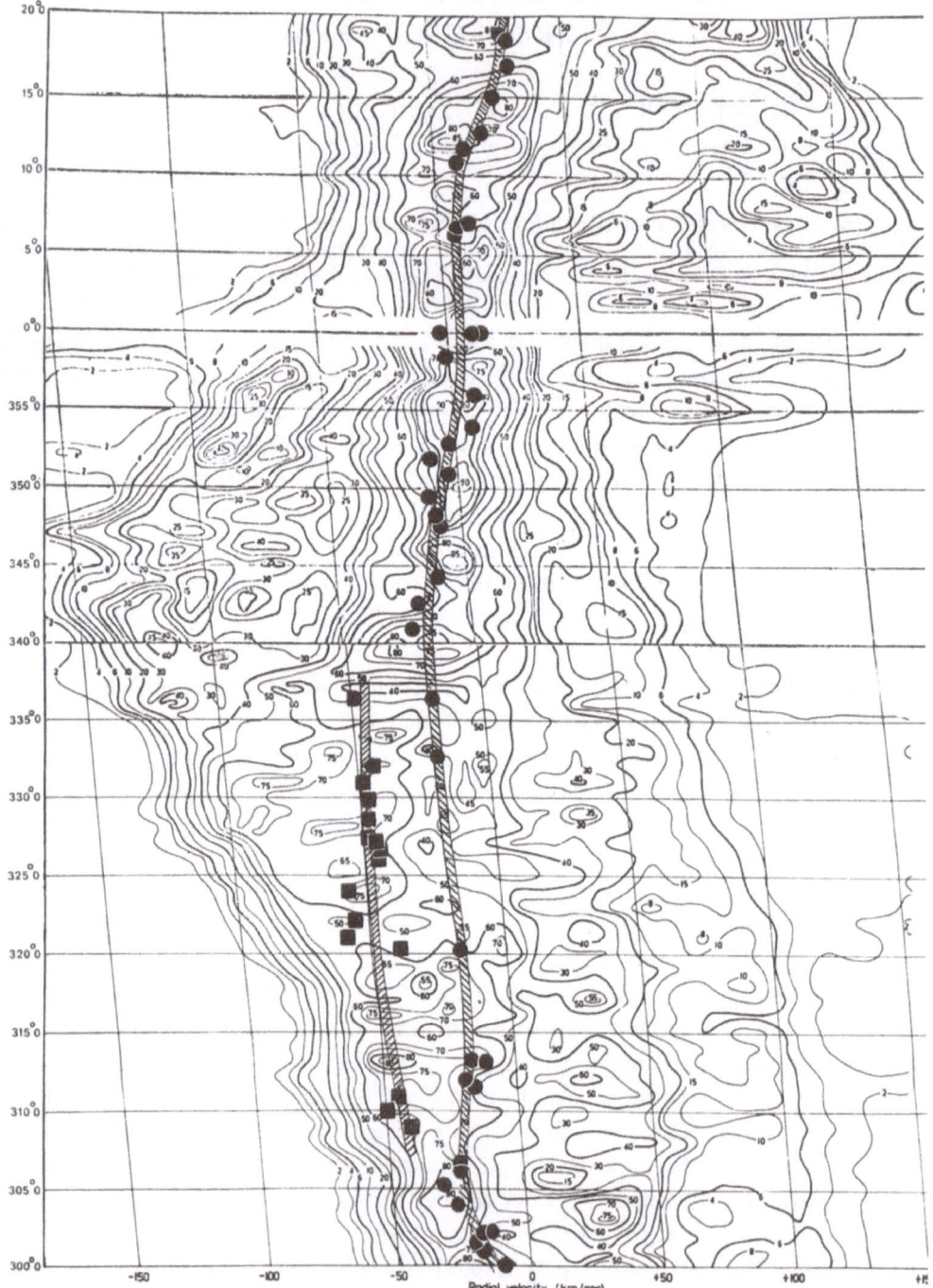

Fig. 1.   Comparison between velocities of HI and HII regions. —— HI data from Kerr and Hindman (1966), Kerr (1969). ●■ HII data from Courtès (1960) Cruvellier (1967) and Georgelin (1969). ● Sagittarius-Carina arm. ■ Norma-Centaurus arm.

On the other hand, spectrophotometric distances for H II regions can be used through the full longitude range. In particular, a clear morphological continuity is obtained for an inner spiral arm reaching from Carina to Sagittarius. Because of this linear continuity, the distance of this arm can also be obtained by the slope of the differential rotation curve in the longitude range 0° – 33° (Cruvellier, 1967) and this distance agrees quite well with the photometric distances.

With this interpretation of 21-cm data, no large discrepancy remains. In the 305–333° range 21-cm observations give a broad H I distribution from 1 to 4 kpc which was interpreted differently by Oort *et al.* (1958) and Kerr and Westerhout (1965). On the other hand, H II regions give a clear separation between two arms at respectively 1.5 and 3.5 kpc.

From those data, four spiral arms can then be drawn with confidence (Table I and Figure 2).

TABLE I

Data on the four nearest spiral arms

| Spiral arm | Longitude range ($l^{II}$) | Distance | Remark |
|---|---|---|---|
| +I | 103°–190° | 3 kpc at 120° | Very conspicuous |
| Perseus | Sharpless 132 Early type stars | | Between 140° and 168° |
| 0 | 59°–254° | 0.5 kpc at 180° | Quite poorly defined |
| Orion | NGC 6820 Rodgers 19 | | The sun is located at the inner edge of the arm |
| −I | 274°–32° | 1.5 kpc at 330° | Well defined |
| Sagittarius | Rodgers 42 Sharpless 69 | | |
| −II | 305°–333° | 3.5 kpc at 330° | Clearly separated |
| Norma | Rodgers 74 Rodgers 106 | | From the Sagittarius arm |

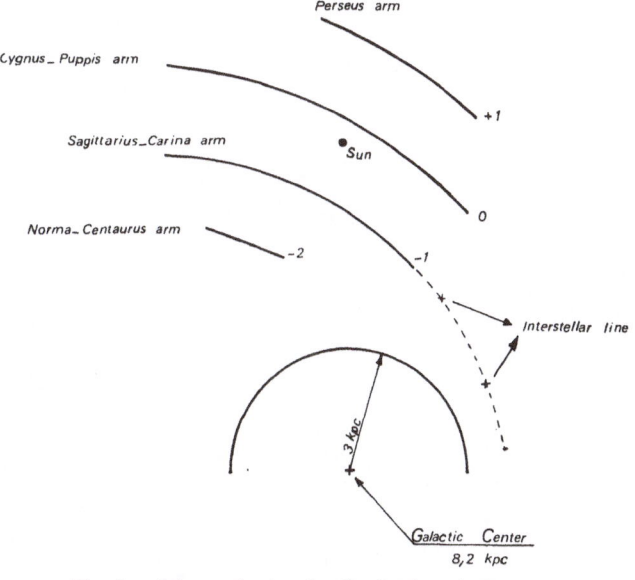

Fig. 2. Schematic sketch of galactic spiral arms.

It is particularly interesting that during this Symposium Dr. Weaver has presented a new interpretation of 21-cm radial velocities, which seems to be in close agreement with the spiral structure presented above – and thus strengthens this result.

The details of this work will be published by Courtès and Georgelin in *Vistas in Astronomy* (1970).

## References

Becker, W.: 1963, *Z. Astrophys.* **57**, 117.
Courtès, G.: 1952, *Compt. Rend. Acad. Sci. Paris* **234**, 506.
Courtès, G.: 1959, *Compt. Rend. Acad. Sci. Paris* **248**, 2953.
Courtès, G.: 1960, *Ann. Astrophys.* **28**, 683.
Courtès, G., Georgelin, Y.P. and Y.M., Monnet, G., and Pourcelot, A.: 1968, in *Interstellar Ionized Hydrogen* (ed. by Y. Terzian), Benjamin, New York.
Courtès, G., Georgelin, Y.P. and Y.M., and Monnet, G.: 1969, *Astrophys. Letters* **4**, 129.
Cruvellier, P.: 1967, *Ann. Astrophys.* **30**, 1059.
Feast, M. W. and Shuttleworth, M.: 1965, *Monthly Notices Roy. Astron. Soc.* **130**, 245.
Georgelin, Y.: 1969, Ph.D. Thesis.
Kerr, F.: 1969, *Australian J. Phys. Astrophys. Suppl. No. 9*, 1.
Kerr, F. and Westerhout, G.: 1965, *Stars and Stellar Systems* **5**, 167.
Kerr, F. and Hindman, J.: 1966, *Symposium on Radio and Optical Studies of the Galaxy*, Mt. Strömlo Obs., Canberra, p. 90.
Kraft, R. P. and Schmidt, M.: 1963, *Astrophys. J.* **137**, 247.
Lyngå, G.: 1964–1965, *Medd. Lunds astr. Obs.*, Ser. II, Nos 139, 140, 141, 142, 143.
Oort, J. H., Kerr, F. J., and Westerhout, G.: 1958, *Monthly Notices Roy. Astron. Soc.* **118**, 319.

# 37. THE DISTRIBUTION OF Hɪɪ REGIONS IN THE LOCAL SPIRAL ARM IN THE DIRECTION OF CYGNUS

H. R. DICKEL, H. J. WENDKER*, and J. H. BIERITZ

*Vermilion River Observatory, University of Illinois, Urbana, Ill., U.S.A.*

**Abstract.** The apparent shapes and orientations of optical nebulae in the Cygnus X complex provide possible evidence for the existence of a symmetry in the local spiral arm which may be related to the structure of the local magnetic field within the spiral arm. We have made a preliminary determination of the distances to about 90 nebulae in the Cygnus X complex by use of the values of interstellar absorption as a function of galactic coordinates. These values of absorption were determined from a comparison of optical and radio data for the nebulae. The more prominent nebulae are clumped at a distance of about 1.5 kpc. The total range in distances is from 1 kpc to at least 4 kpc. We have attempted to fit model spiral arms to this three-dimensional distribution of nebulae by approximating the spiral arm with a truncated cylinder. It has been possible to narrow the range of permissible orientations and sizes etc. for this local section of the Orion arm.

## 1. Introduction

Gaseous nebulae are intimately associated with spiral structure. Thus the distribution and physical properties of the gaseous nebulae in our local spiral arm should help to reveal the structure within the arm. We will present possible evidence for a symmetry about the axis of the arm and also some preliminary results concerning the orientation and geometry of the local region of the Orion arm as deduced from our study of the nebulae in the Cygnus X complex.

The Cygnus X region is approximately bounded by $l^{II} = 70°$ to $90°$ and $b^{II} = -8°$ to $+8°$. Most of the optically-visible nebulae of this complex are concentrated in the western side. So, we began our study by cataloguing 193 optical nebulae in this western section (Dickel *et al.*, 1969). We have followed two main lines of data analysis: the first has been to study the apparent shape and orientation of the optical features in our catalogue. The second aspect of the study has been to determine the distances to about half of these nebulae by use of the values of interstellar absorption as a function of galactic coordinates. These values of absorption were determined from a comparison of optical and radio data for these nebulae.

## 2. Shape and Orientation of the Nebulae

Let us first consider the shape of the nebulae. There are innumerable delicate filaments in addition to the prominent nebulae of IC 1318. One interesting feature of the nebulae is that 75% of those catalogued are highly elongated with axial ratios less than $\frac{1}{2}$ rather than the expected value close to 1. We see from Figure 1 that the axial ratio or

* Now at Max-Planck-Institut für Radioastronomie, Bonn, Germany.

eccentricity decreases with increasing galactic latitude above the plane. The indicated error bars show the extreme values within each latitude range. There are no extremely filamentary nebulae at low latitudes and no circular ones at high latitudes.

It should be pointed out that these highly elongated features that are so prevalent in the optical picture do not seem to appear in the radio data. Furthermore, although the main areas of the optical and radio emission overlap, the peaks of emission do not coincide too well which suggests that the optical appearance of the HII regions is governed to a high degree by the distribution of the absorbing material (Wendker,

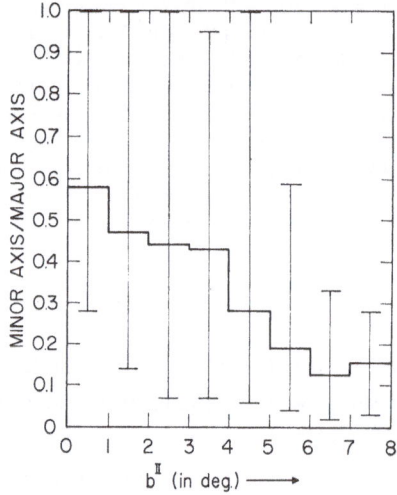

Fig. 1.   The nebular axial ratio vs. galactic latitude.

1970). Thus in the following discussion on the apparent orientation of the optical emission features, probably one is dealing mainly with the orientation of the dust lanes. For this discussion on the orientation, we will consider the following four possibilities:

(i) The orientation of the minor axes of the nebulae is random.

(ii) The nebulae have their minor axes oriented to within $\pm 30°$ of being perpendicular to the galactic plane as suggested by the data.

(iii) The nebulae have their minor axes aligned to within $\pm 30°$ of the radius vector from the apparent center of the spiral arm. Since the center of the Cyg OB 2 association lies conveniently nearly on the axis of the spiral arm, we have chosen it as the reference point in testing for symmetry.

(iv) This case is similar to the third except that the orientations are physically related to the Cyg OB 2 association. However, this is unlikely to be the case since the nebulae are too strung out to be excited by this one association. Also, as previously mentioned, it is probably the dust which is aligned.

Thus we will consider only the first three orientations. Figure 2 shows a histogram of the number of nebulae within each 20° zone of orientation relative to the radius vector from the center of the arm which we will take as Cyg OB 2. From the results of a $\chi^2$ test applied to this histogram to test for deviations from randomness, we conclude that the apparent orientation of the optical nebulae is not random.

In considering the other two possibilities, orientation with respect to the galactic plane or with respect to the axis of the arm, we have grouped the nebulae into 20° sectors of the position angle of the radius vector from the center of the arm as shown in Figure 3. Then we considered the number of nebulae within each sector that have a particular orientation. The results of $\chi^2$ tests applied to these data may be summarized as follows: There seems to be a natural dividing line at a position angle of 70°

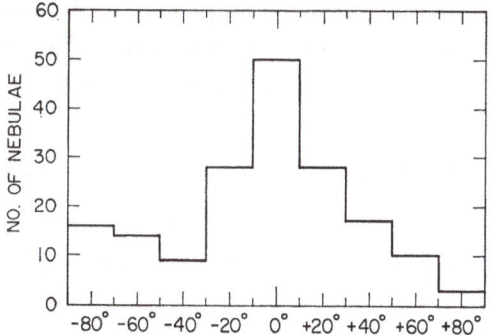

Fig. 2.   The number of nebulae with various orientations relative to the radius vector from the arm center (taken as the center of the Cyg OB2 association).

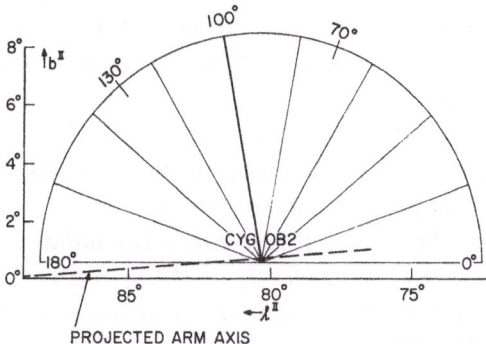

Fig. 3.   Zones of position angles around the Cyg OB 2 association.

which separates a region of random orientations for position angles less than 70° from a region of nearly radial orientations for position angles greater than this value. Most of the large, nearby nebulae such as IC 1318 a, b, c, are at position angles less than 70°. In general the probability for orientation of the minor axis perpendicular to the galactic plane is lower than for orientation with respect to the arm-center. The most pronounced alignment occurs between position angles 100° to 130° which is in the right sense if the arm is tilted above the galactic plane. The projected axis of the arm for a 1° inclination is indicated in Figure 3. Near the axis and in the galactic plane where we can see to further distances along the arm, one might expect any alignment to be smeared out but above the plane at the edge of the arm, the filamentary structure and alignment should be more apparent as we have found.

Struve (1957) has also commented on the outer ring of filaments which seems to surround the main part of the Cygnus X complex (see Figure 4 of Dickel *et al.*, 1969). We applied the $\chi^2$ test to the orientations of the nebulae on the outer ring. The orientations are not random but are again oriented with respect to the center of the arm. Here it may be the nebulae themselves that are so aligned.

In concluding the first part, the apparent orientation of the optical nebulae seems to indicate an alignment for the dust streamers which may be related to the helical structure of the local magnetic field of the spiral arm as given in the recent model by Mathewson (1968). The alignment we have found resembles that found by Hiltner (1951) for the planes of vibration of the polarized starlight.

## 3. Nebular Distances and the Local Spiral Arm

For any nebula, both the observed surface brightness in Hα and the observed radio brightness temperature are directly related to the emission measure by formulae involving known physical constants and the electron temperature. If we adopt a value of 6000 K for this temperature we can then compute the ratio of the observed brightness at the two wavelengths. The observed ratio will differ from that predicted, however, because of the absorption of the optical radiation by interstellar material. The amount of absorption as a function of distance in the Cygnus direction has been determined by Ikhsanov (1959) from measurements of the color excesses of stars. We have therefore used his curves to get the distances to the 90 nebulae for which we have both radio and optical data.

The optical data for the nebulae consist of an overlapping network of two dozen, calibrated plates of the Hα emission covering the western part of the Cygnus X complex. The grid of radio brightness temperatures for the region was obtained from Wendker's observations (Wendker, 1970) with a resolution of 11 min of arc at a wavelength of 11 cm. The microphotometry of all our plates is time consuming and thus we decided to do a preliminary calculation of the distances by estimating the surface brightnesses in Hα of the catalogued nebulae which appear on our plates. The results of these calculations were presented in our paper V (Dickel *et al.*, 1969) where we found that the nebulae are distributed from about 1 kpc to at least 4 kpc with the more prominent nebulae located around 1.5 kpc.

The next concern is where to place the main part of the Orion arm relative to this three-dimensional distribution of nebulae. From an examination of maps of the 21-cm emission it seems possible to approximate the local section of the arm as a truncated cylinder. The center of the arm can be fit reasonably well by a straight line which makes an angle of 77° with the line representing 0° longitude. This simple model gives a good fit from about 0.5 kpc from the sun to about 3.5 or 4 kpc beyond which the curvature of the arm becomes noticeable.

Figure 4 shows the model of the spiral arm which best fits our data. We have tried many models and have been able to significantly narrow the range of possible parameters for this section of the local spiral arm. Our results may be summarized as follows:

(i) The pitch angle – here defined as the angle which the axis of the cylinder makes with the line to the galactic center – lies between 75° and 80° with the best fit giving 77°.

(ii) The angle of tilt above the plane is less than 2° with a best fit of 1°. Putting the arm parallel and above the plane does not improve the fit.

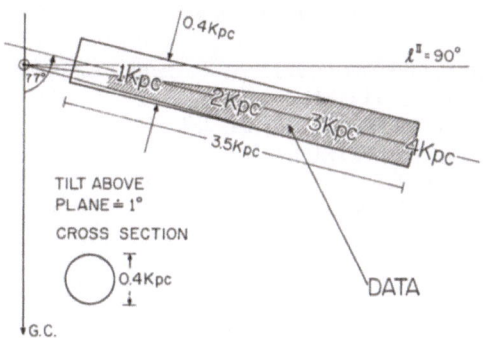

Fig. 4.    Best-fitting model to the local Orion arm.

(iii) The sun is on the inner edge of the arm, roughly $\frac{8}{10}$ of the distance from the center to the edge of the arm.

(iv) The semimajor axis is found to be greater than 0.15 kpc and much less than 0.4 kpc with a best fit of 0.2 kpc.

(v) The ratio of minor and major axes of the cylindrical cross section is more than $\frac{1}{2}$ and close to 1.

## 4. Conclusions

In summary, the data on the apparent filamentary structure of the optical nebulae and of the interstellar absorbing material within the local Orion arm suggest the existence of a symmetry which may be related to the structure of the local magnetic field within the arm. Our preliminary analysis on the distances to the Cygnus X-nebulae indicates that with the refinements in the data and analysis which we are presently incorporating (such as the accurate Hα surface brightnesses, high resolution radio data, etc.) we should get distances to the nebulae which are reliable to within a few tenths of a kpc. The fitting of model spiral arms to these data is a promising technique for determining the geometry of our local arm as defined by the ionized gas. So, by a judicious combination of optical and radio data for the nebulae with stellar color excesses and perhaps additional measurements of polarization and radial velocities at both wavelength ranges, we are beginning to see a picture emerging for the spatial distribution of the components of the local spiral arm. Hopefully, the physical relationships between the gas, stars, dust and magnetic field in the arm will soon be revealed.

## Acknowledgements

We wish to thank R. A. Avner, S. P. Wyatt and K. M. Yoss for helpful discussions and J. H. Cahn and J. R. Dickel for their help in the preparation of the manuscript. This research has been partially supported by the National Science Foundation.

## References

Dickel, H. R., Wendker, H. J., and Bieritz, J. H.: 1969, *Astron. Astrophys.* **1**, 270.
Hiltner, W. A.: 1951, *Astrophys. J.* **114**, 241.
Ikhsanov, R. N.: 1959, *Izv. Krym. Astrofiz. Obs.* **21**, 257.
Mathewson, D. S.: 1968, *Astrophys. J.* **153**, L47.
Struve, O.: 1957, *Sky Telesc.* **16**, 118.
Wendker, H. J.: 1970, *Astron. Astrophys.* **4**, 378.

# 38. REFLECTION NEBULAE AND SPIRAL STRUCTURE

R. RACINE and S. VAN DEN BERGH

*David Dunlap Observatory, Richmond Hill, Ontario, Canada*

**Abstract.** It is shown that stars embedded in reflection nebulosity may be used as spiral-arm tracers.

Surveys of stars embedded in reflection nebulae (Dorschner and Gürtler, 1964, 1965; Van den Bergh, 1966) show that reflection nebulae occur in associations. A detailed photometric and spectroscopic investigation of these associations of reflection nebulae has been carried out by Racine (1968, 1970). Some R associations coincide with known OB associations, others are connected with groupings of T Tauri stars and some are located in regions in which star formation was not previously known to occur.

Racine (1968) and Van den Bergh (1968) were able to show that R associations outline the Orion spiral arm rather better than do OB associations. The reason for this is that the number of R associations per kpc² in the galactic plane is greater than is the number of OB associations. This is so because the late B and early A stars, that dominate the stellar content of R associations, have a much greater space density than do the O and early B stars which populate OB associations.

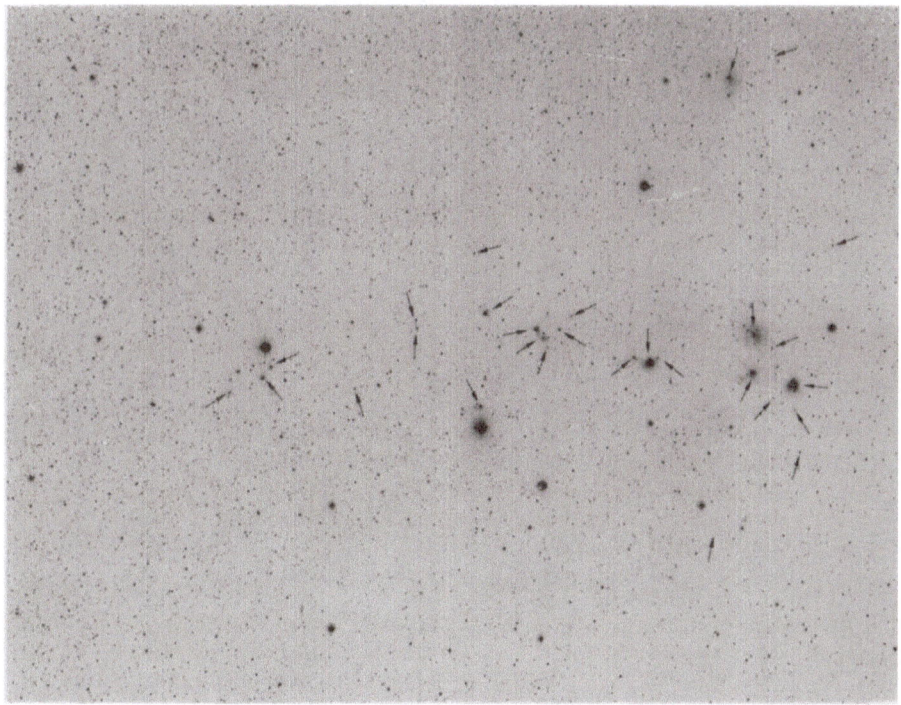

Fig. 1.    Mon R 2 is a good example of an association of reflection nebulae. The earliest spectral type in this R association is B1. The distance of the association Mon R 2 is 950 pc.

*Becker and Contopoulos (eds.), The Spiral Structure of Our Galaxy, 219–221. All Rights Reserved.*
Copyright © 1970 by the I.A.U.

A good example of an association of reflection nebulae is shown in Figure 1. In the figure 27 members of the association Mon R 2 are identified. This association, which contains both early and late B stars, is located at a distance of 950 pc.

Table I lists the associations of reflection nebulae that have so far been studied in detail. The positions of these R associations are plotted in Figure 2. The figure shows that associations of reflections nebulae outline the Orion spiral arm rather well. Figure 2 shows that this arm extends from $l^{II} = 70°$ to $l^{II} = 250°$. In the direction $l^{II} = 160°$ (Perseus-Auriga) the principal axis of the Orion arm, as outlined by reflection nebulae, is located at a distance of approximately 250 pc from the sun. Comparison of Figure 2 with plots of the distribution of clusters, OB associations and H$_{II}$ regions (Becker, 1964) shows excellent agreement between the spiral arms outlined by R associations and those obtained by other methods. The apparent lack of R associations in the Southern Milky Way is entirely due to observational selection. The Curtis Schmidt telescope on Cerro Tololo will shortly be used for a systematic survey of southern reflection nebulae.

One of the most striking features revealed by the distribution of R associations is that the Orion spiral arm is inclined to the galactic plane. In Cygnus and Cepheus the Orion arm is located approximately 50 pc *above* the plane; the Orion arm crosses the galactic plane in Cassiopeia and remains *below* the plane from Perseus to Canis Major.

TABLE I

Data on associations of reflection nebulae

| Association | | $l^{II}$ | $b^{II}$ | $d$(pc) | $z$(pc) | References |
|---|---|---|---|---|---|---|
| Sgr | R1 | 13 | − 0.8 | 1200 | − 17 | 1 |
| Aql | R1 | 30 | + 4 | 380 | + 27 | 4 |
| Vul | R1 | 67 | − 1.1 | 2500 | − 40 | 4 |
| Cyg | R1 | 77 | + 1.5 | 1000 | + 26 | 1 |
| Cep | R1 | 109 | + 4 | 650 | + 46 | 1 |
| Cep | R2 | 111 | +12 | 400 | + 83 | 1 |
| Cas | R1 | 118 | − 3 | 520 | − 27 | 1 |
| Cas | R2 | 133 | + 8 | 310 | + 43 | 4 |
| Cas | R3 | 131 | + 1 | 750 | + 13 | 4 |
| Cam | R1 | 142 | + 2 | 870 | + 30 | 1 |
| Per | R1 | 158 | −18 | 330 | −100 | 1 |
| Tau | R1 | 166 | −24 | 110 | − 45 | 1 |
| Tau | R2 | 171 | −17 | 135 | − 39 | 1 |
| Aur | R1 | 173 | − 3.5 | 1500 | − 90 | 4 |
| Tau-Ori | R1 | 201 | −17 | 360 | −105 | 1 |
| Mon | R1 | 201 | − 1.0 | 1050 | − 18 | 1 |
| Ori | R1 + R2 | 208 | −17 | 600 | −170 | 1 |
| Mon | R2 | 216 | −12 | 950 | −180 | 2 |
| CMa | R1 | 224 | − 2 | 690 | − 30 | 1 |
| CMa | R2 | 236 | −14 | 800 | −190 | 3 |
| Sco | R1 | 354 | +20 | 145 | + 50 | 1 |

1 = Racine (1968); 2 = Racine (1968) revised; 3 = Van den Bergh (1968); 4 = Racine (1970).

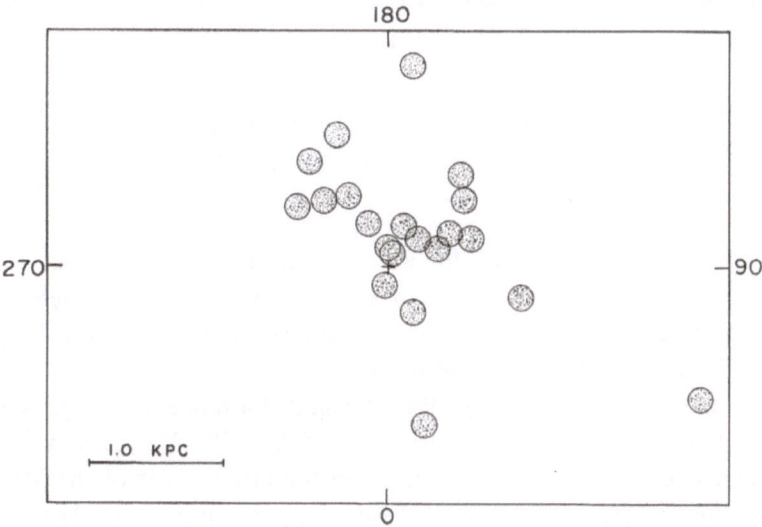

Fig. 2. Plot of the positions of R associations projected on the galactic plane. The position of the sun is marked by a cross.

The most southern R association that has so far been observed in the Orion Spiral arm is CMa R2 (Van den Bergh, 1968) which is situated 190 pc below the galactic plane. Additional southern hemisphere observations will be required to see if the Orion arm peters out at this point or whether it continues on and dips down again into the galactic plane.

## References

Becker, W.: 1964, *Z. Astrophys.* **58**, 202.
Dorschner, J. and Gürtler, J.: 1964, *Astron. Nachr.* **287**, 257.
Dorschner, J. and Gürtler, J.: 1965, *Astron. Nachr.* **289**, 57.
Racine, R.: 1968, *Astron. J.* **73**, 233.
Racine, R.: 1970, in preparation.
Van den Bergh, S.: 1966, *Astron. J.* **71**, 990.
Van den Bergh, S.: 1968, *Astrophys. Letters* **2**, 71.

# 39. THE GALACTIC STRUCTURE AND THE APPEARANCE
# OF THE MILKY WAY

E. D. PAVLOVSKAYA and A. S. SHAROV

*Moscow University, Moscow, U.S.S.R.*

The appearance of the Milky Way for an observer situated within our Galaxy is determined by the spatial distribution of stars and absorbing interstellar matter. Hence it may be hoped that the study of the surface brightness of the Milky Way permits to derive the spiral structure of our Galaxy.

Some years ago Elsässer and Haug (1960) suggested a model of the galactic spiral structure based on an analysis of the distribution of the Milky Way surface brightness along the galactic equator. They have identified the bright parts of the Milky Way with the directions where the lines of sight run along the spiral arms. However Behr (1965) pointed out that, in the presence of interstellar absorption within the arms, the bright Milky Way parts might correspond to the directions between the arms.

We consider this problem in detail using different models of the galactic structure.

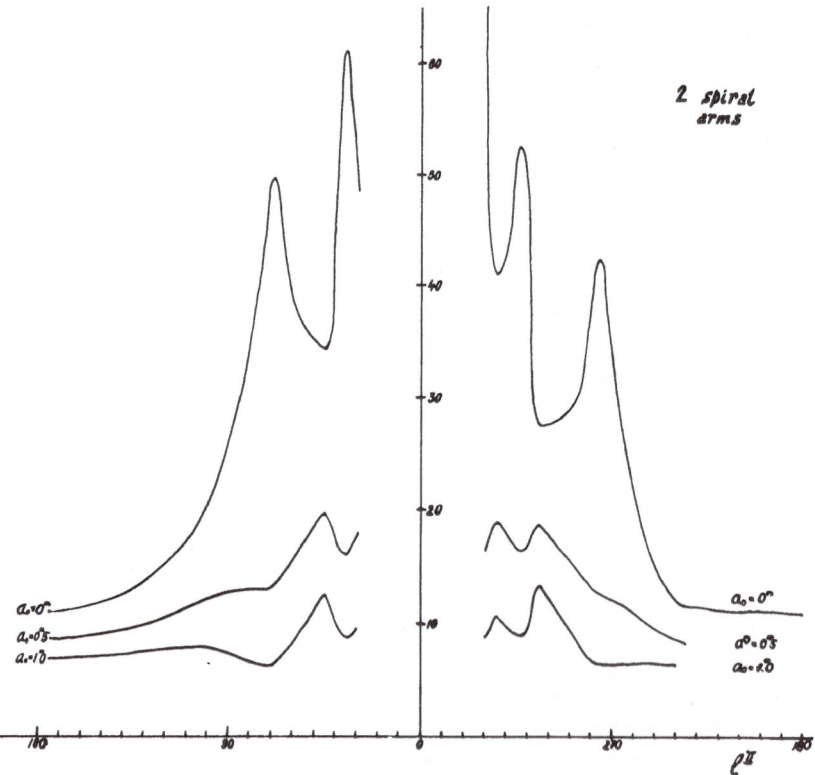

Fig. 1.   The Milky Way brightness along the galactic equator in the 2 spiral arms model of Galaxy with different absorptions in the arms (0$^m$, 0$^m$.5, 1$^m$.0 kpc$^{-1}$).

*Becker and Contopoulos (eds.), The Spiral Structure of Our Galaxy, 222–224. All Rights Reserved.*
*Copyright © 1970 by the I.A.U.*

It is assumed that the model of our Galaxy consists of the disk with a radius of 15 kpc, and of several arms in the form of logarithmic spirals. The sun is situated in an arm at 10 kpc distance from the centre of the system. We considered two geometrical models of the spiral arms – the model of Mills (1959) with 2 arms, and another one, with 14 arms derived by us (Pavlovskaya and Sharov, 1966). Our computational program permits to suppose the absorption either constant or variable across the spirals, and to change other parameters of our models. Results of calculations of the Milky Way brightness along the galactic equator for the two cases with 2 and 14 spiral arms 0.8 kpc wide are represented in Figures 1 and 2. It is supposed that both the disk and the spirals have constant densities, taken to be equal to 1 and 5, respectively.

The upper curves on both figures show the 'photometric profile' of the Milky Way if the interstellar absorption $a_0$ is zero. The two lower curves of each figure give the results of our calculations for constant absorption inside spiral arms: $a_0 = 0^m.5$ and $1^m.0$ kpc$^{-1}$. In the absence of absorption the bright parts of the Milky Way correspond to the directions along the arms. However the presence of absorption leads to a change of the general appearance of the Milky Way. With acceptable values of the absorption the maxima of brightness shift in the directions between the spiral arms.

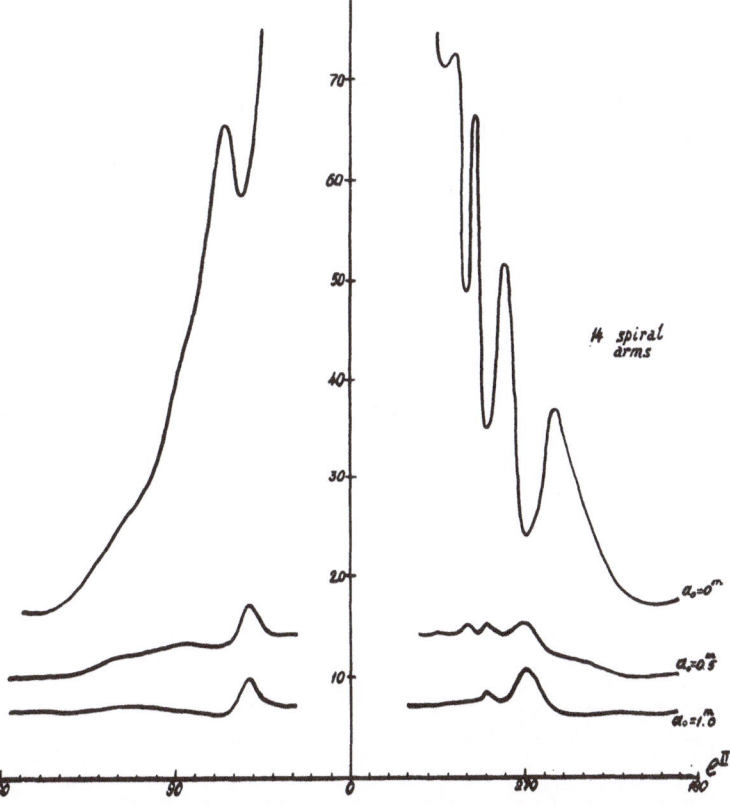

Fig. 2.   The Milky Way brightness along the galactic equator in the 14 spiral arms model of Galaxy with different absorptions in the arms (0$^m$, 0$^m$.5, 1$^m$.0 kpc$^{-1}$).

We considered a number of models with different values of absorption inside and between the arms as well as models with variable absorption across the arms. The analysis of all the results shows that even within the limits of our models the appearance of the Milky Way essentially depends on the value of absorption, and also on other parameters of the system. Of course, the real clouds of dark matter make the general appearance of the Milky Way more complicated. It seems unlikely that the study of surface brightness of the Milky Way might serve as a powerful tool for the investigation of the galactic structure. Hence it should not be supposed that the coincidence of the directions of the spiral arms, derived by other methods, is an essential argument in favour of any model of the galactic structure (Kardashev *et al.*, 1964).

It should be noted that all the regular models do not explain the higher brightness of the southern Milky Way as compared with the northern one. The difference appears to be connected with different values of the mean absorption in the two mentioned areas of the Milky Way (Fernie, 1962; Sharov, 1963).

The complete work will be published in the *Soviet Astronomical Journal*.

### References

Behr, A., 1965, *Z. Astrophys.* **62**, 157.
Elsässer, H. and Haug, U.: 1960, *Z. Astrophys.* **50**, 121.
Fernie, J. D.: 1962, *Astron. J.* **67**, 224.
Kardashev, N. S., Lozinskaya, T. A., and Sleptsova, N. F.: 1964, *Astron. Zh.* **41**, 601.
Mills, B. Y.: 1959, *Publ. Astron. Soc. Pacific* **71**, 267.
Pavlovskaya, E. D. and Sharov, A. S.: 1966, *Astron. Zh.* **43**, 40.
Sharov, A. S.: 1963, *Astron. Zh.* **40**, 900.

# 40. THE STUDY OF THE MILKY WAY INTEGRATED SPECTRUM AND THE SPIRAL STRUCTURE OF THE GALAXY

E. B. KOSTJAKOVA

*Moscow University, Moscow, U.S.S.R.*

The integrated spectrum of the Milky Way can give some information on the composition and large-scale structure of the Galaxy.

A study of the integrated spectrum of the Milky Way was carried out by the author in both hemispheres during several years (Kostjakova, 1952, 1954, 1958, 1963, 1964, 1965; Sharov and Kostjakova, 1967).

The absolute spectrophotometric gradients ($\Phi$) and temperatures ($T_c$) of 20 bright Milky Way fields studied are shown in Figure 1 (filled circles). The solid curve indicates the colour distribution of the Milky Way with the galactic longitude. We can see that over a large part of the Milky Way strip ($120° < l^{II} < 320°$) the spectrophotometric temperature is fairly high, of about 7000–8000 K. Beginning at $l^{II} = 100°$–$110°$, it declines gradually toward the direction of the galactic centre, reaching there its minimum value of about 4000 K. The pattern appears to be an asymmetric one: in the northern Milky Way the decline in temperature sets in at a greater angular distance than in the south.

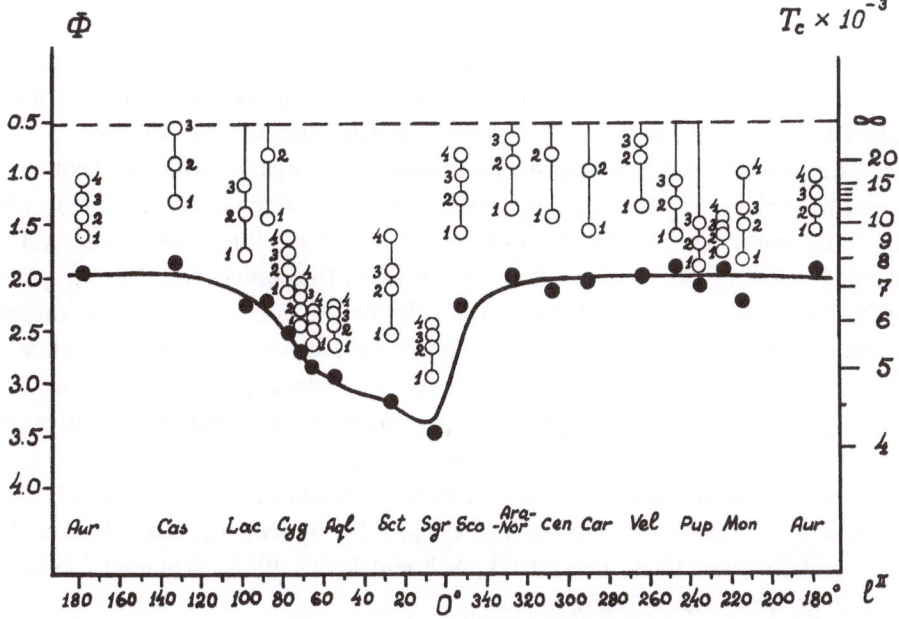

Fig. 1.   The spectrophotometric temperature, $T_c$, of the Milky Way, plotted against galactic longitude, $l^{II}$.

*Becker and Contopoulos (eds.), The Spiral Structure of Our Galaxy, 225–227. All Rights Reserved.*
*Copyright © 1970 by the I.A.U.*

The results obtained were corrected for the interstellar absorption by the method of Parenago (1945), with the data of $a_0$ (absorption per kpc), according to Sharov (1963). The correction procedure is illustrated also in Figure 1, where the open circles with numbers show the values of $T_c$, obtained for four assigned distances: 1, 2, 3 kpc, and $r = \infty$. The results show that the observed reddening of the Milky Way toward the galactic centre is a real effect, not caused by the interstellar absorption. In the 'cool' Milky Way clouds (Sagittarius, Scutum, Aquila), with a plausible $a_0$ adopted, $T_c$ varies over a comparatively small range, while the distance changes greatly. The 'hot' clouds, on the contrary, are very sensitive to the choice of distance. That enables to estimate an upper limit of the distance to individual 'hot' Milky Way clouds. Such estimates were made for several clouds.

The revealed colour distribution of the Milky Way was explained by the difference in stellar composition of the Galaxy in various directions. In regions of high temperature ($T_c$) the main contribution to the integrated radiation evidently comes from hot (blue) giants and supergiants, typical for spiral arm population. By decreasing the angular distance from the galactic centre, the contribution of the 'red populations' – spherical and intermediate – begins to predominate in the total radiation. Such an explanation agrees well with current ideas on the structure and stellar composition of galaxies.

The study of the integrated Milky Way spectrum, together with the upper bound estimates on distances to several clouds, led to certain conclusions as to the galactic structure in the solar neighbourhood.

(1) The colour variations in the northern Milky Way indicate that the region where 'red objects' predominate, has a large angular size. This agrees with the conclusion of Bok (1956, 1967) that the nuclear region of our Galaxy extends to about 5–6 kpc in radius.

(2) The progressive rise of $T_c$ in the direction of Cygnus-Aquila may indicate an actual increase in the number of the flat subsystem objects toward the anticentre; it appears that in this direction we observe the beginning of a spiral arm.

(3) The higher (real) temperature in Scutum as compared with adjacent regions can also be explained by the assumption that in this direction we observe the beginning of an inner spiral arm. In the direction of Sagittarius the line of sight intersects this arm; the remaining radiation would be produced by the coolest stars of the spherical component. In other directions (120°–320°) the main contribution to the integrated radiation evidently comes from flat component objects, causing a high $T_c$ there.

(4) The asymmetry in the galactic longitude dependence of the Milky Way colour can easily be explained by an asymmetric placement of the spiral arms relative to the direction from the sun to the galactic centre.

The spiral arms outlined, agree well with the location of the Population I objects, according to the recent studies of the Galaxy by optical methods (Becker 1964; Becker and Fenkart 1963; Beer, 1961; Schmidt-Kaler 1964; Whiteoak, 1963) (see Figure 2 in Kostjakova, 1965).

A similar analysis, carried out in reference to the results of radio astronomical research, led to the conclusion that in the solar neighbourhood the spiral arms deduced by optical methods and those derived from radio observations, agree neither in position nor in direction. This fact still awaits its explanation.

## References

Becker, W.: 1964, *Z. Astrophys.* **58**, 202.
Becker, W. and Fenkart, R.: 1963, *Z. Astrophys.* **56**, 257.
Beer, A.: 1961, *Monthly Notices Roy. Astron. Soc.* **123**, 191.
Bok, B. J.: 1956, *Vistas in Astronomy* **2**, 1522.
Bok, B. J.: 1967, *Am. Scient.* **55**, No. 4.
Kostjakova, E. B.: 1952, *Soviet Astron. Circ.* No. 129.
Kostjakova, E. B.: 1954, *Izv. Krym. Astrofiz. Obs.* **12**, 118.
Kostjakova, E. B.: 1958, *Soviet Astron. Circ.* No. 192, 13.
Kostjakova, E. B.: 1963, *Astron. Zh.* **40**, 771.
Kostjakova, E. B.: 1964, *Astron. Zh.* **41**, 505.
Kostjakova, E. B.: 1965, *Astron. Zh.* **42**, 537.
Parenago, P. P.: 1945, *Astron. Zh.* **22**, 129.
Schmidt-Kaler, T.: 1964, *Z. Astrophys.* **58**, 217.
Sharov, A. S.: 1963, *Astron. Zh.* **40**, 900.
Sharov, A. S. and Kostjakova, E. B.: 1967, *Astron. Zh.* **44**, 98.
Whiteoak, J. B.: 1963, *Publ. Astron. Soc. Pacific* **75**, 103.

# 41. SPACE DISTRIBUTION OF INTERSTELLAR DUST IN CONNECTION WITH THE GALACTIC SPIRAL STRUCTURE

T. A. URANOVA

*Moscow University, Moscow, U.S.S.R.*

**Abstract.** A maximum of dust density along the inner edge of the Cygnus arm is found. It seems that the same happens along the inner edges of the Perseus and the Sagittarius arms.

Pronik (1962) has examined some large dust clouds, studied by different investigators, and has shown that these clouds form a dust edge near the inner side of the Cygnus-Orion arm.

The present work deals with the general laws of the distribution of the galactic dark matter in connection with its spiral structure. We have studied the regions of the Milky Way Rift and stretches from 32° to 54° and from 67° to 72° in longitude, and from −12° to +10° in latitude. The space distribution of dust was studied with the aid of absorption curves $A_v(r)$. The data of B,V-photoelectric photometry, taken from 80 works, gave the material needed. Our card catalogue contained about 750 stars, one fourth of them having MK spectral classes; the data of one-dimensional spectral classifications were used after transforming them to the MK system with a careful determination of their systematic discrepancies from the latter. The system of absolute magnitudes and normal colors was taken from Boulon's (1963) work, and the value $R = 3.0$ was adopted. The areas studied were divided into regions about $3 \times 3°$ in size. The absorption curves were drawn for each region. The selection effect was controlled by a method analogous to that proposed by Neckel (1966). For other directions we have used the data of photographic photometry from Ikhsanov (1959), Metik (1962), Grigorieva (1965; 1969), Mashnauskas (1968), Brodskaja (1956; 1961), Ampel (1959), Adolfsson (1955), Strajzis (1963), Beer (1961), Pronik (1959) and Apriamashvili (1966), after transforming to the B, V, MK-system.

The optical density of the dust $a(r)$, in mag kpc$^{-1}$, was determined as the derivative of the absorption curve. Values of $a(r)$, with intervals equal to 50 pc, were taken from each curve. These data allowed to study in detail the variations of $a(r)$ with the z-coordinate and with the distance in the galactic plane.

The whole thickness of the absorption layer in the solar neighbourhood was estimated 85 pc above the galactic plane and 160 pc below it. This may be both a real fact and the result of the sun's elevation above the galactic plane.

The variations of the dust density along the line of sight were examined for seven directions in the above mentioned sectors; the regions of latitude $|b| = 2°$ were used. In all directions mentioned the line of sight meets a dense dust front with an absorption of $a(r) \geqslant 5$–20 mag kpc$^{-1}$. The distance of the dust front from the sun increases with the longitude, so that the line of this front is nearly parallel to the edge of the Cygnus arm (Figure 1).

*Becker and Contopoulos (eds.), The Spiral Structure of Our Galaxy, 228–231. All Rights Reserved.*

The results obtained allowed to draw a map of dust density distribution in the galactic plane, with graduations of 1–2 mag $kpc^{-1}$.

This distribution was compared to the positions of hot objects forming the spiral arms. Figure 2 indicates Morgan's aggregates and a part of Becker's O-B2 clusters. This figure shows that the densest parts of the dust layer are arranged along the inner side of the Cygnus arm both in the western and eastern directions.

In the Sagittarius arm direction the density of dust decreases slowly down to zero. The space transparence between the eastern parts of the Cygnus and Sagittarius arms is traced up to 3–8 kpc from the sun.

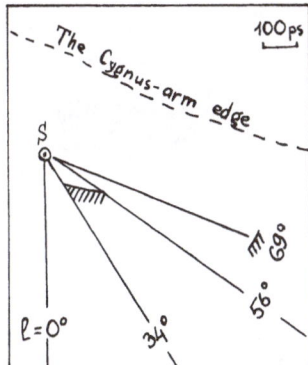

Fig. 1. The position of the dense front of dust in the galactic plane.

One can see more distinctly the concentration of dust between the Cygnus and the Sagittarius arms in cross sections, perpendicular to the arms axes (Figure 3). The x-coordinates in Figure 3 are read out from the inner side of the Cygnus arm to the outer side of the Sagittarius arm; the cross sections refer to distances 200 and 800 pc from the sun. It is seen that the dust is gathered near the edge of the Cygnus arm and its flat extension spreads towards the Sagittarius arm.

Returning to Figure 2 one can see that the central regions of the Cygnus arm as well as of the Sagittarius and Perseus arms, do not contain any dense dust formations.

At the same time, in the interval between the Cygnus and Perseus arms, some dense formations are noticeable near the inner side of the latter, as we can deduce from the absorption curves taken from the work of Brodskaja (1956, 1961), Grigorieva (1965) and Mashnauskas (1968).

The scanty data on dust density beyond the Sagittarius arm suggest the presence of a dust edge on its inner side. (The open clusters absorption curve in the direction $l \sim 310°$ indicates a density increase, as well as some individual distant stars with great colour excesses at $l \sim 45°$.)

It seems that dust is concentrated at the inner sides of the three known arms of Galaxy, whereas their central parts contain considerably smaller quantities of dust.

Certain discrepancies with the conclusions of Neckel (1966) may be explained in our opinion, both by a more detailed graduation of the dust density assumed in the

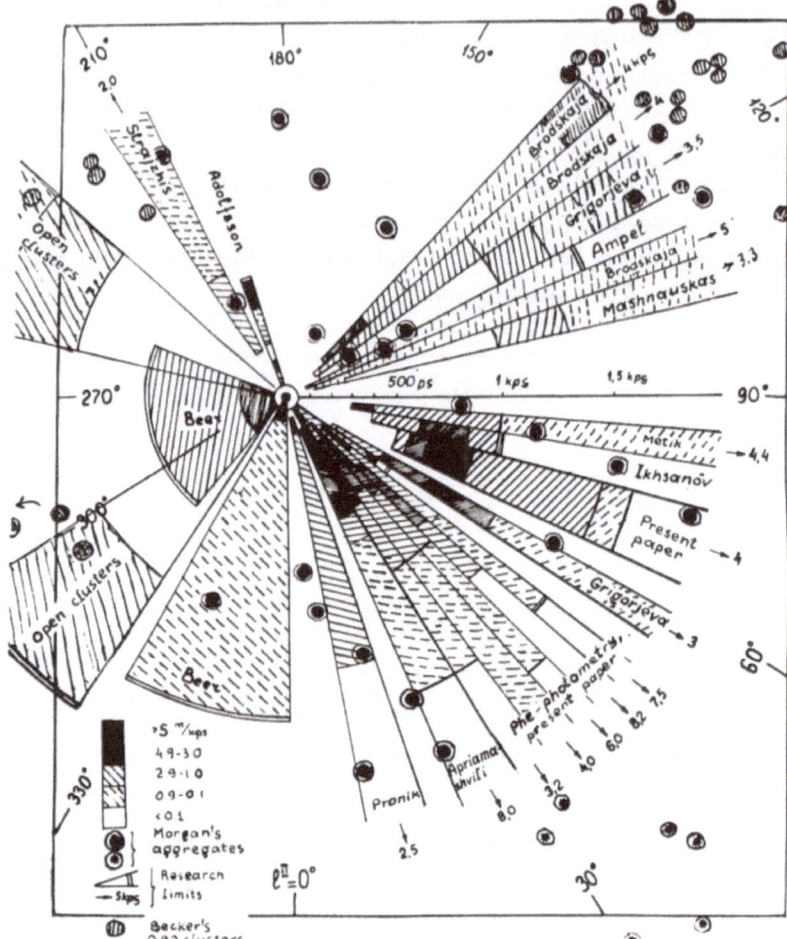

Fig. 2.   The distribution of dust in the galactic plane.

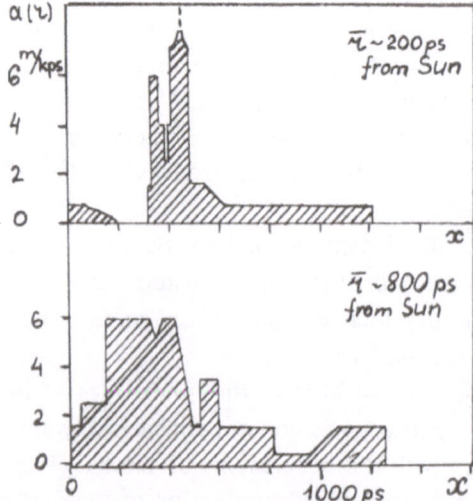

Fig. 3.   The distribution of dust density between the Cygnus and the Sagittarius arms (cross sections).

present work (especially for large values of dust density) and also by the fact that our data are not quite identical.

Detailed results are presented by Uranova (1968a, b). The connection of the Milky Way apparent brightness with the distance of the responsible dust matter is also considered there. It is shown that the Milky Way isophotes in the Aquila direction are due to the dust up to 500 pc from the sun.

## References

Adolfsson, T.: 1955, *Ark. Astron.* **1**, No. 34, 495.
Ampel, R.: 1959, *Bull. Astron. Obs. Univ. N. Copernicus, Torun*, No. 20.
Apriamashvili, S. P.: 1966, *Abastumansk. Astrofiz. Obs. Gora Kanobili Bjull.* **35**.
Beer, A.: 1961, *Monthly Notices Roy. Astron. Soc.* **123**, 191.
Boulon, J.: 1963, *J. Observateurs* **46**, Nos. 10–11.
Brodskaja, E. S.: 1956, *Izv. Krymsk. Astrofiz. Obs.* **16**, 162.
Brodskaja, E. S.: 1961, *Izv. Krymsk. Astrofiz. Obs.* **26**, 382.
Grigorieva, N. B.: 1965, *Izv. Krymsk. Astrofiz. Obs.* **34**, 238.
Grigorieva, N. B.: 1969, *Soobshch. Gos. Astron. Inst. P. K. Sternberga*, No. 162.
Ikhsanov, R. N.: 1959, *Izv. Krymsk. Astrofiz. Obs.* **21**, 257.
Mashnauskas, I.: 1968, *Astron. Obs. Biul. Vilnius* No. 22, 47.
Metik, L. P.: 1962, *Izv. Krymsk. Astrofiz. Obs.* **27**, 283.
Neckel, F.: 1966, *Z. Astrophys.* **63**, 221.
Pronik, I. I.: 1959, *Izv. Krymsk. Astrofiz. Obs.* **21**, 268.
Pronik, I. I.: 1962, *Astron. Zh.* **39**, 362.
Strajzis, V.: 1963, *Astron. Obs. Biul. Vilnius*, No. 8, 18.
Uranova, T. A.: 1968a, *Astron. Zh.* **45**, 1318.
Uranova, T. A.: 1968b, *Soobshch. Gos. Astr. Inst. P. K. Sternberga*, No. 163.

# 42. A COMPARISON OF THE DENSITY GRADIENTS OF MAIN SEQUENCE STARS OBTAINED BY TWO DIFFERENT METHODS IN DIFFERENT GALACTIC LATITUDES

W. BECKER and R. FENKART

*Astronomisches Institut der Universität Basel, Binningen, Switzerland*

The Basel Observatory program of the determination of disc- and halo-density gradients for different intervals of absolute magnitude comprises in addition to Milky Way fields several directions, all pointing to Selected Areas near a plane perpendicular to the galactic equator and passing through the sun and the galactic centre. It was started with SA 51 (Becker, 1965) and continued with Sa 57, 54 and 141 (Fenkart, 1967, 1968, 1969).

The aim of the investigations in the galactic disc is to study stellar clouds and their content of stars of different luminosity, based on the finding that in other galaxies stellar clouds are closely related with spiral structure. In this connection we want above all to check the possibility of applying three-colour photometric methods for this purpose in the simpler case of higher galactic latitudes where interstellar absorption is of no influence.

In each direction we measured photometrically all the 1500 to 2000 stars contained in fields of 0.5 to 2.5$\square°$ on Palomar Schmidt plates, and determined their apparent magnitudes in the RGU system with the help of photoelectric scales comprising between 40 and 60 stars per direction most of them observed by Purgathofer (1969). These magnitudes needed no corrections for interstellar absorption, since the two-colour diagrams showed no perceptible reddening for any of the investigated directions.

The blanketing effect shown in the two-colour diagram allowed the statistical separation of the Population II stars from the disc stars, in which we are interested in this context. The two-colour diagram allows also the attribution of statistical values of absolute magnitudes to each disc star.

Thus the density functions could be derived immediately by counting the stars within a given interval of absolute magnitude falling into consecutive intervals of distance modulus.

In the overlapping distance interval these 'photometric' density gradients can be compared with the 'spectral' gradients, that one can obtain by using the spectral type for the determination of the absolute magnitude. The material of the Bergedorfer Spektraldurchmusterung has been used for this purpose.

Our photoelectric scales and the Schmidt plates allowed us to reach limiting magnitudes of $19^{m}.5$ in G for SA 51, 54 and 57, and $17^{m}.5$ for SA 141. So it is evident that the photometric density functions will extend to much farther distances than the spectral ones, since the Bergedorf classification is not sufficiently reliable for our purposes for stars fainter than $13^{m}.0$ in B.

Therefore the density comparison will be limited to distances within roughly 1 kpc

*Becker and Contopoulos (eds.), The Spiral Structure of Our Galaxy, 232–235. All Rights Reserved.*
*Copyright © 1970 by the I.A.U.*

or less from the sun, depending on what luminosity interval is used. Finally, in order to be able to compare the photometric and the spectral density functions for the same luminosity groups, we had to establish the relationship between our intervals in absolute magnitude $<M_G>$ and the corresponding intervals in Bergedorf spectral type $<S_B>$. That can be done using the relationships:

Bergedorf- to MK-spectral classification (according to Schmidt-Kaler, 1965),

MK-type to $M_V$, B–V and U–B (according to Schmidt-Kaler, 1965 and Johnson, 1966) and the formula:

$$G = V + 0.93(B-V) - 0.08(U-B)$$

(Steinlin, 1968), as our photometric observations are made in the RGU system.

The density functions are given for the two intervals of absolute magnitude: $3 < M_G < 5$ and $5 < M_G < 7$.

The result is shown in Figures 1 and 2 for Selected Areas 54, 57 and 51, 82 respectively. They show in the first line, that the method of three-colour photometry covers distances, which are up to 10 times as large as the distances reached by spectral classification. This is essential for the study of star clouds. Besides, there are pronounced differences in the density values determined by the two different methods. The stars with $M_G$ between 3 and 5 show generally much lower densities if spectral types are used for the determination of absolute magnitude. In the interval between 5 and 7

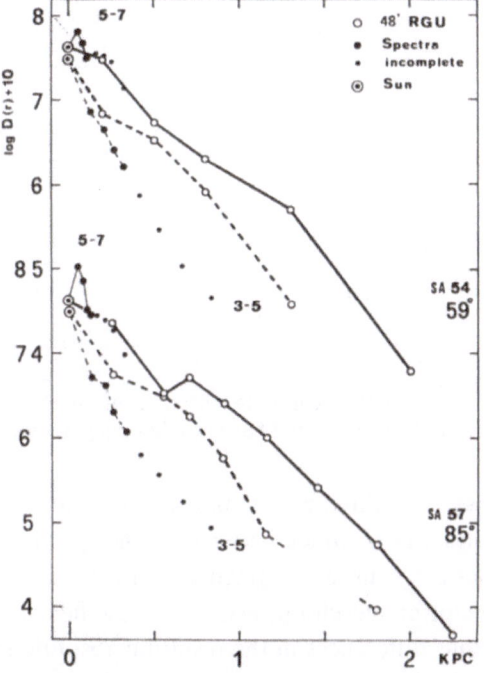

Fig. 1.   Density gradients for Selected Areas No. 54 and 57 for intervals of absolute magnitude from 3 to 5 and from 5 to 7. Strong lines and open circles: RGU photometry; weak lines and points: spectral classification; small points: incomplete data.

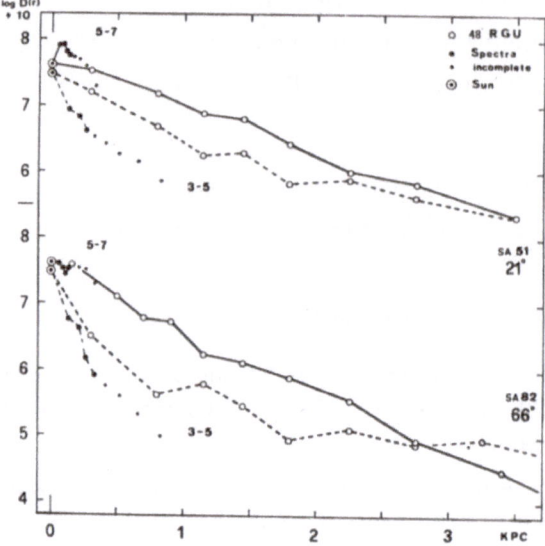

Fig. 2.   The same as Figure 1 for Selected Areas No. 51 and 82.

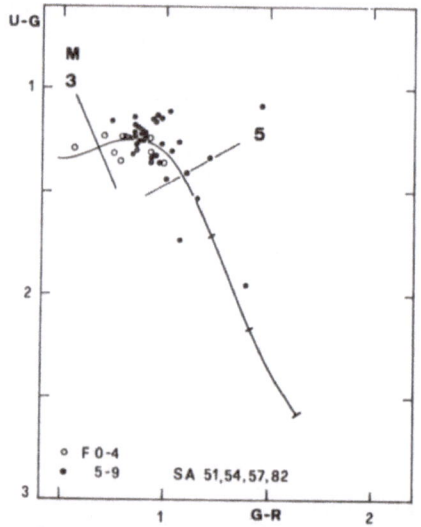

Fig. 3.   The distribution of F stars (Bergedorf classification) in the two-colour diagram. According
to their spectral type the stars should have absolute magnitudes between 3 and 5.

the opposite is the case. The three-colour photometry data match the values in the
vicinity of the sun (Gliese, 1957) much better than the spectral type data. We suspect
that these discrepancies are caused by systematic effects in classification of F and G
type stars in the Bergedorfer Durchmusterung in these fields.

   That there exist serious differences in the resulting absolute magnitudes by spectral
classification and by three-colour photometry can be seen by plotting into two-
colour diagrams those stars in Selected Areas 51, 54, 57 and 82 which are bright enough
to be classified in Bergedorf type.

The result is shown in the Figures 3 and 4, where we have first the stars of Berge-dorf-type F, which correspond to the luminosity interval $3 < M_G < 5$ and then the ones of Bergedorf-type G, corresponding to the luminosity group $5 < M_G < 7$. Whereas the earlier types fall well in the corresponding luminosity interval defined by three-colour photometry, a considerable fraction of the later ones according to three-colour photometry should evidently be classified earlier, if a reasonable relation between spectral type and absolute magnitude is supposed to exist.

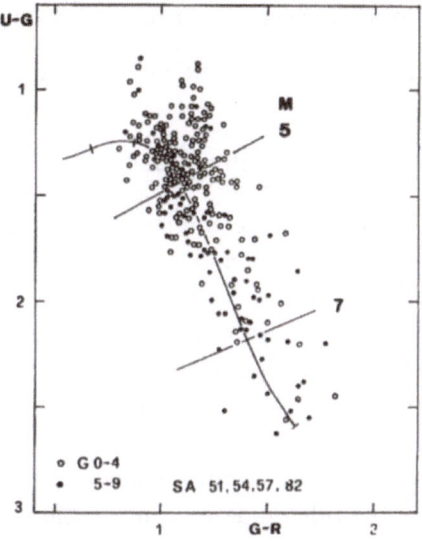

Fig. 4.    The distribution of G stars (Bergedorf classification) in the two-colour diagram. According to their spectral type the stars should have absolute magnitudes between 5 and 7.

This finding gives at least a qualitative hint, as to where to look for the cause of the observed discrepancies between the spectral and the photometric density gradients. We have to take it into account in the discussion of density gradients determined in lower galactic latitudes.

### Acknowledgement

We thank the Schweizerischer Nationalfonds zur Förderung der Wissenschaftlichen Forschung for financial help.

### References

Becker, W.: 1965, *Z. Astrophys.* **62**, 54.
Fenkart, R.: 1967, *Z. Astrophys.* **66**, 390.
Fenkart, R.: 1968, *Z. Astrophys.* **68**, 87.
Fenkart, R.: 1969, *Astron. Astrophys.* **3**, 228.
Gliese, W.: 1957, *Astron. Rechen-Inst. Heidelberg, Mitt. A*, No. 8.
Johnson, H. L.: 1966, *Ann. Rev. Astron. Astrophys.* **4**, 193.
Schmidt-Kaler, T.: 1965, *Landolt-Börnstein, New Series* **1**, 284.
Purgathofer, A.: 1969, *Lowell Obs. Bull.* **147**, 98.
Steinlin, U. W.: 1968, *Z. Astrophys.* **69**, 276.

# 43. THE GALACTIC DISTRIBUTION OF YOUNG CEPHEIDS

G. A. TAMMANN

*Astronomisches Institut der Universität Basel, Binningen, Switzerland*

**Abstract.** The distribution of long-period Population I cepheids is studied. An age comparison of cepheids and galactic clusters shows that cepheids with periods $>11.25$ d, corresponding to ages of $\lesssim 30 \times 10^6$ yr, should be almost as good spiral arm tracers as galactic clusters with earliest types b 2-3. The distances of cepheids are determined from a revised period-luminosity-colour relation and from colour excesses, for the determination of which a new, purely photometric method is given. The resulting distribution of cepheids shows a good correlation with the spiral arms, as traced by young clusters.

## 1. Introduction

Spiral arm tracers have to fulfill two requirements: they must be young enough to be typical for the spiral arm population and it must be possible to determine reliable distances for them. Next to the young galactic clusters and H II regions the Population I cepheids with long periods seem most suitable for this purpose. As for the clusters and exciting stars of H II regions the theory of stellar evolution provides age estimates for the cepheids, and their distances can well be determined from the period-luminosity-colour ($P$-$L$-$C$) relation.

It is then of great interest to decide whether galactic clusters and H II regions on the one side and cepheids on the other side of comparable age define the same spiral arms. The contrary would be most severe: it would mean that there are separate places for stellar formation for different kinds of objects. It would seem easiest to solve the problem in extragalactic spirals whether the loci of young galactic clusters and young cepheids coincide. However, at present there are no complete surveys of cepheids in nearby spirals (M31, M33) available. The detection of cepheids in more distant galaxies is so badly hampered by the discovery chance that a meaningful discussion of their distribution is impossible. It remains only the possibility to study the distribution of galactic cepheids and to compare it with the spiral arms as defined by galactic clusters and H II regions.

The study of the galactic distribution of cepheids has not yet led to clear results, mainly due to the paucity of long period cepheids. Their space density within 1500 pc from the sun is roughly ten times lower than that of young galactic clusters, and their small number does not allow to outline spiral arms with any certainty. Therefore the question whether they are confined to spiral arms can only be settled, if their location is compared with more frequent spiral arm tracers as the young galactic clusters and H II regions are (Becker, 1963; Becker and Fenkart, 1963). Kraft and Schmidt (1963) have found a vague correlation between bright cepheids and OB-associations, and Kraft (1963) a reasonable correspondance between bright cepheids and young clusters, which was confirmed by Schmidt-Kaler (1964) and by Tammann (1968) on the basis of an extended observational material. However, Fernie (1968) has recently denied this correlation; and he has found – not discriminating against cepheids with shorter periods – even an anti-correlation for cepheids with early B stars, galactic clusters, and H II regions.

Becker's (1969a) updated list of cluster and H II region distances and additional observations for long period cepheids as well as a new calibration of the slope and the zero point of the *P-L-C* relation (Sandage and Tammann, 1969) and a better understanding of the age of cepheids as a function of period (Kippenhahn and Smith, 1969) make a new intercomparison desirable.

## 2. The Age of Cepheids

Kippenhahn and Smith (1969) have found from cepheids in galactic clusters an empirical relation between age and period. The relation predicts with surprising probability the age of a cepheid, in spite of the fact that the massive cepheids cross the instability strip several times and hence have different ages at a given period. The spread of age is offset by the fact, that the crossing times are quite different, and that the second crossing is by far the slowest. A least square solution of the data by Kippenhahn and Smith and giving proper weight to the individual crossing times leads to (Tammann, 1969a)

$$\log t (\text{in } 10^7 \text{yr}) = 1.16 - 0.651 \log P \text{ (in d)}. \tag{1}$$

The maximum age of cepheids which are expected to outline well the spiral arms could be determined by Equation (1) and compared to the age of clusters with earliest main sequence type b2-3 ($U-B \lesssim -0^m\!.80$) or earlier, which are known to be spiral arm tracers by Becker's work. But the numerical age of clusters is critically dependent on the intrinsic colour of the brightest unevolved stars and hence on the colour excess assumed. It seems therefore safer to compare directly the periods with the spectral type of the earliest unevolved cluster members.

In Figure 1 the known cluster cepheids (Sandage and Tammann, 1969) are plotted with their periods against the earliest spectral types of their parent clusters. The four cepheids in the h + χ-association are plotted as an error box, because the spread in their periods as well as the work on the association by Wildey (1964) and Schild (1967) seem to indicate a finite formation time of the association. Also included in the diagram are WZ Sgr, SZ Tau, and Anon. Sct, which are believed to be members of the Sgr OB4-association (Tammann, 1969a), of NGC 1647 (Becker, 1969b), and of NGC 6649 (Tammann, 1969b), respectively. There is a clear relation between period and earliest spectral type. The scatter in the diagram is explainable by the non-unique relation between period and cepheid age, as already mentioned, and by observational errors in the determination of the earliest spectral types.

From Figure 1 one finds that all cepheids with $P > 20$ d and practically all cepheids with $P > 15$ d should be as young or younger than b 2-3 clusters and hence be as good spiral arm tracers as the latter. To go to cepheids with periods as low as 11.25 d (average age $= 3 \times 10^7$ yr according to Equation (1)) would apparently mean to

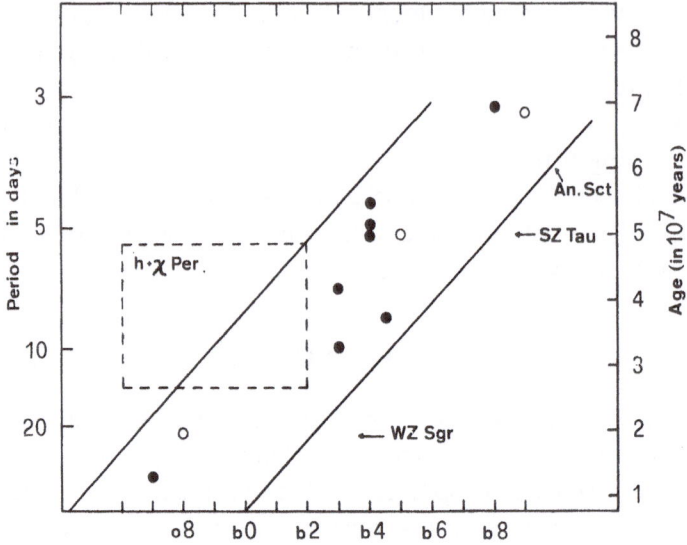

Fig. 1.    The period of cluster cepheids vs. the earliest spectral type of their parent clusters. The scale at the right side gives the age of the cepheids according to Equation (1). Four cepheids in the $h + \chi$-association fall into the dotted square. Three cepheids, whose cluster membership is less reliable, are shown as open circles.

include some cepheids as old as b3-clusters which are known not to be confined to spiral arms.

## 3. The P-L-C Relation

The *P-L-C* relation used here is:

$$M_{\langle V \rangle} = -3.534 \log P + 2.647 (\langle B^0 \rangle - \langle V^0 \rangle) = 2.469 \tag{2}$$

in the range $0.4 < \log P < 1.9$. This is essentially the same form as derived by Sandage and Tammann (1969) from 13 cepheids with known distances except for two slight modifications:

(a) The change of the pulsation constant $Q$ along a constant period line in the $(M_V, B-V)$-plane was considered by Sandage and Tammann (1969) but not taken into account. The best straight line approximation for this change follows from Christy's theory of stellar pulsation (Christy, 1968), which predicts:

$$R^{1.69} / \mathfrak{M}^{0.66} = \text{const. (for } P < 10 \text{ d)} \tag{3}$$

and

$$R^{1.83} / \mathfrak{M}^{0.78} = \text{const. (for } P > 10 \text{ d)}. \tag{4}$$

Using a mass-luminosity relation of $L \propto \mathfrak{M}^{3.3}$ (Sandage and Gratton, 1963) and the same equations for the bolometric corrections and for the $(T_e, B-V)$-relation of supergiants as in Sandage and Tammann (1968) one finds for constant period a relation of:

$$M_V = 2.647 (B - V) - \text{const}. \tag{5}$$

The coefficient of $(B-V)$ is the mean of 2.614 and 2.680, which are the correct values for $P < 10$ d and $P > 10$ d, respectively. (The value assumed before is 2.52).

(b) The mean magnitudes $\langle V \rangle$ and $\langle B \rangle$ of the cluster cepheids are taken from a list of photometric parameters, which were derived by Fourier analysis from all photoelectric observations of cepheids available (Schaltenbrand and Tammann, 1969)

The least square solution leading to Equation (2) is very little changed if EV Sct and VY Per are excluded ($0\overset{m}{.}03$ in the constant term). It is still somewhat doubtful if the parent cluster of EV Sct, NGC 6664, is real (Tammann, 1969c), and VY Per shows the greatest deviation and may not belong to the h + χ-association. The average deviation from Equation (2) is for the 13 cepheids $\pm 0\overset{m}{.}071$ (without VY Per $\pm 0\overset{m}{.}055$), which is somewhat less than in the previous paper (Sandage and Tammann, 1969).

According to the theory of the instability strip the period-luminosity relation is depending on the effective temperature $T_e$, too. This dependence has usually been expressed by the observable quantity $(\langle B \rangle - \langle V \rangle)$. It is evident that instead of $(\langle B \rangle - \langle V \rangle)$ the colour $(\langle U \rangle - \langle B \rangle)$ could be used as well, the latter being almost as sensitive for temperature changes in supergiants. It is therefore attempted here to derive a P-L-C relation in $M_V$, $\log P$, and $(\langle U \rangle - \langle B \rangle)$.

For this purpose 41 cepheids with complete UBV-photometry are taken from the list by Sandage and Tammann (1968). Reliable colour excesses are known for these mainly from the work by Kraft (1961) and Bahner *et al.* (1962); (the cepheids BY Cas, S Nor, and SV Per, which are suspected to have companions, are exluded as well as the peculiar cepheid Y Oph (Evans, 1968)). Mean magnitudes of these cepheids are taken from Schaltenbrand and Tammann (1969). The absolute magnitudes $M_{\langle V \rangle}$ are determined from Equation (2). The intrinsic colours $\langle U \rangle^0 - \langle B \rangle^0$ are derived under the assumption $E_{U-B}/E_{B-V} = 0.80$. By least square solution one then finds:

$$M_{\langle V \rangle} = -3.382 \log P + 1.834(\langle U^0 \rangle - \langle B^0 \rangle) - 1.700. \tag{6}$$

The fact that similar solutions are found if the cepheids are devided into two groups with $E_{B-V} < 0\overset{m}{.}55$ and $E_{B-V} > 0\overset{m}{.}55$, respectively, seems to prove in favour of the near correctness of the adopted ratio $E_{U-B}/E_{B-V}$. The average difference $\Delta M_{\langle V \rangle}$ between Equations (2) and (6) is for 41 cepheids $\pm 0\overset{m}{.}094$.

## 4. The Colour Excess of Cepheids

The colour excesses of relatively few cepheids are well determined by means of spectroscopic or $\Gamma$-photometric observations. Since the colour excesses enter the distance determination of cepheids it is attempted here to derive the $E_{B-V}$-values purely from photometric data.

Equations (2) and (6) can be written in the form

$$M_{\langle V \rangle} = 3.534 \log P + 2.647(\langle B \rangle - \langle V \rangle - E_{B-V}) - 2.469 \tag{7}$$

and

$$M_{\langle V \rangle} = 3.382 \log P + 1.834(\langle U \rangle - \langle B \rangle - 0.80 E_{B-V}) - 1.700. \tag{8}$$

Solving (7) and (8) for $E_{B-V}$ leads to:

$$E_{B-V} = -0.129 \log P + 2.243(\langle B \rangle - \langle V \rangle)$$
$$-1.554(\langle U \rangle - \langle B \rangle) - 0.652. \qquad (9)$$

The colour excesses $E_{B-V}$ were determined from Equation (9) for 41 cepheids with known spectroscopic $E_{B-V}$-values. A comparison of the results of the two methods is shown in Figure 2. There is no indication for a systematic difference. The mean deviation from the 45°-line amounts to $\pm 0\overset{m}{.}08$. Since the average mean error of a spectroscopically determined colour excess is about $\pm 0\overset{m}{.}03$, the colour excesses determined from Equation (9) should be good within $\pm 0\overset{m}{.}07$.

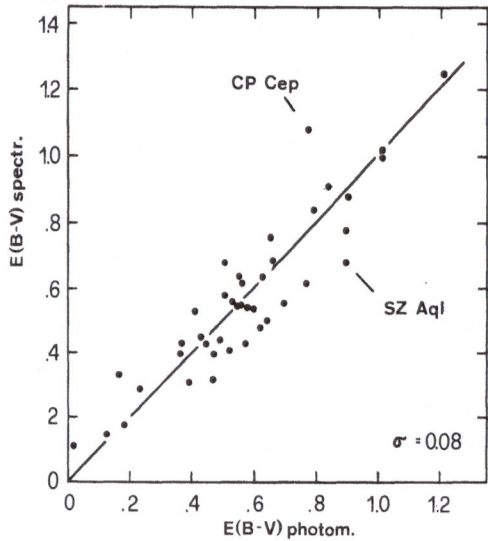

Fig. 2.   Colour excesses $E_{B-V}$ from spectroscopic or $\Gamma$-photometric observations vs. $E_{B-V}$ from
Equation (9). Two cepheids with exceptionally large deviations are indicated.
The drawn line has the slope 1.

For cepheids where spectroscopic excesses are lacking as well as $U$-magnitudes $E_{B-V}$ was taken from the list by Fernie (1967). These values are derived mainly under the assumption that cepheids have the same intrinsic colours at maximum light, which seems to be correct within about $\pm 0\overset{m}{.}10$ (Sandage and Tammann, 1968). The fact that Fernie's intrinsic colours are systematically redder by $\sim 0\overset{m}{.}06$ than those by Kraft can be neglected in the present context.

## 5. The Distances of Cepheids

From Equation (2) and an adopted absorption law $A_V = 3.0\ E_{B-V}$ the distance modulus becomes:

$$(m - M)^0 = \langle V \rangle + 3.534 \log P - 2.647(\langle B \rangle - \langle V \rangle) -$$
$$- 0.353\ E_{B-V} + 2.469. \qquad (10)$$

There are few astronomical formulae so kindly deviced by nature as this one: the coefficients of the observed quantities are the greater the higher the precision is with which they can be observed. The only quantity where an observational error could have any effect on the distance modulus is $E_{B-V}$; even here an error of $0^{m}.10$ affects the modulus by only $0^{m}.035$. The random scatter of cepheids around the adopted $P$-$L$-$C$ relation seems also to be quite small as indicated by the calibrating cepheids.

It remains the question whether systematic errors have entered the $P$-$L$-$C$ relation. The problem of the reliability of cluster distances based on the Hyades shall not be discussed here (see e.g. Sandage and Tammann, 1969), because any – quite unlikely – change of the Hyades distance would equally affect the clusters and cepheids, whose distances are to be compared here. In determining the slope of the $P$-$L$-$C$ relation, RS Pup, the only calibrating cepheid with very long period, enters with great weight. However the slope found here seems to be quite reliable because in the range $0.4 < \log P < 1.9$ it nearly coincides with the best linear fit to the ridge line of the $P$-$L$-$C$ relation derived from all well observed extragalactic cepheids (Sandage and Tammann, 1968) and agrees also very nearly with the slope Gascoigne (1969) found from the combined solution for cepheids in galactic clusters and in the Magellanic Clouds.

A list of distances from Equation (10) for cepheids with $P > 9$ d is given in Table I. It is attempted to include all galactic cepheids of Population I with sufficient observational data; however, cepheids of doubtful population and with suspected companions are excluded.

For a number of long period cepheids photoelectric photometry is missing. Distance estimates for these are taken from the list by Fernie and Hube (1968); they are not shown in Table I. The distances of cepheids in common in Table I and in the list by these authors agree reasonably well; this is somewhat surprising because they use a quite different period-luminosity relation.

## 6. Galactic Distribution of Cepheids

The distances of 32 cepheids with $P > 15$ d and of 20 cepheids with $15 > P > 11.25$ d are shown in Figure 3 projected into the galactic plane; also drawn are 14 cepheids of the long period group and 7 of the shorter period group with lacking photoelectric observations. The young clusters and the H II regions according to the latest results (Becker, 1969a) are drawn as open circles.

From Figure 3 the following conclusions can be drawn:

(a) The cepheids with $P > 15$ d either fall into the slightly extended boundaries, which are outlined as spiral arms by young cluster and H II regions, or lie at distances, where no young clusters and H II regions are known. The same holds for 25 out of 27 cepheids with $15 > P > 11.25$ d, only 2 of them fall between reliably determined spiral arms (TT Aql and Z Sct.). This result conforms with the view that long period cepheids, young clusters and the exciting stars of H II regions were formed essentially at the same places.

(b) The exception of TT Aql is not surprising. It lies mid-way between the spiral

## TABLE I

Distances of Cepheids with $P > 9d$

| (1) | (2) | (3) | (4) | (5) | (6) | (7) | (8) | (9) | (10) | (11) |
|-----|-----|-----|-----|-----|-----|-----|-----|-----|------|------|
| Cepheid | $l^{II}$ | $b^{II}$ | $\log P$ | $\langle U \rangle$ | $\langle B \rangle$ | $\langle V \rangle$ | $E_{sp}$ | $E_{ph}$ | $E_{Fernie}$ | $r$ (pc) |
| SZ Aql | 35.6 | − 2.3 | 1.234 | 11.06 | 10.09 | 8.66 | 0.68 | | | 1950 |
| TT Aql | 36.0 | − 3.1 | 1.138 | 9.44 | 8.44 | 7.14 | 0.55 | | | 1000 |
| FN Aql | 38.5 | − 3.1 | 0.977 | 10.53 | 9.62 | 8.38 | | 0.59 | | 1450 |
| RX Aur | 165.8 | − 1.3 | 1.065 | 9.33 | 8.65 | 7.68 | | 0.32 | | 1760 |
| SY Aur | 164.7 | + 2.1 | 1.006 | 10.83 | 10.12 | 9.06 | 0.44 | | | 2660 |
| YZ Aur | 167.3 | + 0.9 | 1.260 | 12.74 | 11.79 | 10.38 | | 0.87 | | 4480 |
| AN Aur | 164.9 | − 1.0 | 1.012 | 12.49 | 11.66 | 10.44 | | 0.68 | | 4010 |
| RW Cam | 144.9 | + 3.8 | 1.215 | – | 10.06 | 8.67 | | | 0.70 | 1990 |
| SS CMa | 239.2 | − 4.2 | 1.092 | 11.97 | 11.11 | 9.87 | | 0.62 | | 3510 |
| $l$ Car | 283.2 | − 7.0 | 1.551 | 6.04 | 5.00 | 3.72 | | 0.39 | | 430 |
| U Car | 289.1 | + 0.1 | 1.588 | 8.35 | 7.46 | 6.27 | | 0.43 | | 1630 |
| VY Car | 286.6 | + 1.2 | 1.278 | 9.36 | 8.58 | 7.45 | | 0.51 | | 1790 |
| WZ Car | 289.3 | − 1.2 | 1.362 | – | 10.46 | 9.29 | | | 0.33 | 4730 |
| XX Car | 291.3 | − 4.9 | 1.196 | – | 10.39 | 9.35 | | | 0.30 | 4290 |
| XY Car | 291.4 | − 3.9 | 1.095 | – | 10.50 | 9.30 | | | 0.49 | 2880 |
| XZ Car | 290.3 | − 0.8 | 1.221 | – | 9.84 | 8.59 | | | 0.55* | 1680 |
| YZ Car | 285.6 | − 1.4 | 1.259 | 10.63 | 9.83 | 8.71 | | 0.45 | | 3160 |
| AQ Car | 285.8 | − 3.3 | 0.990 | 10.44 | 9.78 | 8.84 | | 0.29 | | 2800 |
| CR Car | 285.7 | − 0.4 | 0.990 | 13.82 | 12.90 | 11.58 | | 0.75 | | 5750 |
| FI Car | 287.8 | + 0.7 | 1.129 | – | 13.24 | 11.65 | | | 0.88 | 5170 |
| FO Car | 290.5 | − 2.1 | 1.015 | – | 12.05 | 10.78 | | | 0.64 | 4440 |
| FR Car | 291.1 | + 0.6 | 1.030 | 11.65 | 10.82 | 9.68 | | 0.51 | | 3260 |
| RW Cas | 129.0 | − 4.6 | 1.170 | 11.41 | 10.46 | 9.21 | 0.41 | | | 2980 |
| RY Cas | 115.3 | − 3.3 | 1.084 | 12.34 | 11.31 | 9.94 | 0.76 | | | 2980 |
| SZ Cas | 134.8 | − 1.2 | 1.134 | 12.39 | 11.32 | 9.83 | 0.88 | | | 2546 |
| CH Cas | 112.9 | + 1.6 | 1.179 | – | 12.61 | 10.96 | | | 0.94: | 3830 |
| CY Cas | 113.9 | + 2.0 | 1.158 | – | 13.35 | 11.65 | 1.10 | | | 4640 |
| DD Cas | 116.8 | + 0.5 | 0.992 | 12.00 | 11.08 | 9.86 | 0.56 | | | 3020 |
| TX Cen | 315.2 | − 0.6 | 1.233 | 13.50 | 12.23 | 10.53 | | 1.04 | | 3130 |
| VW Cen | 307.6 | − 1.6 | 1.177 | 12.53 | 11.57 | 10.22 | | 0.73 | | 4010 |
| XX Cen | 309.5 | + 4.6 | 1.040 | 9.47 | 8.79 | 7.82 | | 0.34 | | 1790 |
| KK Cen | 294.2 | + 2.7 | 1.086 | – | 12.82 | 11.50 | | | 0.57 | 6620 |
| KN Cen | 307.8 | − 2.1 | 1.532 | 12.50 | 11.40 | 9.84 | | 0.96 | | 4430 |
| V339 Cen | 313.5 | − 0.5 | 0.976 | 10.73 | 9.90 | 8.70 | | 0.64 | | 1740 |
| CP Cep | 100.4 | + 1.1 | 1.252 | 13.53 | 12.18 | 10.54 | 1.08 | | | 3480 |
| SU Cru | 299.2 | − 0.6 | 1.109 | 12.82 | 11.56 | 9.79 | | 1.21 | | 1640 |
| X Cyg | 76.9 | − 4.3 | 1.215 | 8.43 | 7.55 | 6.40 | 0.45 | | | 970 |
| SZ Cyg | 84.4 | + 4.0 | 1.179 | 12.19 | 10.95 | 9.42 | | 0.70 | | 2250 |
| TX Cyg | 84.4 | − 2.3 | 1.168 | 12.77 | 11.37 | 9.49 | 1.25 | | | 1380 |
| VX Cyg | 82.2 | − 3.5 | 1.304 | 13.18 | 11.81 | 10.07 | | 0.96 | | 2730 |
| BZ Cyg | 84.8 | + 1.4 | 1.006 | 12.98 | 11.83 | 10.22 | 1.00 | | | 2140 |
| CD Cyg | 71.1 | + 1.4 | 1.232 | 11.33 | 10.29 | 8.97 | 0.64 | | | 2580 |
| EZ Cyg | 67.1 | + 0.6 | 1.067 | 13.70 | 12.51 | 11.06 | | 0.63 | | 4420 |

Table I (continued)

| Cepheid | $l^{II}$ | $b^{II}$ | $\log P$ | $\langle U \rangle$ | $\langle B \rangle$ | $\langle V \rangle$ | $E_{sp}$ | $E_{ph}$ | $E_{Fernie}$ | $r$ (pc) |
|---|---|---|---|---|---|---|---|---|---|---|
| β Dor | 271.7 | − 32.8 | 0.993 | 5.13 | 4.56 | 3.75 | | 0.14 | | 320 |
| ζ Gem | 195.7 | + 11.9 | 1.007 | 5.30 | 4.71 | 3.89 | 0.15 | | | 350 |
| AA Gem | 184.6 | + 2.7 | 1.053 | 11.56 | 10.81 | 9.71 | | 0.53 | | 3600 |
| Z Lac | 105.8 | − 1.6 | 1.037 | 10.33 | 9.58 | 8.43 | 0.48 | | | 1860 |
| T Mon | 203.6 | − 2.6 | 1.432 | 8.28 | 7.33 | 6.14 | 0.43 | | | 1180 |
| SV Mon | 203.7 | − 3.7 | 1.183 | 10.11 | 9.30 | 8.25 | | 0.28 | | 2540 |
| S Mus | 299.6 | − 7.5 | 0.985 | 7.52 | 6.96 | 6.14 | | 0.22 | | 920 |
| UU Mus | 296.8 | − 3.2 | 1.066 | 11.84 | 10.91 | 9.78 | | 0.29 | | 3840 |
| S Nor | 327.8 | − 5.4 | 0.989 | 8.01 | 7.36 | 6.41 | 0.21 | | | 910 |
| U Nor | 325.6 | − 0.2 | 1.102 | 11.96 | 10.83 | 9.23 | | 1.04 | | 1570 |
| SY Nor | 327.5 | − 0.7 | 1.102 | 11.72 | 10.85 | 9.50 | | 0.89 | | 2470 |
| SV Per | 162.6 | − 1.5 | 1.047 | 10.55 | 9.99 | 8.95 | 0.44 | | | 2800 |
| VX Per | 132.8 | − 3.0 | 1.037 | 11.38 | 10.52 | 9.30 | | 0.61 | | 2500 |
| X Pup | 236.1 | − 0.8 | 1.414 | 10.70 | 9.76 | 8.54 | | 0.45 | | 3350 |
| RS Pup | 252.4 | − 0.2 | 1.617 | 9.54 | 8.44 | 7.01 | | 0.65 | | 1710 |
| VZ Pup | 243.4 | − 3.3 | 1.365 | 11.56 | 10.77 | 9.61 | | 0.57 | | 5290 |
| AD Pup | 241.9 | − 0.0 | 1.133 | – | 10.97 | 9.88 | | | 0.42* | 4620 |
| AQ Pup | 246.2 | + 0.1 | 1.475 | 11.09 | 10.15 | 8.79 | 0.62 | | | 3370 |
| VY Sgr | 10.1 | − 1.1 | 1.132 | – | 13.50 | 11.53 | | | 1.30* | 2920 |
| WZ Sgr | 12.1 | − 1.3 | 1.339 | 10.56 | 9.41 | 8.02 | 0.68 | | | 1820 |
| YZ Sgr | 17.8 | − 7.1 | 0.980 | 9.13 | 8.36 | 7.34 | | 0.32 | | 1230 |
| RY Sco | 356.5 | − 3.4 | 1.308 | 10.49 | 9.44 | 7.99 | | 0.79 | | 1560 |
| V500 Sco | 359.0 | − 1.4 | 0.969 | 11.03 | 10.03 | 8.74 | | 0.55 | | 1610 |
| Y Sct | 24.0 | − 0.9 | 1.015 | 12.29 | 11.19 | 9.63 | 0.80 | | | 1810 |
| Z Sct | 26.8 | − 0.8 | 1.111 | 11.88 | 10.94 | 9.60 | | 0.74 | | 2760 |
| RU Sct | 28.2 | + 0.2 | 1.294 | 12.61 | 11.21 | 9.50 | | 0.84 | | 2200 |
| TY Sct | 28.1 | + 0.1 | 1.044 | 13.80 | 12.52 | 10.79 | | 1.11 | | 2490 |
| UZ Sct | 19.2 | − 1.5 | 1.169 | 14.78 | 13.17 | 11.28 | | 0.94 | | 3220 |
| RY Vel | 282.6 | + 1.5 | 1.449 | 10.71 | 9.71 | 8.36 | | 0.64 | | 2690 |
| RZ Vel | 262.9 | − 1.9 | 1.310 | – | 8.23 | 7.11 | | | 0.30 | 1700 |
| SV Vel | 286.0 | + 2.4 | 1.149 | – | 9.65 | 8.57 | | | 0.35 | 2650 |
| SW Vel | 266.2 | − 3.0 | 1.371 | 10.13 | 9.28 | 8.13 | | 0.44 | | 2790 |
| SX Vel | 265.5 | − 2.2 | 0.980 | – | 9.14 | 8.26 | | | 0.33 | 2240 |
| DD Vel | 271.5 | − 1.4 | 1.120 | – | 14.13 | 12.47 | | | 0.98* | 6860 |
| DR Vel | 273.2 | + 1.3 | 1.049 | – | 11.10 | 9.54 | | | 0.85 | 1810 |
| Ex Vel | 274.1 | − 2.2 | 1.122 | – | 13.31 | 11.72 | | | 0.88 | 5340 |
| SV Vul | 63.9 | + 0.3 | 1.654 | 9.85 | 8.69 | 7.22 | 0.64 | | | 1930 |

Column 1–4: name of cepheid with galactic coordinates and logarithm of period.
Column 5–6: mean magnitudes from Schaltenbrand and Tammann (1969).
Column 8:   $E_{B-V}$ from spectroscopic or $\Gamma$-photometric observations (Sandage and Tammann, 1968).
Column 9:   $E_{B-V}$ from Equation (9).
Column 10:  $E_{B-V}$ from Fernie (1967). Stars* indicate that $E_{B-V}$ was determined from the period-colour relation for galactic cepheids (Sandage and Tammann, 1968, Equation (7)).
Column 11:  the distance in pc from Equation (10).

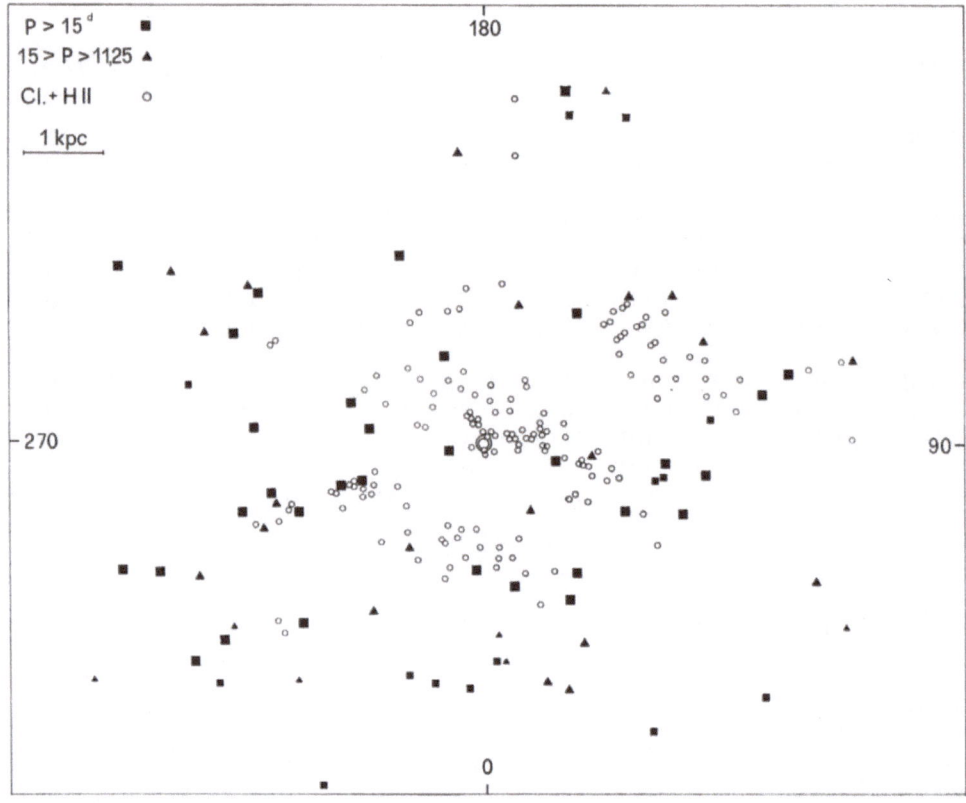

Fig. 3. The galactic distribution of young objects projected into the galactic plane. Big squares (■): cepheids with $P > 15$ days; big triangles (▲): cepheids with $15 > P > 11.25$ days. Small squares (■) and small triangles (▲): cepheids with uncertain distances in the corresponding period intervals; open circles (○): clusters with earliest type b2-3 and H II regions (compare Becker and Fenkart, this volume, p. 205, Figure 1).

arms 0 and −I (the designation of spiral arms according to Becker, (1963)). Its age is from Equation (1) $2.7 \times 10^7$ yr. If one assumes that it has moved in the galactic plane with 1.5 times the random speed of cepheids of 15 km s$^{-1}$ (Kraft and Schmidt, 1963; Oort, 1964), and that its velocity is perpendicular to the axis of a spiral arm, it has travelled during its life-time very nearly the distance between the ridge line of one of the neighbouring spiral arms and its present location.

The case of Z Sct, lying between −I and −II, is less clear. There is a slight indication from Figure 3 that the distance between these two spiral arms decreases if one goes from $l^{II} = 310°$ to 0° to 30°. This possible merging of the spiral arms in the direction of Scutum might be supported by the distribution of early Be-stars (Schmidt-Kaler, 1964). If this was correct, Z Sct would fit well into the spiral pattern.

(c) The agreement between young clusters, H II regions and long period cepheids is especially tight in the spiral arm −I; it is still quite reasonable for the spiral arms 0 and +I. The existence of the spiral arms −II and +II, barely indicated by young clusters and H II regions, seems to be confirmed by the cepheids. There is hope that the spiral arm −II can be traced quite well from cepheids, when photoelectric UBV-observations of all long period cepheids in this direction become available.

(d) If one goes to shorter periods ($P < 11.25$ d) more and more cepheids fall inbetween the spiral arms. Eventually the correlation between young clusters, HII regions and cepheids is smeared out and turns into the anti-correlation found by Fernie (1968).

## Acknowledgements

I would like to thank Prof. W. Becker for many discussions and for letting me use his unpublished list of data for the clusters and HII regions. I am also indebted to Prof. R. F. Christy and Dr. S. C. B. Gascoigne for unpublished material. I acknowledge that part of this work was supported by the Swiss National Science Foundation.

## References

Bahner, K., Hiltner, W. A., and Kraft, R. P.: 1962, *Astrophys. J. Suppl. Ser.* **6**, 319.
Becker, W.: 1963, *Z. Astrophys.* **57**, 117.
Becker, W.: 1969a, unpublished.
Becker, W.: 1969b, private communication.
Becker, W. and Fenkart, R. P.: 1963, *Z. Astrophys.* **56**, 257.
Christy, R. F.: 1968, private communication.
Evans, T. L.: 1968, *Monthly Notices Roy. Astron. Soc.* **141**, 109.
Fernie, J. D.: 1967, *Astron. J.* **72**, 422.
Fernie, J. D.: 1968, *Astron. J.* **73**, 995.
Fernie, J. D. and Hube, J. O.: 1968, *Astron. J.* **73**, 492.
Gascoigne, S. C. B.: 1969, *Monthly Notices Roy. Astron. Soc.* **146**, 1.
Kippenhahn, R. and Smith, L.: 1969, *Astron. Astrophys.* **1**, 142.
Kraft, R. P.: 1961, *Astrophys. J.* **134**, 616.
Kraft, R. P.: 1963, *Stars and Stellar Systems* **5**, 157.
Kraft, R. P. and Schmidt, M.: 1963, *Astrophys. J.* **137**, 249.
Oort, J. H.: 1964, IAU Symposium No. 20, p. 8.
Sandage, A. and Gratton, L.: 1963, *Proc. Int. School of Physics "Enrico Fermi"*, Academic Press, New York, p. 11.
Sandage, A. and Tammann, G. A.: 1968, *Astrophys. J.* **151**, 531.
Sandage, A. and Tammann, G. A.: 1969, *Astrophys. J.* **157**, 683.
Schaltenbrand, R. and Tammann, G. A.: 1969, unpublished.
Schild, R.: 1967, *Astrophys. J.* **148**, 449.
Schmidt-Kaler, Th.: 1964, *Z. Astrophys.* **58**, 217.
Tammann, G. A.: 1968, *Colloque Obs. Genève*, May 16–18, 1968, chapter 18.
Tammann, G. A.: 1969a, Cepheiden als Mitglieder offener Sternhaufen und Sternassoziationen, Habilitationsschrift, Basel.
Tammann, G. A.: 1969b, *Astron. Astrophys.* **3**, 308.
Tammann, G. A.: 1969c, IAU Inform. Bull. Variable Stars, No. 351.
Wildey, R. L.: 1964, *Astrophys. J. Suppl. Ser.* **8**, 439.

# 44. A PROGRESS REPORT ON THE CARINA SPIRAL FEATURE

B. J. BOK, A. A. HINE, and E. W. MILLER

*Steward Observatory, Tucson, Ariz., U.S.A.*

**Abstract.** The existing data on the distribution of O and B stars, of optical and radio HII sources, of HI and of cosmic dust have been assembled for the Carina-Centaurus Section of the Milky Way, $l^{II} = 265°$ to 305°. The published data have been supplemented by recent photoelectric UBV data and new photographic material. Two Working Diagrams (Figures 9 and 10) of the Carina Spiral Feature have been prepared. The Feature is sharply bounded at $l^{II} = 282°$ and again at $l^{II} = 295°$ in the range of distance from 1.5 to 6 kpc from the sun. Its outer rim is observed from the sun almost tangentially to a distance of 8 kpc from the sun. The Feature is found to bend at distances greater than 9 to 10 kpc from the sun, a result shown by both radio HI and radio HII data.

Figure 9 presents our basic data for the stellar, gas and dust components of the Feature. The O and early B stars and the HII Regions are closely associated and within 6 kpc of the sun they are concentrated in the range $285° < l^{II} < 295°$. The distribution in longitude of HI is broader and spills over on both sides of the O and B and the HII peak distributions. Long period cepheids yield a concentration similar to that shown by O and B stars and HII Regions. The visual interstellar absorption between $l^{II} = 282°$ and 295° is represented by a value $A_V = 0.5$ mag kpc$^{-1}$, or less, applicable to distances of 4 to 5 kpc. Much higher absorption is present on the outside of the Carina Spiral Feature, $265° < l^{II} < 280°$, where total visual absorptions as great as 3.5 mags. are found at distances of the order of 2 kpc. Even heavier absorption is indicated for these longitudes at 4 kpc from the sun, thus suggesting that the heavy obscuration on the outside of the Carina Spiral Feature is a phenomenon of general structural relevance (see Figures 8 and 9). Only small values of $A_V$ are found at the inside of the Spiral Feature.

The Working Diagram (Figure 10) shows that the O and B star peak and the HII peak have a width of 800 parsecs (12°) at 4 kpc from the sun, whereas the HI width is at least 1500 parsecs at the same distance. The peak of the O and B star distribution and of the HII distribution lies at about 600 parsecs (8° at 4 kpc) within the outer edge of the spiral feature. The heaviest interstellar absorption is on the outside of the Feature.

## 1. Preamble

In the early 1950s Baade and Mayall (1951) found from their studies of external galaxies that emission nebulae and O and early B stars are the best tracers of spiral arms. Morgan *et al.* (1952) accepted the challenge that these same objects should be used to trace the spiral structure of our Galaxy. Thus, the first diagrams of the local spiral structure were obtained. Their work was extended later by Becker and Fenkart (1963), Becker (1963, 1964) and by Schmidt-Kaler (1964). The presence of three local spiral features was suggested: the Perseus Arm, the Orion Arm, and the Sagittarius Arm. The Carina HII regions and OB stars were provisionally linked to the Sagittarius Arm, but they did not receive the attention they deserved. The pitch angle, the angle between the direction of an arm and the direction of rotation, for the three arms averaged about 25°.

The 21 cm line of neutral atomic hydrogen was discovered by Ewen and Purcell (1951) and the results of the first extensive studies of spiral structure based on 21 cm profiles by the Leiden and Sydney observers became available shortly after the publication of the Morgan-Sharpless-Osterbrock results. They revealed an overlying near-circular spiral pattern of HI distribution. The Leiden-Sydney hydrogen map by Oort *et al.* (1958), and more recently the studies by Westerhout (1968), Kerr

*Becker and Contopoulos (eds.), The Spiral Structure of Our Galaxy, 246–261. All Rights Reserved.*
*Copyright © 1970 by the I.A.U.*

(1969b), and Hindman (1969), show the spiral arms to be circular, with pitch angles of only 5° to 6°. These pitch angles are much smaller than those found by Becker and Schmidt-Kaler.

To reconcile the optical and radio pictures of the Galaxy, Bok (1959) suggested that the Carina Spiral Feature is part of a spiral arm separate from the Sagittarius Arm, and that this feature extends possibly through the sun to connect with the Cygnus Arm. The Orion Arm is then considered to be a second-class spiral feature, a spiral spur rather than a spiral arm. Bok's emphasis on the Carina Feature goes back to his doctoral thesis (1932). The importance of the Carina Feature was stressed especially by him in 1937 in *The Distribution of the Stars in Space* (University of Chicago Press), notably in the concluding section of the book. A Carina-Cygnus spiral arm would possess a pitch angle in good agreement with the pitch angles found from 21 cm studies.

Recent investigations in Carina have strengthened the importance of the concentrations of OB stars and interstellar gas in Carina as a major spiral feature. Becker (1956) and Bok (1956) have stressed the high concentration of O and B stars in Carina

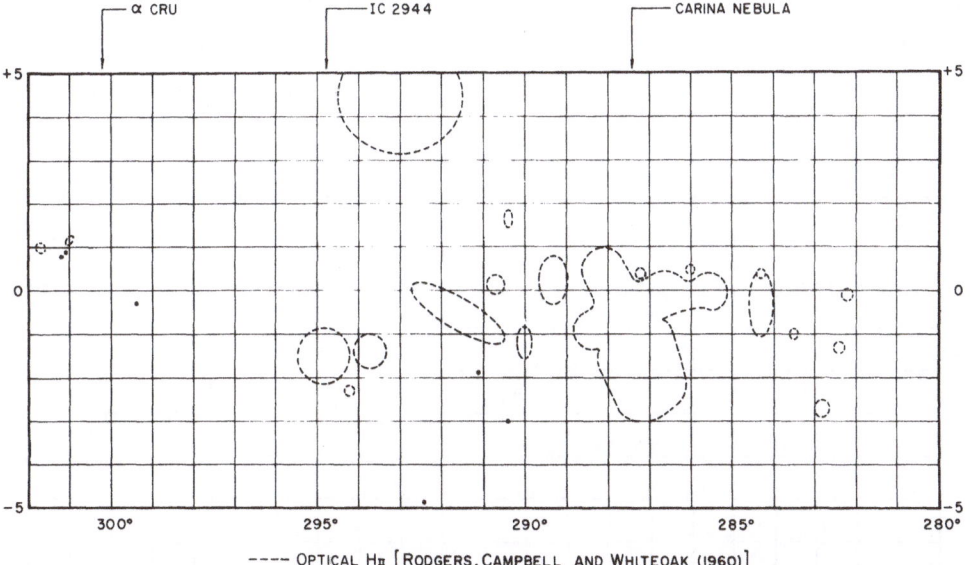

---- OPTICAL Hɪɪ [RODGERS, CAMPBELL AND WHITEOAK (1960)]

Fig. 1.   The distribution of optical H ɪɪ regions in Carina. Dotted lines indicate the optical H ɪɪ regions in the list of Rodgers, *et al.* (1960a). The coordinates are $l^{II}$, $b^{II}$. The positions of the three optically bright objects are indicated: the Carina Nebula, IC 2944, and the first magnitude binary star α Crucis. In order to show the apparent H ɪɪ distribution schematically, dimensions of the regions (as given by Rodgers *et al.* (1960a)) are used as the axes of the ellipses. Relative brightnesses are not shown. *The Mt. Stromlo H-Alpha Emission Atlas* (Rodgers *et al.* 1960b) was consulted for confirmation of the position, orientation, and shape of each object. Since the outline of RCW 54 (a large emission region between $l^{II} = 289°$ and 291°) could not be ascertained, the four concentrations within the region, given in the notes to Tables II and III of Rodgers *et al.* (1960a) are shown in the figure. Many optical H ɪɪ regions appear in the longitude range $l^{II} = 282°$ to 295°; there are very few H ɪɪ regions outside these limits. Note also that the H ɪɪ regions appear generally south of the galactic plane. The single large H ɪɪ region north of $b^{II} = 0°$ is RCW 59, which is probably a nearby object because of its relatively high galactic latitude $(+4°.5)$ and its radial velocity $(-12.2 \text{ km s}^{-1})$ as observed by Courtès (1969); its probable distance is 1.7 kpc.

with distances ranging from 1 to 4 kpc. More recent observations by Feinstein (1969) have revealed the presence of OB stars to distances of 6 kpc and Graham (1970) has observed OB stars in Carina to distances of 8–10 kpc. Extensive H II catalogues by Hoffleit (1953) and by Rodgers *et al.* (1960a) show the Carina region to be abundant in emission nebulae.

Radio studies of H II are providing additional valuable information. The continuum studies of Mathewson *et al.* (1962) and of Hill (1968) have revealed the presence of many thermal H II regions in the Carina-Centaurus section. There have been studies recently by Wilson (1969) by the newly-developed hydrogen 109 α techniques. Radial velocities have been obtained for most H II regions in this section. From these radial velocities, kinematical distances are found for many optical H II regions and also for H II regions which are too distant to be observed optically. The Wilson results clearly show that we are looking tangentially along a spiral feature to distances of 8–9 kpc and that the feature curves at distances of 9–10 kpc. Further support for the presence of a major Carina Spiral Feature is provided by the long-period cepheids. In the most recent paper on the subject, Fernie (1968) has found that the cepheids near $l^{II} = 280°$ to 290° extend to distances of 10 kpc; the Carina Feature is shown beautifully in his Figure 2.

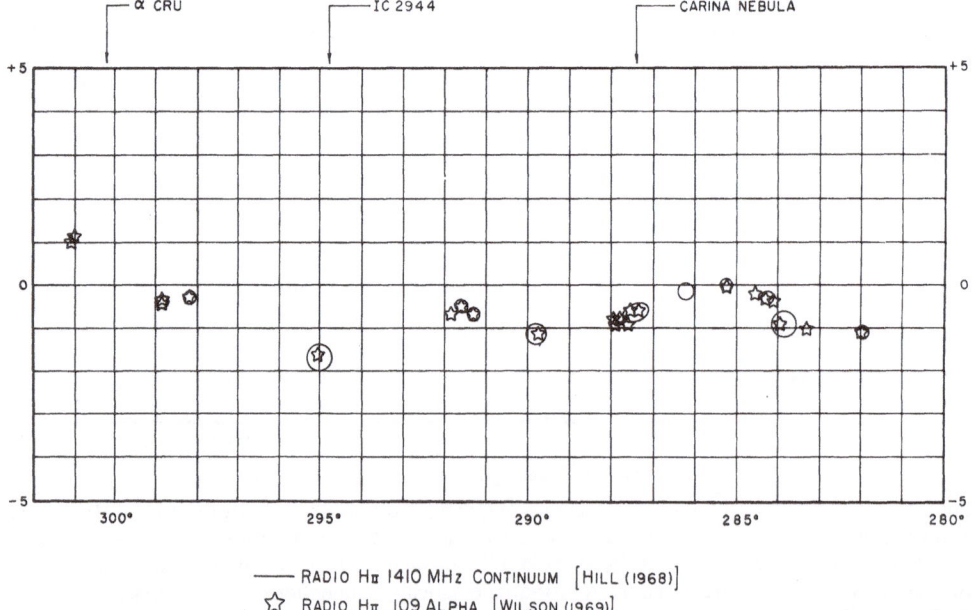

Fig. 2. The distribution of radio H II sources in Carina. Solid lines indicate schematically the 1410 MHz continuum sources observed by Hill (1968). Stars mark the positions of the H II 109α line observations of Wilson (1969). Sources which have been found to be nonthermal by Mathewson *et al.* (1962) or by Wilson (1969) are not shown. For the continuum sources, the positions and the dimensions at half intensity are taken from Hill's contour map and are necessarily approximate. Relative brightnesses are not shown. The radio H II sources, like the optical sources, are concentrated in Carina between $l^{II} = 282°$ and 295° and south of the galactic plane. The two sources between 298° and 299° are very distant, and hence are not part of the nearer concentration between 282° and 295°; they are discussed in the caption for Figure 3.

Because of the volume of evidence pointing to a major spiral feature in Carina which runs to great distances along the line of sight, we have undertaken a study in Carina of the distribution of OB stars, emission nebulae, H I and cosmic dust as a function of galactic longitude. Such a study should enable us to make cross cuts through the spiral feature at various distances and thus determine where the OB stars, emission nebulae, H I and dust are located. This paper is a progress report on our work to date.

The results are shown in terms of 9 figures each with a lengthy descriptive caption. We have brought together in one place all the data accessible to us in print, or made available to us in advance of publication. B. J. Bok and P. F. Bok are publishing their standard sequences and UBV magnitudes and colors for individual OB stars in a separate paper (Bok and Bok, 1969), but the derived absorptions are given in the figures that follow. A section entitled Conclusions and Recommendations will be found at the end of the Progress Report.

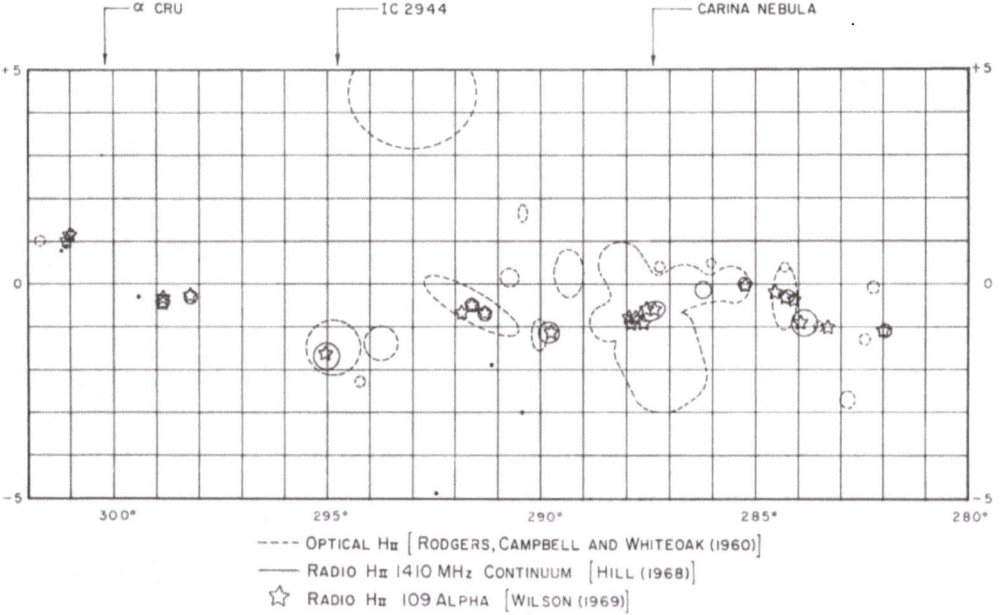

Fig. 3. Optical and radio H II. The optical and radio H II distributions of Figures 1 and 2 are here combined. Most of the radio sources correspond to optical H II regions. The sources observed between $l^{II} = 298°$ and 299° have no optical counterparts, but this is not surprising, since their kinematical distances from H II 109α data are of the order of 11–12 kpc. At this distance the Carina Spiral Feature has bent over, and we are no longer viewing the feature tangentially. An analysis of the combined radio and optical data permits us to place a boundary on the Carina Spiral Feature. If we exclude the distant sources between $l^{II} = 298°$ and 299°, we find in the figure a region of low H II concentration between $l^{II} = 295°$ and 302°. Hill's work at 1410 MHz (1968) shows that this gap extends to $l^{II} = 305°$. This lack of emission nebulae strongly suggests an interarm direction. For the other side of the Carina Spiral Feature the Mathewson et al. (1962) 1440 MHz observations show no H II sources between $l^{II} = 265°$ and 282°, again suggesting an inter-arm direction. We place the boundaries of the Carina Spiral Feature between $l^{II} = 282°$ and 295° for distances up to 10 kpc from the sun. Beyond 10 kpc the feature is no longer observed tangentially.

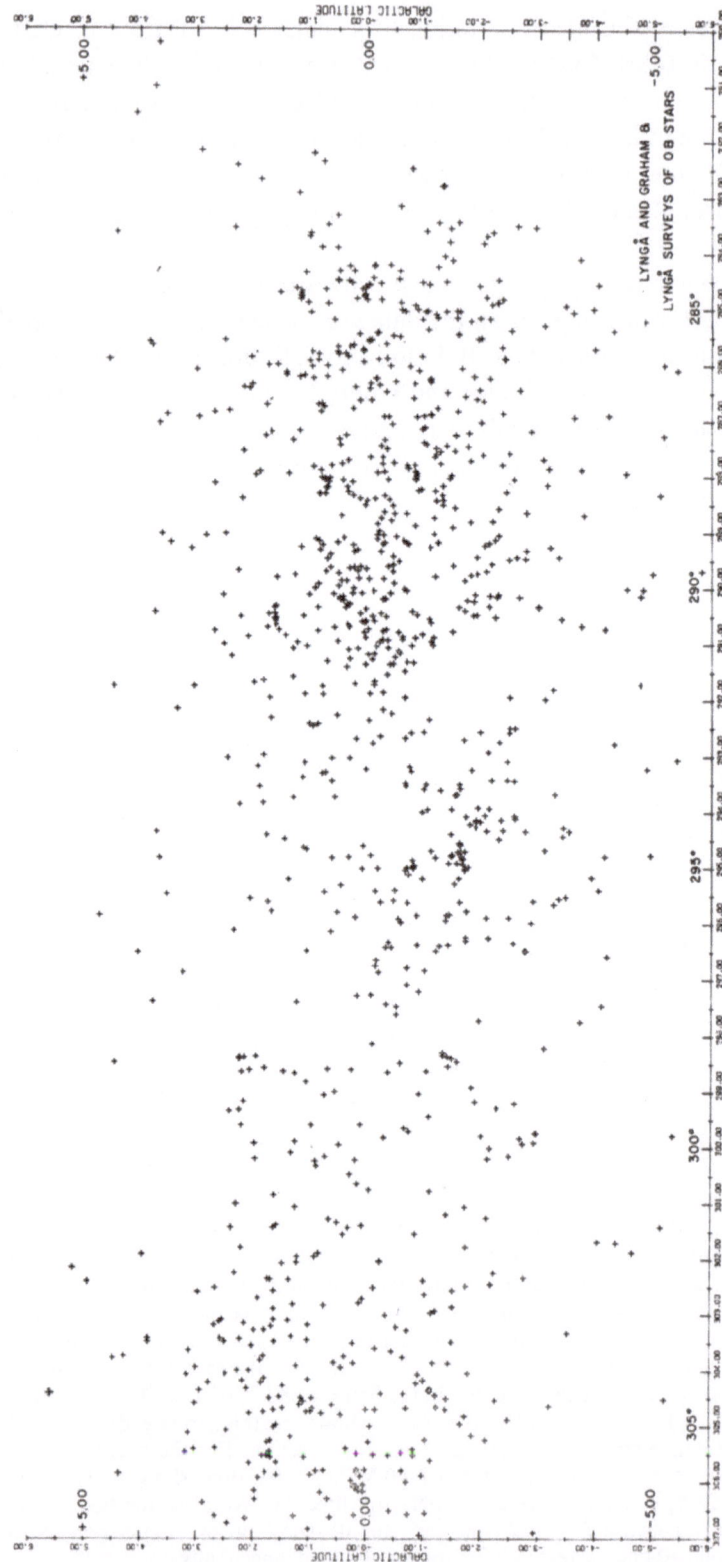

Fig. 4. Graham-Lyngå (1965) OB stars, Lyngå Carina-Centaurus (1968a) OB stars and Klare and Szeidl OB stars (1966) between $l^{II} = 280°$ and $307°$. Three extensive Catalogs of OB stars have been used in the preparation of Figure 4, which shows the distribution over the sky of known OB stars to a limiting apparent magnitude between 11 and 12. We draw attention to the following features: (1) The concentration of OB stars is greatest between $l^{II} = 284.5°$ and $l^{II} = 291°$. (2) There is a marked secondary concentration near IC 2944, at $l^{II} = 295°$, $b^{II} = -1°.5$. (3) There is evidence for the presence of an obscuring cloud near $l^{II} = 292°$, $b^{II} = -1°.5$. (4) Some faint OB stars make an appearance at positive galactic latitudes near $l^{II} = 296°$, but the numbers at comparable negative galactic latitudes are still much larger than those at positive latitudes. The computer card catalog used in the preparation of Figure 4 was copied for us by Westerlund from an unpublished catalog compiled by Lyngå.

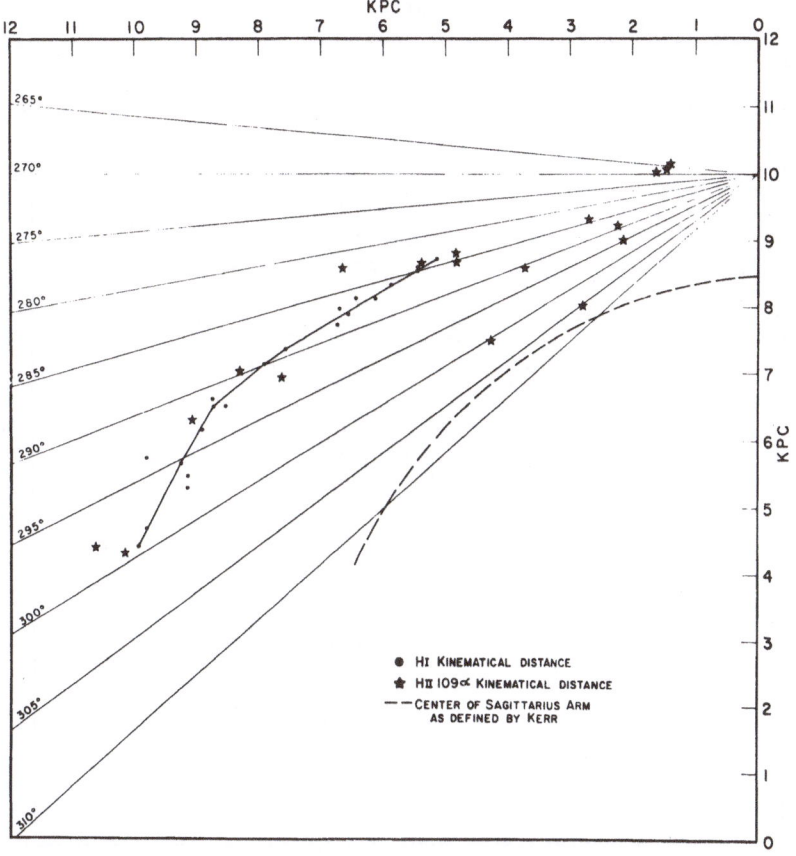

Fig. 5. The spatial distribution of HI and HII 109α sources in Carina. To obtain the kinematic distances of the HI and the HII 109α sources, the Schmidt model of the Galaxy was used (Schmidt, 1965). Kerr and Hindman (1969) provided the HI velocity profiles and Wilson (1969) provided the radial velocities for the 109α sources. Because of the poor distance resolution between $l^{II} = 270°$ and 295° and because of the double-valued distances obtained inside the solar circle, the radio HI data have not been used to define the Carina Spiral Feature within 5 kpc of the sun. At distances greater than 5 kpc, the radial velocities resulting from galactic rotation are all greater than zero and the observed radial velocities yield unique distances. The dots in the diagram represent the kinematical distances of the neutral hydrogen beyond 5 kpc. A solid line has been drawn to connect the dots and defines the spiral feature as determined from the HI. Because of deviations from circular motion of ± 10 km s$^{-1}$ a distance error of 1 kpc may result. The stars represent the kinematical distances of the HII 109α sources. At distances greater than 5 kpc from the sun the HII 109α sources agree well with the HI positions and strengthen the manner in which we have drawn the spiral feature. At distances less than 5 kpc from the sun, where double valued distances are derived, the radio HII 109α sources can be identified with optical HII sources from position correspondence. Whenever this is the case, that kinematical distance of the radio source is chosen which best agrees with the optical distance of the observed HII region. No radio HII 109α sources are observed between $l^{II} = 270°$ and 282°. The source at $l^{II} = 282°$ has a kinematical distance of 7 kpc and lies near the faint Rodgers *et al.* (1960a) optical source RCW 46. The position correspondence is not exact, and we may be looking at a close optical source near $l^{II} = 282°$ in front of a more distant source at 7 kpc from the sun. Optical counterparts to the HII 109α sources are observed between $l^{II} = 285°$ and 295° with some near and some very distant sources superimposed near $l^{II} = 290°–291°$. Beyond $l^{II} = 295°$, however, no large optical HII regions are observed. At $l^{II} = 298°$ a very distant radio HII region is observed at 11–12 kpc distance from the sun (see Figure 3). At $l^{II} = 301°$ a visible HII region is observed which corresponds to an HII 109α source. It is small and we will neglect it as insignificant; it corresponds to RCW 65. At $l^{II} = 305°$ a strong optical HII region is observed which corresponds to a strong HII 109α source. At this longitude we are probably encountering the outer regions of the Sagittarius Arm. We conclude then from the radio HII 109α data that the Carina Spiral Feature is confined between $l^{II} = 282°$ and 295°, with only extremely distant HII sources ($d > 10$ kpc) located between $l^{II} = 295°$ and 305°. At $l^{II} = 305°$ we encounter the next inner arm. Figure 5 is a basic diagram and will be used to discuss the distribution of young clusters, optical HII regions and cosmic dust.

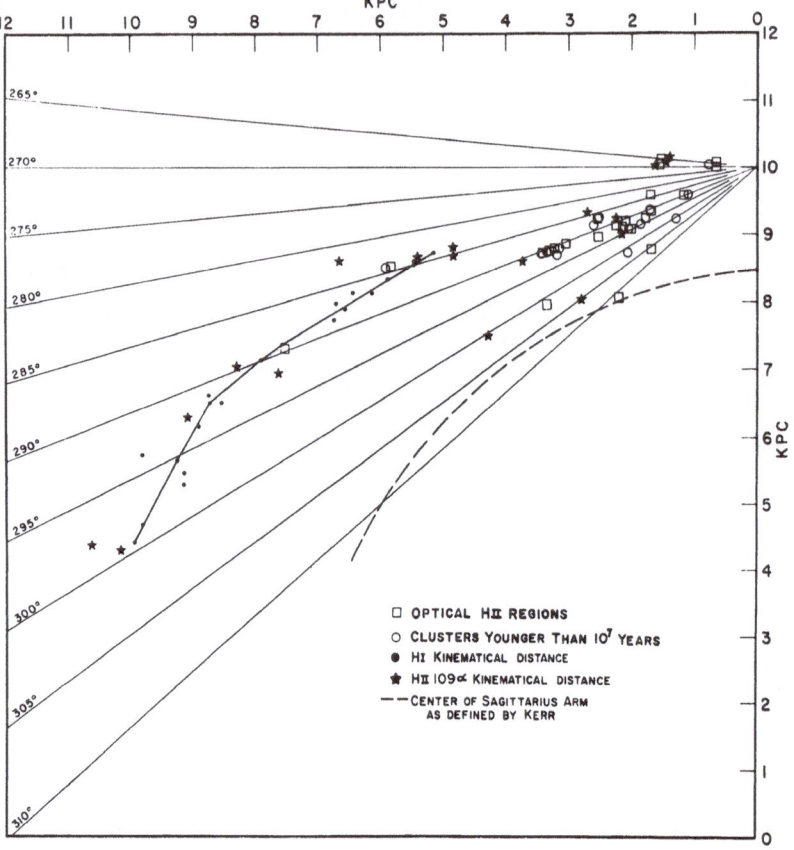

Fig. 6. Young clusters and optical H II regions in Carina. The H II distances indicated by the squares in Figure 6 come principally from Sher's (1965) paper. A few additional H II distances come from Courtès *et al.* (1968). The distances plotted are those of the exciting stars. One H II source, RCW 54 (1), at $l^{II} = 289°.8$, $b^{II} = -1°.2$ has been plotted at a distance of 7.9 kpc, Lindsey Smith's (1966) assigned distance for the exciting star. The star is a Wolf-Rayet, WN8, with an OB companion and appears in her list as LS 30. The distance is not unreasonable, as Wilson (1969) finds an H II $109\alpha$ source at the same position as the Lindsey Smith H II source at a distance of 9 kpc. Two distant optical H II regions ($> 5$ kpc) fall along the spiral feature as defined by the H I. These H II regions are Wd 1 (Westerlund, 1960) at 6 kpc and the Lindsey Smith 7.9 kpc region. They support the way in which we have drawn the more distant part of the arm. The Rodgers *et al.* (1960a) optical H II catalogue lists no H II regions between $l^{II} = 270°$ and 282°. The radio data show that the lack of H II sources between these longitudes is real and not caused by absorption. The most conspicuous optical sources in the RCW catalogue are located between $l^{II} = 285°$ and 295° and at distances between 2 and 4 kpc. These H II regions provide a means of defining the Carina Spiral Feature at distances nearer than 5 kpc from the sun. An outer envelope (towards smaller longitudes) can be drawn around the H II regions. This envelope appears to link up with the H I arm near $l^{II} = 284°$ at a distance of 5 kpc from the sun. If we draw the total spiral arm in this manner, then it is clear that the prominent H II regions are on the inside of the Carina Spiral Feature as defined mostly by H I. If the H II regions were to be placed on the outside of the spiral feature, then we would require the gaseous H I arm to suddenly jut inward near $l^{II} = 284°$, which seems rather unlikely. Additional information, presented later in the present paper, leads to the conclusion that the H II regions are indeed on the inside of the spiral feature. All but three of the young clusters indicated by circles come from Sher's work (1965). The other three clusters have been included from Lindoff's (1968) paper on ages of open clusters. There are some 140 listed clusters between $l^{II} = 265°$ and 310°. Many are older than $10^7$ years, but a number are believed to be young. Data on the young clusters would certainly improve our analysis of the spatial distribution of clusters in Carina. If one draws the near portion of the spiral feature (within 5 kpc of the sun) as the outer envelope of the H II optical and radio sources, then the clusters appear on the inside of the Carina Spiral Feature.

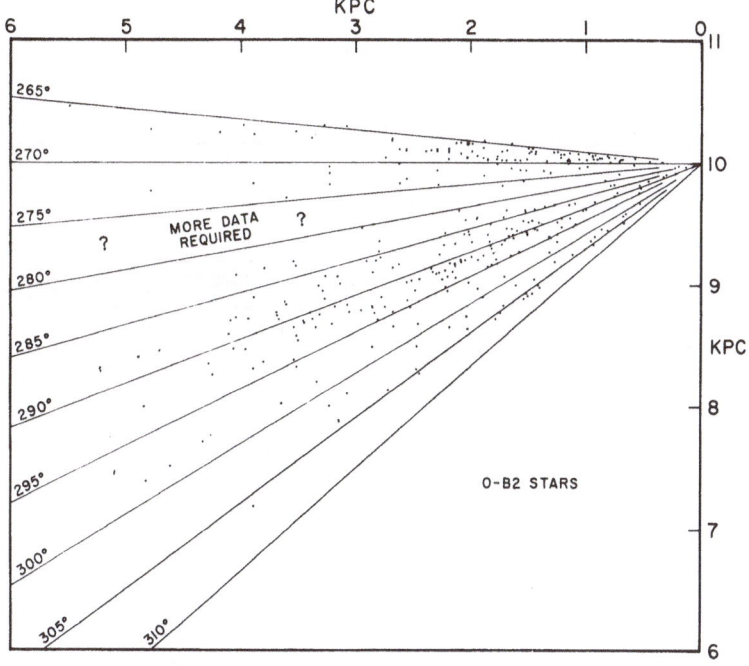

Fig. 7. O-B2 stars. O-B2 stars, which have observed UBV colors, or which have B–V colors and spectral types, are plotted in Figure 7. The distances have been corrected for reddening. All distances from the sun for the stars in this diagram have been placed on a homogeneous basis by using in the reductions the intrinsic colors and absolute magnitudes given by Schmidt-Kaler (1965). For each star the color excess is defined as the difference between the observed (B–V) color and the value of (B–V)$_0$ of Schmidt-Kaler. The total absorption $A$v is found by multiplying the (B–V) color excess by a factor 3. The absolute magnitudes for the stars in Figure 7 are assigned on the basis of the published luminosity classification. When no luminosity classification exists for the star, the star is assumed to be of class V and the corresponding absolute magnitude for a class V star is then assigned. The stars between $l^{II} = 265°$ and 273° are from the unpublished work of Velghe, who kindly made his results available to us in advance of publication. Stars at other longitudes are from the lists given in the references at the end of this paper. The most striking feature in the diagram is the apparent lack of stars between $l^{II} = 275°$ and 285°. Bok and Van Wijk (1952) observed the same feature and concluded that the deficiency in early B stars in Vela near $l^{II} = 283°$ is real and not due to absorption alone. This region is now being studied by Graham (this volume, p. 262) who has observed stars between $l^{II} = 282°$ and 292°, and by Velghe and Denoyelle (this volume, pp. 278 and 281) who are working on O and B stars in Vela. Graham finds from his H$\beta$ study of 454 OB stars between $l^{II} = 282°$ and 292° that the OB stars do not appear at large distances until near $l^{II} = 285°$. At this and greater longitudes he finds OB stars present to great distances along the spiral feature. Figure 7 supports Graham's conclusion in that a large number of O-B2 stars are observed between $l^{II} = 285°$ and $l^{II} = 295°$. We would hesitate to draw any definite conclusion from the O-B2 stars plotted in Figure 7 regarding an edge to the spiral feature between 280° and 285° as defined by O-B2 stars. Our sample of stars is too small. However, the Warner and Swasey Observatory (WSO) OB star prism survey is more homogeneous and extends to 12th magnitude. From the (WSO) OB star counts between $l^{II} = 265°$ and 305° a sharp peak is found in the star numbers between $l^{II} = 285°$ and 290° indicating that the stellar arm begins somewhere between these longitudes. Figure 7 shows a decrease in the numbers of O-B2 stars from $l^{II} = 295°$ to 305°. This decrease is also noted in the number of OB stars found between these longitudes in the Warner and Swasey Observatory survey. Absorption might be suggested as a possible cause for the drop in star numbers at $l^{II} = 295°$–305°. However, from the absorption studies we have made between $l^{II} = 295°$ and 305° and from studies by Lyngå (1968b) for distant stars between $l^{II} = 298°$ and 306°, the absorption is shown to be low in the regions not obviously affected by local obscuration. In agreement with Graham, we conclude from the OB stars that an outer edge to the Carina Spiral Feature is found near 285°, and we find an inner edge occurs near $l^{II} = 295°$.

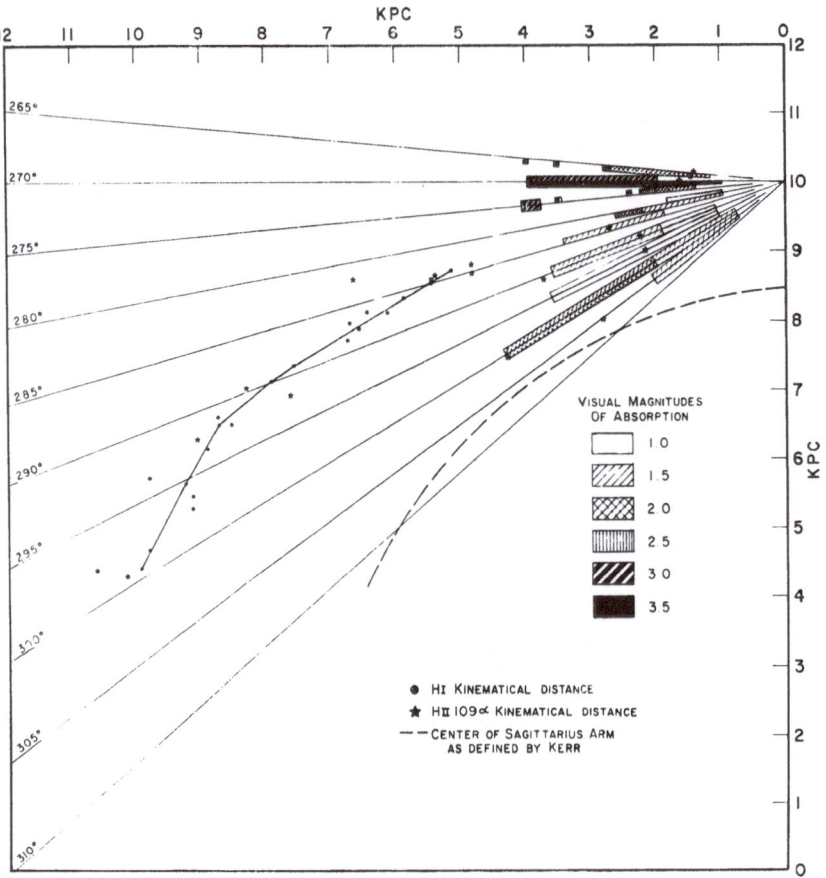

Fig. 8.    Absorption in Carina-Centaurus. The magnitudes and colors for the stars that are plotted
in Figure 7 form the basis for Figure 8. The only additional data that we have used are the results
obtained by Neckel (1967) for the range $275° < l^{II} < 280°$, where we have data for only very few stars.
High absorption values are found for the range $265° < l^{II} < 280°$ and generally low values for the
range $282° < l^{II} < 305°$; in the latter sector the average visual coefficient of absorption for the areas
not obviously affected by local obscuration is $Av = 0.5$ mag kpc$^{-1}$. The low average coefficient of
absorption for $282° < l^{II} < 305°$ renders it possible to penetrate optically to great distances, with
stars at distances of 6 to 8 pc becoming accessible to observation. The highest absorption occurs in the
range $265° < l^{II} < 275°$. Total absorptions in visual light as great as 3 to 4 magnitudes are found in
this sector. Our data give information only to a distance of 4 kpc. Without further study we cannot
say anything about the total absorptions affecting the stars at greater distances for this critical sector.
We take special note of one important structural feature: in the Carina-Centaurus section of the
Milky Way, the heaviest absorption is found just on the outside of the Carina Spiral Feature as it is
defined by OB stars, clusters and interstellar hydrogen. As of now, this does not appear to be
a local phenomenon.

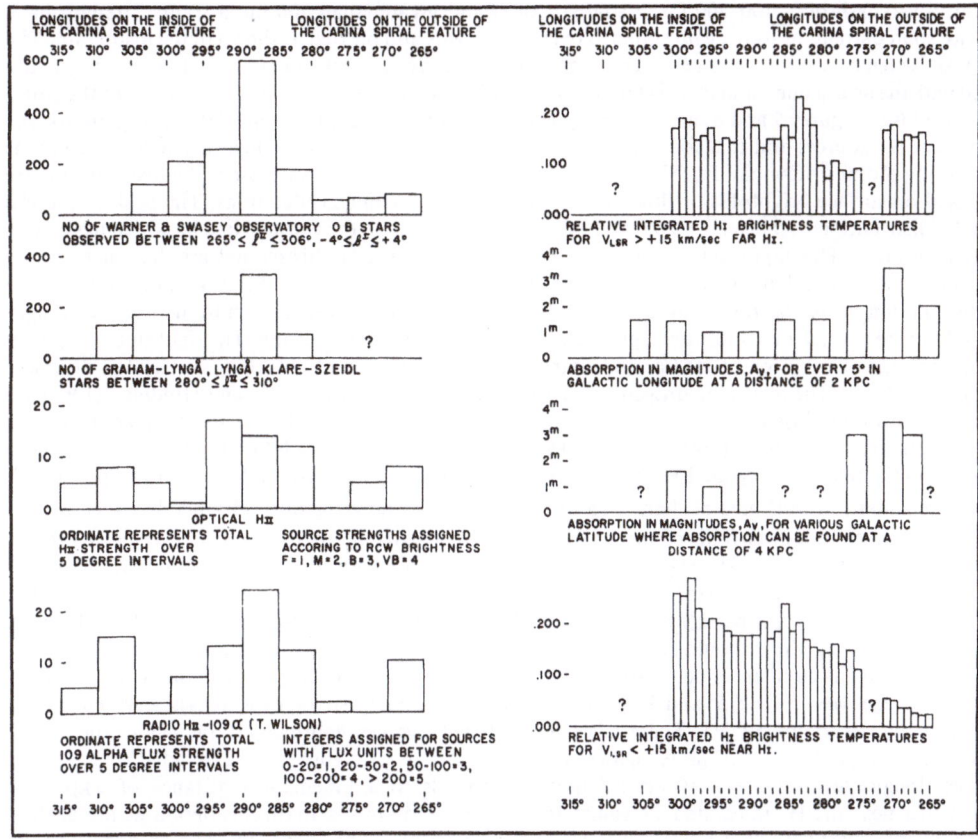

Fig. 9.   The distribution of OB stars, optical H II sources, radio H II 109α sources, H I (far), cosmic dust and H I (near). This diagram collects in a single figure all data obtained thus far. At the top, and on each half of the figure, the galactic longitudes, $l^{II}$, are indicated. The question marks indicate areas for which no data are presently available to us. The distribution of OB stars is seen to peak between $l^{II} = 285°$ and 290° in both the Warner and Swasey Observatory Survey and the combined Graham-Lyngå (1965), Lyngå (1968a), and Klare-Szeidl (1966) OB star searches. Graham has called to our attention the fact that in the Graham-Lyngå OB star search the region near the Carina Nebula was purposely omitted and that, if it had been included, the number of OB stars between 285° and 290° would have been substantially greater than appears in the figure. Warner and Swasey Observatory observations of OB stars for $l^{II} > 305°$ are available, but they are not included in this progress report. Lyngå has also published OB star data for $l^{II} > 311°$ (Lyngå, 1964) which we have not used as the data are beyond the galactic longitude limits of our Carina-Centaurus program. The optical H II distribution was determined by assigning to the Rodgers *et al.* (1960a) H II objects source strengths of 1, 2, 3, and 4 depending on their RCW assigned brightnesses. The faintest sources were assigned source strengths of 1, and the very bright H II sources were assigned a value of 4. The ordinate is the sum of the source strengths over 5° intervals of galactic longitude. The peak is seen to occur between longitudes 290° and 295°. The peak between 290° and 295° in the optical H II histogram must be judged with some care, since some faint sources may be distant. If they were close to us, they would be listed as brighter by RCW; the total source strength in 5° longitude intervals might be affected and hence the position of the peak. The Carina Nebula was assigned a source strength of only 4. It could easily be assigned a strength of 40, which would cause the optical H II to peak in the range 285°–290°. The important point is that there is a peak in the optical H II distribution between $l^{II} = 285°$ and 295° and that it agrees with the OB star peak distribution as shown in the first two diagrams of Figure 9. The distribution of H II 109α sources studied by Wilson (1969) is shown in the last diagram on the left side of Figure 9. Again source strengths were assigned on the basis of Wilson's published fluxes. 0–20 flux units (f.u.) were assigned a source strength of 1, 20–50 f.u. were assigned a source strength of 2, 50–100

f.u. a strength of 3, 100–200 f.u. a strength of 4 and greater than 200 f.u. a source strength of 5. The source strengths were again summed over 5° intervals and plotted as shown in the histogram. The peak is seen to occur between $l^{II} = 285°$ and 290°. Some very distant sources will obviously be included with the nearer ones, and there is bound to be an effect in this histogram of the curvature of the spiral feature (see Figures 5 and 6) at distances greater than 9 kpc. A comparison of the histogram for the optical H II sources and that for the radio H II 109α sources in Figure 9 shows the two distributions peaking at different longitudes. The discrepancy is not of great importance, since the distributions are based on surface brightness or flux without any distance (size) considerations. The peak in the H II 109α distribution between $l^{II} = 285°$ and 290° is primarily due to the Carina Nebula, which has a large H II 109α flux. The important conclusions to be drawn from the H II histograms are that the emission nebulae reach a peak between $l^{II} = 285°$ and 295° and that on either side of this longitude interval the distribution drops sharply. Between $l^{II} = 305°$ and 310° the H II sources begin to increase in strength and number, which may indicate a spiral feature closer to the galactic center. The right side of Figure 9 presents the distribution of H I (far and near) and the distribution of absorbing material at distances of 2 and 4 kpc. The neutral hydrogen profiles which we obtained from Kerr and Hindman (1969) are all for positions close to $b^{II} = 0$. These profiles were simply integrated by means of a planimeter and the planimeter recordings plotted as a function of longitude. The first plot on the right side of Figure 9 shows the relative integrated H I brightness temperatures for $V_{LSR}$ (the velocity with respect to the local standard of rest) greater than $+15$ km s$^{-1}$, $\int_{15}^{\infty} T_b dv$. The value of $+15$ km s$^{-1}$ was chosen in order to place the lower limit of the integration well away from any local H I. The velocity profiles reveal a large amount of H I near the solar circle ($V_{LSR} = 0$) between $l^{II} = 265°$ and 295°, and we wanted to avoid an unnecessary mixing of near and distant H I concentrations. $V_{LSR} = +15$ km s$^{-1}$ corresponds to a kinematical distance of 5–6 kpc at $l^{II} = 280°$ for strictly circular velocities; smaller values of $V_{LSR}$ should refer generally to H I concentrations within 5 kpc, unless marked deviations from circular velocity are present. The striking feature in the diagram is the low integrated $T_b$ between $l^{II} = 275°$ and 280°. At 281° the integrated $T_b$ suddenly increases, as we encounter a large concentration of H I. These large integrated $T_b$ values continue all the way to the end of our H I data at $l^{II} = 300°$. The H I between 265° and 271° is probably local. The radial velocity over these longitudes averages $+10$ km s$^{-1}$ and corresponds to a kinematical distance of 2 kpc. We consider next the H I integrated $T_b$ values for $V_{LSR} < +15$ km s$^{-1}$. In this histogram we are dealing for the most part with neutral hydrogen which is local, within 3 kpc of the sun. The broadness of the longitude histograms for H I is very striking when a comparison is made with the histograms for the OB stars and for H II. The question marks in the far H I and near H I diagrams are at longitudes where we have no H I profiles. Kerr has velocity profiles for these longitudes, but we have not used them in this progress report. The observed longitude distribution of absorbing material has been plotted for distances of 2 and 4 kpc from the sun. The diagrams refer to average values of the visual absorption, $A_V$, for areas within 3° of the galactic equator which are free from easily recognized irregular local obscuration. Both diagrams show a small amount of absorption in Carina between $l^{II} = 280°$ and 305°. The maximum amount of absorption occurs between $l^{II} = 265°$ and 275°, where it reaches 3.5 magnitudes at some longitudes.

## 2. Conclusions and Recommendations

A. CONCLUSIONS

The primary conclusion of the present investigation for the Carina-Centaurus Section between $l^{II} = 280°$ and $l^{II} = 305°$ is that we are observing at these longitudes lengthwise along a major Spiral Feature, stretching in distance between 1 kpc and 6 kpc from the sun, and probably beyond.

The OB stars and the H II regions, optical and radio, exhibit a sharply-bounded Feature in the range 283° to 295°. Our results are in full agreement with those presented at this Symposium by Graham, who finds a sharp straight edge at $l^{II} = 285°$ for the distribution in depth of the OB stars with distances between 3 kpc and 9 kpc

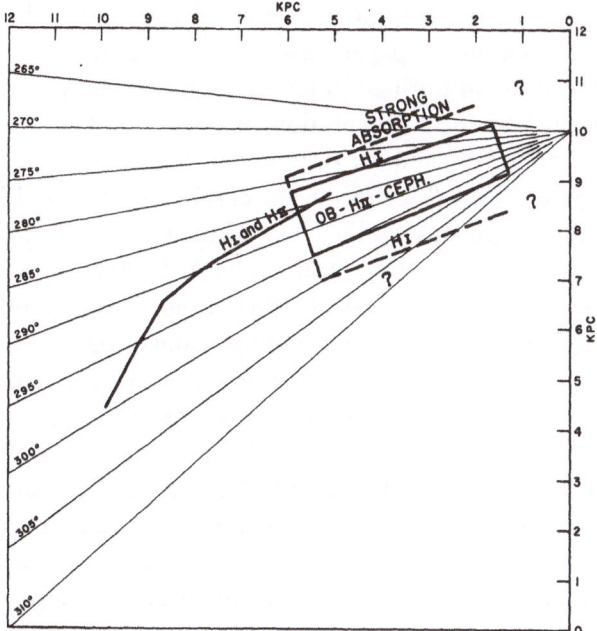

Fig. 10.    1969 working diagram of the Carina spiral feature.

from the sun. Graham finds no super-giant OB stars in the range $282° < l^{II} < 285°$. In a current, as yet unpublished, survey of the distribution in longitude of emission-line stars, Wray and Wackerling find that their distribution exhibits a peak for $285° < l^{II} < 295°$. The H II regions mimic the principal features of the distributions in longitude of the OB stars – as is to be expected – but the neutral hydrogen spills over on both sides of the limited longitude range of the H II concentrations, especially within 4 kpc of the sun. The classical cepheids, studied by Fernie (1968), exhibit precisely the same pattern as the OB stars; this is shown most clearly by Fernie's Figure 2. The density of absorbing matter increases very markedly on the outside of the edge near $l^{II} = 283°$, with very high values of $A_V$ found for the range $265° < l^{II} < 275°$.

We have noted above that the Spiral Feature appears to be broader in H I than in H II, in OB stars or in cepheids. A 1969 working diagram of the Carina Spiral Feature is shown in Figure 10. At a distance of 4 kpc from the sun, the OB and H II peaks in the longitude distribution are about 12° in width, which implies a linear width of the spiral feature of about 800 pcs. The H I feature has about twice the width, 1500 pcs across, more or less. The peak of the H II and OB star distribution is approximately 8° within the outer edge of the spiral feature shown in H I, hence within 600 pcs of the outer boundary of the H I spiral feature.

We note further that the Spiral Feature in Carina lies surprisingly far below the traditional galactic circle. OB stars and H II regions are found in greatest concentration in the range $-2° < b^{II} < 0°$. Kerr (1969a) has found that the greatest H I concentration in Carina is well below the plane, near $b^{II} = -1°$ to $-1.5°$ for $l^{II} = 265°$ to 310°. Unfortunately the data available to us for the H I profiles near $b^{II} = -1.5°$ are very scanty. The consequences of the peculiar latitude distribution of the spiral tracers are fully discussed in Dr. Graham's paper (this volume, p. 262).

The general absorption in Carina-Centaurus is quite small, at least for the regions not obviously affected by local obscuration. $A_v = 0.5$ mag kpc$^{-1}$ is a good average for the range $280° < l^{II} < 300°$ and for distances to 4 kpc from the sun. The heaviest absorption is found at $265° < l^{II} < 275°$ on the *outside* of our Spiral Feature. Some of the heavy absorption originates within 1 kpc of the sun, but there is good evidence for further heavy absorption setting in at distances between 3 and 4 kpc from the sun. Hence the heavy absorption on the *outside* of the Carina Spiral Feature is not a local phenomenon. However, it is not beyond the realm of possibilities that this heavy concentration of cosmic dust is really at the inside of the next spiral arm, visible in Puppis and Vela! This is a question beyond the scope of our present study.

We refer throughout our Paper to the *Carina Spiral Feature* rather than to the *Carina Spiral Arm*. The Feature that we have described may either be a part of a major spiral arm, or it may represent a very strong link between the Sagittarius and Orion Arms. We shall require studies comparable to ours for several other sections of galactic longitude before we can hope to fit the Carina Spiral Feature into an overall pattern.

## B. RECOMMENDATIONS

In the course of our researches on the Carina Spiral Feature, we have found several gaps in our basic observational data which, if possible, should be filled in the years to come. We list six projects that seem most in need of attention.

### (a) H I *Survey*

We need most urgently a high resolution 21 cm survey for each degree of longitude in the range $265° < l^{II} < 310°$, with profiles for each degree of latitude between $b^{II} = 0°$ and $\pm 5°$.

### (b) *A Search for Faint OB Stars*

Since adequate UBV sequences are now available to $V = 15$, we are undertaking the search for OB stars in the range $12 < V < 15$, possibly to somewhat fainter limits. The techniques are quite straightforward (Bok, 1966) and simple to apply. Not only will we thus obtain better data on the stellar distribution of young stars in the Carina Spiral Feature, but improved information on interstellar absorption should become available through such studies. We either have, or can readily obtain, the photographic plates required for these studies.

### (c) *Open Clusters*

There are approximately 140 open clusters listed for the Carina-Centaurus section, of which so far only 20 have been studied in some detail. There are several clusters with probable ages of the order of $10^7$ years or less that await further study. Distance determinations for young clusters with an associated H II region would assist greatly in the study of the distances of the H II regions.

(d) *Optical Radial Velocities*

With image tubes and other specialized auxiliary apparatus tested and available, it becomes imperative to measure optical radial velocities for a large sample of OB stars and clusters and for all optical H II regions. There is at present little new information available to supplement the fine analysis of stellar and nebular radial velocities carried out some years ago by Feast and Shuttleworth (1965). Spiral theory demands as much information as is obtainable about differences of velocity between stars and gas. A useful by-product of optical radial velocity data for H II regions will be that these should enable us to list without doubt whether or not a given H II region shown optically is the same as one in the same direction observed by the radio H II 109 α-techniques. High dispersion studies of interstellar absorption lines will obviously be of great importance.

(e) *Polarization*

We have as yet only very incomplete information on the interstellar polarization for the Carina-Centaurus section. The new data by Mathewson, and by Klare and Neckel reported in this volume should prove very helpful but much work remains to be done.

(f) *Long-Period Cepheids*

The Harvard and Leiden variable star surveys have yielded very useful information on the long-period cepheid variables in this section, but the interpretation of the data in relation to spiral structure is confused by the statistical inhomogeneity of the searches made to date. As Kraft (1965) has stressed, the long-period cepheids show great promise as optical spiral tracers for large distances from the sun. Color studies, comparable to those of Olmsted (1966), should be undertaken to check on the interstellar absorption to distances up to 10 kpc from the sun.

The Carina-Centaurus section of the Southern Milky Way is rich and ready for harvest.

## Acknowledgements

We wish to acknowledge advice and assistance received from many individuals and organizations. The financial backing for the Project has come from the National Science Foundation, Grant No. GP-7882. We much appreciate the cooperation received from Dr. V. M. Blanco and the Staff of the Cerro Tololo Interamerican Observatory. Following the Chilean visit from the Boks, Dr. J A. Graham and Dr. R. E. White rounded out some of the incomplete photoelectric and photographic programs. We should record that we have profited greatly from our continued close collaboration with Dr. Graham.

Several colleagues made new material available to us in advance of publication. We wish in this connection to express our gratitude to the following individual astronomers:

B. F. Burke, W. Buscombe, G. Courtès, T. Denoyelle, A. Feinstein, S. Garzoli, D. Goniadski, J. A. Graham, J. V. Hindman, D. Hoffleit, P. M. Kennedy, F. J. Kerr, G. R. Knapp, S. L. Knapp, C. C. Lin, G. Lyngå, S. W. McCuskey, D. S. Mathewson,

P. G. Mezger, G. Monnet, W. W. Roberts Jr., N. Sanduleak, D. Sher, C. B. Stephenson, C. M. Varsavsky, A. G. Velghe, L. R. Wackerling, G. Westerhout, B. Westerlund, T. L. Wilson, J. D. Wray, C. Yuan.

In conclusion we should mention the persons at Steward Observatory who have been deeply involved in the Project, Mrs. Lynn Glaspey, K. Ebisch, D. B. Daer, and E. D. Howell.

## References

Baade, W. and Mayall, N. U.: 1951, *Problems of Cosmical Aerodynamics* Central Air Documents Office, Dayton, Ohio, p. 165.
Becker, W.: 1956, *Vistas in Astronomy* **2**, 1515.
Becker, W.: 1963, *Z. Astrophys.* **57**, 117.
Becker, W.: 1964, *Z. Astrophys.* **58**, 202.
Becker, W. and Fenkart, R.: 1963, *Z. Astrophys.* **56**, 257.
Beer, A.: 1961, *Monthly Notices Roy. Astron. Soc.* **121**, 191.
Bok, B. J.: 1932, *Harvard Repr.* No. 77.
Bok, B. J.: 1956, *Vistas in Astronomy* **2**, 1522.
Bok, B. J.: 1959, *Observatory* **79**, 58.
Bok, B. J.: 1966, IAU Symposium No. 24, p. 228.
Bok, B. J. and P. F.: 1960, *Monthly Notices Roy. Astron. Soc.* **121**, 531.
Bok, B. J. and P. F.: 1969, *Astron. J.* **74**, 1125.
Bok, B. J. and van Wijk, U.: 1952, *Astron. J.* **57**, 213.
Courtès, G.: 1969, private communication.
Courtès, G., Georgelin, Y. P. and Y. M., Monnet, G., and Pourcelot, A.: 1968, in *Interstellar Ionized Hydrogen* (ed. by Y. Terzian), Benjamin, New York, p. 571.
Cousins, A. W. J. and Stoy, R. H.: 1963, *Roy. Obs. Bull.*, No. 64.
Ewen, H. I. and Purcell, E. M.: 1951, *Nature* **168**, 356.
Feast, M. W., Stoy, R. H., Thackeray, A. D., and Wesselink, A. J.: 1961, *Monthly Notices Roy. Astron. Soc.* **122**, 239.
Feast, M. W. and Shuttleworth, M.: 1965, *Monthly Notices Roy. Astron. Soc.* **130**, 245, and corrigenda **134**, 107.
Feinstein, A.: 1969, *Monthly Notices Roy. Astron. Soc.* **143**, 273.
Fernie, J. D.: 1968, *Astron. J.* **73**, 995.
Graham, J. A.: 1970, IAU Symposium No. 38, p. 262.
Graham, J. A. and Lyngå, G.: 1965, *Mem. Mt Stromlo Obs.*, No. 18.
Hill, E. R.: 1968, *Australian J. Phys.* **21**, 735.
Hindman, J. V.: 1969, in preparation.
Hoffleit, D.: 1953, *Harvard Ann.* **119**, 37.
Kerr, F. J.: 1969a, *Ann. Rev. Astron. Astrophys.* **7**, 39.
Kerr, F. J.: 1969b, *Austr. J. Phys. Astrophys. Suppl.* **9**, 1.
Kerr, F. J. and Hindman, J. V.: 1969, in preparation.
Klare, G. and Szeidl, B.: 1966, *Veröff. Landessternwarte Heidelberg-Königstuhl* **18**, 9.
Kraft, R. R.: 1965, *Stars and Stellar Systems* **5**, 157.
Lindoff, U.: 1968, *Ark. Astron.* **5**, 1.
Lodèn, L. O.: 1968, *Ark. Astron.* **5**, 161.
Lyngå, G.: 1964, *Medd. Lund Obs.*, Series II, No. 39.
Lyngå, G.: 1968a, *Ark. Astron.* **5**, 161.
Lyngå, G.: 1968b, *Proc. Astron. Soc. Austr.* **1**, 92.
Mathewson, D. S., Healey, J. R., and Rome, J. M.: 1962, *Austr. J. Phys.* **15**, 354.
Morgan, W. W., Sharpless, S., and Osterbrock, D. E.: 1952, *Astron. J.* **57**, 3.
Neckel, T.: 1967, *Veröff. Landessternwarte Heidelberg-Königstuhl* **19**.
Olmsted, M.: 1966, *Astron. J.* **71**, 916.
Oort, J. H., Kerr, F. J., and Westerhout, G.: 1958, *Monthly Notices Roy. Astron. Soc.* **118**, 379.
Rodgers, A. W., Campbell, C. T., and Whiteoak, J. B.: 1960a, *Monthly Notices Roy. Astron. Soc.* **121**, 103.

Rodgers, A. W., Campbell, C. T., Whiteoak, J. B., Bailey, H. H., and Hunt, V. O.: 1960b, *An Atlas of H-Alpha Emission in the Southern Milky Way*, Mt Stromlo Obs., Australian National Univ., Canberra.

Schmidt, M.: 1965, *Stars and Stellar Systems* **5**, 513.

Schmidt-Kaler, T.: 1964, *Z. Astrophys.* **58**, 217.

Schmidt-Kaler, T.: 1965, *Landolt-Börnstein, New Series* **1**, 284.

Sher, D.: 1965, *Quart. J. Roy. Astron. Soc.* **6**, 299.

Smith, L. F.: 1966, Ph. D. Thesis, Australian National University.

Velghe, A. G.: 1969, unpublished.

Westerhout, G.: 1968, *Maryland-Green Bank Galactic 21 cm Line Survey*, 2nd ed., Univ. of Maryland.

Westerlund, B. E.: 1960, *Ark. Astron.* **2**, 419.

Wilson, T. L.: 1969, Ph.D. Thesis, M.I.T.

# 45. THE SPACE DISTRIBUTION OF THE OB-TYPE STARS IN CARINA

J. A. GRAHAM

*Cerro Tololo Inter-American Observatory, La Serena, Chile\**

**Abstract.** Distances have been determined for 436 OB-type stars in Carina. A sharp outside edge is found to the OB star distribution with respect to the center of the Galaxy. The layer of OB stars follows the galactic plane to distances of the order of 3 kpc, but at distances greater than 4 kpc it appears to bend away from the plane to negative latitudes by 2° or 3° out to distances of the order of 10 kpc.

The Carina section of the Milky Way is at present perhaps the most promising of all fields for the optical study of a single large scale feature of the Galaxy. An increasing amount of evidence confirms that the very many young stars of population I are spread out over distances of several kiloparsecs along the line of sight. Young open clusters and associations, H II regions, OB-type stars, cepheid variables and Wolf-Rayet stars are all found here over a very wide range of apparent magnitude. On the other hand, the interstellar absorption is not, on the average, very heavy and it does not appear to increase rapidly with distance. This contrasts with the situation in most other directions in the galactic plane where there are strong concentrations of population I and it is this characteristic more than any other which allows us to study the Carina region optically over great distances with the presently available observational techniques.

In this short report, the preliminary results of a study of the OB star distribution are reported. The observing list was the objective prism survey made in this part of the sky by Graham and Lyngå (1965). In this publication 454 stars of type OB are listed. They are mainly stars with apparent magnitudes between 8.5 and 11.5 and are spread over an area of approximately 75 sq. deg. centered on the star $\eta$ Carinae. Because of crowding problems, the stars contained in several young open clusters are not included in the Graham-Lyngå survey. Magnitudes and colors for the OB survey stars have now been published. These were measured with the Leiden Observatory's 5-color photoelectric photometer using the system of Walraven and Walraven (1960). On the basis of the intrinsic color relations published by Walraven (1966), these data have been used to derive absorption corrected visual magnitudes. A ratio of total to selective absorption corresponding to 3.2 on the UBV system is assumed. Absolute magnitudes are derived mainly from the H$\beta$ photometry which has recently been obtained at Cerro Tololo Inter-American Observatory and is now being prepared for publication. The calibration relation for converting the observed H$\beta$ indices into absolute magnitudes is the mean of the two relations given by Fernie (1965) and by Graham (1967). For the spectral type OB, there is good agreement with the new calibration

\* Operated by the Association of Universities for Research in Astronomy, Inc., under contract with the National Science Foundation.

relation reported at this symposium by Crawford. Using the absolute magnitudes derived in this way, distance moduli and distances are then computed. The main advantage of the Hβ method lies in the simple yet efficient technique used to obtain a limited amount of information for a large number of stars. By nature of its derivation, the Hβ index is independent of both atmospheric and interstellar reddening. The observation and reduction procedure is very straightforward although the uncertainties of the absolute magnitudes do place a limit on the utility of the method. The contributors to the scatter in absolute magnitude include Hβ line emission (even though the Hδ surveys of The (1966) and Wray (1967) identify nearly all the Be type stars in the field) as well as stellar multiplicity rotation and evolutionary effects. It is difficult to see how the uncertainty of an absolute magnitude derived in this way can be appreciably reduced from its present value of about 0.6 magnitudes (standard error) without a detailed study of every program star to assess the contribution of each of the above effects. A full discussion of these errors will be given in the paper now in preparation.

In the preliminary study of the overall distribution of the present sample of OB stars, two diagrams have been made. Distance is plotted in the first diagram against galactic longitude and in the second diagram against distance, z, from the new galactic plane. The longitude (new) range of the OB star survey is 282° to 292° and the many stars with distances of the order of 2 kpc are fairly evenly distributed over the whole longitude range. On the other hand, the stars with distances greater than 5 kpc are only found in the longitude sector 285° to 292°. The simplest and most likely interpretation of the results is that we are seeing a real outside edge to the OB star concentration in this direction. The edge is apparently in the direction 287°. There is a suggestion that a tangential point may be observed at longitude 283°.5 distance 1.8 kpc. This result is still uncertain and needs confirmation. There appears to be a less certain extension of this edge out to 10 kpc. Both Hβ photometry and spectra agree in showing that several of the stars which have been observed must have distances of this order. We seem to be slightly inside this edge and therefore its distance from the sun varies with galactic longitude. If, as seems likely, the normal absorption law is valid, then there is the following relation between galactic longitude and the distance to the edge:

| $l^{II}$ | d(kpc) |
|------|--------|
| 283.5 | 1.8 |
| 284.0 | 2.0 |
| 284.5 | 4.0 |
| 285.0 | 5.0 |
| 285.5 | 6.0 |

With the second diagram, in which the distance, z, from the galactic plane is plotted against the distance from the sun for each star, a number of conclusions can be drawn about the distribution of the stars perpendicular to the galactic plane. At distances less than 4 kpc, the stars scatter more or less symmetrically about the galactic plane with a dispersion |z| of about 40 pc. The z dispersion of the necessarily young sample of supergiant stars is significantly less than the $z$ dispersion of the sample of less luminous stars which contains stars several times as old. At distances greater than 4 kpc the data are consistent with a general bending or distortion of the galactic plane by 2° or 3° to negative latitudes out to distances of the order of 10 kpc. In the region between longitudes 288° and 291° there is some support for this picture from the 21 cm Hɪ data presented by Kerr at this symposium. Plans are now being made to obtain radial velocities for a sample of these distant stars in order to make detailed comparisons with the velocity of the neutral hydrogen in this part of the sky.

## References

Fernie, J. D.: 1965, *Astron. J.* **70**, 575.
Graham, J. A.: 1967, *Monthly Notices Roy. Astron. Soc.* **135**, 377.
Graham, J. A. and Lyngå, G.: 1965, *Mem. Mt Stromlo Obs.* **4**, 16.
The, Pik-Sin: 1966, *Contr. Bosscha Obs.*, **35**.
Walraven, T.: 1966, IAU Symposium No. 24, p. 274.
Walraven, T. and Walraven, J. H.: 1960, *Bull. Astron. Inst. Netherl.* **15**, 67.
Wray, J. R.: 1967, Ph.D. Thesis, Northwestern University.

# 46. PHOTOMETRIC INVESTIGATION OF THE
# ASSOCIATION CAR OB 2

W. SEGGEWISS

*Observatorium Hoher List der Universitäts-Sternwarte Bonn, Germany*

**Abstract.** A photometric investigation of the association Car OB 2 was carried out with a photo-electric standard sequence observed with the ESO photometric telescope and plates taken with the ADH telescope of Boyden Observatory. The distance of Car OB 2 is 2.04 kpc, the color excess $E_{B-V}$ is $0^m.38$. The main sequence of this 'OB' association reaches at least to late A-type stars. The luminosity function was determined and compared with the luminosity functions of very young open clusters observed by Walker (1957). The distance of a group of 13 early-type stars in the area of Car OB 2 ('group c') was found to be 2.5 kpc, the reddening $0^m.68$. The relation of Car OB 2 and group c to the structure of the Carina arm is regarded.

The association Car OB 2 is marked by a group of brilliant blue stars which are distributed over a nearly elliptical ring. The major axis of the ellipse is approximately 30 min, the minor axis 23 min of arc. The center lies a half degree above the galactic plane near the galactic longitude of 290°. The precise coordinates are

$$\alpha = 11^h03^m.7 \qquad l^{II} = 290°.0$$
$$\delta = -59°31' \quad (1950) \quad b^{II} = +0°.4.$$

This remarkable compact group of early-type stars has been the subject of some photometric and spectroscopic studies: In 1952 Bok and van Wijk (1952) observed four stars of the group in the PV-system (HD 96446, 96622, 96638, 96670). They also were the first who qualified the group as an association. Unfortunately they had only Henry Draper spectral types at their disposal with which to derive a distance of 1.4 kpc. Therefore Hoffleit (1956) determined MKK spectral types for five stars in Car OB 2 (HD 96248 B 1 I, 96446 B 2 V, 96622 O 9.5 IV, 96638 O 8, 96670 O 8). Together with the photoelectric magnitudes and colors in the Bok-van Wijk system she derived a distance of 1.7 kpc. But if one excludes HD 96446 ('exceptionally nearby for group membership', Hoffleit) one derives 2.0 kpc as the mean distance.

Feinstein (1964) observed 27 stars in the UBV-system. He used the main-sequence fitting-method and found a distance of 1.2 kpc and a color excess $E_{B-V} = 0^m.40$. He also noticed a second group ('group b') of stars in the field of the association with a distance of 0.4 kpc and a reddening of $0^m.15$.

Early in 1969 the author observed a photoelectric standard sequence of 25 stars in the UBV-system with the 1m-telescope of the European Southern Observatory in Chile. The limiting magnitude was $14^m.2$ in V. The plates had been taken by H. Haffner during his stay at Boyden Observatory in 1958. 480 stars were measured on four plates in each color with a Becker-type iris photometer of Hoher List Observatory. The more interesting blue parts of the two-color diagram and the color-magnitude diagrams are shown in Figures 1 and 2. The main-sequence of the association is clearly recognizable. The lefthand dashed line in Figure 1 and the solid lines in Figure

*Becker and Contopoulos (eds.), The Spiral Structure of Our Galaxy, 265–269. All Rights Reserved.*
Copyright © 1970 by the I.A.U.

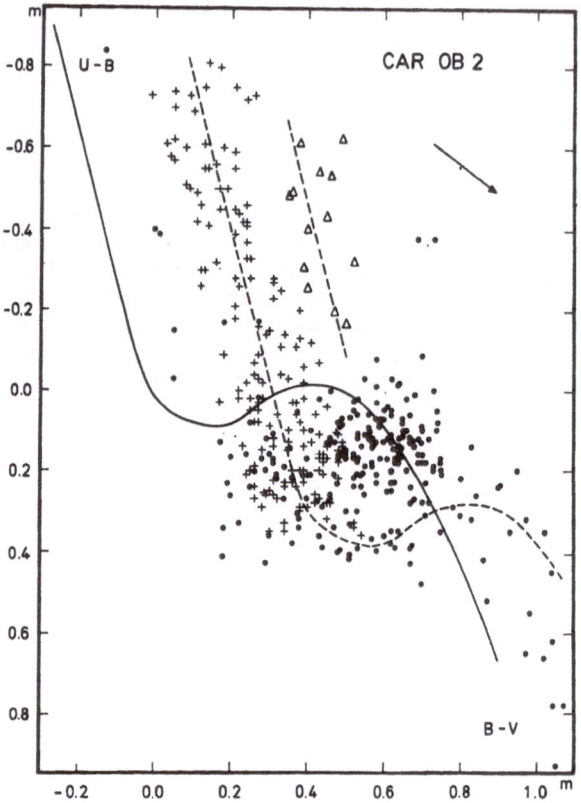

Fig. 1.  Two-color diagram of the association Car OB 2.  + probable members,  ● field stars,
△ stars of group c.

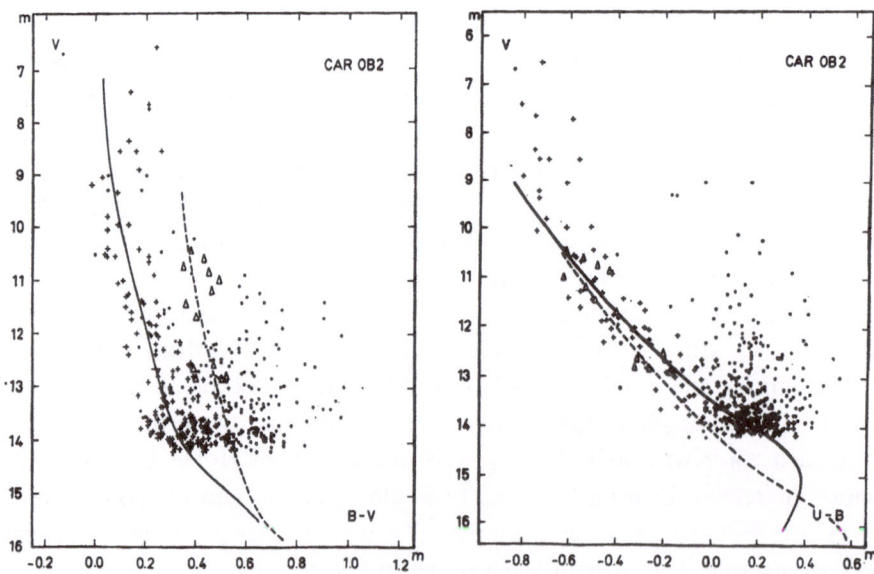

Fig. 2.  Color-magnitude diagrams of the association Car OB 2.  + probable members,  ● field
stars, △ stars of group c.

2 are the fitted ZAMS. In order to separate the association members from non-members, all stars nearer to the superimposed ZAMS than two times the mean internal error of the photometry ($\sigma = +0^m\!.04$ in V, $0^m\!.05$ in B–V, and $0^m\!.07$ in U–B) were selected. These probable members are indicated by crosses. The color excess $E_{B-V}$ equals $0^m\!.38$, the distance modulus not corrected for absorption equals $12^m\!.65$. Under the assumption $R = 3.0$ the distance was determined to be 2.04 kpc, the same value that Hoffleit derived. The linear dimensions of Car OB 2 are therefore 18 by 13 pc. The discrepancy between our distance and that of Feinstein is easily explained: If one has only those stars brighter than the 11th mag. and only the long wavelength color-magnitude diagram at one's disposal it is impossible to fit the main-sequence in a correct way. UBV data for fainter stars as well and the short wavelength color-magnitude diagram with its more steeply inclined upper part of the main-sequence are needed.

It is worthwhile to point out that Car OB 2 not only consists of O and early B stars – as the name 'OB' association may suggest – but that there is a continuous transition to later B- and A-type stars. The same was found by Seggewiss (1968) in the northern part of the association Sco OB 1 ($=$ Tr 24 and its subgroups).

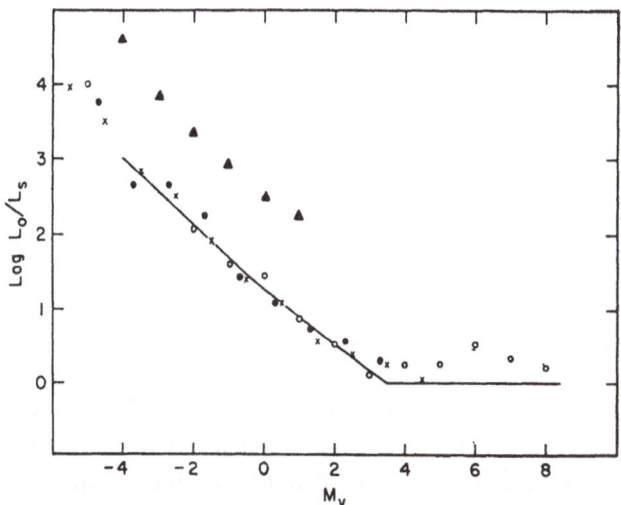

Fig. 3.    Luminosity function of Car OB 2 compared with the luminosity functions of young open clusters as derived by Walker (1957). ▲ Car OB 2, ● NGC 6530, ○ NGC 2264, × Orion nebula cluster.

The luminosity function of Car OB 2 has the same slope as the initial luminosity function derived by Salpeter (1955) and the luminosity functions of very young open clusters observed by Walker (1957). Figure 3 was taken from Walker's paper. The ordinate is the logarithm of the ratio of the observed luminosity functions $L_o$ to the luminosity function $L_s$ of main-sequence stars in the solar neighborhood. The line indicates Salpeter's initial luminosity function. Dots refer to NGC 6530, open circles to NGC 2264, crosses to the Orion nebula cluster (star numbers adjusted). The luminosity function of Car OB 2 is represented by triangles (star numbers not adjusted).

A group of 13 highly reddened stars, marked by triangles, can be distinguished in Figure 1. It is surprising to notice that all 13 stars lie within a small area, only 8 min of arc diameter, in the field of the association. One does not find any indication that the interstellar absorption in the field of Car OB 2 is non-uniform. Therefore one may conclude that these early-type stars form a separate group c, more reddened than Car OB 2 ($E_{B-V} = 0^m\!.68$) and more distant (2.5 kpc, fitted ZAMS as dashed lines in Figure 2).

Probable field stars in the area of Car OB 2, mostly weak reddened A- and F-type stars, are indicated by dots (Figures 1 and 2). Only very few stars form the upper part of Feinstein's group b (Figure 1) which therefore consists probably only of field stars. The bluest star is the helium star HD 96446 investigated by Jaschek and Jaschek (1959).

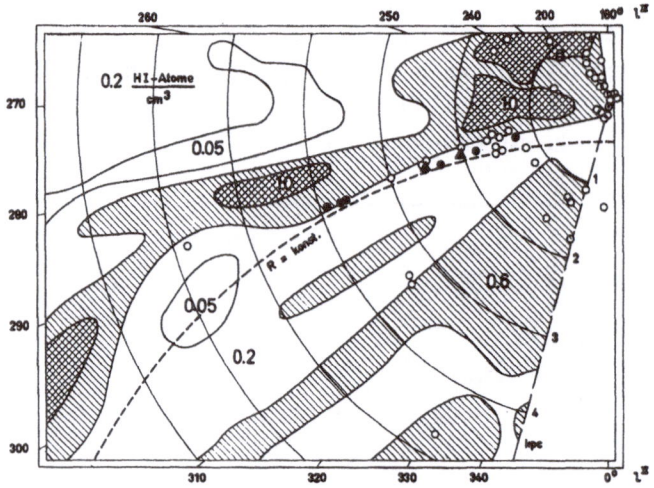

Fig. 4. Galactic plane around $l^{II} = 290°$. ▲ Car OB 2 and group c, ● young open clusters and stellar groups investigated by Schmidt and Diaz Santanilla (1964), ○ open clusters and H II regions.

Figure 4 shows the galactic plane after a drawing by Beer (1961). The density of neutral hydrogen atoms is indicated by different types of hatching. Small circles denote open clusters and H II regions. Car OB 2 and group c are represented by triangles. The dots indicate 6 young open clusters and stellar groups in the vicinity of Car OB 2, which were investigated by Schmidt and Diaz Santanilla (1964) of Hoher List Observatory with the plates now used for Car OB 2. The earliest spectral types, distances and reddening of these clusters together with the corresponding values of Car OB 2 and group c are given in the Table I.

Two points are to be stressed: First, the chain of spiral arm tracers does not coincide with the maximum density of neutral hydrogen. Second, the interstellar reddening is surprisingly low, only $0^m\!.5$ to a distance of 4 kpc, though we are looking lengthwise through a spiral arm.

Finding charts and UBV data of all 480 stars are available upon request from Hoher List Observatory.

## TABLE I

Earliest spectral types, distances and color excesses of young open clusters and stellar groups in the vicinity of Car OB 2

| Object near $l = 290°$ | Sp. Type | $d$ kpc | $E_{B-V}$ m |
|---|---|---|---|
| NGC 3572 a1 | B 1 IV–V | 1.24 | 0.26 |
| 3572 a2 | B 1 IV–V | 3.65 | 0.51 |
| 3572 b | O 9 V | 3.90 | 0.52 |
| 3590 a | B 5 V | 1.82 | 0.19 |
| 3590 b | B 2 IV | 3.70 | 0.44 |
| Tr 18 | B 4 V | 2.33 | 0.37 |
| Car OB 2 | O 8 | 2.04 | 0.38 |
| 'Group c' | O 7 | 2.50 | 0.68 |

## References

Beer, A.: 1961, *Monthly Notices Roy. Astron. Soc.* **123**, 191.
Bok, B. J. and van Wijk, U.: 1952, *Astron. J.* **57**, 213.
Feinstein, A.: 1964, *Observatory* **84**, 111.
Hoffleit, D.: 1956, *Astrophys. J.* **124**, 61.
Jaschek, M. and C.: 1959, *Publ. Astron. Soc. Pacific* **71**, 465.
Salpeter, E. E.: 1955, *Astrophys. J.* **121**, 161.
Schmidt, H. and Diaz Santanilla, G.: 1964, *Veröffent. Astron. Inst. Univ. Bonn*, **71**.
Seggewiss, W.: 1968, *Z. Astrophys.* **68**, 142.
Walker, M. F.: 1957, *Astrophys. J.* **125**, 636.

# 47. PHOTOMETRIC AND SPECTROSCOPIC DATA
## FOR SOME DISTANT O AND B STARS

G. LYNGÅ

*Lund Observatory, Lund, Sweden*

**Abstract.** The open cluster Stock 16 and some distant OB stars in Crux have been studied. Lack of absorption between spiral arms −I and −II is apparent. Negative residual radial velocities are found at distances of about 6 kpc at $l^{II} = 298°$.

The first group of stars that I am about to describe is situated in Crux at $l^{II} = 298°.5$; $b^{II} = +2°.1$ and the second group is essentially the cluster St. 16 at $l^{II} = 306°.1$, $b^{II} = 0°.2$ (Stock in Alter *et al.*, 1959). Data for these stars are given in Table Ia and Ib respectively. The numbers refer to my catalogue (Lyngå, 1968) of faint OB stars between Carina and Centaurus; identifications with CPD numbers are also given. The photoelectric UBV data were obtained at various telescopes at Mount Stromlo Observatory and the spectral plates were taken with the 100 cm reflector at Siding Spring Observatory equipped with a Meinel spectrograph (dispersion 119 Å/mm). The spectral classification (column 3 of Table I) was made using a number of standard spectra taken with the same instrument, and radial velocities were measured using the Abbe comparator with automatic setting device (Gollnow, 1962) at Mount Stromlo Observatory. The radial velocities with their mean errors are entered in column 4. Where variable radial velocity is suspected individual plate results are listed. The estimates of interstellar K line strength (column 5) were made on an arbitrary scale from 0 to 4. Column 6 gives absolute magnitude, colour excess and distance, spectroscopically determined. Colour excesses in brackets are photometrically determined on the assumption of luminosity class V. Residual radial velocities, using the same parameters of circular galactic motion as assumed by Feast (1967), are given in column 7 of Table I. In Table Ib, column 7 also contains remarks on membership of St. 16.

Using the unreddened colours given by Johnson (1963) I derive a reddening in $B–V$ for the stars in Crux, the mean value of which is $0^m.66$ with an rms scatter of $0^m.07$, all stars having nearly the same reddening. The reddening ratio $E_{U-B}/E_{B-V}$ varies with spectral class in much the same way as the data treated by Lindholm (1957). The distances given in column 7 depend largely on the estimated luminosity classes and the calibration given by Blaauw (1963). From these it is concluded that the stars do not form a physical group.

To discuss the absorption conditions in more detail, I shall divide the stars into two groups. The three closest stars Nos. 111, 112 and 115 have a mean distance of 3.1 kpc and a mean colour excess of $0^m.66$. The mean distance of the four remaining stars is 5.8 kpc with the same mean colour excess $0^m.66$. If the reddening is proportional to a density of interstellar matter

$$d(z) = d(0)\, e^{-z/125}$$

*Becker and Contopoulos (eds.), The Spiral Structure of Our Galaxy, 270–275. All Rights Reserved.*
*Copyright © 1970 by the I.A.U.*

TABLE Ia

Data for seven OB stars in Crux

| Star No. CPD No. | $V$ $B-V$ $U-B$ $n$ | MK class $N_{sp}$ | Stellar r.v. (km s$^{-1}$) | K-line int. r.v. (km s$^{-1}$) | $M_{sp}$ $E_{B-V}$ $r_{sp}$(kpc) | Remarks |
|---|---|---|---|---|---|---|
| 110 $-59°4119$ | $10^m.22$ $+^m.26$ $-^m.60$ $5$ | B0 III $3$ | $-33$(var.) $-39$ $+16$ | $2.0$ $+30$ | $-5^m.0$ $^m.56$ $5.0$ | |
| 111 $-59°4127$ | $10^m.73$ $+^m.39$ $-^m.57$ $5$ | B1 IV $3$ | $-3$(var.) $-84$ $+21$ | $3.0$ $-18$ | $-4^m.1$ $^m.65$ $3.6$ | |
| 112 $-59°4128$ | $11^m.00$ $+^m.39$ $-^m.59$ $4$ | B2 III $3$ | $+12$(var.) $-19$ $+35$ | $2.3$ $-12$ | $-3^m.6$ $^m.63$ $3.5$ | |
| 113 $-59°4130$ | $10^m.42$ $+^m.36$ $-^m.67$ $5$ | O7 $3$ | $-34 \pm 7$ | $2.0$ $+39$ | $-5^m.4$ $^m.68$ $5.8$ | res. vel. $-12$ km s$^{-1}$ |
| 114 $-$ | $11^m.48$ $+^m.48$ $-^m.51$ $5$ | B0 III $3$ | $-43 \pm 6$ | $1.7$ $+8$ | $-5^m.0$ $^m.78$ $6.9$ | res. vel. $-25$ km s$^{-1}$ |
| 115 $-59°4147$ | $9^m.40$ $+^m.43$ $-^m.59$ $6$ | B1 III $4$ | $-34 \pm 4$ | $3.0$ $-17$ | $-4^m.4$ $^m.69$ $2.2$ | res. vel. $-19$ km s$^{-1}$ |
| 116 $-59°4152$ | $10^m.71$ $+^m.31$ $-^m.73$ $5$ | O9 V $2$ | $+34$: $-36$ | $2.0$ $-$ | $-4^m.8$ $^m.60$ $5.5$ | |

where $z$ is height above the galactic plane in parsec, then the further group is expected to be more reddened than the nearer one by $0^m.26$. I suggest that there is a lack of absorbing matter at distances 3–6 kpc in this direction, $l^{II} = 298°.5$, which also agrees with the expected gap between spiral arms $-$I and $-$II.

The strength of the interstellar K-line varies from star to star, but is not correlated with distance for the stars studied, which indicates that the K-line absorption takes place close to the sun. This agrees with the fact that the K-line velocities are not systematically negative, which would have been required by a large distance for this galactic longitude.

## TABLE Ib

Data for stars in the field of St. 16

| Star No. CPD No. | $V$ $B-V$ $U-B$ $n$ | MK class $N_{sp}$ | Stellar r.v. (km s$^{-1}$) | K-line int. r.v. (km s$^{-1}$) | $M_{sp}$ $E_{B-V}$ $r_{sp}$(kpc) | Remarks |
|---|---|---|---|---|---|---|
| 233 $-61°3549$ | 10$^m$.06 1$^m$.08 $-^m$.11 2 | O9 III 2 | $-61$ (var.) $+1$ | 1.0 $-18$ | $-5^m$.7 1$^m$.39 2.2 | non-member |
| 235 $-61°3558$ | 10$^m$.24 $+^m$.30 $-^m$.52 2 | B1 Vn 1 | $-63$ | $-$ | 3$^m$.6 $^m$.56 2.7 | possible member |
| 236 $-61°3566$ | 9$^m$.29 $+^m$.23 $-^m$.75 2 | B0 IIIn 4 | $-36$ (var.) $-147$ $-52$ $-47$ | 1.0 $+24$ | $-5^m$.0 $^m$.53 3.5 | possible member |
| 238 $-62°3544$ | 10$^m$.65 $+^m$.49 $-^m$.46 2 | $-$ | $-$ | $-$ | $-$ ($^m$.78) $-$ | non-member; multiple star; in nebulosity |
| 239 $-61°3575$ | 7$^m$.95 $+^m$.19 $-^m$.82 4 | O9 V 5 | $-53 \pm 4$ | 1.4 $-37$ | $-4^m$.8 $^m$.50 1.8 | member res. vel. $-31$ km s$^{-1}$ |
| 240 $-61°3576$ | 9$^m$.48 $+^m$.23 $-^m$.71 4 | B1 Vn 4 | $+51$ (var.) $-111$ $+98$ $-18$ | 2.0 $+22$ | $-3^m$.6 $^m$.49 2.2 | member |
| 241 $-61°3579$ | 10$^m$.44 $+^m$.25 $-^m$.71 3 | B2 V 3 | $-92$ (var.) $+14$ $-33$ | 2.0 $-3$ | $-2^m$.5 $^m$.49 2.1 | member |
| 241 A | 11$^m$.62 $+^m$.26 $-^m$.51 1 | $-$ | $-$ | $-$ | $-$ ($^m$.49) $-$ | member |
| 241 B | 11$^m$.43 $+^m$.34 $-^m$.30 1 | $-$ | $-$ | $-$ | $-$ ($^m$.52) $-$ | member |
| 241 C | 12$^m$.50 $+^m$.33 $-^m$.16 1 | $-$ | $-$ | $-$ | $-$ ($^m$.46) $-$ | member |

*Table Ib (continued)*

| Star No.<br>CPD No. | $B$<br>$B-V$<br>$U-B$<br>$n$ | MK class<br>$N_{sp}$ | Stellar r.v.<br>(km s$^{-1}$) | K-line int.<br>r.v. (km s$^{-1}$) | $M_{sp}$<br>$E_{B-V}$<br>$r_{sp}$ (kpc) | Remarks |
|---|---|---|---|---|---|---|
| 242<br>$-61°3581$ | $10^m.09$<br>$+^m.27$<br>$-^m.66$<br>3 | B0: V<br>2 | $-78$ (var.)<br>$+1$ | 1.0<br>0 | $-4^m.4$<br>$^m.57$<br>3.6 | member |
| 246<br>$-61°3587$ | $10^m.73$<br>$+^m.28$<br>$-^m.58$<br>3 | B1 V<br>3 | $-21$ (var.)<br>$-50$<br>$+3$ | 0.7<br>$+39$ | $-3^m.6$<br>$^m.54$<br>3.5 | possible member;<br>in nebulosity |
| 247<br>$-61°3598$ | $10^m.28$<br>$+^m.83$<br>$-^m.17$<br>3 | B2 Ib<br>4 | $-22\pm3$ | 4.0<br>$+6$ | $-5^m.7$<br>$1^m.00$<br>4.0 | non-member<br>res. vel.<br>$+13$ km s$^{-1}$ |
| HD 115071<br>$-61°3544$ | $6^m.64$<br>$+^m.22$<br>$-^m.74$<br>1 | O9 Vn<br>3 | $-147$ (var.)<br>$-124$<br>$-49$ | 1.0<br>$-$ | $-4^m.8$<br>$^m.53$<br>0.9 | non-member |

Stars Nos. 113, 114 and 115 have constant radial velocities. Of the other stars, No. 111 has probably variable radial velocity and the remaining may or may not have variable radial velocities. If one or several of these stars were undisclosed binaries their distances would have been somewhat underestimated but the above discussion would not be significantly altered.

The area of the open cluster Stock 16 is shown in Figure 1. Numbers marked refer to Table Ib. It is clear from this table that the cluster stars have common colour excesses $E_{B-V}=0^m.53$ and $E_{U-B}=0^m.32$. The slope of the reddening trajectory is then 0.60. Figure 2 shows the colour-magnitude diagram $V/(U-B)_0$ and gives an apparent distance modulus of $13^m.4$. Assuming $R=3.0$ we derive a true distance modulus of $11^m.8$, which agrees with the mean of the spectroscopic distance moduli of Table Ib, thus confirming our choice of absorption to reddening ratio. We also conclude that duplicity of cluster stars has not greatly affected our photometric distance determination, as such an effect would tend to increase the observed distance modulus.

The cepheid V 378 Cen is situated 8 min of arc from the cluster centre. If this star is a member of the cluster, then it is $0^m.7$ more luminous than was estimated from its period by Kraft and Schmidt (1963).

Four stars in Table Ib have been considered as non-members because of deviating colour excesses and because of deviating positions in the colour-magnitude diagram. None of these is situated close to the cluster (compare Figure 1). Star No. 238 is a multiple star involved in nebulosity and may be considered as a very compact cluster.

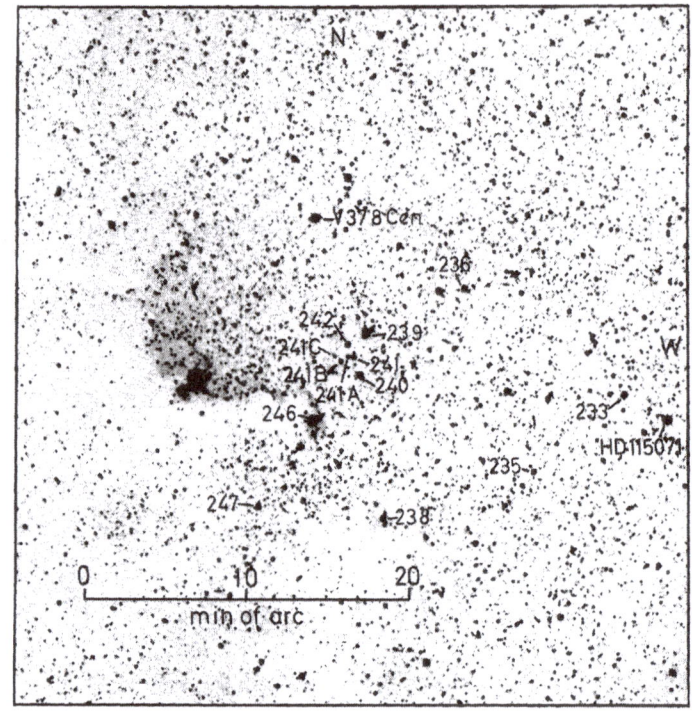

Fig. 1.   Finding chart for St. 16. Nos. 233, 238, 247 and HD 115071 are considered as non-members.

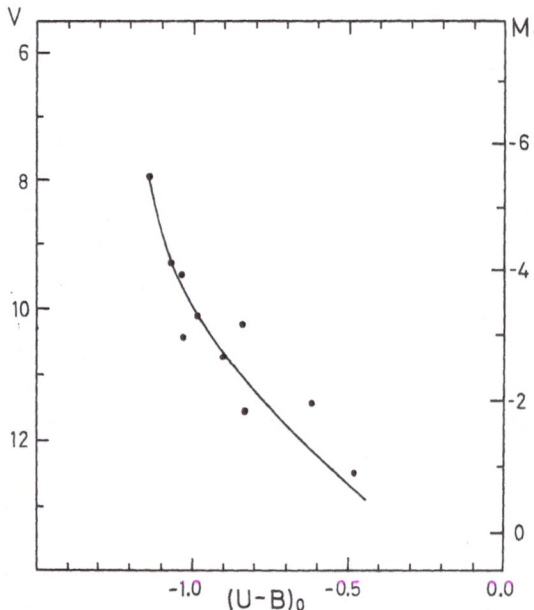

Fig. 2.   Colour-magnitude diagram for members and possible members of St. 16 assuming $E_{U-B} = 0^m.32$. The right-hand scale and the curve assume an apparent distance modulus of $13^m.4$.

On St. 16 we can further remark that it is a very young cluster since the O9V star No. 239 is on the main sequence.

It is of interest to compare the residual velocities listed in Table I with the compilation of residual velocites for OB stars and cepheids made by Feast (1967). Stars Nos. 113 and 114 are situated beyond the region in which residuals are predominantly negative thus extending this region. Nos. 115, 239 and 247 are situated in or slightly east of Feast's regions and confirm the sign of the residual in all cases. The mean value of the radial velocity for all members and possible members of St. 16 is $-38$ km s$^{-1}$ from 22 plates. This gives a residual of $-16$ km s$^{-1}$ as compared to $-31$ km s$^{-1}$ measured from No. 239 alone.

## Acknowledgements

This work was supported by the Swedish Natural Science Research Council under contract No. 2309-5. The use of Mount Stromlo instruments and computer is gratefully acknowledged.

## References

Alter, G., Hogg, H. S., Ruprecht, J., and Vanýsek, V.: 1959, *Bull. Astron. Inst. Csz.* **10**, Appendix, p. 1.
Blaauw, A.: 1963, *Stars and Stellar Systems* **3**, 383.
Feast, M. W.: 1967, *Monthly Notices Roy. Astron. Soc.* **136**, 141.
Gollnow, H.: 1962, *Monthly Notices Roy. Astron. Soc.* **123**, 391.
Johnson, H. L.: 1963, *Stars and Stellar Systems* **3**, 204.
Kraft, R. P. and Schmidt, M.: 1963, *Astrophys. J.* **137**, 249.
Lindholm, E. H.: 1957, *Astrophys. J.* **126**, 588.
Lyngå, G.: 1968, *Medd. Lunds Astron. Obs.*, Ser. I, No. 238.

# 48. PROGRESS REPORT ON THE CLEVELAND-CHILE SURVEY FOR SOUTHERN OB STARS

C. B. STEPHENSON and N. SANDULEAK

*Warner and Swasey Observatory, Cleveland, Ohio, U.S.A.*

From some standpoints it should seem surprising that not all of the OB stars of our Galaxy that can be identified to a feasible limiting magnitude (say 12–13) by conventional objective prism techniques have yet been so identified, but this is in fact the state of our published data. Some people here will recall that such a systematic survey for the northern Milky Way was carried out several years ago jointly by the Hamburg and Warner and Swasey Observatories. Now at the Warner and Swasey Observatory we are extending this northern survey into the southern Milky Way that could not be reached from the north, and we are making this extension as homogeneous as possible with respect to the northern survey.

We expect our goal of homogeneity to be achieved because of two factors: Stephenson was extensively involved in the northern survey, and the southern extension is being accomplished with equipment nearly identical to that used in Cleveland, viz. the Michigan Schmidt telescope (now at Cerro Tololo) and the same objective prism as used at Cleveland. The chief difference between the new and old plate material is that the new plates are of superior quality to the Cleveland ones, mostly because the Michigan Schmidt is optically superior to the Warner and Swasey one and partly because of the good seeing at Cerro Tololo; however, the southernmost Chile plates are quite variable in limiting magnitude, for reasons partly known and partly unknown.

We expect to publish our catalogue and identification charts in 1970, and a major purpose of our presenting this report is to solicit opinions as to the most useful publishing format. We will probably adopt at least the following changes with respect either to the northern catalogues or to common southern-hemisphere practice: (1) We shall arrange the stars by right ascension throughout, instead of by declination zones. (2) We shall give CD numbers in preference to CPD numbers, even though most southern workers for some reason are doing the opposite. Our reason for this choice is that CD charts exist for the whole southern sky, while the same is not true of the CPD. (3) We are considering including an index that will cross-reference our star numbers against those of previous surveys. (4) The charts will be prepared from direct plates, probably of the scale of the *Lick Sky Atlas*, rather than from objective prism plates as was done for the Cleveland zones (no charts were ever published for the Hamburg zones).

Finally we should mention that we are receiving numerous requests for pre-publication data from this survey. Because of this we should point out that such data exist at present only in the form of handwritten material, which is not yet computer-compatible and is thus very time-consuming to prepare for transmission to a potential user, and this will remain true almost up to the time when we submit our manuscripts for publication. For this reason we would like to apologise for our having to decline requests for such material for about another year, unless the need for it should be truly exceptional.

# 49. A STUDY OF O AND B STARS IN VELA, ALONG THE GALACTIC EQUATOR

A. G. VELGHE

*Koninklijke Sterrenwacht van België, Ukkel, Belgium*

The spiral structure of our Galaxy, as defined by optical evidence, shows a striking gap around $l^{II} = 270°$. Intrigued by this feature, the author started, in 1955, a survey of early type stars based on objective-prism plates taken with the ADH–Schmidt-telescope of the Boyden Observatory (South-Africa); dispersion of the spectra: 240 Å/mm at Hγ.

On the basis of the principle of natural groups (Morgan, 1951), 196 OB stars were segregated. The limiting magnitude is 12.5. For all these stars $UBV$-photometry was carried out with the 60-inch Rockefeller telescope of the Boyden Observatory (1961 and 1962).

The survey deals with the region along the galactic equator, form $l^{II} = 262°$ to 273°, and $b^{II} = -4°.6$ to $+2°.0$. This region shows a conspicuous concentration of early-type stars and several clusters. In earlier work it was supposed to contain the so-called association I Vel, but later on the reality of this association was questioned (Alter *et al.*, 1958). On long-exposure plates (cf: *Georgetown Atlas of the Southern Milky Way*, 1952), a dark cloud narrows the Milky Way boundaries in this area, and filaments of obscuring matter can be traced within the region concerned.

TABLE I

Visual absorption at a distance of 1.3 kpc for the various regions, indicated by capitals in Figure 1

| Region | $A_v$ | Region | $A_v$ |
|--------|-------|--------|-------|
| A | 0ᵐ.45 | G | 1ᵐ.95 |
| B | 1ᵐ.15 | H | 2ᵐ.30 |
| C | 1ᵐ.30 | I | 2ᵐ.75 |
| D | 1ᵐ.35 | J | 3ᵐ.15 |
| E | 1ᵐ.50 | K | 3ᵐ.80 |
| F | 1ᵐ.95 | L | 4ᵐ.40 |

Figure 1 shows: (a) the surface distribution of the OB stars segregated from the Boyden objective-prism plates; (b) the distribution of the O stars and of those having intrinsic colors $(B-V)_0 < -0.30$; and (c) the boundaries of the bright starfield and of the dark regions around and within it. Capitals in Figure 1 (c) refer to regions of different obscuration; the corresponding $A_v$-values, given in Table I, are based on data of the stars for which MK types are available (102 stars).

*Becker and Contopoulos (eds.), The Spiral Structure of Our Galaxy, 278–280. All Rights Reserved.*
*Copyright © 1970 by the I.A.U.*

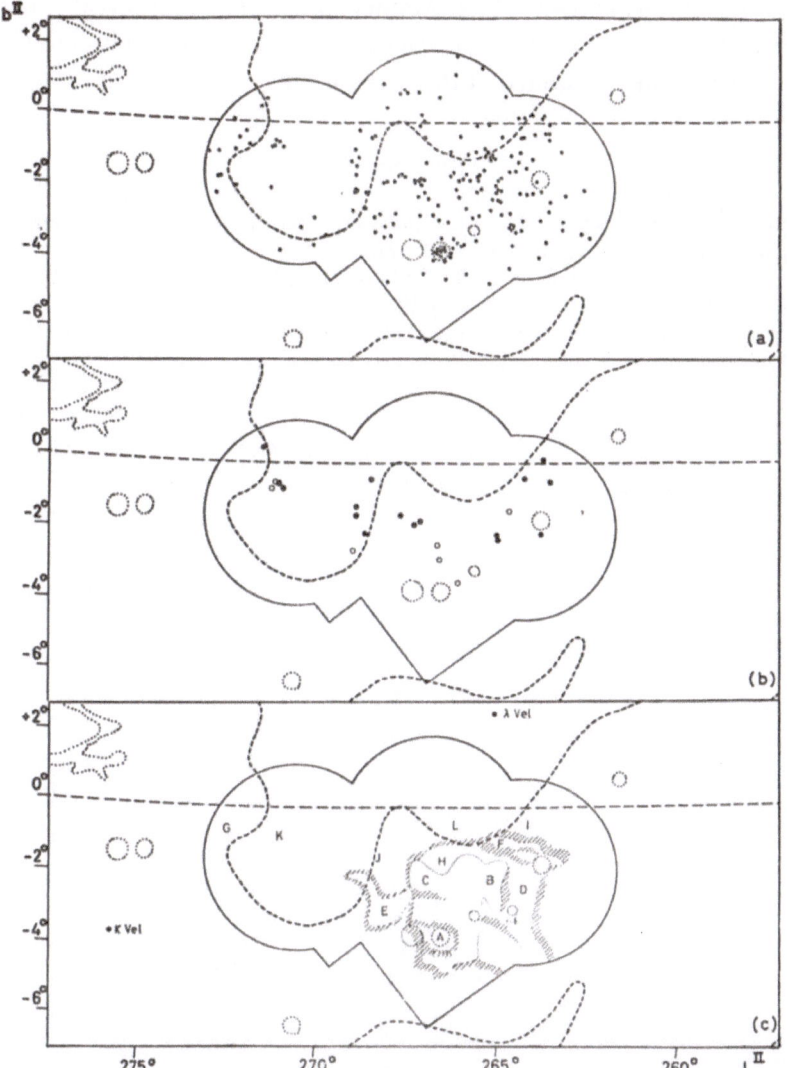

Fig. 1.  (a): The surface distribution of the OB stars in the investigated field; (b): the distribution of the O stars and stars with $(B-V)_0 < -0.30$; and (c): boundaries of the bright star cloud and of the obscuring matter; capitals indicate specific regions of different absorption as given in Table I. The background material for the drawings is taken from the Skalnaté Pléso *Atlas of the heavens*.

From the discussion of the data appears: (1) that the assumption concerning the reality of an association in the Vela-region has to be rejected; (2) that a spiral feature at right angles to the galactic center extends up to large distances (more than 5 kpc); this feature could be a linkage between the known Carina-arm and some spiral structure in Puppis, but most of it is hided by obscuring matter. In this connection one should recall that a study of the surface-photometry of the Milky Way by Elsässe-and Haug (1960), indicates the existence of an arm in this region, and that the Austrar lian radio-observations at 1440 MHz (Mathewson *et al.*, 1962) reveal a not unimportant peak in the same direction. An extension of the present study is dealt with in the paper by Denoyelle, presented at this symposium.

## References

Alter, G., Ruprecht, J., and Vanýsek, V.: 1958, *Catalogue of Star Clusters and Associations*, Prague.
Elsässer, H. and Haug, U.: 1960, *Z. Astrophys.* **50**, 121.
Mathewson, D. S., Healey, J. R., and Rome, J. M.: 1962, *Australian. J. Phys.* **15**, 354, 369.
Morgan, W. W.: 1951, *Publ. Obs. Univ. Michigan* **10**, 33.
*Photographic Atlas of the Southern Milky Way*, 1952, Georgetown Coll. Obs. and Nat. Geogr. Soc., Washington.

# 50. PROGRESS REPORT OF THE CURRENT RESEARCH ON THE GALACTIC STRUCTURE IN VELA

J. DENOYELLE

*Koninklijke Sterrenwacht van België, Ukkel, Belgium*

In order to find evidence for supporting or rejecting the idea of a major spiral feature in the direction of $l^{II}=270°$, as proposed by Velghe (1969), it is necessary to extend the limits of the survey both in longitude and in distance. The extension towards smaller longitudes was placed arbitrarily at $l^{II}=257°$, while in the direction of Carina, the investigation was confined to $l^{II}=285°$. In doing so, sufficient overlap was made with the studies by Velghe and by Graham and Lyngå (1965), in order to get a homogeneous and complete material in a large, but important section of the southern Milky Way. Following the same technique as outlined in the preceding paper, about 360 young type stars were selected from objective-prism plates, taken with the ADH-telescope of the Boyden Observatory. Later on (in 1966) a first series of *UBV* photoelectric measurements was carried out at the Boyden Observatory, however with insufficient accuracy. Before finishing the reductions, a second series was made with the 1m-photometric telescope of the ESO (La Silla, Chile) at the end of 1968. These (ESO)-values, in the Johnson *UBV*-system, will be available before long. In the definite form, they will be combined with the Boyden-measurements. The surface distribution of the stars in the whole section is shown in Figure 1. Many of the stars between $l^{II}=273°$ and 282° are of rather late B-type, but they were included for the purpose of determining the absorption in this field. The space distribution will enable

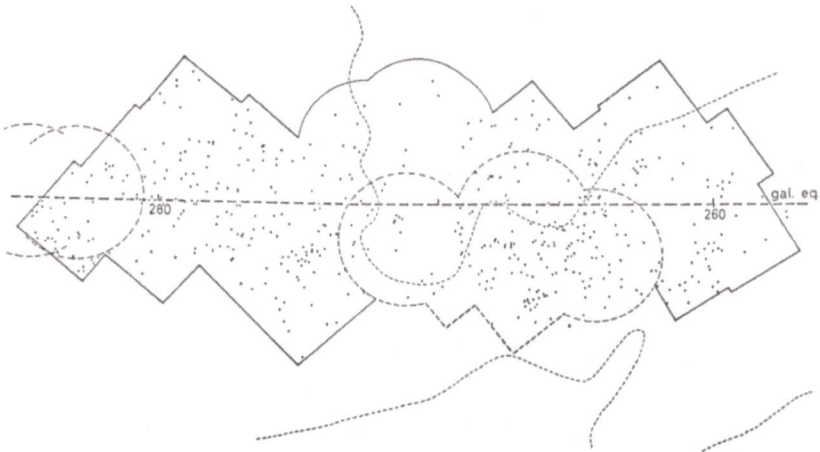

Fig. 1.   Surface distribution of O- and B-type stars in and near Vela. Limits of the survey: (———) Denoyelle; (– – – –) Velghe and (for $l^{II}>280°$) D. Hoffleit; (- - - - - - - -) Boundary of the Milky Way (Skalnaté Pléso).

us to decide whether the edge of the Carina spiral feature is seen in this direction (Bok *et al.*, 1969) and if there is any indication for a possible linkage from Carina to Puppis. To extend the distance limit, a search for faint blue stars, applying the idea of Bok (1966), is planned and some photoelectric sequences will be measured early in 1970 as standards for photographic photometry. As another approach to the galactic structure in this field a complete set of red objective-prism plates has been obtained since 1966, to pick out the objects with H $\alpha$ in emission. As for the reality of the association I Vel, some preliminary material on Radial Velocities will be collected also in 1970, using the Fehrenbach astrograph of the ESO at La Silla (Chile).

## References

Bok, B. J.: 1966, IAU Symposium No. 24, p. 228.
Bok, B. J., Hine, A. A., and Miller, E. W.: 1969, IAU Symposium No. 38, p. 246.
Graham, J. A. and Lyngå, G.: 1965, *Mem. Mt. Stromlo Obs.* No. 18.
Velghe, A. G.: 1969, IAU Symposium No. 38, p. 278.

# 51. CALIBRATION OF 4-COLOR AND Hβ PHOTOMETRY FOR B- AND A-TYPE STARS

D. L. CRAWFORD

*Kitt Peak National Observatory\*, Tucson, Ariz., U.S.A.*

Observing has been underway for several years, at Kitt Peak and Cerro Tololo, of the bright B and A stars, and of the brighter and nearer open clusters, to supply data necessary for calibration of 4-color and Hβ indices in terms of intrinsic color and absolute magnitude, as well as for use in studies of galactic structure. To date, data are on hand – some published, most being prepared for publication – for all O- to G0-type stars brighter than $V = 5^{m}0$, for most O- to B5-type to $6^{m}5$, and for most northern hemisphere O- to G0-type to $6^{m}5$. In addition, the following open clusters have been studied: Hyades, Coma, Ursa Major, Praesepe, α Persei, Pleiades, IC 2602, IC 2391, h and χ Persei, NGC 6231, NGC 752, and IC 4665.

The following parameters are derived for the $u\,v\,b\,y\,\beta$ photometry:

(1) $V$, the visual magnitude;

(2) $(b–y)$, a color index, which has 70% of the scale of $(b–v)$;

(3) $m_1$, a measure of 'blanketing' in the $\lambda4100$ Å region. $E(m_1) = -0.3\ E(b–y)$. For the A and F stars, a standard $m_1$ vs. $\beta$ relation is used to derive $\delta m_1 = m_1$ (std) $-m_1$ (obs.).

(4) $c_1$, a measure of the Balmer discontinuity. $E(c_1) = 0.2\ E(b–y)$. For the A- and F-type stars $\delta c_1 = c_1$ (obs.) $- c_1$ (std).

(5) $(u–b)$, a color index. $E(u–b) = 1.7\ E(b–y)$.

(6) $\beta$, a measure of the Hβ line strength.

The definitive calibrations are being prepared for publication soon, but I would like to present a simple preliminary set of calibrations at this time:

(a) For the A-type stars $(2.890 < \beta > 2.720,\ \delta c_1 < 0.280)$:

(1) $(b–y)_0 = 2.943 - 1.0\ \beta - 0.1\ \delta c_1 - 0.1\ \delta m_1$. Cosmic scatter $\pm 0^{m}011$ (mean error, one star).

(2) $M_v = M_v(\text{Z.A.M.S.}) - 8\ \delta c_1$.

(b) For the B stars $(-0.1 < c_1 < 1.000)$:

(1) $(b–y_0) = -0.116 + 0.097\ c_1$. Cosmic scatter $\pm 0^{m}01$.

(2) $M_v = M_v(\text{Z.A.M.S.}) - 8\ \delta\beta$. Cosmic scatter $\pm 0^{m}2$ at B9, $\pm 0^{m}7$ at B0.

---

\* Operated by the Association of Universities for Research in Astronomy, Inc., under contract with the National Science Foundation.

## 52. WOLF-RAYET STARS, RING-TYPE HII REGIONS, AND SPIRAL STRUCTURE

TH. SCHMIDT-KALER

*Astronomisches Institut, Ruhr-Universität, Bochum, Germany*

At the Prague IAU meeting (Isserstedt and Schmidt-Kaler, 1967) we presented some results of investigations on stellar rings, a new kind of stellar association and powerful spiral tracer. We have since kept looking for similar features of gas or dust which might be early stages in the evolution of stellar rings.

We have found about a dozen regular ellipsoidal HII shells with very sharp filamentary outer boundaries (appearing as ring nebulae), and about half a dozen ring-type dark clouds for which a distance estimate was possible (Schmidt-Kaler, 1968, 1969). The minor diameter of the ring-type dark clouds is on the average about 6.5 pc. The exciting stars of 8 sharp HII ring nebulae are Wolf-Rayet stars. All of these are broad-lined, definitely single stars of the classes WN 5, 6, 8 – sequence B (following the definition of Hiltner and Schild, 1966). This is remarkable since WR-stars are a very rare class of stars, and about half of them are binaries. The minor diameter of the HII rings is in the average 6.8 pc. Only one object (NGC 7635) is definitely deviating from the average; it is peculiar also regarding the filaments, the exciting star WN4 p or 07 f, and its location on the rim of the HII filaments rather than near the centre. The observations of the other HII rings are compatible with the assumption of a unique minor diameter of 7.3 pc.

We decided to study the ring nebula NGC 6888 around HD 192163 in detail since a slight concentration of stars inside and particularly on the rim of the nebulae is apparent. Star counts confirmed that the density of stars brighter than $B = 17^m$ is about 30% higher inside and about twice as high on the rim (the shape of the rim is completely determined by gas filaments, and *not* by the stars lined up along the edge). Photographic UBV photometry down to $V = 16^m$, based on a photoelectric sequence set up in NGC 6888, has been obtained. Preliminary results for the stars on the rim show the colour-magnitude diagram of a very young cluster. The position of the ZAMS as defined by the reddening and distance modulus of the two high luminosity stars in the area with known MK class, HD 192163 WN6-B and BD+37° 3827 F3 Ib, gives a perfect fit to the distribution of the fainter stars in the colour-magnitude diagram. The distance is 1.9 kpc, the minor diameter 6.3 pc. If the supergiant is a member, it should be a contracting star. Cooperating with Mr. Schwartz (Bonn) continuum measurements of the northern ring nebulae at 2.7 GHz have been obtained, measurements at 1.4 and 11 GHz are under way. The results confirm the HII rings as purely thermal sources.

In a recent survey (with Dr. Haupt) of the 23 well-established X-ray sources on the Palomar and the Ross Calvert Milky Way Atlas we found again a connection with

*Becker and Contopoulos (eds.), The Spiral Structure of Our Galaxy, 284–286. All Rights Reserved.*
Copyright © 1970 by the I.A.U.

gaseous shells appearing as rings. Excluding one extragalactic radio source and three well-known supernovae remnants there are 15 X-ray sources for which charts were available. In four cases an X-ray star has already been identified:

GX3+1 lies on the rim of an H II ring, the geometric distance estimated from the ring diameter agrees exactly with the completely independent distance estimate of Freeman *et al.* (1968).

Sco X-1 lies on the outer edge of a ring-type dark cloud, the geometric distance estimate puts it at 300 pc to be compared with the distance of approximately 500 pc given by Westphal *et al.* (1968).

Cen X-2 lies again on the outer edge of a ring-type dark cloud, its geometric distance of 170 pc would relate it to the immediately adjoining coal-sack. For Cyg X-2 no conspicuous optical feature has been noted on the photographs. In the remaining 11 cases where no X-ray star has yet been identified the error circle contains in 4 cases a ring-type dark cloud and in 6 cases a schmetterling H II region (filamentary nebulae of a typical butterfly shape like NGC 6302, often containing elliptical ring filaments).

Cyg X-3 lies in the same region as Cyg X-2 and no optical feature can possibly be associated with it. In the neighbourhood of the schmetterlings always dark cloud rings are seen.

The distances of the X-ray sources have been estimated either by assuming a unique minor diameter of the associated ring type feature and/or from the photometric distance modulus of the associated schmetterling. They are located in the spiral arms as defined by early-type clusters and H II regions (Becker and Fenkart, 1970) or stellar rings (Isserstedt, 1970). The H II ring nebulae are also located in the spiral arms.

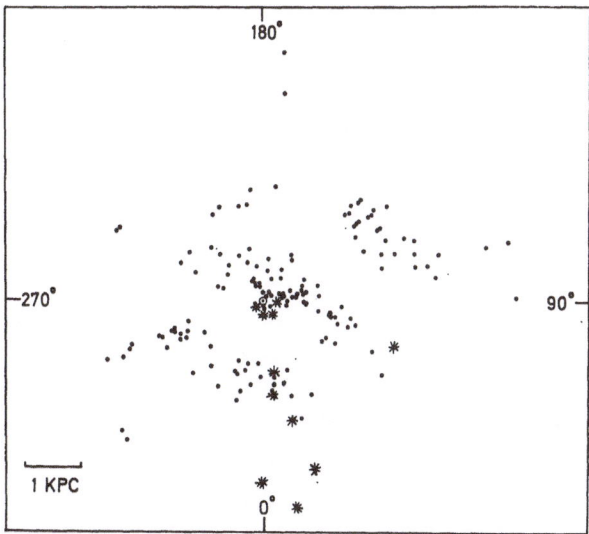

Fig. 1.   Distribution of X-ray sources projected on the galactic plane (stars) and early-type clusters and H II regions (dots) according to Becker and Fenkart (1970).

In conclusion I would like to summarize:

(1) Star formation seems to be going on in dense filamentary shells of gas of ellipsoidal shape, connected with non-binary WR-stars.

(2) The galactic X-ray sources which may be called non-synchotron may be very young objects showing signs of heavy mass loss, and are associated to ring-type dark clouds or H II regions.

(3) The characteristic diameter of these shells is about 6.9 pc.

(4) The X-ray sources are located on the spiral arms; the X-ray sources and the WR-stars (Smith, 1968) seem to be the farthest-reaching spiral tracers of the Galaxy, apart from the stellar rings.

A detailed paper will appear elsewhere.

## References

Becker, W. and Fenkart, R. P.: 1970, IAU Symposium No. 38, p. 205.

Freeman, K. C., Rodgers, A. W., and Lyngå, G.: 1968, *Nature* **219**, 251.

Hiltner, W. A. and Schild, R. E.: 1966, *Astrophys. J.* **143**, 770.

Isserstedt, J.: 1970, IAU Symposium No. 38, p. 287.

Isserstedt, J. and Schmidt-Kaler, T.: 1967, Preprint, see also *Sky Telesc.* **34**, 299.

Schmidt-Kaler, T.: 1968, *Veröff. Bochum* No. 1, 80.

Schmidt-Kaler, T.: 1969, *Colloque Internat. Liège*, No. 16.

Smith, L. F.: 1968, *Monthly Notices Roy. Astron. Soc.* **141**, 317.

Westphal, J. A., Sandage, A., and Kristian, J.: 1968, *Astrophys. J.* **154**, 139.

# 53. STELLAR RINGS AND GALACTIC STRUCTURE

J. ISSERSTEDT

*Astronomisches Institut, Ruhr-Universität, Bochum, Germany*

**Abstract.** Stellar rings are shell type prolate ellipsoidal stellar aggregates. The most important property is the constancy of the minor diameters which allows precise geometric distance determinations up to great distances. New photoelectric and photographic UBV-observations of two stellar rings confirm the reality and the diameter constancy of these objects. All A- and F-stars of a nearby ring in Aquila exhibit an UV-excess, similar to that observed in the very young open cluster NGC 2264.

Star formation seems to go on in certain gaseous nebulae (Schmidt-Kaler, 1970). In NGC 6888 we are apparently observing the evolution from a gaseous shell to a stellar ring. A few years ago such stellar rings were discovered in a later stage of their evolution without correlated H II regions (Isserstedt, 1968a).

Most important criteria for the definition of the stellar rings are the symmetry to the major diameter, the small thickness of the ring, which is between $\frac{1}{40}$ of the minor diameter, and $\frac{1}{10}$ of this diameter in the worst cases, and a sharp outer boundary. The ratio of the axes is always smaller than 2.0. The rings contain between 25 and 200 stars; an average value is 70.

In a systematic search for stellar rings over most of the sky with the Lick Observatory Sky Atlas, and near the galactic plane with the red prints of the Palomar Observatory Sky Survey (incl. the Whiteoak Extension) a total of 1070 rings have been found. The minor diameters are between 0.7 and nearly 4°. The rings have been catalogued with their coordinates, diameters and the position angles of the major axes.

Some characteristics of the stellar rings make it most unlikely that they should be nothing but accidental configurations of stars. There is, for example, a great difference between the distributions of stellar rings and normal field stars up to 21$^m$. The rings are much more concentrated to the plane, and the absorption layer, recognizable in the distribution of the rings, indicates that the average distance of the rings is greater than the average distance of the field stars.

Another conspicuous fact is the observation that in the greatest rings which were found in the Palomar Atlas, the density of the luminous stars of the Hamburg-Cleveland Spectral Survey is a dozen times the average density in the field near the galactic plane.

A few stars with known MK-classification or with spectra from the Henry Draper Catalogue were found in stellar rings. With these data it became possible to determine a few distances of stars in rings and even three Hertzsprung-Russell diagrams for the nearest objects. All these observations seemed to confirm the reality of the rings and indicated that the minor diameters of the rings are scattering only a little around a mean value of 7.1 pc.

In analogy to the shell type gaseous nebulae the stellar rings are shell type ellipsoidal

*Becker and Contopoulos (eds.), The Spiral Structure of Our Galaxy, 287–289. All Rights Reserved.*

stellar associations whose symmetry axis is the major diameter. The minor diameter of these prolate ellipsoidal shells therefore appears always without perspective shortening in projection to the sphere.

To test whether the rings really have always the same minor diameter, the distance of the rings was determined only from the apparent diameters with the assumption that the true minor diameters are always 7.1 pc. This hypothesis led to a picture of the galactic structure up to a distance of 15 kpc (Isserstedt, 1968a, 1969b). This structure, almost free of heliocentric effects, proves the reality of the rings and the constancy of the minor diameters. The scatter of the diameters is not larger than 5%; otherwise the far distant details in this picture would be smeared out. It should be stressed that we do not see any possibility to explain this observational result in another way. The Galaxy is of type Sb, perhaps Sb-Sc. The scale in the picture of the galactic structure is still rather uncertain and depends on the value for the minor ring diameter. With a minor diameter of 7.1 pc the distance to the galactic centre is 10.8 kpc.

At first glance the distance range seems surprisingly large. This is possible, because most rings at great distances from the sun are far distant from the galactic plane. This means, that the light has to pass the thin absorption layer only for 2 or 3 kpc. The problem is, therefore, not the *visibility* of these rings but the fact, that there *are* rings so far from the galactic plane and that these rings conform to the structure in the plane. Most rings which define the structure are less than 1000 pc distant from the plane, but the rings between 500 pc and 1000 pc contribute an important part of the picture, and there are moreover many rings above 1000 pc, which are at least partly real objects and not accidental configurations. In spite of the great scatter around the galactic plane the rings define a flat plane in good agreement with the results from 21 cm-line observations (Isserstedt, 1968b).

The galactic structure in the solar neighbourhood, as shown by the distribution of H II regions and young open clusters (Becker and Fenkart, 1970) is very similar to the local distribution of stellar rings (Isserstedt, 1969b).

To investigate the reality of individual stellar rings and to determine the mean value of the minor diameter more accurately, we began a program of photoelectric and photographic UBV-photometry for several objects. The first results for ring No. 373 show that it is real and that the minor diameter is within the errors of the method identical with the mean value of 7.1 pc. The distance of this ring is nearly 3 kpc. A detailed analysis is difficult because the absorption is strong and not homogeneous in the field (Isserstedt and Schmidt-Kaler, 1969).

To determine properties of stellar rings, it is necessary to observe the nearest objects with small reddenings and bright stars. Therefore, the Aquila ring, one of the nearest objects, was observed photoelectrically in UBV. The reddening is $0^m.08$ only and homogeneous all over the ring. The distance is 240 pc, the minor diameter 7.4 pc. There is a clear UV-excess for all A- and F-stars, indicating that the stars are possibly still contracting to the main sequence. The same effect can be seen in the two colour diagrams of the central region of the very young open cluster NGC 2264 (Walker, 1956) and of the nearby stellar ring in Orion (Isserstedt, 1969a).

We think that these UV-excesses are further evidence for the reality of the stellar rings because such systematic effects are not expected in accidental stellar configurations.

## References

Becker, W. and Fenkart, R. P.: 1970, IAU Symposium No. 38, p. 205.
Isserstedt, J.: 1968a, *Veröff. Bochum*, No. 1, 1.
Isserstedt, J.: 1968b, *Veröff. Bochum*, No. 1, 121.
Isserstedt, J.: 1969a, *Astron. Astrophys.* 3, 210.
Isserstedt, J.: 1969b, in preparation.
Isserstedt, J. and Schmidt-Kaler, T.: 1969, in preparation.
Schmidt-Kaler, T.: 1970, IAU Symposium No. 38, p. 284.
Walker, M. F.: 1956, *Astrophys. J. Suppl. Ser.* 2, 365.

## Discussion

*B. J. Bok:* Why are no globular clusters observed on the other side of the galactic center while the ring-spiral system comes through apparently undisturbed by cosmic dust?

What is your response to the paper by David Crampton, who claims that the dispersions in radial velocity in the Aquila and Orion rings are too great to make them stable?

*Isserstedt:* (1) Nearly no rings are found behind the galactic centre. From 87 rings with distances $R > R_0$ between longitudes 0° and 10° only 9 objects are less than 500 pc distant from the plane. No ring at all could be found near the galactic centre up to a distance $R = 500$ pc. The structure of the other side of the galaxy can be observed only because those rings are far distant from the plane.

(2) I cannot discuss Crampton's letter, because no individual data are given.

# 54. RADIAL VELOCITIES OF EARLY-TYPE STARS

J. P. KAUFMANN

*Technische Universität, Berlin, Germany*

With the Fehrenbach objective prism radial velocities of about 700 stars of type B0 to A0 were determined in two fields of the Southern Milky Way ($l^{II}=295°$ $b^{II}=-0.6°$; $l^{II}=320°$ $b^{II}=-2.5°$), with a mean error of $\pm 20$ km s$^{-1}$. An additional photographic *UBV*-photometry with plates of the ADH-telescope at Boyden Observatory was accomplished. Minimum distances for the stars resulted from absorption-corrected magnitudes and a MK-spectral classification. About 200 stars lay at distances greater than 1.5 kpc from the sun. The largest distances determined were 5 kpc. From the radial velocities and distances circular velocities were derived and plotted against galactocentric distances $R$. Even within the possible error limits a positive velocity gradient showed up in the range 8 kpc $< R <$ 9.5 kpc, which French authors had already found for the region 10.5 kpc $< R <$ 12.5 kpc. If there do not exist significant deviations from circular motion for these stars, a conformity with Schmidt's 1965 model cannot be obtained.

*Becker and Contopoulos (eds.), The Spiral Structure of Our Galaxy,* 290. *All Rights Reserved.*
Copyright © 1970 *by the I.A.U.*

# 55. THE VELOCITY DISPERSIONS OF O- AND B-STARS WITHIN A FEW KPC

R. B. SHATSOVA

*Moscow University, Moscow, U.S.S.R.*

**Abstract.** The results of a study of the dispersion of velocities of O-B5 stars up to a distance of 3 kpc are described.

The modern catalogues of motions of OB stars enable us to pass, in the study of stellar kinematics, from the solar vicinity to more distant regions. In their works Rubin and Burley (1964), Feast and Shuttleworth (1965), Petrie (1963), Petrie and Petrie (1968), Bonneau (1967a) and others are already using new observational data for obtaining the velocity field up to 3–4 kpc distance.

To each volume element of the Galaxy one may attribute as kinematic characteristics the velocity of the centroid and the distribution function of peculiar velocities or, what amounts to the same, the velocity dispersion, the asymmetry, etc. In the present paper an attempt is made to derive within the accessible volume of the Galaxy the distribution of velocity dispersions.

As observational basis we used the radial velocities and distances taken from the catalogues of Rubin *et al.* (1962) and Bonneau (1967b). The volume of space covered by these catalogues was divided into heliocentric rings, or semirings, and sectors. The average number of stars in a single section for $r<3$ kpc amounts to 56 and 53 in Rubin's and Bonneau's catalogues respectively.

The observed velocities of stars were corrected for standard solar motion, the Oort's term of galactic rotation and the average residual radial velocity corresponding to the volume element to which the star belongs. A special examination has shown that the average velocity consists of the errors in Oort's terms (mainly due to systematic errors in the distances), the higher order terms of galactic rotation, the radial motion (if any) in the Galaxy, as well as any local motion.

The errors in the mean distances in the Oort's term and in the velocity dispersion were determined by an analytic method which was worked out by us following the ideas of Feast and Shuttleworth (1965). The following values of the parameters of the differential motions were assumed:

$$A = 15 \text{ km s}^{-1} \text{ kpc}^{-1}, \quad \omega_0'' = 0.7 \text{ km s}^{-1} \text{ kpc}^{-3},$$
$$\varepsilon_0 = 0.5 \text{ km s}^{-1} \text{ kpc}^{-1}, \quad \varepsilon_0' = -0.6 \text{ km s}^{-1} \text{ kpc}^{-2}.$$

These values figure among other solutions in the paper of Petrie and Petrie (1968). The local motions in the majority of space elements do not exceed the errors of average residual velocities. The maximum velocities appeared in the sections $l^{II} = 120°-150°$ and $r>2$ kpc ($V_r=7-11$ km s$^{-1}$) and $r=1-2$ kpc, $R<R_0$ ($V_r=-6$ km s$^{-1}$).

*Becker and Contopoulos (eds.), The Spiral Structure of Our Galaxy, 291–294. All Rights Reserved.*
Copyright © 1970 by the I.A.U.

The velocity dispersion, defined by

$$\sigma^2 = \frac{1}{n} \sum_{i=1}^{n} (V_i - \bar{V})^2,$$ (1)

was corrected for the influence of accidental errors of radial velocities and of distance moduli, as well as for the differential motions of the centroids corresponding to different parts of the section with respect to the centre of the whole section.

The final results, given in Tables I and II, allow us to formulate some inferences:

(1) The dispersions of $V_r$ velocities near the sun are approximately the same for the O-B5, O-B1 and B2-B5 spectral groups and are equal to 13–14 km s$^{-1}$.

(2) The dispersion increases with the distance from the sun in the direction of the galactic centre for the bulk of OB stars, while it is almost constant for the B2-B5 (according to Rubin's catalogue). In the anticentre direction the dispersion either slightly decreases or, within the limits of uncertainty, remains constant.

(3) For $R < R_0$ there exists no symmetry with respect to galactic longitude zero.

TABLE I

The dispersion and the parameter $C$ of the peculiar radial velocities of stars from the catalogue of Rubin *et al.*

| $l^{II}$ | $r$ kpc | O-B5 | | | B3-B5 | | |
|---|---|---|---|---|---|---|---|
| | | $\sigma$ | $C$ | $n$ | $\sigma$ | $C$ | $n$ |
| 330°–30° | 0 –0.5 | $13.8 \pm 1.7$ | $59 \pm 7$ | 32 | $14.5 \pm 2.1$ | $60 \pm 9$ | 24 |
| | 0.5–1 | $14.2 \pm 1.3$ | $63 \pm 6$ | 59 | $11.8 \pm 1.3$ | $72 \pm 8$ | 42 |
| | 1 –2 | $19.1 \pm 1.4$ | $70 \pm 5$ | 96 | $14.7 \pm 2.7$ | $68 \pm 12$ | 15 |
| | 2 –3 | $18.3 \pm 1.8$ | $70 \pm 7$ | 53 | | | |
| 30°–120° | 0 –0.5 | $10.0 \pm 0.7$ | $48 \pm 3$ | 99 | $9.7 \pm 0.8$ | $44 \pm 4$ | 80 |
| | 0.5–1 | $11.1 \pm 0.9$ | $43 \pm 4$ | 71 | $10.7 \pm 1.7$ | $49 \pm 8$ | 19 |
| | 1 –2 | $19.1 \pm 1.6$ | $76 \pm 6$ | 72 | | | |
| | 2 –3 | $21.3 \pm 2.5$ | $74 \pm 9$ | 36 | | | |
| 120°–150° | 0 –0.5 | $14.9 \pm 2.4$ | $63 \pm 10$ | 20 | $14.7 \pm 2.9$ | $60 \pm 12$ | 13 |
| | 0.5–1 | $20.0 \pm 3.0$ | $93 \pm 14$ | 23 | $10.9 \pm 2.7$ | $45 \pm 11$ | 8 |
| | 1 –2 | $12.2 \pm 1.7$ | $77 \pm 11$ | 26 | | | |
| | 2 –3 | $8.6 \pm 1.1$ | $41 \pm 5$ | 29 | $5.0 \pm 1.6$ | $33 \pm 10$ | 10 |
| 150°–240° | 0 –0.5 | $13.9 \pm 1.1$ | $64 \pm 5$ | 84 | $13.0 \pm 1.2$ | $58 \pm 5$ | 61 |
| | 0.5–1 | $14.8 \pm 1.5$ | $59 \pm 6$ | 51 | $11.2 \pm 1.8$ | $58 \pm 9$ | 20 |
| | 1 –2 | $13.3 \pm 1.0$ | $62 \pm 5$ | 91 | $11.4 \pm 1.6$ | $93 \pm 13$ | 26 |
| | 2 –3 | $12.2 \pm 1.8$ | $66 \pm 10$ | 22 | | | |
| 240°–330° | 0 –0.5 | $13.4 \pm 0.8$ | $56 \pm 4$ | 129 | $13.2 \pm 0.9$ | $51 \pm 4$ | 107 |
| | 0.5–1 | $18.7 \pm 2.0$ | $64 \pm 7$ | 45 | $6.6 \pm 0.9$ | $61 \pm 8$ | 26 |
| | 1 –2 | $14.6 \pm 1.7$ | $119 \pm 14$ | 38 | | | |
| | 2 –3 | $11.4 \pm 1.1$ | $53 \pm 5$ | 51 | | | |

TABLE II

The dispersion and the parameter $C$ of the peculiar radial velocities of stars from the catalogue of Bonneau

| $l^{II}$ | $r$ kpc | O-B2 $\sigma$ | $C$ | $n$ | B3-B5 $\sigma$ | $C$ | $n$ |
|---|---|---|---|---|---|---|---|
| $< 120°$ | 0 –0.5 | $9.3 \pm 1.6$ | $50 \pm 9$ | 17 | $7.6 \pm 0.9$ | $32 \pm 4$ | 35 |
| | 0.5–1 | $15.5 \pm 1.3$ | $55 \pm 5$ | 67 | $3.2 \pm 0.8$ | $16 \pm 4$ | 9 |
| | 1 –2 | $19.9 \pm 1.6$ | $86 \pm 7$ | 75 | $8.8 \pm 2.1$ | $36 \pm 8$ | · 9 |
| | 2 –3 | $18.1 \pm 2.3$ | $89 \pm 11$ | 31 | | | |
| | 3 –4 | $28 \pm 6$ | $130 \pm 28$ | 11 | | | |
| $120°–180°$ | 0 –0.5 | $17.2 \pm 3.0$ | $63 \pm 11$ | 16 | $8.9 \pm 1.2$ | $41 \pm 6$ | 26 |
| | 0.5–1 | $22.5 \pm 2.7$ | $93 \pm 11$ | 35 | $14.6 \pm 3.0$ | $56 \pm 11$ | 12 |
| | 1 –2 | $12.8 \pm 1.2$ | $89 \pm 9$ | 52 | $12.6 \pm 3.1$ | $119 \pm 30$ | 8 |
| | 2 –3 | $12.6 \pm 1.6$ | $48 \pm 6$ | 32 | | | |
| | 3 –4 | $18.0 \pm 4.3$ | $165 \pm 39$ | 9 | | | |
| $> 180°$ | 0 –0.5 | $13.2 \pm 1.5$ | $66 \pm 7$ | 39 | $10.8 \pm 1.1$ | $45 \pm 5$ | 48 |
| | 0.5–1 | $12.5 \pm 1.5$ | $63 \pm 8$ | 34 | $12.1 \pm 2.5$ | $111 \pm 23$ | 12 |
| | 1 –2 | $10.7 \pm 2.0$ | $79 \pm 8$ | 48 | $4.1 \pm 1.0$ | $20 \pm 5$ | 8 |
| | 2 –3 | $10.7 \pm 2.0$ | $50 \pm 9$ | 15 | | | |
| | 3 –4 | $16.0 \pm 4.0$ | $143 \pm 36$ | 8 | | | |

(4) In the region of the local spiral arm (in the first quadrant) the dispersion has its minimum value for $r < 1$ kpc and its maximum value for $r > 1$ kpc. This fact has been established only tentatively, owing to the largeness of the sections of space which contain spiral arms as well as regions between them. It is to be noted that an exclusion of a few stars with $V_r > 40$ km s$^{-1}$ reduces the value of the dispersion to circumsolar values. Small values of $\sigma$ we observe also in the Perseus arm in distinction to the inner arm. Small values in the Orion and Perseus arms are exhibited also by the first absolute moments as may be seen from Petrie's (1963) data.

At the same time the $C$ parameter of the Planck distribution function of peculiar velocities has been considered. The Planck distribution functions for the moduli of space velocities and for their radial components in the simplest form are:

$$f(V) = \frac{15}{\pi^4} \frac{C^4}{V^5(e^{C/V} - 1)},$$ (2)

$$f(V_r) = \frac{15}{2\pi^4 C} \int_0^{C/|V_r|} \frac{x^4 dx}{e^x - 1}.$$ (3)

The theoretical curves represent well the observed frequencies in the whole range of velocities, including very large ones, for stars of different types of spectra and luminosity. Theoretical moments of the distributions (2) and (3) truncated at the largest velocity in the sample $V^*$, coincide practically with (1).

In the range of the smallest velocities (for the OB-stars, $V < 5$ km s$^{-1}$) the curve (2) lies lower than the observed one. This defect may be removed by the introduction of a small parameter $\delta$ so that $V^2/C^2 + \delta$ is used as argument instead of $V^2/C^2$. For small values of $V$ this distribution is transformed into a Maxwellian one, while for large values into the Rayleigh-Jeans distribution.

The dispersion $\sigma$ and the parameter $C$ are connected by the relation

$$\sigma = S(V^*/C)\,C \tag{4}$$

and $S(V^*/C)$ is tabulated by Shatsova (1965a, b).

The advantage of $C$, as compared with $\sigma$, consists in its almost full independence of the truncation value. Because of that $C$ may be calculated either from the whole distribution curve of velocities, or from any of its parts, in particular from its 'tail' as well.

The range of the C-values in its main features resembles the range of dispersions. The relative differences of $C$ for different sections are somewhat smaller than for the $\sigma$.

## References

Bonneau, M.: 1967a, *Bull. Astron.* **2**, 13.
Bonneau, M.: 1967b, *J. Observateurs* **50**, 237.
Feast, M. W. and Shuttleworth, M.: 1965, *Monthly Notices Roy. Astron. Soc.* **130**, 245.
Petrie, R. M.: 1963, *Publ. Astron. Soc. Pacific* **75**, 354.
Petrie, R. M. and Petrie, J. K.: 1968, *Publ. Dom. Astrophys. Obs. Victoria* **13**, 253.
Rubin, V. C., Burley, J., Kiasatpoor, A., Klock, B., Pease, G., Ruthscheidt, E. and Smith, C.: 1962, *Astron. J.* **67**, 491.
Rubin, V. C. and Burley, J.: 1964, *Astron. J.* **69**, 80.
Shatsova, R. B.: 1965a, *Astron. Zh.* **42**, 160.
Shatsova, R. B.: 1965b, Planck's Distribution of Stellar Velocities near the Sun, Rostov-on-the-Don University.

# 56. ON THE DISTRIBUTION OF SPACE VELOCITIES
# OF OB STARS

L. V. MIRZOYAN and M. A. MNATSAKANIAN

*Bjurakan Astrophysical Observatory, Erevan, Armenia, U.S.S.R.*

OB stars, as the most typical members of O-associations, are of great interest for the problems of stellar evolution. Their concentration in spiral arms indicates the importance of the O-associations for the formation of the spiral arm population in galaxies.

In this report we present briefly the results of a study of the space velocities of O-B1 stars in stellar associations which can give some information on the internal motions in the spiral arms.

We have considered the so called synthetic association formed by superposing the subsystems of O-B1 stars around the nuclei of all known O-associations (Mirzoyan, 1961). If we assume that the associations are expanding, as has been predicted by Ambartsumian (1949, 1954) then we have to expect an increase of the mean velocity of expansion $V$ with the distance $r$ from the centre of the synthetic association as a result of the large dispersion of runaway velocities and non-simultaneous formation of stars in associations. We can assume as well that the intensity of star formation in the synthetic association, that is in the observed volume of the Galaxy during the last tens of millions of years has remained almost constant for all O-associations (Mirzoyan, 1965).

The first rough estimates of the velocities of O-B1 stars situated at different distances from the centre of the synthetic association confirmed the expected increase of the mean velocity with distance from the centre (Mirzoyan, 1961). Here we have applied a more correct method for the derivation of the function $V(r)$.

The general solution of the problem is presented by a formula which needs the use of the differentials of the observed distribution functions. Therefore the accuracy given by this formula is not high enough.

However, the problem becomes much simpler when we study the unknown function $V(r)$ only quantitatively. In this case one may express the mean values of $\bar{V}$ and $\bar{r}$ for the stars situated in plane-parallel layers in the projection of the association by two functions which can be determined from the observations: the number of stars and the sum of absolute radial velocities of stars in a circle of a given radius around the centre. Changing the widths and the positions of the layers in the association one may determine $\bar{V}$ for different $\bar{r}$ and, thus, the function $\bar{V}(\bar{r})$.

For the determination of $\bar{V}(\bar{r})$ we have used the radial velocities of 290 O-B1 stars from the Wilson (1953) catalogue. The distribution of these stars around the nuclei of stellar associations has been determined on the basis of their distances from the nearest nuclei. The distances of O-B1 stars have been taken from Hiltner's

*Becker and Contopoulos (eds.), The Spiral Structure of Our Galaxy, 295–296. All Rights Reserved.*

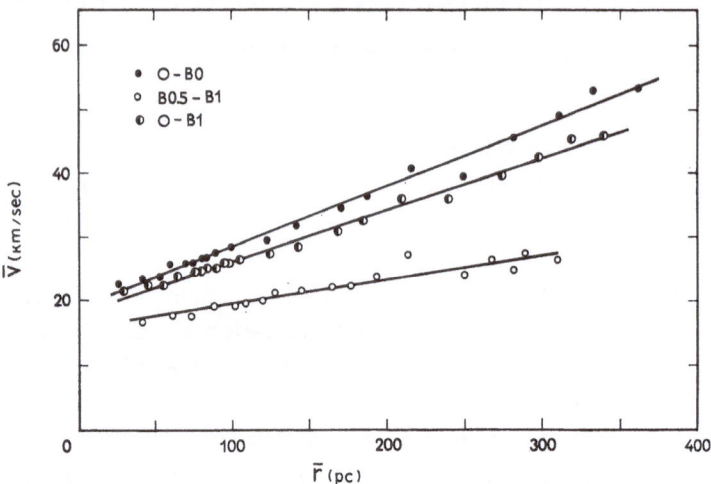

Fig. 1. The dependence $\bar{V}(\bar{r})$ in the synthetic association: for O-B0 (full dots), B0.5-B1 (open circles) and O-B1 (semifilled circles) stars. The lines are the solutions by the least-square method.

(1955) list. For the stellar associations the data of Ruprecht's (1966) catalogue have been used.

The results of calculations for O-B0 stars (222), B0.5-B1 stars (68) and for all O-B1 stars (290) are presented in Figure 1, where the lines are obtained by the least-square method.

It is well seen in Figure 1, that the dependence of mean velocity from the distance to the centre is almost linear in all three cases.

It can be shown, that in this particular case, when the function $\bar{V}(\bar{r})$ is linear, the unknown function $V(r)$ must be linear as well. Moreover, in this case both functions must be identically equal: $V(r) \equiv \bar{V}(\bar{r})$.

Thus the analysis of the distribution of velocities of O-B1 stars based on the radial velocities and the distribution of these stars around the nuclei of associations confirm in full the conclusion concerning the increase of the mean velocities of stars with their distances from the giving birth nuclei (Mirzoyan, 1961). This fact is a strong evidence in support of the expansion of stellar associations.

Therefore the existing distribution of O-B stars around the nuclei of stellar associations and in the spiral arms of our Galaxy is caused by the intensity of star formation during the life-time of these systems and by the process of star aging with running away from the parent nuclei (Mirzoyan, 1965).

## References

Ambartsumian, V. A.: 1949, *Astron. Zh.* **26**, 3.
Ambartsumian, V. A.: 1954, *Trans. IAU* **8**, 665.
Hiltner, W. A.: 1955, *Astrophys. J. Suppl.* **2**, 389.
Mirzoyan, L. V.: 1961, *Soobshch. Bjurak. Obs.* **29**, 81.
Mirzoyan, L. V.: 1965, *Astrofiz.* **1**, 109 = 1965, *Astrophys.* **1**, 70.
Ruprecht, J.: 1966, *Trans. IAU* **13A**, 830.
Wilson, R. E.: 1953, *General Catalogue of Stellar Radial Velocities*, Washington.

# 57. QUESTIONS CONCERNING THE USEFULNESS AND FEASIBILITY OF LARGE-SCALE STELLAR STATISTICS

W. SEITTER

*Observatorium Hoher List der Universitäts-Sternwarte Bonn, Germany*

**Abstract.** A proposal for a cooperative effort to secure three-color photometry of all stars up to the 18th magnitude of a 10° field around the galactic equator has been made.

PART III

# THEORY OF SPIRAL STRUCTURE

# 58. GRAVITATIONAL THEORIES OF SPIRAL STRUCTURE

G. CONTOPOULOS

*University of Chicago, Chicago, Ill., U.S.A.\**

**Abstract.** The basic ideas and some of the most important recent developments of the gravitational theories of spiral structure are described. A separation between linear and non linear effects is made. The linear self consistent problem consists of the problem of modes and of the initial value problem, which is discussed here in some detail. More emphasis is put on the non linear problem near resonances and in particular the inner Lindblad resonance. The linear density response to a slightly growing spiral potential (trailing or leading) near the inner Lindblad resonance is always trailing, while non linear effects form a density distribution with a roughly quadruple symmetry.

## 1. Introduction

One of the main early difficulties of the gravitational theory of spiral structure was the problem of differential rotation. If the spiral arms were composed always of the same stars then, after a few revolutions, the spiral arms would be wound very tightly and practically disappear. The way out of this difficulty is the notion of spiral arms as *waves*. Stars move through the spiral arms, but they stay there longer, on the average, so that the spiral arms are, at every moment, the maxima of density; they are not material arms, but spiral waves.

The idea of spiral waves is due to B. Lindblad. In a series of papers, starting in the early forties, he developed many of the elements of the present day theory of spiral waves (Lindblad, 1941, 1942, 1948, 1950; Lindblad and Langebartel, 1953; see also Coutrez, 1947). This is particularly remarkable, in view of the fact that no experience from similar problems in plasma physics was available at that time.

However, although B. Lindblad is the father of the gravitational theory of spiral structure, his views were never widely accepted by astronomers in general. Two reasons, I think, were responsible for that. First the fact that Lindblad's papers are difficult to follow, introducing many assumptions and approximations at every step, and second Lindblad's insistence on *leading* spirals (except in his last papers before his death). In his paper with Langebartel (1953) it is pointed out that spiral patterns can be both leading or trailing. The authors stress that their "general conclusions ... do not depend on which one of the two alternatives will ultimately prove to be most important". This statement is known to-day to be correct in the general case away from resonances, and, in fact, most of the recent work in spiral wave theory does not discriminate between leading and trailing spirals. However, after making this statement, Lindblad and Langebartel go on discussing in detail only leading spirals.

Thus, the credit goes to C. C. Lin, who not only developed the theory of spiral waves in much more detail, but also presented it in a relatively simple form that made it acceptable to the rest of the astronomical world. The response to the work of Lin

---

\* Present address: University of Thessaloniki, Greece.

*Becker and Contopoulos (eds.), The Spiral Structure of Our Galaxy, 303–316. All Rights Reserved.*
*Copyright © 1970 by the I.A.U.*

and his associates has been an ever growing wave of research in this area, that has produced many important new results.

In the present Report I will describe some of the main recent developments in this field, including my own recent work.

## 2. Outline of the Theory of Spiral Waves

One can divide the theories of spiral waves into two basic categories: local and global.

Local theories deal with relatively small regions of the galaxy, considering the center to be at distance large with respect to the local dimensions.

A local theory, referring to the gas, was developed by Goldreich and Lynden-Bell (1965) and another one, referring to stars, by Julian and Toomre (1966). Such theories explain the almost universal appearance of small, broken waves in spiral galaxies, which are trailing and may have a rather large inclination angle. Even in quite regular galaxies one sees such 'wavelets' as branches or bridges between the main arms.

However, most of the recent work deals with global theories, which aim at explaining the *grand design* of the more or less regular two-armed spirals that we see in abundance in the sky. Any irregularities in the spiral pattern are considered, in this approach, as higher order effects, to be introduced, eventually, at a later stage of the theory.

The kinds of problems considered by the theory of spiral structure are given in terms of increasing difficulty, in Table I.

TABLE I

Theory of spiral waves

|  |  |
|---|---|
| A. Linear | 1. Given spiral potential |
|  | 2. Self consistent problem |
|  | 2a. Modes |
|  | 2b. Initial value problem |
| B. Non-Linear | 1. Far from resonances |
|  | 2. Near resonances |

The first step in developing a theory of spiral waves is the linearization of the collisionless Boltzmann equation

$$\frac{\partial f}{\partial t} + \mathbf{v} \cdot \frac{\partial f}{\partial \mathbf{x}} - \frac{\partial V}{\partial \mathbf{x}} \cdot \frac{\partial f}{\partial \mathbf{v}} = 0. \tag{1}$$

This equation is known also as Vlasov's equation in plasma physics. Here $f$ is the distribution function, while $V$ is the potential of the spiral galaxy. We consider, first, a two-dimensional model of the galaxy; thickness effects can be introduced later.

We assume that $V$ is composed of an unperturbed, axisymmetric, part, $V_0(r)$, and a spiral part, $V_1$, which is small with respect to $V_0$. Thus we can write

$$V = V_0 + V_1,$$

and
$$f = f_0 + f_1 + f_2 + \cdots,$$  (2)

where $f_0$ is the distribution function of the axisymmetric substratum. If we write Equation (1) as

$$D(f, V) = 0$$  (3)

we notice that the operator $D$ is linear in $f$ and in $V$, therefore, if we introduce the value (2) and equate to zero the terms of various orders of Equation (3), we find, first,

$$D(f_0, V_0) = 0;$$  (4)

this means that $f_0$ is a function of the (isolating) integrals of motion of the unperturbed problem. In general, i.e. away from resonances, the only isolating integrals of motion are the angular momentum $J_0 = r^2 \dot{\vartheta}$ and the energy $E_o = \frac{1}{2}(\dot{r}^2 + J_0^2/r^2) + V_0$, therefore

$$f_0 = f_0(E_0, J_0).$$  (5)

The next equation, derived from Equation (3), is

$$D(f_1, V_0) + D(f_0, V_1) = 0;$$  (6)

this is the basic linearized collisionless Boltzmann equation, used extensively in galactic dynamics.

The first problem now consists in finding the response of a galactic disk to a *given* potential of the form (2).

If the spiral part of the potential, $V_1$, is *given*, the solution of Equation (6) can be written explicitly for every $f_0$. In fact Equation (6) is a partial differential equation, with characteristics the unperturbed orbits of the axisymmetric field $V_0(r)$, and its solution is

$$f_1 = \int P(f_0, V_1) \, d\tau,$$  (7)

where $P$ is an operator linear in $f_0$ and $V_1$, and the integration is along the unperturbed orbits; $\tau$ is an auxiliary parameter, namely the time along unperturbed orbits, appearing only in trigonometric terms (except in resonances) and after the integration it is expressed in terms of the coordinates.*

In a similar way one can find $f_2$, etc.; thus the distribution function $f$ can be found, step by step, as a formal series; $f$ is an integral of motion of the same form as the 'third' integral, found in other galactic problems. In particular if $f_0 = E_0$ we find a 'generalized energy' $E = E_0 + E_1 + \cdots$, and if $f_0 = J_0$ we find a 'generalized angular momentum' $J = J_0 + J_1 + \cdots$ (Contopoulos, 1967). In the case of a spiral pattern

---

* This method of solution of the linearized collisionless Boltzmann equation is used extensively in stellar dynamics and plasma physics (see, e.g., Contopoulos, 1960; Shu 1968).

rotating as a rigid body with angular velocity $\Omega_s$ the Hamiltonian

$$H = E - \Omega_s J = \tfrac{1}{2}(\dot{r}^2 + J_0^2/r^2) + V - \Omega_s J_0 \tag{8}$$

is known to be an analytic integral of motion.

A more difficult problem is the self-consistent (or self-gravitating) problem. In this case $V_1$ has to be found, together with $f_1$, through Equation (6) and Poisson's equation

$$\nabla^2 V_1 = 4\pi G \delta(z)\, \sigma_1 = 4\pi G \delta(z) \int f_1 d\mathbf{v}, \tag{9}$$

where $\delta(z)$ is Dirac's delta function, and $\sigma_1$ the perturbed surface density.

Lin and his associates (Lin 1966a, b, 1967a, b; Lin and Shu 1964, 1966, 1967; Lin et al., 1969) have considered in detail the problem of *modes*. This problem deals with spiral solutions of Equations (7) and (9), of the form

$$\begin{aligned}
V_1 &= V_1^* \exp\left[i(\omega t - m\vartheta)\right], \\
f_1 &= f_1^* \exp\left[i(\omega t - m\vartheta)\right], \\
\sigma_1 &= \sigma_1^* \exp\left[i(\omega t - m\vartheta)\right],
\end{aligned} \tag{10}$$

where $V_1$ is the spiral component of the potential, $f_1$ the corresponding distribution function, and $\sigma_1$ the surface density; here $m$ is the number of spiral arms (usually $m=2$), $\vartheta$ the angle in an inertial frame, and

$$\omega = m\Omega_s. \tag{11}$$

The functions $V_1^*$, $\sigma_1^*$ depend only on $r$, while $f_1^*$ depends on $r$ and the velocities.

If we integrate the solution (7) of Equation (6) over all velocities we find the surface density $\sigma_1$. At the same time we replace $V_1$ in Equation (7) by a solution of Poisson's Equation (9). Then, if we eliminate the factor $\exp\left[i(\omega t - m\vartheta)\right]$, we find an integral equation of the form

$$r\sigma_1^*(r) = \int K_{m,\,\omega}(r, r')\, r'\sigma_1^*(r')\, dr', \tag{12}$$

where the kernel $K$ is a complicated function, depending on $m$ and $\omega$. This is the basic integral equation of galactic dynamics. It is of the general form of an homogeneous Fredholm equation of the second kind (the difference is that the dependence on the parameter $\omega$ is not linear, and the equation may be singular). Its eigenvalues $\omega$ are the modes of the self consistent problem and its eigenfunctions $r\sigma_1^*$ give the corresponding perturbed surface density. This integral equation was given first by Kalnajs (1965) and then in the formalism used in Lin's theory by Shu (1968).

The general solution of this equation is extremely difficult. Thus Lin introduced an 'asymptotic' approximation that simplifies the problem considerably and makes it tractable. The 'asymptotic' approximation consists in assuming the radial wavelength $\lambda$ of the spiral pattern as small and omitting all higher order terms in $\lambda$. Then the integral equation is reduced to an algebraic relation between $\omega$, the wave number

$k = 2\pi/\lambda$, and $r$,

$$D(\omega, k, r) \equiv 1 - \frac{2\pi G\sigma_0}{|k| \langle \dot{r}^2 \rangle} \left[ 1 - \frac{\nu\pi}{\sin \nu\pi} \mathfrak{G}_\nu(\chi_*) \right] = 0. \tag{13}$$

Here $\sigma_0$ is the surface density of the basic, axisymmetric, distribution, $\langle \dot{r}^2 \rangle^{\frac{1}{2}}$ the velocity dispersion, $\nu$ is the 'relative frequency' defined by

$$\nu = (\omega - 2\Omega)/\kappa, \tag{14}$$

with $\Omega$ the angular velocity of the galactic rotation and $\kappa$ the 'epicyclic frequency' at distance $r$,

$$\chi_* = k^2 \langle \dot{r}^2 \rangle/\kappa^2 \tag{15}$$

and

$$\mathfrak{G}_\nu(\chi_*) = \frac{1}{2\pi} \int_{-\pi}^{\pi} \cos \nu\gamma \exp\left[ -\chi_*(1 + \cos\gamma) \right] d\gamma. \tag{16}$$

The relation (13) is Lin's dispersion relation (Lin, 1966a) for a Schwarzschild distribution of unperturbed velocities.

By solving this equation for a given $\omega$ we find $k = k_\omega(r)$. Then the spiral arms, in a frame of reference rotating with the spiral field, are given by

$$\vartheta' = \frac{1}{2} \int k_\omega(r) \, dr + \text{const} \, (+ \pi), \tag{17}$$

where

$$\vartheta' = \vartheta - \Omega_s t. \tag{18}$$

Lin found which spirals of the form (17) fit best the spiral arms of our Galaxy and in this way he derived a value of $\Omega_s = \omega/2$ near 13 km s$^{-1}$ kpc$^{-1}$.

A confirmation of this value of the angular velocity of the spiral pattern, $\Omega_s$, came from a rather different approach by Fujimoto at Columbia University. Fujimoto (1968) studied in particular gaseous spiral arms, using the hydrodynamic equations instead of the collisionless Boltzmann Equation. He solved the linearized equations numerically and found the density response to various imposed spiral potentials. Imposing self-consistency, i.e. agreement between the phase and amplitude of the response with the density responsible for the spiral field he could find a value of $\Omega_s$ similar to that of Lin, and, further, a relation between the inclination of spiral arms and the proportion of gas in them; more open spirals contain more gas.

Accurately speaking a linear theory cannot give the absolute value of the amplitude of the wave, because if $\sigma_1^*$ is a solution of Equation (12), or of a similar linear equation, so is also $c\sigma_1^*$, where $c$ is an arbitrary constant. Thus Fujimoto considered only the relative variations of the response with the radius $r$. The absolute value of the amplitude can be found only by a non-linear theory. This has been done recently by Vandervoort and will be reported during this Symposium.

Kalnajs has recently solved numerically the integral Equation (12) and has found values of $\Omega_s$ of the order of 30 km s$^{-1}$ kpc$^{-1}$. His spirals are rather open and cannot be treated by an 'asymptotic' theory, like Lin's. Most of Kalnajs' original work is contained in his Thesis (1965), while his recent work will be reported during the present Symposium. The problem of open spirals is quite difficult and more effort should be turned in this direction.

Let us now turn to the initial value problem. This is the problem of the evolution of a *given* initial perturbation (given at time $t=0$).

This problem is well known in plasma physics. In the case of an homogeneous plasma its solution by Landau (1946) is by now classical. However in the galactic case only Kalnajs (1965) mentioned it briefly.

One can solve, in principle, the initial value problem after the problem of modes has been solved. The solution is found by a variation of Landau's method of Fourier transforms in space and a Laplace transform in time. Namely we perform a Fourier analysis in the angle $\vartheta$, followed by a Laplace transform in time, omitting a Fourier transform in $r$.

It is obvious that any perturbation of an axisymmetric galaxy, being a periodic function in the angle $\vartheta$, with period $2\pi$, can be Fourier analyzed into a one-armed perturbation, a two-armed perturbation, etc. Let us consider only two-armed perturbations. Then we can write

$$
\begin{aligned}
V_1 &= V_{11} \exp(-2i\vartheta), \\
f_1 &= f_{11} \exp(-2i\vartheta), \\
\sigma_1 &= \sigma_{11} \exp(-2i\vartheta).
\end{aligned}
\tag{19}
$$

The initial perturbation, at $t=0$, is also written

$$
f_{1;0} = f_{1;0}^* \exp(-2i\vartheta),
\tag{20}
$$

where $f_{1;0}^*$ is a function of $r$ and the velocities. We can write

$$
f_{1;0}^* = f_0 a \exp(i\phi),
\tag{21}
$$

where $f_0$ is the unperturbed distribution function, $a$ a relative amplitude and $\phi$ a phase angle.

Then we take the Laplace transforms of $V_{11}, f_{11}$ and $\sigma_{11}$

$$
f_1^* = \int_0^\infty f_{11} \exp(-i\omega t)\, dt
\tag{22}
$$

and similar expressions for $V_1^*$ and $\sigma_1^*$. Thus Equation (6) gives a differential equation for $f_1^*$

$$
i\left(\omega - \frac{J_0}{r^2}\right) f_1^* + \dot{r}\frac{df_1^*}{dr} = \frac{\partial f_0}{\partial E_0}\dot{r}\frac{dV_1^*}{dr} - 2iV_1^*\left(\frac{\partial f_0}{\partial E_0}\frac{J_0}{r^2} + \frac{\partial f_0}{\partial J_0}\right) + f_{1;0}^*,
\tag{23}
$$

under the restriction

$$
\lim_{t\to\infty}\left[f_{11}\exp(-i\omega t)\right] = 0.
\tag{24}
$$

Equation (23) is the same as the equation for the modes (Shu, 1968) except for the last term $f_{1;0}^*$. If we solve it for $f_1^*$ and integrate over all velocities we find an integral equation very similar to Equation (12), namely

$$r\sigma_1^*(r) = \int K_{m,\omega}(r, r')\, r'\sigma_1^*(r')\, dr' + s_{10}(r),$$  (25)

Where $s_{10}(r)$ is a known function, depending on the initial conditions. This is of the general form of a *nonhomogeneous* Fredholm equation of the second kind, and its solution can be given once the solution of the homogeneous Equation (12) is known.

In the asymptotic case the solution can be given by using the formalism of Lin and his associates. In the Appendix we derive the solution for the perturbed density $\sigma_1$ in the form

$$\sigma_1 = \sigma_0 \, a \exp\left[i(\phi - \pi/2 - 2\vartheta)\right] \frac{1}{2\pi\kappa} \int\limits_{-\infty + i\omega_I}^{\infty + i\omega_I} \frac{1}{\sin\nu\pi} \frac{\mathfrak{G}_\nu(\chi_{*;0})}{D} \exp(i\omega t)\, d\omega,$$  (26)

where $D$ is the function (13) and $i\omega_I$ is the imaginary part of $\omega = \omega_R + i\omega_I$ along the line of integration; this line is drawn below all singularities of the integrand, which are assumed to be poles.

Following Landau we move the line of integration in the integral (26) parallel to itself, so that it comes at $i\omega'$ with $\omega' > 0$. Then it is known that for large $t$ the only contribution to $\sigma_1$ comes from the poles of the integrand. If we are not at a resonance (i.e. when $\sin\nu\pi \neq 0$) the only poles are the roots of the equation $D = 0$, i.e. Lin's modes. Thus the problem of the origin of spiral waves becomes, in some sense, trivial. Because practically any two-armed initial perturbation excites Lin's modes. (Of course the appearance of other modes also is not excluded, especially in open spirals.)

However, there is a difference between this result and Lin's original picture of the modes. The value of $\omega$, which is a solution of Equation (13), is not unique. In fact the solution $\sigma_1$ (Equation (26)), gives the wave number $k$ as a function of $r$ and $t$,

$$k = k(r, t),$$  (27)

thus Equation (13) gives also $\omega$ as a function of $r$ and $t$,

$$\omega = \omega(r, t).$$  (28)

Relations of this form have been used as the starting point of Toomre's recent work (1969) on the evolution of the density waves in a galaxy. The fact that $\omega$ is not constant along a spiral wave produces a differential rotation, affecting the wave itself, which tends to produce the ultimate dissolution of spiral waves.

Toomre finds that the 'group velocity' of the spiral waves, $d\omega/dk$, is directed inwards in the main part of the galaxy and of the order of 10 km s$^{-1}$. Any information contained in the spiral waves moves inwards with this group velocity. Toomre found that certain quantities, like the wave number and the action density (energy density divided by the relative frequency $v$) are preserved, as they are transmitted with the group velocity. Thus the whole wave pattern moves towards the inner Lindblad resonance. Toomre suspects that the energy of the wave is transformed there to thermal energy of the stars, and the wave is damped. In a numerical example he found that the amplitude of the wave decreases as the wave moves inwards, and tends to zero near the inner Lindblad resonance. This damping of the wave happens in about $10^9$ years, therefore a mechanism is needed to regenerate the spiral waves. Various suggestions of 'exterior' forcing mechanisms were proposed by Toomre, including the Magellanic Clouds or a small bar near the center of the Galaxy.

However Toomre's conclusion is not generally accepted. An alternative assumption, invoking a *reflection* of the wave near the inner Lindblad resonance, will be presented by Lin during this Symposium. At any rate near the inner Lindblad resonance strong non-linear phenomena take place and as we will see presently, any linear theory is not applicable there except for short times.

## 3. Non-Linear Theory

Away form resonances non-linear effects appear only as small corrections in $f_1$ (higher order terms $f_2$, etc.). Their main function is in stopping the growth of a finite, growing, wave, so that it reaches a stationary state. Then one can calculate the *amplitude* reached by the wave (Vandervoort).

Non-linear effects are extremely important near resonances. There the whole linear theory is inapplicable for long times. In fact in the response integral (7), as given by Lin and his associates, there is a denominator $\sin v\pi$, and this becomes very small if $v$ is near an integer. Then $f_1$ is larger than $f_0$ and the whole approximation scheme implied by the linearization is not valid. In particular, if $v$ is exactly an integer, $f_1$ contains a secular term; therefore the linear theory can be applied only for a short time.

Whenever $v$ is an integer (or a rational number) we have a resonance between the frequency of rotation in a frame of reference rotating with the spiral pattern, and the epicyclic frequency; the unperturbed orbits in the rotating frame are closed, periodic, orbits. The most important resonance in a galaxy is the inner Lindblad resonance, where

$$v = -1. \tag{29}$$

In order to solve the collisionless Boltzmann equation in this case we cannot start with $f_0$ a function of the energy $E_0$ and the angular momentum $J_0$, because then $f_0$ contains secular terms, except in the quite special case that $f_0$ is a function of the Hamiltonian only.

However, in the case $v = -1$ we have one more isolating integral of the axisymmetric problem, namely the initial phase difference between the motion around the center of the galaxy and the radial (epicyclic) motion. We use this integral in the form

$$\frac{S_0}{C_0} = \frac{\sin}{\cos}(2\vartheta_1 - \omega\tau_1), \tag{30}$$

where $\vartheta_1$ and $\tau_1$ are the angle and the corresponding time of a pericentron passage.

We can now find a function $f_0$ of $E_0$, $J_0$ and $S_0$ (or $C_0$), such that it does not produce secular terms in $f_1$. This means that $f_0$ cannot be a Schwarzschild distribution function, but has a more complicated form. The main result (Contopoulos, 1970a) is that $f_0$ has a $4\vartheta'$ dependence, where $\vartheta'$ is the angle (17) in the rotating frame. Namely the main term of $f_0$ contains $\cos[4\vartheta' + \Phi_1(r)]$, where $\Phi_1$ is a phase function.

This quadruple symmetry of $f_0$ is evident in the form of the orbits near the inner Lindblad resonance. Away from resonances the orbits fill rings around the center (Figure 1); for every given value of the Hamiltonian (8) there is one almost circular

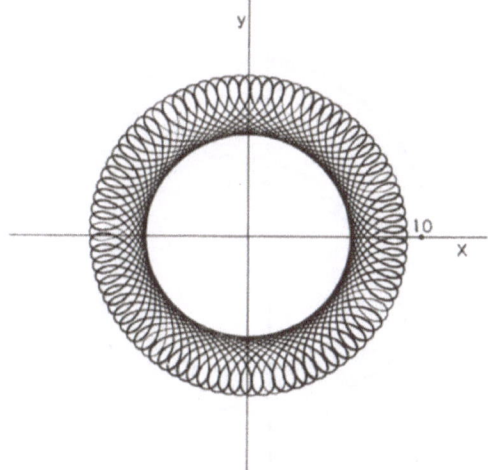

Fig. 1. A ring-type orbit.

periodic orbit, and all the rest form rings with boundaries on each side of the periodic orbit. Near the inner Lindblad resonance, however, for every value of the Hamiltonian there are two resonant periodic orbits, like ellipses with their center at the center of the Galaxy. Orbits near the resonance (e.g. starting at a distance up to 2 kpc from the resonance, with appropriate values of the Hamiltonian) form *tube orbits* around the two resonant periodic orbits (Figure 2). Orbits starting further away from the resonance form rings, as in Figure 1, but of larger width; the set of such orbits, however, is small.*

* The other resonances, besides the inner Lindblad resonance, are less important, because they involve much smaller sets of orbits. A particularly interesting resonance is the particle resonance, where $v = 0$, which will be discussed by Barbanis.

Resonances are encountered also in many problems of stellar dynamics and celestial mechanics. Tube orbits in a meridian plane of an axisymmetric galaxy were found by Torgard and Ollongren (1960) (see also Ollongren, 1965). Their theoretical explanation, as resonance phenomena, was given by means of the 'third' integral (Contopoulos, 1965) and follows the lines discussed above. Excellent agreement was found between theory and numerical experiments, by using an extra integral, like $S_0$ (or $C_0$), in the resonant case. The analogy with the present problem is discussed in more detail in a forthcoming paper (Contopoulos, 1970a).

In celestial mechanics the same problem is known as the problem of 'small divisors'. Resonances appear in the gaps of the asteroids, in satellite orbits at the 'critical inclination', etc.

The superposition of all the orbits near the inner Lindblad resonance gives the density response. This has also a rough quadruple symmetry. In fact, in some of the galaxies shown by Morgan during this Symposium, one could see nuclei showing a rough quadruple symmetry.

It is obvious that if the spiral field is infinitesimal resonant effects are also infinitesimal. In order to see the growth of a resonance we have calculated numerically orbits in a *growing* wave (Contopoulos, 1970a). Initially the spiral potential $V_1$ is zero and the stars start moving along their unperturbed (epicyclic) orbits. However as the wave gradually grows, tending to its maximum amplitude, the orbits deviate gradually,

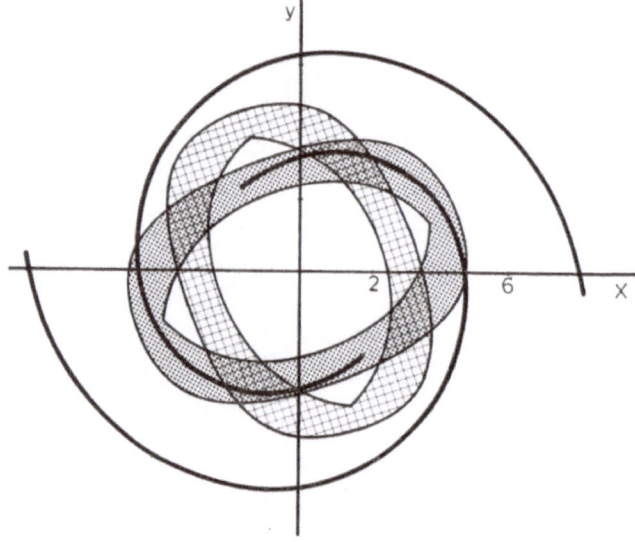

Fig. 2.    The areas covered by two tube orbits near the inner Lindblad resonance. The thick line gives the minimum of the spiral potential.

tending to their final tube form around one or the other of the two resonant periodic orbits. In a particular example the field reaches 0.9 of its maximum amplitude in about $2 \times 10^9$ years, while the orbits reach almost exactly their final tube forms after at most 10 revolutions. Therefore the growth of resonances is an important factor in the evolution of a spiral wave.

We must stress that any 'absorption' or 'reflection' of the waves near the inner Lindblad resonance is different from the corresponding linear phenomena, known from plasma physics, which happen away from resonances. Lin will mention the possibility of the excitation of a long outgoing wave near the inner Lindblad resonance which may be considered as a non-linear reflection. If such a wave exists it may also have important consequences for the dynamics of spiral waves.

## 4. Preference of Trailing Waves

It is generally believed to-day that spiral waves are trailing rather than leading. This is well known for local, sheared, wavelets (Goldreich and Lynden-Bell, 1965; Julian and Toomre, 1966), but in the case of the grand design of spiral waves leading and trailing waves appear as equivalent. Some 'indications' for the amplification of trailing waves (Lin and Shu 1966; Lin 1967a) have not proved working (Toomre, 1969). The only evidence for the preference of trailing waves has been provided by Kalnajs (1965), who states that the response to a trailing wave is trailing, while the response to a leading wave is both a leading and a trailing wave. Kalnajs' argument is correct, but difficult to follow, and applies to the general case of open spirals.

Thus we made some calculations following Lin and Shu's formalism, for a slightly growing wave near the inner Lindblad resonance. Namely we calculated the linear response, near the inner Lindblad resonance, of a spiral wave of the form

$$V_1 = A(r) \exp\{i[\Phi(r) + \omega t - 2\vartheta]\}, \tag{31}$$

where $\omega = \omega_R + i\omega_I$ has a small negative imaginary part, which gives a slightly growing wave (because $V_1$ contains the factor $\exp(-\omega_I t)$). The value of $|\omega_I|$ was taken equal to $0.03\ \omega_R$. The results of the calculation are shown in Figure 3. It is seen that near resonance the response to a trailing wave is trailing, while the response to a leading wave is also trailing.

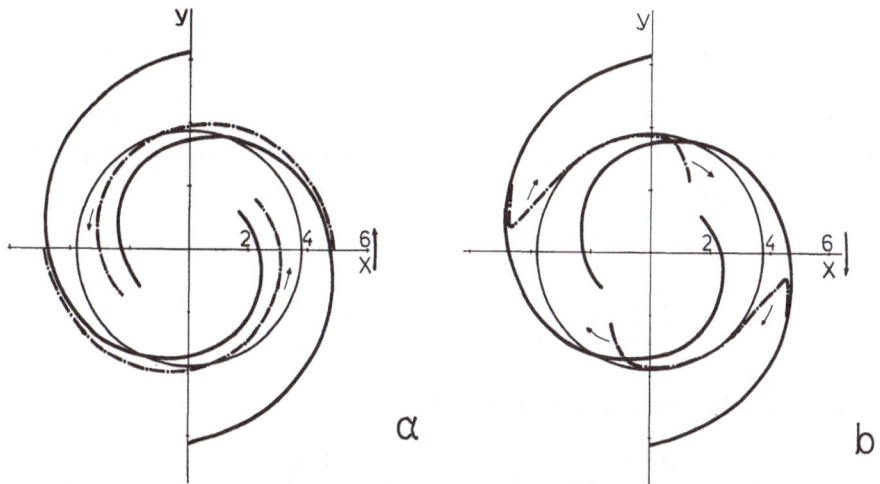

Fig. 3.   Linear density response to a growing imposed field near the inner Lindblad resonance (thin circle). The thick solid line represents the minimum of the potential, which is trailing in (a) and leading in (b). The response (dash-dotted line) is trailing in both cases.

It is known from Lin's theory that outside the inner Lindblad resonance the response is almost exactly in phase with the imposed potential, i.e. the maxima of density almost coincide with the minima of potential. Inside the Lindblad resonance the response is 90° out of phase. Therefore there cannot be a self consistent spiral wave inside the inner Lindblad resonance, at least in the linear theory. This can be well understood in terms of the initial value problem. As the Equation (13) has no solution inside the inner Lindblad resonance, the integrand of Equation (26) has no poles, therefore the solution (26) does not tend to a stationary spiral form after a long time.

This is consistent with observations of the nuclear region of galaxies and with the numerical experiments of Miller and Prendergast.

In the case of an imposed growing spiral field it is not at all evident how the in-phase response outside the inner Lindblad resonance is joined to the out-of-phase response inside it. However an exact linear calculation shows that in both cases of a trailing and of a leading wave the response precedes the imposed potential, and more strongly so in the case of the leading wave (Figure 3). Thus a leading wave is strongly distorted near the inner Lindblad resonance tending to become trailing. This effect discriminates strongly in favor of trailing waves. More details will be given in a future paper (Contopoulos, 1970b). It appears that permanent leading waves are impossible in galaxies possessing an inner Lindblad resonance.

The above review of the recent developments in the theory of spiral structure shows that much progress has been made in recent years. However there are still some basic unsolved problems. Perhaps the most urgent at this moment is to find what is the fate of the groups of waves moving towards the inner Lindblad resonance with Toomre's group velocity. The complete exploration of the non-linear effects near resonances may give the answer to the basic problems of the origin and persistence of spiral waves.

### Appendix. The Initial Value Problem in the Asymptotic Approximation

The solution of Equation (23) of the initial value problem can be found in the same way as the solution of the corresponding equation for modes (Shu, 1968). It is

$$f_1^* = \frac{\partial f_0}{\partial E_0} V_1^* - \frac{1}{2 \sin(\omega \tau_0 - 2\vartheta_0)}$$

$$\times \int_{-\tau_0}^{\tau_0} \left\{ V_1^* \left( 2 \frac{\partial f_0}{\partial J_0} + \omega \frac{\partial f_0}{\partial E_0} \right) + i f_{1;\,0}^* \right\} \cos[\omega \tau - 2\vartheta(\tau)] \, d\tau. \quad (A1)$$

Then, using a Schwarzschild distribution function

$$f_0 = \frac{\Omega_0 \sigma_0(r_0)}{\pi \kappa_0 \langle \dot{r}^2 \rangle_0} \exp\left\{ -\frac{[\dot{r}^2 + \kappa_0^2(r_0 - r)^2]}{2 \langle \dot{r}^2 \rangle_0} \right\}, \quad (A2)$$

(where a zero subscript in $\Omega$, $\kappa$, $\langle \dot{r}^2 \rangle$ means quantities calculated at $r = r_0$) and the approximate relation (Shu, 1968)

$$2 \frac{\partial f_0}{\partial J_0} + \omega \frac{\partial f_0}{\partial E_0} = -\frac{f_0 v_0 \kappa_0}{\langle \dot{r}^2 \rangle_0} \quad (A3)$$

we find, using the same approximations as Shu,

$$f_1^* = \frac{f_0}{\langle \dot{r}^2 \rangle_0} \left\{ -V_1^* + \frac{v_0}{2 \sin v_0 \pi} \int_{-\pi}^{\pi} \left[ V_1^* - \frac{i \langle \dot{r}^2 \rangle_0}{v_0 \kappa_0} a \exp(i\phi) \right] \cos v_0 \gamma \, d\gamma \right\}.$$

$$(A4)$$

Integrating over all velocities we find the density response. If we assume that $a$ and $\phi$ depend only on $r$ and not on the velocities, we find

$$\sigma_1^* = \frac{-\sigma_0 V_1^*}{\langle \dot{r}^2 \rangle} \left\{ 1 - \frac{v\pi}{\sin v\pi} \mathfrak{G}_v(\chi_*) \right\} - \frac{i\pi\sigma_0 a \exp(i\phi)}{\kappa \sin v\pi} \mathfrak{G}_v(\chi_{*;0}), \qquad (A5)$$

where $\mathfrak{G}_v(\chi_*)$ is given by Equation (16), and $\chi_{*;0} = (\phi')^2 \langle \dot{r}^2 \rangle / \kappa^2$, where $\phi' = k$ at $t = 0$. If the last assumption about $a$ and $\phi$ is not made the last term of Equation (A5) is slightly more complicated.

If we replace $V_1^*$ by the lowest order solution of Poisson's equation, $V_1^* = -2\pi \times G\sigma_1^*/|k|$, and omit the second term of Equation (A5) we find Lin's dispersion relation (Equation (13)).

In the initial value problem $D$ is not zero, in general, because of the second member in (A5). The solution of Equation (A5) is

$$\sigma_1^* = \frac{\pi\sigma_0 a \exp[i(\phi - \pi/2)]}{\kappa \sin v\pi} \frac{\mathfrak{G}_v(\chi_{*;0})}{D}. \qquad (A6)$$

Then, inverting the Laplace transform (22), and using Equation (19), we find the solution of the initial value problem in the form (26).

If we move the line of integration above the real axis of $\omega$, then, for large $t$, only the poles of the integrand give a contribution to $\sigma_1$.

Let us consider the main root of the dispersion relation $D = 0$, say $\omega = \omega\,(r, k)$. Then we find

$$\sigma_1 = \frac{\sigma_0 a \exp[i(\phi - 2\vartheta)]}{\sin v\pi} \frac{\mathfrak{G}_v(\chi_{*;0})}{\partial D/\partial v} \exp(i\omega t), \qquad (A7)$$

where it is assumed that $\sin v\pi \neq 0$.

Equation (A7) gives the form of the perturbed density distribution for every (large) $t$, therefore it gives also the wave number $k$. We know that initially (for $t = 0$) it is

$$k(r) = \phi'(r). \qquad (A8)$$

Let us disregard the transition period needed for $\sigma_1$ to reach the form (A7) and consider a new 'initial time', such that (A7) is satisfied approximately for every $t \geqslant 0$. Then we have two possibilities.

(a) If the initial perturbation satisfies Lin's dispersion relation for a fixed $\omega$, i.e. if $k(r)$, defined by Equation (A8), satisfies the equation $D(\omega, k(r), r) = 0$, then this mode is stationary, except for resonance effects.

(b) If the initial perturbation does not satisfy the above condition, then, for $t = 0$, $\omega$ is not constant, but a function of $r$. Equation (A7) gives at time $\Delta t$

$$k(r) = \phi'(r) + \omega'(r)\,\Delta t, \qquad (A9)$$

therefore the wave number changes in time. Then the dispersion relation gives $\omega$ as a function of $r$ and $t$, i.e. the spiral *pattern* has a differential rotation.

# References

Contopoulos, G.: 1960, *Z. Astrophys.* **49**, 273.

Contopoulos, G.: 1965, *Astron. J.* **70**, 526.

Contopoulos, G.: 1967, *Proceedings 14th Liège Colloquium,* p. 213.

Contopoulos, G.: 1970a, *Astrophys. J.,* in press.

Contopoulos, G.: 1970b, *Astrophys. J.,* to be published.

Coutrez, R.: 1947, *Stockholm Obs. Ann.* **15**, No. 3.

Fujimoto, M.: 1968, *Astrophys. J.* **152**, 391.

Goldreich, P. and Lynden-Bell, D.: 1965, *Monthly Notices Roy. Astron. Soc.* **130**, 125.

Julian, W. H. and Toomre, A.: 1966, *Astrophys. J.* **146**, 810.

Kalnajs, A.: 1965, Ph.D. Thesis, Harvard University.

Landau, L.: 1946, *J. Phys. U.S.S.R.* **10**, 25.

Lin, C. C.: 1966a, *SIAM J. Appl. Math.* **14**, 876.

Lin, C. C.: 1966b, in *Galaxies and the Universe,* Columbia University Press, New York and London, p. 33.

Lin, C. C.: 1967a, in *Relativity Theory and Astrophysics, 2: Galactic Structure,* Amer. Math. Soc., Providence, p. 66.

Lin, C. C.: 1967b, *Ann. Rev. Astron. Astrophys.* **5**, 453.

Lin, C. C. and Shu, F. H.: 1964, *Astrophys. J.* **140**, 646.

Lin, C. C. and Shu, F. H.: 1966, *Proc. Nat. Acad. Sci. U.S.A.* **55**, 229.

Lin, C. C. and Shu, F. H.: 1967, IAU-URSI Symposium No. 31, p. 313.

Lin, C. C., Yuan, C., and Shu, F. H.: 1969 *Astrophys. J.* **155**, 721.

Lindblad, B.: 1941, *Stockholm Obs. Ann.* **13**, No. 10.

Lindblad, B.: 1942, *Stockholm Obs. Ann.* **14**, No. 1.

Lindblad, B.: 1948, *Stockholm Obs. Ann.* **15**, No. 4.

Lindblad, B.: 1950, *Stockholm Obs. Ann.* **16**, No. 1.

Lindblad, B. and Langebartel, R.: 1953, *Stockholm Obs. Ann.* **17**, No. 6.

Ollongren, A.: 1965, *Ann. Rev. Astron. Astrophys.* **3**, 113.

Shu, F. H.: 1968, Ph.D. Thesis, Harvard University.

Toomre, A.: 1969, *Astrophys. J.* **158**, 899.

Torgard, I. and Ollongren A.: 1960, NUFFIC Intern. Summer Course in Science, Part X.

# 59. THE THEORY OF SPIRAL STRUCTURE OF GALAXIES

## L. S. MAROCHNIK

*Astrophysical Institute of the Academy of Sciences of the Tadjik S.S.R.,*
*Dushanbe, U.S.S.R.*

**Abstract.** A Landau-type instability mechanism for generating spiral waves is suggested.

Two populations of stars, Populations I and II, are considered, the second one with mean rotational velocity zero. Then a dispersion relation is derived which is reduced to the Lin-Shu dispersion relation in the case of vanishing Population II. The amplification of the wave is of the same type as the two-stream instability. It occurs if the angular velocity of the spiral pattern $\Omega_s$ is smaller than the angular velocity of the Population I stars. A value of $\Omega_s = 22$–$25$ km s$^{-1}$ kpc$^{-1}$ was found, as well as the growth parameter. Spiral arms are formed in $10^8$–$10^9$ yr, while trailing and leading waves grow at the same rate.

A quasi-linear theory is developed to account for the limited growth of the spiral waves.

Detailed accounts of the theory and of its implications are contained in recent publications (Marochnik, 1969; Marochnik and Suchkov, 1969a; 1969b; Marochnik and Ptitzina, 1969; Marochnik *et al.*, 1969).

## References

Marochnik, L. S.: 1969, *Astrofiz.* **15**, 487.
Marochnik, L. S. and Suchkov, A. A.: 1969a, *Astron. Zh.* **46**, 319, 524.
Marochnik, L. S. and Suchkov, A. A.: 1969b, *Astrophys. Space Sci.* **4**, 317.
Marochnik, L. S. and Ptitzina, N. G.: 1969, *Astron. Zh.* **46**, 762.
Marochnik, L. S., Pomagaev, S. G., Sagdeev, R. Z., and Suchkov, A. A.: 1969, *Dokl. Akad. Nauk S.S.S.R.,* in press.

# 60. SMALL AMPLITUDE DENSITY WAVES ON A FLAT GALAXY

A. J. KALNAJS

*Harvard College Observatory, Cambridge, Mass., U.S.A.*

**Abstract.** By numerical methods we have found an unstable two-armed density wave on a flat galactic model. We present the results in a form of four plots, and briefly discuss the observational implications as well as the uncertainties involved in the models and the calculations.

## 1. Brief Description of the Calculations

We have investigated the stability of a flat model galaxy derived from the rotation curve of M31 (Van de Hulst *et al.*, 1957). From 4 kpc outward sufficiently high random motions were incorporated in the model to make it stable against local axisymmetric collapse. To stabilize the inner part rather large orbital eccentricities are required. Rather than extrapolate our epicyclic orbits to eccentricities larger than 0.2, we reduced the response by supposing that only a fraction of the stars participated in the collective modes. Such a procedure appears to be qualitatively correct and is consistent with the results reported below. The perturbations or modes are assumed to be small in amplitude and are found by solving the Poisson and linearized Vlasov equations. The solutions of these equations can be written in the form $\mathrm{Re}\left[A(r)\exp i(m\theta+\omega t)\right]$ and are density waves which rotate around the axis of the galaxy with a pattern speed $=\mathrm{Re}(-\omega/m)$. The amplitude $A(r)$ and the frequency $\omega$ are obtained by solving an integral equation (Kalnajs, 1965).

The two armed or $m=2$ modes are special for the reasons pointed out by Lindblad (1958). We have obtained numerically the largest discrete $m=2$ mode. By largest we mean that the gravitational interactions associated with it are strongest as measured by the shift of the pattern speed ($=30$ km s$^{-1}$ kpc$^{-1}$) from the kinematical value ($\Omega-\kappa/2\approx10$ km s$^{-1}$ kpc$^{-1}$) which would obtain if the gravitational effects of the perturbation were neglected (Lindblad, 1958). The calculation itself involved the replacement of the kernel of the integral equation by a $60\times60$ complex matrix. The eigenvector solution was numerically stable.

After finding a self-consistent mode it is possible to calculate the response of any subsystem of the galaxy to it. Since the spiral structure is seen most prominently in the gas and objects with the lowest velocity dispersion, we have calculated the density response and velocity fields of zero velocity dispersion objects. This is a reasonable approximation since the pressure effects in the gas are negligible on the scale of kiloparsecs.

The results are graphically summarized in Figures 1–4. The rotation of the galaxy as well as of the pattern is counterclockwise. The amplitude of the latter grows at the rate of two powers of $e$ in $10^9$ yr. For the sake of clarity only the positive values of the perturbed quantities have been printed. The numbers in the figures must be multiplied by the indicated scale factors to obtain a consistent set of values. The

*Becker and Contopoulos (eds.), The Spiral Structure of Our Galaxy, 318–322. All Rights Reserved.*
*Copyright © 1970 by the I.A.U.*

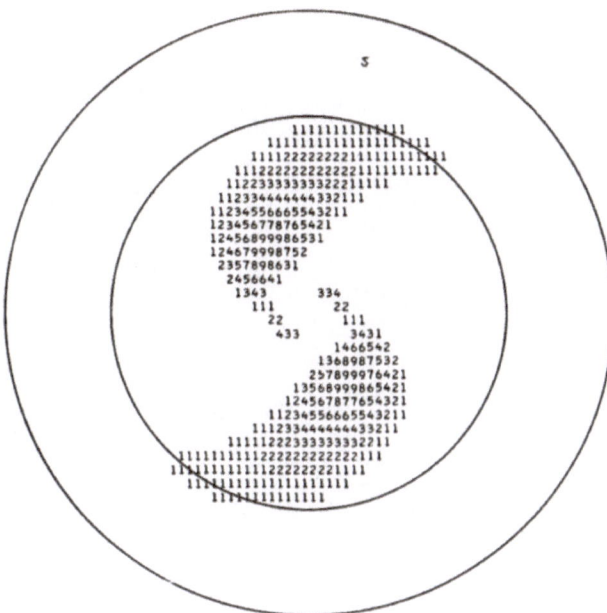

Fig. 1. Excess mass density. Scale factor is 6.17.

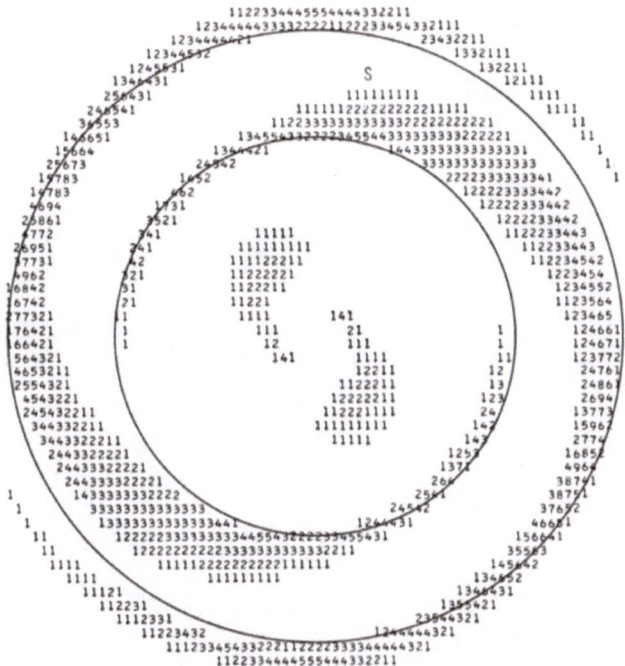

Fig. 2. Excess gas density. Scale factor is 0.0557.

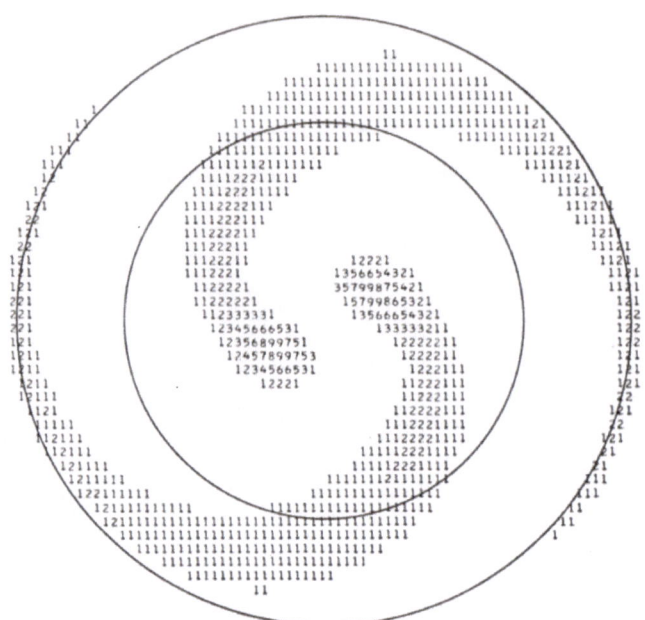

Fig. 3.    Tangential velocity. Scale factor is 2.32.

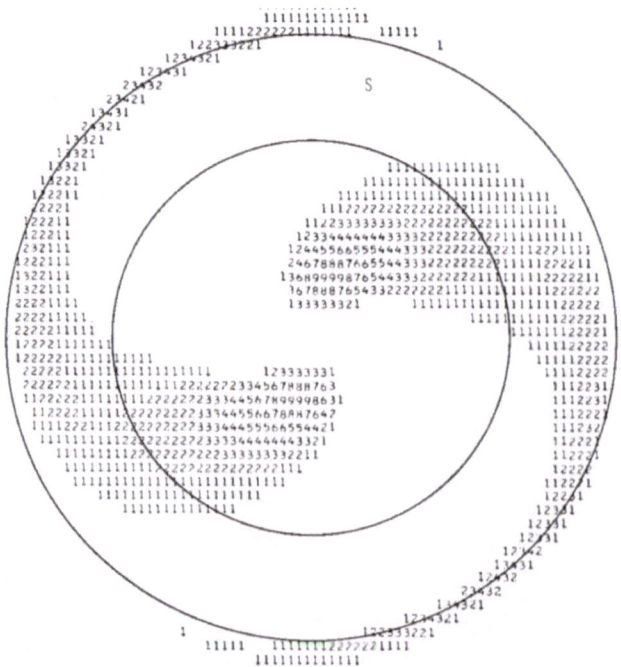

Fig. 4.    Radial velocity. Scale factor is 2.42.

surface mass density is in solar masses pc$^{-2}$, the velocities are in km s$^{-1}$, and the surface gas density is normalized by its equilibrium value. In the calculation the latter was assumed to be constant, however a smooth variation with radius will not change the pattern significantly.

The high frequency ($\omega = -60 - 2i$) gives rise to two resonances. At the particle resonance (inner circle, $r = 9$ kpc) the stars travel with the wave, whereas at the outer resonance (outer circle, $r = 14$ kpc) the stars see the force field varying at their epicyclic frequency. Energy and angular momentum are conserved over the whole disk but they are redistributed outward: the stars inside the particle resonances, which move faster than the wave, lose both to the stars on the outside of the resonance. An angular momentum transfer is already apparent from the trailing nature of the pattern.

## 2. Discussion

The interesting result of these calculations is the strong dependence of the perturbed densities on the velocity dispersion of the subsystem. The density wave is essentially a bar-like distortion of the central region of the galaxy which drives the gas. The tightly wound pattern and large density contrast in the latter are due to the presence of the resonances. The position of the resonance is determined by the inner part of the model, whereas the growth rate of the pattern depends on the mass density chiefly at the outer resonance. A decrease of the latter means a slower growth rate which tends to break up the gas pattern into nearly circular arcs around the resonance radii.

Insofar as the model is entirely determined by the rotation curve, the above results should apply to our galaxy since the rotation curves are similar (at least from 4 kpc outward). The sun's radial position would fall half-way between the two resonances. If we take the observed north-south asymmetry in the rotation curve (Kerr, 1964) as evidence for the presence of such a mode, then we can explain it if the sun is placed in the position $s$, and the local standard of rest (LSR) has a radial motion which is one-half that of the gas, or $-4$ km s$^{-1}$. The asymmetry also determines the amplitude of the mode which is twice as large as that quoted in the figures. If we further assume that the LSR moves with the gas in the tangential direction (at 6.5 km s$^{-1}$) we can fill in most of the dip in the rotation curve which occurs at 6.7 kpc. There are other features such as the 50 km s$^{-1}$ maximum in the radial velocity field in the direction of the galactic center which would appear as an outward moving 'arm', the general asymmetry with respect to the center, and the fact that the mode is unstable, which encourage further investigation.

There are also many deficiencies, notably those in the equilibrium model and non-linearities in the gas distribution, which have to be corrected. Small corrections in the model arise from the fact that the presence of the mode increases the Oort constants of the gas, $A$, $B$, $C$, and $K$ by 2.7, 0.7, $-1.7$, and $-3.5$ km s$^{-1}$ kpc$^{-1}$ respectively. A larger uncertainty comes from the central region. The model we used has no central condensation which would lead to an inner resonance.

The interpretation that leads to the 1965 Schmidt model, which is the commonly accepted one, tends to overestimate the central rotation rates. We have repeated the above calculation using this model and find that the mode is not very much affected in the outer regions ($r > 5$ kpc), but is significantly modified in the vicinity of the inner resonance point at 2.3 kpc. If an amplitude is chosen to match the outer parts, we find non-circular velocities at 2.3 kpc in the range of 100–150 km s$^{-1}$. Such amplitudes severely strain the linear theory. The credibility of the result is further lessened by our orbit approximations which in the presence of a resonance do become suspect. However, viewed as an order of magnitude calculation, the result suggests that a bar-like mode of the disk can produce large non-circular motions if a central condensation exists, and that the nature of the motions will have to be understood in order to obtain a quantitative description of the center. It is conceivable that the wide wings on the line profiles near the center are associated with an inner Lindblad resonance, which would imply a smaller central concentration than suggested by the Schmidt model.

Another interesting feature of the inner resonance is the coincidence of the maxima in gas densities with the maxima in the inward radial velocities.

A more detailed description of these calculations will be published later.

## Acknowledgement

This work has been supported by the National Science Foundation.

## References

Kalnajs, A. J.: 1965, Ph.D. Thesis, Harvard University.
Kerr, F. J.: 1964, IAU-URSI Symposium No. 20, p. 81.
Lindblad, B.: 1958, *Stockholm Obs. Ann.* **20**, No. 6.
Van de Hulst, H. C., Raimond, E., and Van Woerden, M.: 1957, *Bull. Astron. Inst. Netherl.* **14**, 1.

# 61. THE PROPAGATION AND ABSORPTION OF SPIRAL DENSITY WAVES

F. H. SHU

*State University of New York at Stony Brook, N.Y., U.S.A.*

**Abstract.** An 'anti-spiral theorem' holds with limited validity for the neutral modes of oscillation in a stellar disk – namely, whenever the effects of stellar resonances can be ignored. In the regions between Lindblad resonances, a group of spiral waves will propagate in the radial direction with the group velocity found by Toomre. This propagation occurs with the conservation of 'wave action', wave energy, and wave angular momentum.

## 1. Permissibility of Modes of Spiral Form

It is known from the work of Lynden-Bell and Ostriker (1967) that an 'anti-spiral theorem' holds for all neutral modes of oscillation associated with a differentially rotating, self-gravitating, gaseous system. However, galaxies are composed mostly of stars, and the dynamics of stars differs from that of gas (and dust) in that stars can resonate with an oscillating gravitational field without continual interruptions from collisions. These resonances arise whenever the stars feel the same phase of the perturbation gravitational field in each cycle of their peculiar motions. An exact formulation of the linearized problem, including appropriate boundary conditions, shows that an analogous 'anti-spiral theorem' holds for neutral modes in a stellar disk (modelled with infinitesimal thickness) when the effects of stellar resonances can be ignored (Shu, 1968, 1970a).

While such results pose obvious difficulties for the density wave theory, they also suggest the permissibility of spirals to depend on the prevalence of one or a combination of the following conditions:
  (a) the existence of strong stellar resonances,
  (b) the effects of finite amplitude,
  (c) the existence of overstabilities,
  (d) the driving of the disk by other agencies.

It is difficult to accept the hypothesis that resonances can *by themselves* generate extensive spiral patterns. Indeed, in the linear theory, the analogy with plasma physics would lead one to suspect that, except for special circumstances (possibly such as those described by Kalnajs), resonances act to absorb density waves (by mechanisms similar to Landau damping). Of course, the conversion of wave energy into energy of particle motions cannot proceed indefinitely, and non-linear effects may eventually alter these considerations.

Any expectation for waves to develop spontaneously is related to the possibility for instability. Thus, it is important that (when stellar resonances are ignored) recent investigations – summarized in the rest of this paper – reveal no short-scale instabilities in addition to the Jeans instability discussed for this geometry by Toomre (1964).

*Becker and Contopoulos (eds.), The Spiral Structure of Our Galaxy, 323–325. All Rights Reserved.*

This result leads us to consider seriously that spiral structure may be the result of forcing by a yet unspecified agency.

## 2. Local Properties of Density Waves

To investigate the properties of density waves, it is convenient to adopt the WKBJ approximation advocated by Lin (1966), but we carry the approximation one order higher than previous treatments. We treat leading and trailing spirals separately, admitting, however, the possibility that boundary conditions may ultimately require superposition.

### A. THE DISPERSION RELATION FOR SPIRAL DENSITY WAVES

In the lowest order of approximation, we recover the dispersion relation of Lin and Shu (1966). Wavenumber 'information' propagates radially inside the principal range with the group velocity found by Toomre (1969). Toomre's analysis, thus, provides an important insight into the nature of 'anti-spiral theorems'. Even existing spiral waves must eventually disappear if they are neither replenished nor returned.

### B. THE PROPAGATION OF THE DENSITY OF WAVE ACTION

In the next order of approximation, we obtain a relation governing the variation of amplitude with galactocentric radius (Shu, 1968, 1970b). Toomre (1969) showed that the amplitude relation of Shu can be recovered from physical considerations providing the 'density of wave action' is propagated with the group velocity. Kalnajs (private communication) further clarified this interpretation by demonstrating that the principle is equivalent to the conservation of wave energy and wave angular momentum when viewed by an observer in an inertial frame. In this approximation, then, spiral density waves of short scale show neither a tendency to be overstable nor a tendency to be damped.

Thus, the Jeans instability appears to be the only instability mechanism locally operative in the plane of a stellar disk (whose distribution function for small peculiar velocities is taken locally to be Schwarzschild's distribution). Hence, a point of great importance and some debate, is whether the Galaxy is everywhere more than marginally stable.

### C. POSSIBLE FORCING BY GRAVITATIONAL CLUMPING

For a combined star-gas disk, one region, at least, appears unlikely to be completely stabilized. Depletion of interstellar gas by star formation is yet relatively incomplete in the outer regions of the Galaxy, and the continuous dissipation of turbulent velocities in the interstellar gas will almost certainly lead to effective values of the mean random velocities which are below that required for stability. In these regions may still occur the process of gravitational clumping of interstellar gas described by Goldreich and Lynden-Bell (1965).

Lin (1970) has proposed that gaseous condensations so produced – aided by the excitation of density 'wakes' in the stellar sheet (Julian and Toomre, 1966) – may serve as a source for trailing spiral waves which propagate into the interior. Even though the 'forcing' of the disk originates in the outer and more rarified regions, an impression can be made on the interior because the wave energy density tends to 'pile up' in the interior where the group velocity is small. Furthermore, because the waves are initiated as nearly co-rotating disturbances in the outer parts of the galaxy, pattern speeds characteristic of the material rotation in the outer parts would automatically result. A partial test of these ideas is possible with external galaxies since a pattern speed chosen to match the material rotation near the 'outer edge' of the observed spiral structure should yield an accurate theoretically-deduced spiral pattern for the interior.

## References

Goldreich, P. and Lynden-Bell, D.: 1965, *Monthly Notices Roy. Astron. Soc.* **130**, 125.
Julian, W. H. and Toomre, A.: 1966, *Astrophys. J.* **146**, 810.
Lin, C. C.: 1966, *SIAM J. Appl. Math* **14**, 876.
Lin, C. C.: 1970 IAU Symposium No. 38, p. 377.
Lin, C. C. and Shu, F. H.: 1966, *Proc. Nat. Acad. Sci. U.S.A.* **55**, 229.
Lynden-Bell, D. and Ostriker, J. D.: 1967, *Monthly Notices Roy. Astron. Soc.* **136**, 293.
Shu, F. H.: 1968, Ph.D. Thesis, Harvard University.
Shu, F. H.: 1970a, *Astrophys. J.* **160**, in press.
Shu, F. H.: 1970b, *Astrophys. J.* **160**, in press.
Toomre, A.: 1964, *Astrophys. J.* **139**, 1217.
Toomre, A.: 1969, *Astrophys. J.* **158**, 899.

# 62. LARGE SCALE OSCILLATIONS OF GALAXIES

## C. HUNTER

*Massachusetts Institute of Technology, Cambridge, Mass., U.S.A.*

**Abstract.** The observational evidence that may indicate the presence of large scale modes of oscillation in galaxies is reviewed, and some results of theoretical modal calculations are described.

## 1. Observational Evidence

Although to a first approximation the matter in galaxies rotates in circular orbits about a center, there are also indications of large scale systematic and non-random departures from purely circular motion. A natural interpretation of these motions is that they indicate the presence of various free modes of oscillation of the galaxy that have been excited at some time in the past, either during the formation of the galaxy or as a result of a tidal interaction with another galaxy. Three important pieces of evidence of large scale non-circular motions are listed below.

(i) Kerr (1962) found that the Northern and Southern hemisphere 21 cm maps of our Galaxy fitted together better if there is a general radial outflow of 7 km s$^{-1}$ in the solar neighborhood.

(ii) It has been noticed in many instances that the derived rotation curves of external galaxies are not symmetrical about a center. One example is M31 for which Burke *et al.* (1964) found different rotation curves for the opposite NF and SP sections. Such differences can be explained by the presence of modes of odd angular wave number $m$. Presumably $m=1$ modes are the most fundamental of these and are the most likely to be prominent.

(iii) Roberts' (1966) detailed map of the observed radial velocity field for M31 (his Figure 2) is markedly different from that which would be observed for pure circular motion. In particular the minor axis is not a line of constant radial velocity (the systemic radial velocity) but shows an inflow of matter towards the center at distances less than 9 kpc, and an outflow at greater distances.

This interpretation assumes that M31 is flat. An alternative explanation is possible in terms of a bending of the plane of M31. Reasons for rejecting this explanation are that substantial bending is required on account of the high inclination of M31, whereas the theoretical analysis of Hunter and Toomre (1969) showed that bending of the central regions of a galaxy is hard to maintain. Our Galaxy, for example, is bent substantially only in the outer parts (Kerr *et al.*, 1957).

Roberts remarked on the similarity of opposite quadrants of his radial velocity map, and this indicates the predominance of motions of even angular wave number. Figures 1 and 2 show the isovel maps of the upper half of M31 that result from adding a prescribed radial velocity in the plane of the galactic disk of M31 to the circular velocity derived by Roberts. The form of the radial velocity used was chosen to agree

*Becker and Contopoulos (eds.), The Spiral Structure of Our Galaxy, 326–330. All Rights Reserved.*

roughly with that shown in Roberts' map along the minor axis, and was then extended over the whole plane of the disk. Figure 1 results from the addition of an axisymmetric non-circular motion. Isovel lines in the left hand quadrant have the more involved forms away from the central regions, and in some areas slope counter to the direction they would for pure circular motion (Roberts' Figure 3). This is in qualitative agreement with Roberts' Figure 2. The features noted above are more marked in my

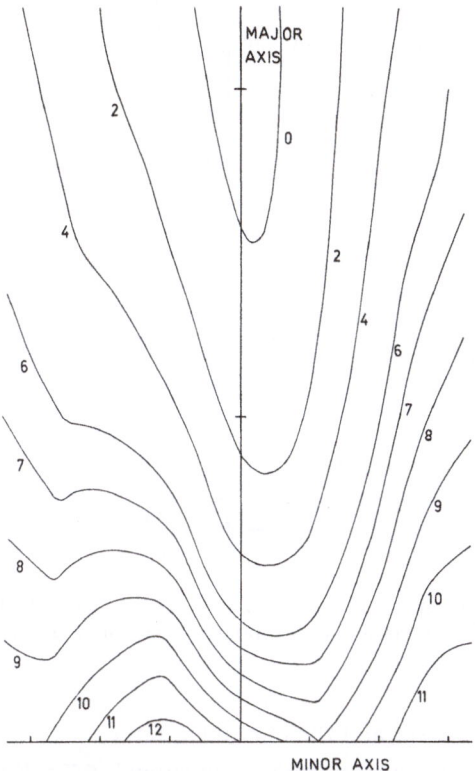

Fig. 1. Curves of constant radial velocity for an axisymmetric radial motion added to the circular velocity field of M31. A unit change between isovels corresponds to a jump of $-21$ km s$^{-1}$, and 11 represents the systemic velocity of $-310$ km s$^{-1}$. The marks along the axes are at 4 kpc intervals.

Figure 2 in which the added radial motion has an $m=2$ angular variation. Greater radial motions are present in this case, and the figure changes considerably with the orientation of the motion. Since the beam width used for the observations was relatively large, the true isovel map is still somewhat uncertain, but the indications are that it is consistent with the presence of organized large scale non-circular motions in the plane of M31.

## 2. Theoretical Calculations of Modes

The most straightforward calculations of free modes of oscillation are those obtained using the 'cold disk' model. A particular equilibrium model is selected, typically by

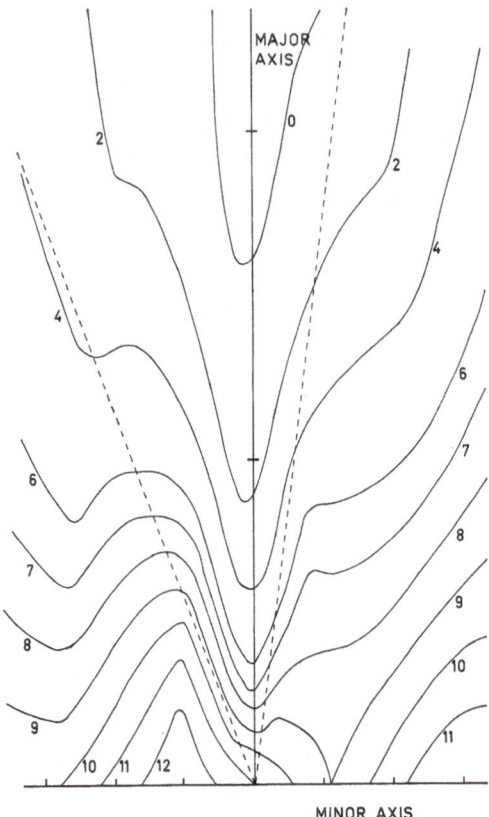

Fig. 2.   Curves of constant radial velocity for an $m = 2$ radial motion added to the circular velocity field of M31. The dashed lines show the directions of maximum amplitude of the radial motion.

taking a given rotation curve and then finding a finite mass distribution whose gravitational attraction provides the necessary centrifugal force. Possible free oscillations are determined by solving in a fully self-consistent manner the zero pressure hydrodynamic equations that govern departures from the unperturbed state of circular motion (Hunter, 1965). Although such an analysis produces unstable modes of short wavelength where there is nothing to prevent the growth of Jeans instability, a few stable large scale modes are also found which remain discrete when the shorter wavelength modes tend to form continua (Hunter, 1969). Figure 3 shows the radial and circular velocity components of the first two axisymmetric modes of a model of M31 derived by fitting Roberts' combined $n = \frac{3}{2}$ rotation curve within 25 kpc by a suitable mass distribution. The first mode is highly concentrated towards the center and has a higher frequency of 37.5 km s$^{-1}$ kpc$^{-1}$ than that of 10.8 km s$^{-1}$ kpc$^{-1}$ of the second mode. The latter also has an associated radial velocity field more like that observed for M31. Non-radial modes may also by computed though, in this theory, steady modes show no spiral structure. Figure 4 of Hunter (1969) shows the circular velocity field associated with a large scale and centrally concentrated $m = 1$

mode for another galactic model which would cause an asymmetrical rotation curve to be observed for it.

Recently I have been calculating modes of oscillation by using a more realistic set of hydrodynamic equations for a galaxy of stars that takes account of stellar random velocities. The latter are supposed small compared with circular velocities, and the collisionless Boltzmann equation is expanded in terms of an appropriate small parameter. This expansion leads to a set of equations for the moments of the perturbed distribution function. An early truncation of the expansion leads to the equations of zero pressure hydrodynamics, but the continuation of the expansion one stage further gives a more complicated set of hydrodynamic equations with general stress terms. These stresses can be related to the systematic velocities so that a closed set of equations is obtained.

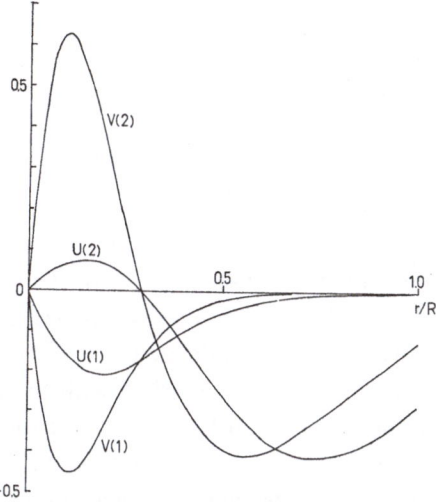

Fig. 3.   The spatial variation of the radial velocity $U$ and circular velocity $V$ for the first two axisymmetric modes of M31. Here $r$ measures radial distance from the galactic center, and the outer radius $R$ is at 25 kpc.

The distribution of stellar random radial velocities in the unperturbed galaxy must be specified in this theory. Toomre (1964) has shown via a local analysis of the full collisionless Boltzmann equation that all short wavelength axisymmetric oscillations are stable for a Schwarzschild distribution of random velocities provided the mean radial random velocity $c > 3.36\, G\sigma/\kappa$, where $G$ is the constant of gravitation, $\sigma$ is the surface density and $\kappa$ is the epicyclic frequency. A similar stability condition can be established for the stellar hydrodynamic equations; the only difference being that the coefficient 3.36 is replaced by 3.0.

Modal calculations have been performed with the distribution of mean radial random velocity given by $c^2 = 4\pi^4 \beta (G\sigma/\kappa)^2$. Here $\beta$ is a parameter that is varied between calculations, $\beta = 0$ corresponding to a cold disk, and $\beta = 0.023$ corresponding to the theoretical stability limit described above. The results of axisymmetric calculations confirm that the theoretical stability limit, derived strictly only for short wavelength disturbances, is more generally valid, though the critical value of $\beta$ may be more like 0.025 than 0.023. The precise point of stabilization is hard to determine because of the sensitivity of the calculations to numerical errors. Non-axisymmetric calculations, principally for $m=2$, show that these modes are not so readily stabilized even at considerably larger values of $\beta$ where instabilities still persist. The instabilities of large scale do not have any clear spiral form.

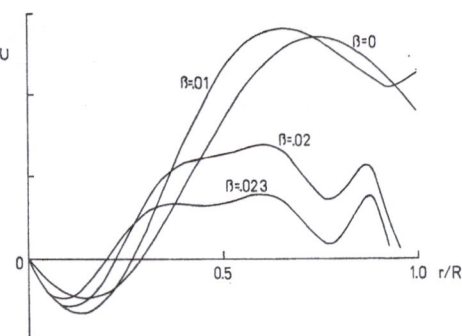

Fig. 4.   The radial velocity $U$ for the second axisymmetric mode of M31 for increasing random velocities as specified in the text.

The increase of random velocities causes changes in the shape of modes. Figure 4 shows how the radial velocity of the second axisymmetric mode of the present M31 model is affected. Generally there is a tendency for the smooth shapes of cold disk modes to break down. This is much more marked with the first axisymmetric mode which disintegrates for values of $\beta$ much less than the critical one. Thus the slower second axisymmetric mode seems from the calculations to be the more fundamental mode of a galaxy and the rough agreement between the non-circular velocities observed on the minor axis of M31 and the calculated shapes of Figure 4 support the idea that this mode is present in M31.

## Acknowledgements

The work described here was supported by both the National Science Foundation and the National Aeronautics and Space Administration.

## References

Burke, B. F., Turner, K. C., and Tuve, M. A.: 1964, IAU-URSI Symposium No. 20, p. 99.
Hunter, C.: 1965, *Monthly Notices Roy. Astron. Soc.* **129**, 321.
Hunter, C.: 1969, *Studies in Appl. Math.* **48**, 55.
Hunter, C. and Toomre, A.: 1969, *Astrophys. J.* **155**, 747.
Kerr, F. J.: 1962, *Monthly Notices Roy. Astron. Soc.* **123**, 327.
Kerr, F. J., Hindman, J. V., and Carpenter, M. S.: 1957, *Nature* **180**, 677.
Roberts, M. S.: 1966, *Astrophys. J.* **144**, 639.
Toomre, A.: 1964, *Astrophys. J.* **139**, 1217.

# 63. GENERATING MECHANISMS FOR SPIRAL WAVES

D. LYNDEN-BELL

*Royal Greenwich Observatory, Sussex, England*

To maintain the spiral structure over long periods in spite of the chaotic disturbances so often seen in galaxies it is probably necessary to have a large energy store which is gradually tapped to generate and maintain the spiral wave. Now in a sense stable axially symmetrical model galaxies are configurations of minimum energy for a given angular momentum structure. By a given angular momentum structure I mean that the function $\mu(h)\,dh$ giving the total mass with specific angular momentum between $h$ and $h+dh$ is given and in making the minimum energy statement I have assumed that there is sufficient 'random' motion to satisfy Toomre's local stability criterion. To tap a store of energy one must either tap the energy of the whole rotational structure by finding disturbances that may change $\mu(h)$, or one may tap the smaller energy store of the 'random' motions. Clearly the larger energy store is the more attractive. Now $\mu(h)$ is conserved for any axially symmetrical time dependent motions so the only disturbances that can change $\mu(h)$ are non-axially symmetrical. If one considers a single star moving in nearly circular motion and interacting with a weak disturbance in the gravitational potential which may be Fourier analysed into components of the form $S(R)\exp[i(m\phi+\omega t)]$ then one finds that the energy and angular momentum of the star oscillate but do not change in the mean. This statement is untrue when the force due to the wave does not average to zero or when it resonates with the natural oscillation of the star about the circular motion. The angular frequency of the force seen by the star is $\omega+m\Omega$, where $\Omega(R)$ is the rotational angular velocity of the galaxy, so the exceptions occur at the resonances

$$\omega + m\Omega = 0, \pm\kappa$$

Here $\kappa(R)=(4B(B-A))^{1/2}$, the epicyclic angular frequency, and $A(R)$ and $B(R)$ are Oort's constants. At the radius where $\omega+m\Omega(R)=0$ the pattern of the disturbance moves around at the same velocity as the circular motion of the stars epicentre. The other resonances that were discovered by B. Lindblad lie one within and one outside that circle.

I wish to point out here the strong connection between the physical picture of Landau damping and Landau excitation in the electrostatic waves of an encounterless plasma and the physics of the resonances in the theory of spiral structure. The physical picture of the Landau problem is that a wave that can propagate in the bulk of the plasma will resonate with particles travelling very close to the wave velocity. These special particles must be treated separately. Those that travel just faster than the wave, so little faster that they are unable to climb over the wave crest, together produce a net pushing on the wave in the direction of its propagation through the

*Becker and Contopoulos (eds.), The Spiral Structure of Our Galaxy, 331–333. All Rights Reserved.*
*Copyright © 1970 by the I.A.U.*

plasma. Conversely those moving just below the wave speed are laggards holding the wave back. Normally there are more laggards than pushers so the wave loses energy and momentum and is damped. If, however, there are more pushers than laggards, as when two streams of particles interpenetrate, the wave is amplified leading to Landau excitation, which is another name for the two stream instability.

Compare this picture with what is happening close to the circle where $\omega + m\Omega(R) = 0$. Stars with epicentres just within that circle have mean angular velocity just greater than the wave – they are the pushers; stars with epicentres just outside have mean angular velocity just less than the wave – they are the laggards. If the angular momentum imparted to the wave by the pushers is greater than that extracted from the wave by the laggards then we may expect that the wave will be enhanced. This will clearly happen if the surface density of epicentres falls off rapidly with $R$ close to the resonance. If such a rapid fall occurs at some radius disturbances moving around the galaxy with a pattern speed equal to the circular velocity these will be excited. Outside such a ring we must have an outward going wave since its source will be within. If we write the wave locally in the form $S' \exp[i(m\phi + \omega t - kR)]$ then $k/\omega$ must be positive to get an outward moving wave; however $m\Omega$ must have the opposite sign to $\omega$ by the resonance condition. Hence $k/m$ has the same sign as $-\Omega$. Now locally the spiral has the form $\phi = (k/m)R$. Hence $\phi$ increases with $R$ when the picture of the galaxy is drawn with the galactic rotation in the sense of $\phi$ decreasing. In the outer parts such arms will therefore trail. Within the resonant circle the same argument would prove that the wave leads but for the fact that any wave propagating inwards will be reflected in the central parts. Thus within the resonant circle there will be a standing wave with a cartwheel or, in the simplest case, a barred structure. The rather vague outer 'edge' of the galaxy does not likewise reflect a wave for the same reason that a pond with a gradually sloping shore does not reflect water waves. The generating of spiral waves is analogous to the forcing of a semi-infinite stretched string fixed at one end and forced at one point. Between the forcing point and the fixed end a standing wave develops but a propagating wave occurs on the outside.

The resonant circle driving the most prominent spiral structure is not necessarily that with the largest density gradient across it because that circle may be poorly matched to the reflected wave from the centre. However assuming that the matching is never too bad it is interesting to see how the changes in the angular momentum of the stars close to resonance will secularly enhance the generating mechanism. Since stars just inside the resonant circle lose angular momentum they move inwards, while those just outside acquire angular momentum and move outwards. Thus the resonant circle is evacuated with density enhancements on either side. The density enhancement just inside the resonant circle followed by the dearth at the resonant circle gives a very strong gradient of epicentre density just inside the old resonance. We deduce that there is a secular movement of the strongest excitation circle inwards and that it sweeps before it a growing ring of matter. Thus in external galaxies we should see a pronounced ring surrounding a bar structure and the spiral structure should start from the ring.

I have not investigated the physical mechanisms that may occur at the Lindblad resonances but it is not unlikely that these too can be made to generate spiral structures. However it is not clear to me that these resonances give direct access to the main energy store of the galactic rotation rather than to the smaller energy of epicyclic motion. I remark however that the mechanism I have described seems to lead inevitably to galaxies with some cartwheel, bar-like or ring structure such as de Vaucouleurs's type SABrs and there is obviously room for mechanisms that generate structures devoid of rings or bars. One might speculate that the two main classes of spirals, SB and S are generated from the resonant circle and from the Lindblad resonance respectively.

None of the statements in this paper have been properly proved by detailed calculations, rather they are a physicist's attempt at a mechanistic explanation.

# 64. SPIRAL WAVES CAUSED BY A PASSAGE OF THE LMC?

A. TOOMRE

*Massachusetts Institute of Technology, Cambridge, Mass., U.S.A.*

This progress report on the deduction of an almost grazing orbit of the Large Magellanic Cloud from the warped shape of our Galaxy focused on two issues of special relevance to this Symposium:

(i) *Distance $R_0$*. – An implausibly large mass ($> 3 \times 10^{10}\, M_\odot$) of the LMC is needed to account for the observed warp even with the most optimal orbit, using Schmidt's (1965) disk model of our Galaxy, or any close variant thereof. Although as yet only tentative, the inference seems to be that our distance $R_0$ from the galactic center has been overestimated; instead of the now 'standard' $R_0 = 10$ kpc, these disk-bending calculations suggest a value more like the older $R_0 \cong 8$ kpc.

(ii) *Forced spiral waves*. – Even with the revised $R_0$, the *sense* of the LMC orbit remains ambiguous (though not the perigalactic distance of roughly 20 kpc nor the low absolute inclination of that orbit relative to the galactic plane). Neither the detailed shapes of the vertically bent model disks, nor any computed tidal effects of the Galaxy upon the LMC and SMC, either separately or as a system, rule out an LMC passage that is *retrograde* with respect to the galactic rotation, but all favor a *direct* passage as being the more plausible. The latter kind of passage, however, presents an embarrassment of riches. This is best illustrated by Figure 1.

Like the earlier but less specific diagrams of Pfleiderer (1963), it shows the present

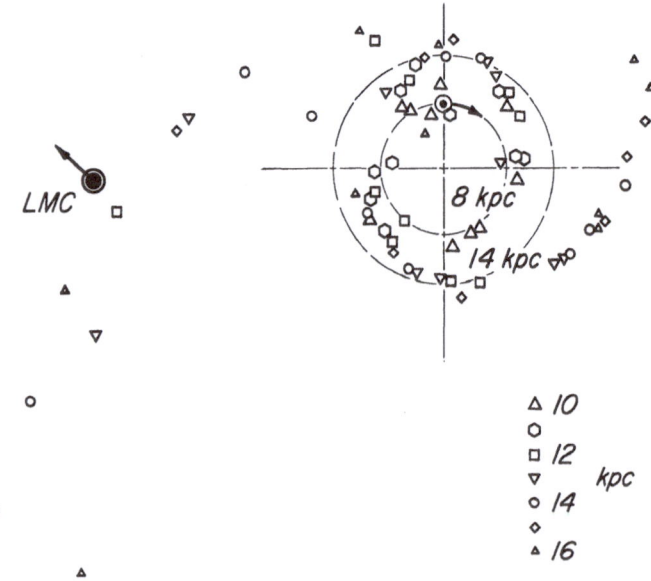

Fig. 1.

locations (projected onto the galactic plane) of 70 test particles which orbited initially with uniform angular spacing in circles of 10, 11, ..., 16 kpc radius from a point mass of $8 \times 10^{10} \ M_\odot$ representing the bulk of the Galaxy, and which were then perturbed by the direct, 20 kpc passage of another $1.5 \times 10^{10} \ M_\odot$ particle representing the LMC. Obviously this diagram contains a spectacular 'bridge' of former galactic material towards the LMC, and an opposite feature which resembles (even in latitude) the 'Outer Arm' studied especially by Habing (1966). But it is the vehemence of these phenomena that is alarming: Although it now seems clear from group velocity considerations (Toomre, 1969) that the neglected interactions between the actual disk particles should have caused some of these outer vibrations to propagate inward as a *transient* spiral density wave, and would thereby have reduced their amplitudes, the amount or rate of that reduction can be estimated only from properly full-scale calculations. Hence we will probably have to wait until some large *n*-body computations such as those of Prendergast and Miller or Hohl have been adapted to this purpose, before we can either dismiss or accept this suggestion that the present spiral structure of our Galaxy may be significantly indebted to external tidal forces from the not too distant past.

## References

Habing, H. J.: 1966, *Bull. Astron. Inst. Netherl.* **18**, 323.
Pfleiderer, J.: 1963, *Z. Astrophys.* **58**, 12.
Schmidt, M.: 1965, *Stars and Stellar Systems* **5**, 513.
Toomre, A.: 1969, *Astrophys. J.* **158**, 899.

# 65. NON-LINEAR DENSITY WAVES IN PRESSURELESS DISKS

C. L. BERRY and P. O. VANDERVOORT

*University of Chicago, Chicago, Ill., U.S.A.*

**Abstract.** A method is described for the construction of a density wave of finite amplitude and in the form of a tightly wound spiral in a pressureless, self-gravitating disk of infinitesimal thickness in an external gravitational field. Waves of one kind are found for systems in which the law of rotation departs only slightly from that of a solid body. Waves of a second kind are found in systems possessing appreciable differential rotation; however, such waves can occur only if the self-gravitation of the disk is small compared with the external gravitation. The latter waves are to be identified with the waves described by linear theories.

## 1. Introduction

This is a report on our first results in a non-linear theory of density waves. We have considered the hydrodynamics of a pressureless, self-gravitating disk of infinitesimal thickness in a given external gravitational field which is time-independent and axi-symmetric; and we have constructed spiral waves of finite amplitude in that disk.

The system envisaged is a model of a galaxy in which the disk represents the sub-systems of low-velocity stars and interstellar gas which can participate appreciably in a density wave whereas the external field is attributed to those subsystems of high-velocity stars which cannot participate appreciably in the wave. In *linear* theories of density waves a basis for this model is provided by the manner in which the peculiar motions of the stars (or the pressure of the gas) tend to inhibit the participation of a subsystem in a density wave (see, e.g., Lin *et al.*, 1969).

## 2. Construction of Density Waves of Finite Amplitude

We consider a wave which has a stationary spiral pattern in a uniformly rotating frame of reference. In that frame the disk is in a steady state. Its structure and the pattern of flow are governed by the hydrodynamical equations of continuity and motion and Poisson's equation. Solutions of the linearized form of this problem have been described by Fujimoto (1968).

We assume that the pattern is tightly wound. Our solutions of the basic equations are asymptotic solutions of the form of series in powers of a parameter $\lambda$ ($\ll 1$) whose smallness characterizes the tightness of the winding. In order to explain certain aspects of our results, we must describe our solution of Poisson's equation in some detail. This analysis is a generalization of the asymptotic solution of Poisson's equation given by Lin and Shu (1964) in their linear theory of density waves.

In the absence of the density wave, let the surface density and gravitational potential of the disk be $\sigma^{(\infty)}(\varpi)$ and $\mathfrak{B}^{(\infty)}(\varpi, z)$, respectively, where the system of cylindrical polar coordinates $(\varpi, \theta, z)$ is defined in the frame of reference rotating uniformly with the angular velocity $\Omega$ of the pattern, the $z$-axis is the axis of rotation, and the plane

*Becker and Contopoulos (eds.), The Spiral Structure of Our Galaxy, 336–340. All Rights Reserved.*

$z=0$ contains the disk. To construct the wave, we write the potential of the perturbed disk in the form

$$\mathfrak{B}(\varpi, \theta, z) = \mathrm{Re}\left[\mathfrak{B}^{(\infty)}(P, z)\right], \tag{1}$$

where

$$P = \varpi + \lambda F(\varpi, z, \chi), \tag{2}$$

$$\chi = m\theta + \lambda^{-1} u(\varpi, z), \tag{3}$$

and $m$ is an integer equal to the number of arms in the pattern. The functions $F$ and $u$ are complex in general. The condition that $\mathfrak{B}$ is a single-valued function of position implies that $F$ is periodic in the real part of $\chi$ with period $2\pi$.

To make the geometry of the disk that of a tightly wound spiral, we require $F$ and $u$ to be slowly varying functions of $\varpi$ and $z$, and we require $u$ to be real in the plane $z=0$. In that plane, the curves $\chi=$ constant are the spirals. The spiral component of the field diminishes rapidly with distance from the disk. Accordingly, the function $F$ is required to vanish at infinity.

Outside the plane $z=0$, the potential is governed by Laplace's equation. We have obtained a formal solution by letting

$$u(\varpi, z) = u(\varpi \pm iz), \quad \text{an arbitrary function}, \tag{4}$$

and writing $F(\varpi, z, \chi)$ as a series

$$F = F^{(0)} + \lambda F^{(1)} + \cdots \tag{5}$$

in powers of $\lambda$. The requirement that the potential must satisfy Laplace's equation in each order in $\lambda$ separately leads to a hierachy of equations governing $F^{(0)}$, $F^{(1)}$, etc. The solution of the first member of the hierarchy is

$$F^{(0)}(\varpi, z, \chi) = [\varpi^{1/2}\mathfrak{B}_\varpi^{(\infty)}(\varpi, z)]^{-1} G(\varpi \pm iz, \chi), \tag{6}$$

where $G(\varpi \pm iz, \chi)$ is an arbitrary function, and the subscript denotes differentiation with respect to $\varpi$. On each side of the plane $z=0$, the ambiguity of sign in Equations (4) and (6) can be removed with the aid of the condition that $F$ vanishes at infinity.

The density distribution which gives rise to this potential is given by

$$\sigma(\varpi, \theta, z) = \frac{1}{2\pi G} \lim_{z \to 0+} \mathfrak{B}_z(\varpi, \theta, z)$$

$$= \sigma^{(\infty)}(\varpi) \mp \frac{1}{2\pi G} \lim_{z \to 0+} \left[\mathfrak{B}_\varpi^{(\infty)}(\varpi, z)\,\mathrm{Im}\,(u_\varpi F_\chi^{(0)})\right] \tag{7}$$

$$+ O(\lambda),$$

where subscripts again denote differentiation. Equation (7) expresses the density as a superposition of the unperturbed density and the density of the wave.

In the lowest order the structure of the wave is characterized by the 'amplitude' $F^{(0)}(\varpi, z, \chi)$ and a wave number

$$k(\varpi) = \lambda^{-1} u_\varpi(\varpi). \tag{8}$$

The indeterminacy of these functions, arising from the arbitrariness of the functions $u(\varpi \pm iz)$ and $G(\varpi \pm iz, \chi)$, is reduced in the solution of the hydrodynamical equations. We encounter an integrability condition in the form of an equation involving $\Omega$, $k(\varpi)$, and $F^{(0)}(\varpi, z, \chi)$. This integrability condition determines the amplitude of the wave once the angular velocity of the pattern and the dependence of the wave number on $\varpi$ have been chosen. These choices remain arbitrary.

We have been able to construct solutions of only two kinds along these lines.

## 3. Density Waves of the First Kind

A necessary condition for the construction of a solution of the first kind is that the pattern of the wave must rotate more slowly than the interior of the disk. The wave extends from the origin to the radial distance at which the pattern and the disk are in corotation; and the amplitude of the wave vanishes at both the origin and the co-rotation point. In practice, we have been able to construct such a wave only over a central region of the disk in which the angular velocity of the unperturbed rotation departs only slightly (of the order of 10%) from a constant. Thus, a wave of the first kind can extend over a large region of the disk only if the law of rotation in that region departs but slightly from that of a solid body. These waves do not possess inner Lindblad resonances.

## 4. Density Waves of the Second Kind

Solutions of a second kind have been found for waves in disks in which differential rotation is appreciable. A necessary condition for the construction of these solutions is that

$$\mathfrak{B}^{(\infty)}(\varpi, z) = O\left[\lambda \mathfrak{B}^{(B)}(\varpi, z)\right], \tag{9}$$

where $\mathfrak{B}^{(B)}(\varpi, z)$ is the external potential. When we interpret the external field in terms of subsystems of high-velocity stars which cannot participate in a density wave, this conditions implies that a density wave of finite amplitude can occur in a galaxy in differential rotation only if the subsystems of gas and low-velocity stars which can support the wave make only a small contribution, of order $\lambda$, to the total mass of the galaxy. It further implies that the smaller the fraction of the total mass of a galaxy in subsystems which can support a density wave, the more tightly wound will be the spiral pattern of the wave. Waves of the second kind are to be identi-fied with the density waves described by linear theories; when we consider a wave of infinitesimal amplitude and linearize our integrability condition in that amplitude, we recover the dispersion relation for $\Omega$ and $k(\varpi)$ which was given by Lin and Shu (1964) in the earliest version of their theory.

An example of a density wave of the second kind is illustrated in Figure 1. Here the external field is that of a disk of surface density

$$\sigma^{(B)}(\varpi) = \frac{\mathfrak{M}a}{2\pi}\left(a^2 + \varpi^2\right)^{-3/2}; \tag{10}$$

and the disk which supports the wave has an unperturbed surface density

$$\sigma^{(\infty)}(\varpi) = \frac{3\lambda\mathfrak{M}}{2\pi R^2}\left(1 - \frac{\varpi^2}{R^2}\right)^{1/2}, \quad \varpi \leqslant R,$$

(11)

$$\sigma^{(\infty)}(\varpi) = 0, \quad \varpi > R,$$

where $\mathfrak{M}$, $a$, and $R$ are constants. In this example, $\lambda = 0.15$ and $R = 3a$. We have constructed a wave of angular velocity $\Omega$ given by

$$\Omega^2 = 0.08944\, G\mathfrak{M}/a^3 ;$$

(12)

the corotation point is then $\varpi = 2a$. Outside an annulus in which

$$v^2 \equiv m^2(\Omega - \Omega_c)^2/\kappa^2 < v_0^2 = 0.49456,$$

(13)

Fig. 1.   Pattern of the density wave in the example described in the text.

where $\Omega_c$ is the angular velocity of the unperturbed disk and $\kappa$ is the epicyclic frequency, we let $k(\varpi)$ be given by the dispersion relation for linear waves. In accordance with the remark at the end of the preceding paragraph, the consequence of this choice is that the amplitude of the wave vanishes. Within the annulus, we permit the wave to have a finite amplitude by choosing $k(\varpi)$ in the manner

$$|k| = \frac{\kappa^2}{2\pi G\sigma^{(\infty)}}\left(1 - v^2\right)\left[1 - \alpha v^2\left(v_0^2 - v^2\right)\right],$$

(14)

with $\alpha = 0.5$. (With $\alpha = 0$, Equation (14) is the dispersion relation for linear waves.) Figure 1 shows the spiral pattern of the wave; the boundary of the shaded region is a contour of constant density (equal to the central density) in the disk which supports the wave. The amplitude of the wave is typically of the order of 30–40% of the unperturbed density. This wave does not possess an inner Lindblad resonance, and the outer Lindblad resonance lies outside the annulus to which the wave has been confined.

## 5. Concluding Remarks

A number of problems remains to be investigated in this work. We conclude this report by commenting on two of them.

(1) The role of resonances in non-linear waves is not clear. Within the framework of the present analysis, we can avoid the effects of resonances by constructing the wave in such a way that it is confined, as in the preceding example, to an annulus which does not contain resonance points. In doing so, however, we are probably exploiting an artificial property of the asymptotic approximation. For this reason, we intend to investigate the role of resonances.

(2) Leading and trailing spiral patterns are allowed without distinction in the present theory. It seems, therefore, that to account for a preference for trailing patterns, we would have to distinguish leading and trailing patterns in terms of the circumstances of their origin or in terms of their stability.

## Acknowledgement

This research has been supported in part by the Office of Naval Research under Contract N00014-67-C-0260 with the University of Chicago.

## References

Fujimoto, M.: 1968, *Astrophys. J.* **152**, 391.
Lin, C. C. and Shu, F. H.: 1964, *Astrophys. J.* **140**, 646.
Lin, C. C., Yuan, C., and Shu, F. H.: 1969, *Astrophys. J.* **155**, 721.

## Discussion

*R. Graham:* D. J. Carson has carried out an analysis of static non-linear axisymmetric waves in a differentially rotating thin stellar sheet. A local approximation is used; it is only valid for moderately large amplitude waves. He finds that these static waves have a greater velocity dispersion than static (i.e. zero frequency) linear waves of the same wavelength, which suggests that non-linear waves are less stable than linear ones. He has also carried out a similar analysis for a differentially rotating incompressible fluid sheet, and finds that static non-linear waves exist for sheets thicker than those of the corresponding linear waves, which produces a similar suggestion.

# 66. DENSITY WAVES IN GALAXIES OF FINITE THICKNESS

P. O. VANDERVOORT

*University of Chicago, Chicago, Ill., U.S.A.*

**Abstract.** The effect of the finite thickness of a galaxy on the propagation of density waves of the type described by Lin and his collaborators has been calculated. The calculated effect does not differ appreciably from what has been estimated previously on the basis of heuristic arguments.

In investigations of the equilibria and stability of galaxies in the context of the problem of spiral structure, it has been customary to neglect the finite thicknesses of galaxies. However, in the application of these theoretical developments to the interpretation of observed spiral structure, as in the work of Lin *et al.* (1969), corrections for finite thickness are significant. Attempts to estimate these corrections by Shu (1968) and others are usually based on heuristic arguments formulated in order to avoid a detailed consideration of the dynamics of galaxies perpendicular to their principal planes.

The work reported here is part of a program to investigate the equilibria and departures from equilibrium of rapidly rotating galaxies of small but finite thickness. These investigations are based on simultaneous solutions of Liouville's equation (the encounterless Boltzmann equation) and Poisson's equation. Although the solution of these equations involves several features which do not appear when the effects of finite thickness are neglected, only the essential feature of the present treatment of the dynamics of a galaxy in the direction perpendicular to its principal plane will be described here.

It can be shown that in a galaxy which is flattened sufficiently, the frequency of the perpendicular oscillation of a star is large compared with the other frequencies of its motion. Under this condition, the perpendicular oscillation is characterized by the approximate constancy of an adiabatic invariant. In a system of cylindrical polar coordinates $(\varpi, \theta, z)$ oriented so that the plane of the galaxy is the $(\varpi, \theta)$-plane, the energy of the perpendicular oscillation is

$$E_z = \tfrac{1}{2}Z^2 + \mathfrak{B}(\varpi, \theta, z, t) - \mathfrak{B}(\varpi, \theta, 0, t), \tag{1}$$

where $Z$ is the component of the velocity in the perpendicular direction, $\mathfrak{B}$ is the gravitational potential, and $t$ is the time. The adiabatic invariant is the action integral

$$J = \sqrt{2} \oint [E_z - \mathfrak{B}(\varpi, \theta, z, t) + \mathfrak{B}(\varpi, \theta, 0, t)]^{1/2} \, dz, \tag{2}$$

where the integration extends over one period of the perpendicular oscillation. The methods of solving Liouville's equation in the present work are based on the approximate constancy of $J$.

Part of this work is an investigation of the effects of the finite thickness of a galaxy on the propagation of density waves of the type considered by Lin *et al.* (1969; see this paper for references to the other publications of these authors). In considering this problem, Shu (1968) writes the dispersion relation in a form equivalent to

$$\kappa^2 - m^2 (\Omega_p - \Omega)^2 = 2\pi G \sigma |k| \mathfrak{F} \mathfrak{T}, \tag{3}$$

where $\kappa$ is the epicyclic frequency, the integer $m$ is the multiplicity of arms in the spiral pattern of the wave, $\Omega_p$ and $\Omega$ are the angular velocities of the pattern and the galaxy, respectively, $G$ is the constant of gravitation, $\sigma$ is the surface density of the galaxy, $k$ is the radial wave number of the wave, $\mathfrak{F}$ is a reduction factor representing the effect of the peculiar motions of stars in directions parallel to the plane of galaxy, and $\mathfrak{T}$ is a reduction factor intended to represent the effects of the finite thickness. The reduction factor $\mathfrak{T}$ depends on the model adopted for the unperturbed structure of the galaxy. Shu has made estimates of the values of $\mathfrak{T}$ for a self-consistent model which leads, in the lowest order of approximation, to a density distribution

$$\varrho(\varpi, z) = \varrho_0(\varpi) \operatorname{sech}^2(z/\Delta), \tag{4}$$

where $\varrho_0(\varpi)$ and $\Delta(\varpi)$ are arbitrary functions. The present calculations of $\mathfrak{T}$ are based on substantially the same model. For typical conditions in a galaxy, for example those in the solar neighborhood, the values of $\mathfrak{T}$ calculated in the present work are 5–10% larger than the values estimated by Shu. It does not appear that these differences are large enough to alter significantly the discussion by Lin *et al.* (1969) of their interpretation of the spiral structure of the Galaxy.

A detailed account of this work will soon be submitted to the *Astrophysical Journal*.

## Acknowledgements

This work was performed at the Leyden Observatory during the tenure of a National Science Foundation Senior Postdoctoral Fellowship (1967–68). I am grateful to Professor Oort and his colleagues for their hospitality and for their interest in the work.

## References

Lin, C. C., Yuan, C., and Shu, F. H.: 1969, *Astrophys. J.* **155**, 721.
Shu, F. H.: 1968, Ph.D. Thesis, Harvard University, §§ 12–14.

# 67. PARTICLE RESONANCE IN A SPIRAL FIELD

B. BARBANIS

*Dept. of Astronomy, University of Patras, Greece*

**Abstract.** Plane galactic orbits near the particle resonance of a spiral field have been calculated numerically. Besides the ring type orbits, i.e. orbits filling a ring around the galactic center, there are orbits librating near the potential maxima at the co-rotation distance (Lagrangian points). This effect tends to disrupt the spiral arms at the co-rotation distance.

A large number of particle plane orbits has been calculated in a typical logarithmic spiral potential superimposed on an axisymmetric potential which represents the Contopoulos-Strömgren (1965) model of the force field on the galactic plane. The spiral force field used is about 7% of the axisymmetric field (Barbanis and Woltjer, 1967; Barbanis, 1968a, b).

Expressing the distance to the galactic center $\varpi$ in kpc and the unit of time in $10^7$ yr the axisymmetric potential is

$$\phi_0 = 1.0459 \left( -7.334\varpi^{-1} - 0.15818\varpi + 0.1721015\varpi^2 - \right.$$
$$\left. - 0.0134207\varpi^3 + 0.000323505\varpi^4 \right). \tag{1}$$

The adopted logarithmic spiral potential, consisting of two arms and rotating as a rigid body with an angular velocity $\Omega_s$, is

$$\varepsilon\phi_s = \varepsilon\varpi^{1/2}(16 - \varpi)\cos\{2\varphi - 2\Omega_s t + 20\ln((\varpi + 4)/8)\}, \tag{2}$$

where $\varpi$, $\varphi$ are polar coordinates with respect to an inertial frame and $\varepsilon = 0.001$.

The angular velocity of the spiral pattern used is $\Omega_s = 25$ km s$^{-1}$ kpc$^{-1}$ and it is equal to the angular velocity $\Omega$ of the axisymmetric model at 10 kpc. This value is larger than Lin's value (Lin *et al.*, 1969) by a factor of 2, but close to the values given by Kalnajs during this Symposium (this volume, p. 318) and Marochnik and Suchkov (1969). However, the main purpose of this paper is to investigate the particle resonance phenomena wherever they appear.

The equations of motion of a mass point with coordinates $(x,y)$ in a reference frame rotating with angular velocity $\Omega_s$ are

$$\frac{d^2x}{dt^2} = \Omega_s\left(2\frac{dy}{dt} + \Omega_s x\right) - \frac{\partial\phi}{\partial x}, \tag{3}$$

$$\frac{d^2y}{dt^2} = -\Omega_s\left(2\frac{dx}{dt} - \Omega_s y\right) - \frac{\partial\phi}{\partial y}, \tag{4}$$

where $\phi = \phi_0 + \varepsilon\phi_s$; the argument of the cosine in $\phi_s$ is

$$2\theta = 2\tan^{-1}(y/x) + 20\{\ln[(\varpi + 4)/8] - \ln[(\varpi_0 + 4)/8]\}, \tag{5}$$

where $\varpi$, $\theta$ are polar coordinates in the rotating frame with origin the galactic center; the x-axis intersects the spiral arm at a distance $\varpi_0 = 10$ kpc.

The Jacobi integral is

$$H = \tfrac{1}{2}(X^2 + Y^2 - \Omega_s^2 \varpi^2) + \phi, \tag{6}$$

where $X = dx/dt$, $Y = dy/dt$, and $\varpi^2 = x^2 + y^2$.

If $H$ is smaller than 1.67 (kpc/$10^7$ yr)$^2$ the curve of zero velocity consists of two near circles (dotted curves) around the galactic center G.C. (Figure 1). The velocity is real inside the inner curve or outside the outer curve. The co-rotation point CR on the x-axis lies inside the ring, where the motion is not permitted. The width of this ring gradually decreases as the value of $H$ increases and tends to the value 1.67.

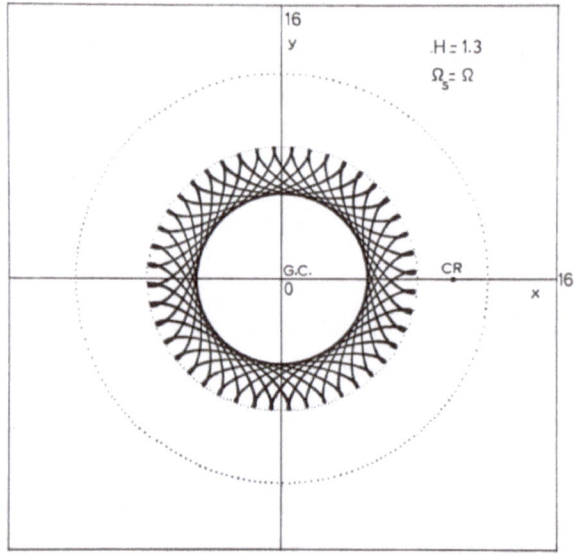

Fig. 1 A ring type orbit for $H = 1.3$, $\Omega_s = \Omega = 25$ km s$^{-1}$ kpc$^{-1}$, and initial conditions: $x_0 = 5$ kpc, $y_0 = 0$, $X_0 = 0$. G.C. is the galactic center and CR the co-rotation point on the x-axis. The curve of zero velocity (dotted curves) consists of two near circles; the motion is not permitted inside the ring.

Orbits 'near' a circular orbit, starting inside the inner curve, fill rings around the origin with their external boundary inside the curve of zero velocity. The corresponding invariant curves on the plane $xX$, for $x > 0$, are inside a boundary (Figure 2). Each invariant curve of Figure 2 is marked by two numbers giving the initial distance and radial velocity (the initial point is always on the x-axis). Orbits starting near the periodic orbit have closed invariant curves, but the invariant curves of orbits further away are open and terminate on the boundary. The terminal points correspond to those loops of an orbit (Figure 1) which are tangent to the x-axis.

For $H$ larger than 1.67 the zero velocity curve disappears and the motion is now permitted everywhere on the xy-plane. However, some of the orbits fill again a ring around the galactic center. This leads to the conclusion that besides the energy integral there is another integral which restricts the motion in a finite area.

The most important effect near the co-rotation point is a new type of orbits, which appear for $H > 1.67$ and a certain range of initial conditions. These orbits librate up and down with respect to the $x$-axis. Figure 3 gives two such librating orbits for $H = 1.7$ and initial conditions: $x_0 = \pm 8.5$ kpc $y_0 = X_0 = 0$. The co-rotation points, on

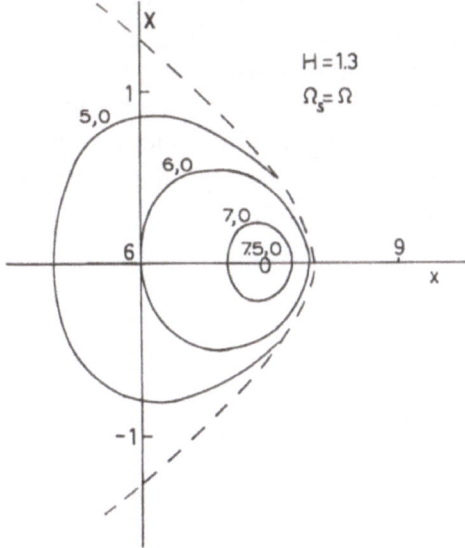

Fig. 2. Invariant curves of various orbits for $H = 1.3$. The dashed line gives the boundary of the invariant curves on the $xX$-plane for orbits starting inside the inner zero velocity curve. The two numbers on each invariant curve give the initial position in kpc and the radial velocity in km s$^{-1}$ of the moving point.

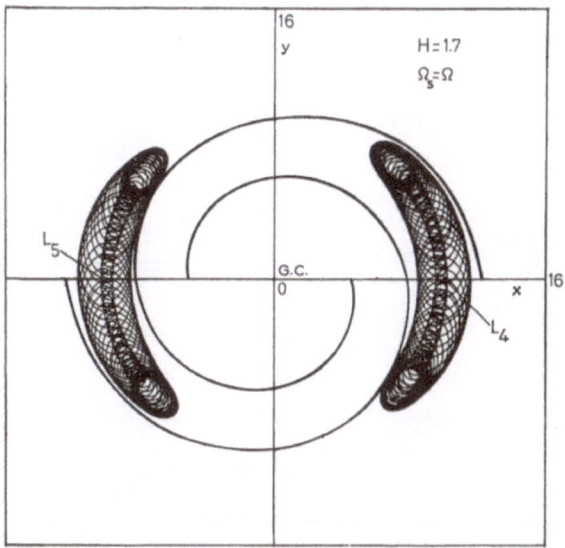

Fig. 3. Two librating orbits for $H = 1.7$ and $x_0 = \pm 8.5$ kpc, $y_0 = X_0 = 0$. The two spiral arms give the density maxima. $L_4$ and $L_5$ are the co-rotation points on the $x$-axis, where the potential is maximum; these points correspond to the Lagrangian points $L_4$, $L_5$ of the restricted three-body problem.

the x-axis, lie near the middle between the extreme intersections of the x-axis by each orbit. At these points the spiral potential is maximum. These points correspond to the Lagrangian points $L_4$, $L_5$ of the restricted three-body problem. The librating orbits correspond to the Trojan orbits near $L_4$ and $L_5$.

Such librating orbits appear if $8.3 < |x_0| < 9.8$ kpc. One orbit near each Lagrangian point $L_4$, $L_5$ is periodic. Orbits near it fill an elongated ring surrounding the periodic orbit (Figure 4). For $|x_0| \geqslant 9.8$ kpc the orbits fill again rings around the galactic center.

For $H = 1.7$ the boundaries of the invariant curves (Figure 5; dashed lines) do not intersect the x-axis. For each orbit, filling a ring around the galactic center, correspond two open invariant curves, one with positive $x$ and the other with negative $x$. On the other hand, the invariant curves of librating orbits are closed curves or two open curves situated on one side of the X-axis.

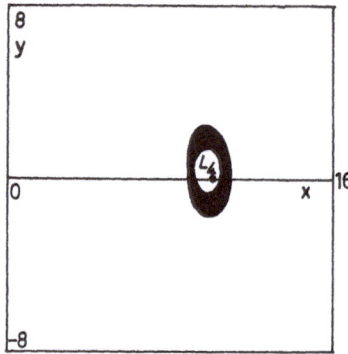

Fig. 4. An orbit for $H = 1.7$ and $x_0 = 9.1$ kpc, $y_0 = X_0 = 0$. Such an orbit fills an elongated ring around $L_4$.

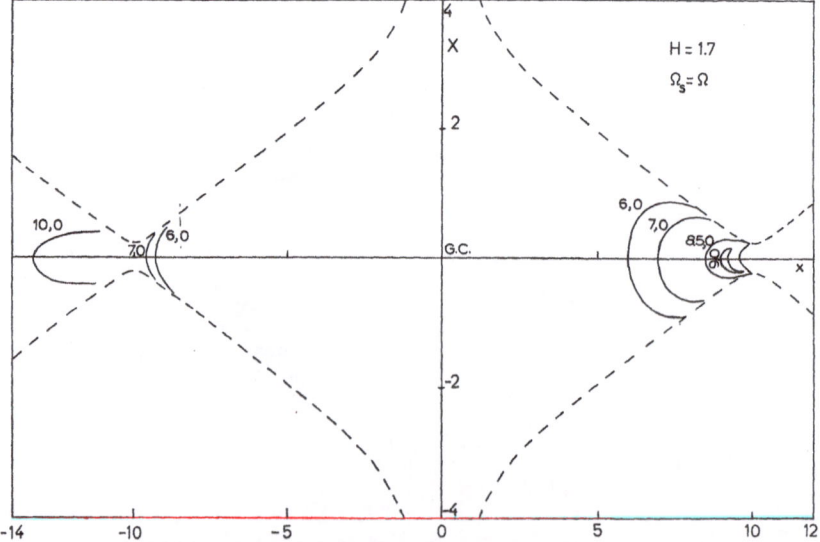

Fig. 5.   Invariant curves of various orbits calculated for $H = 1.7$. Ring type orbits produce two open invariant curves one at the positive and the other at the negative part of the x-axis. Invariant curves of librating orbits are closed, or two open curves situated on the same side of the X-axis.

Similar results were found for $H = 2$. Therefore librating orbits appear for a substantial range of initial conditions near the co-rotation distance. We conclude that an appreciable amount of matter in this neighbourhood is concentrated near the Lagrangian points $L_4$, $L_5$. Matter goes away from the minima of potential (the outermost points of intersection of the spiral arms of Figure 3 by the $y$-axis) and forms two density maxima near the Lagrangian points; this phenomenon is well known in celestial mechanics.

The consequence for the Galaxy is a disruption of the observed spiral arms at the co-rotation distance. Self-consistent spiral waves can exist inside or outside the co-rotation point, but near this point non-linear effects produce their breaking. This effect may be connected with the observed broken form of the spiral arms in the outer parts of many galaxies.

## References

Barbanis, B.: 1968a, *Astrophys. J.* **153**, 71.

Barbanis, B.: 1968b, *Astron. J.* **73**, 784.

Barbanis, B. and Woltjer, L.: 1967, *Astrophys. J.* **150**, 461.

Contopoulos, G. and Strömgren, B.: 1965, *Tables of Plane Galactic Orbits*, Inst. Space Studies, New York.

Lin, C. C., Yuan, C., and Shu, F.: 1969, *Astrophys. J.* **155**, 721.

Marochnik, L. S. and Suchkov, A. A.: 1969, *Astrophys. Space Sci.* **4**, 317.

# B. BARRED SPIRALS

# 68. BARRED SPIRAL GALAXIES

K. C. FREEMAN

*Australian National University, Canberra, Australia*

**Abstract.** We point out some properties of barred spiral galaxies which are important for the theory of their formation and spiral structure, and describe some theoretical work on the dynamics of these systems.

## 1. Some Observed Properties

What are barred spirals? They are members of the family of disk-like galaxies which includes all spirals and lenticulars (SO), and they have an inner bar structure; many examples are shown in *The Hubble Atlas of Galaxies* (Sandage, 1961). Apart from this bar and associated features, such as the dustlanes, they are similar in structure and content to non-barred systems.

Two remarks: (a) There is no clear division between barred and non-barred systems; the transition SA-SAB-SB at any stage is smooth (de Vaucouleurs, 1959). (b) Barred spirals are not rare oddities of nature. Among 994 bright spirals, about $\frac{2}{3}$ show bar structure (31% SA, 28% SAB, 37% SB, 4% S; see de Vaucouleurs, 1963a). If one plans to discuss the dynamics of disk galaxies, then it must be clear that axisymmetry is the exception, not the rule.

On most reproductions, the bar looks dominant in size and luminosity, but this impression is not usually correct. Surface photometry shows very clearly that the bar is a relatively small feature immersed in an approximately axisymmetric disk. Table I gives the NGC number, the morphological type, the ratio $L_B/L_T$ of the bar luminosity to the total luminosity, the ratio $a_B/a_{27.0}$ of the bar length to the disk diameter at a surface brightness of 27.0 magnitudes arc sec$^{-2}$, and the source of the photometry, for three barred galaxies.

TABLE I

Data for three barred spirals

| NGC | Type | $L_B/L_T$ | $a_B/a_{27.0}$ | Source |
|------|----------|------|------|----------------------------|
| 1313 | SB(s)d | 0.12 | 0.15 | de Vaucouleurs (1963b) |
| 6744 | SAB(r)bc | 0.10 | 0.10 | de Vaucouleurs (1963c) |
| 1433 | SB(r)a | 0.15 | 0.30 | de Vaucouleurs (unpublished) |

I emphasise that, in the luminosity distribution, the asymmetry associated with the bar is limited to the inner part of the system; most of the light comes from the approximately axisymmetric disk. First I discuss this disk, and then the bar.

The disks of SB systems have in general an exponential radial luminosity distribution $I(R) = I_0 \exp(-\alpha R)$, and differ in no way from the disks of SA systems. For example, the parameters $I_0$ and $\alpha$ change with classification stage ($a, b, ..., m$) in the same way for the disks of SA and SB galaxies (see Freeman, 1970). As in SA systems, the spiral arms in blue light are defined mainly by HII regions, hot stars, and dust. Little is known about the involvement of the *old* disk population (which usually contains most of the mass) in the spiral structure. The author's infrared plates of M83, and Garrison's visual plates of many spirals (see Morgan, this volume, p. 9) show relatively amorphous spiral arms underlying the highly resolved blue light structure. Are these 'red arms' just a slightly older population I feature, or do they involve the old disk itself? In Baade's picture, is the spiral structure just frosting on the cake, or do we have a spiral cake? It seems important for understanding spiral structure that this point be cleared up: unfortunately it is not obvious that it can be done observationally, even in principle.

What is the nature of the *bar*, i.e. to what population does it belong? In some earlier type systems (e.g. NGC 6744), the bar follows the luminosity distribution $\log I(R) \propto R^{1/4}$, which appears to be characteristic of the spheroidal component in SA galaxies (de Vaucouleurs, 1959). There is evidence that the bar is imbedded in the spheroidal component for systems earlier than about SBc; the bar colours and spectra are consistent with this picture. For later type galaxies, the bar nature is not clear; in NGC 55, a nearly edge-on SBm system (de Vaucouleurs, 1961), the bar is no more extended in the direction perpendicular to the galactic plane than is the disk. At least for the earlier type systems, we identify the bar photometrically and spectroscopically with the old spheroidal population.

Two remarks about the bar: (a) Note the characteristic dust lanes along the bar in earlier type SB systems like NGC 1300. These probably have some dynamical importance which is not yet understood. (b) Systems earlier than about SBc have the bar central in the disk. At about SBd a *second* asymmetry appears: the bar becomes displaced from the disk centre. This effect is discussed by de Vaucouleurs and Freeman (this volume, p. 356).

Finally I discuss the velocity field in barred galaxies. Apart from rotation, the HII velocity field shows noncircular motions in many SB systems, at least in the nuclear regions. In some late type barred spirals, there appears to be a gas flow *outwards* along the bar, with velocities of order 100 km s$^{-1}$ and mass flux of order 1 $M_{\odot}$ yr$^{-1}$. NGC 4027 (de Vaucouleurs *et al.*, 1968) is a particularly good example, because its bar is almost exactly perpendicular to its line of nodes; it is then easy to distinguish streaming motions along the bar from the general rotation field.

To summarise this section: The main difference between the A and B families of spirals is the departure from axisymmetry in the *inner* parts of B galaxies. It seems that the basic problems for the two families are similar: the formation and dynamics of the old background disk + spheroidal (+ bar) components, and the formation and maintenance of the spiral structure.

## 2. Theoretical Work

I describe a gravitational theory which appears to have some success in explaining features of SB galaxies, although its relevance is limited to these systems. I do not consider the disk here, because its problems are common to all disk galaxies. First I discuss the bar.

The formation of the bar may not be a fundamental problem. Several authors have shown how the collapse of uniform density spheroids (rotating or not) tends to produce prolate figures if the ratio (gravitational energy)/(internal energy) is fairly large. If it is true that the bar is part of the old spheroidal component, then this ratio may determine whether the spheroidal component stays axisymmetric, as in SA galaxies, or develops a bar structure.

Little is known about the stellar dynamics of realistic steady state bar-like systems. However the study of unrealistic models seems to be very instructive. Consider a general uniformly rotating bar, in a steady state referred to the rotating frame. It follows from the collisionless Boltzmann equation that the distribution function $f$ depends only on the only isolating (Jacobi) integral $J = \frac{1}{2}c^2 - \Phi_1$, where $\Phi_1$ is the potential for the centrifugal + gravitational force field. The average velocity $\bar{c}$ of stars, referred to the rotating frame, is then identically zero because

$$\int \mathbf{c} f(J)\, \mathrm{d}^3 c \equiv 0$$

by the symmetry of $J$ in $\mathbf{c}$.

Now consider a special bar model. It is possible to construct uniform density ellipsoidal stellar systems, rotating rigidly about a principal axis (Freeman, 1966). These are completely self-consistent exact solutions of the collisionless Boltzmann equation and Poisson's equation. (It is not known whether they are stable.) Because of the properties of uniform ellipsoids, $J$ is not the only isolating integral, so $\bar{c}$ is not identically zero. It turns out that the mean stellar motion shows a strong circulation *counter* to the sense of rotation, just as in the fluid Dedekind ellipsoids. The circulation is so strong that near the minor axes of the bar, the tangential velocity of the stars in the bar is in the *opposite sense* to the rotation of the figure of the bar. In NGC 4027, which is ideally orientated (minor axis of the bar along the line of nodes), this remarkable predicted effect has been observed (de Vaucouleurs *et al.*, 1968). The effect was evident in 7 stellar absorption lines in 3 independent spectograms, so it is probably real. The relevance of this effect to the problem of the Symposium is that, for constructing zero-order background models of barred systems, it is probably incorrect to take $f = f(J)$ alone.

I now discuss the spiral arms and ring structures (e.g. NGC 1433, de Vaucouleurs, 1959) of SB systems. Most of the theoretical work peculiar to barred galaxies depends on the properties of test particle orbits in the gravitational field of a rotating bar.

(i) *The spiral arms.* Several authors have described spiral structure formation based on the gas outflow observed along the bars of some SB systems. There are some interesting results: (a) Model the bar by a rapidly rotating ellipsoid of typical mass and dimensions. The gravitational torque of this bar can support an outflow along the bar, against the Coriolis force, of order 100 km s$^{-1}$, as observed. (b) Follow numerically these outflowing particles after they leave the bar: they define quite realistic spiral arms as long as the outflow continues (see e.g. Freeman, 1965).

Julian and Toomre (1966) suggest that a bar rotating in a differentially rotating disk of stars may set up a forced trailing density wake response in the disk from the bar ends. This suggestion raises again the question of the involvement of the old disk in the spiral structure.

(ii) *The ring structure*, observed in about 25% of SB systems. It seems possible to account for this structure from the properties of particle orbits in the gravitational field of a rotating bar. Consider a prolate spheroidal bar alone, axial ratio $\simeq 5$, mass $\mathscr{M}$, semi-length $L$, and uniform angular velocity $\Omega$. There is one dimensionless number, $Q = 3G\mathscr{M}/(\Omega^2 L^3)$. (The breakup angular velocity corresponds to $Q \simeq 1$.) For a fairly slowly rotating bar ($Q \simeq 15$), there exists a family of approximately circular periodic orbits, encircling the bar near bar ends. These orbits lead the bar slightly, are highly stable, and cover a fairly wide annulus (width $\simeq L/4$) around the bar (see de Vaucouleurs and Freeman, 1970, for full details). It can be shown numerically, from surface of section studies, that it is easy to trap enough matter near these stable periodic orbits to produce the apparent increase in luminosity associated with the observed rings. However, as $Q$ is decreased, the annulus covered by the stable circular periodic orbits becomes less wide, and the matter trapping efficacy decreases. For $Q \lesssim 8$, the periodic orbits no longer exist, and the trapping of matter to form the ring structure is no longer possible. In this picture, the SB(r) systems are slower rotators, in a dimensionless sense, than the SB(s) systems. There is no real observational evidence yet either for or against this inference.

The particle orbit theory for the spiral structure and rings in SB galaxies, and for some of the features of the highly asymmetric SBm systems (see de Vaucouleurs and Freeman, this volume, p. 356) seems to have some success, and probably points to some relevant dynamical process. However I am uneasy about any theory for galactic structures that does not work for all galaxies, SA or SB; the transition SA-SAB-SB is so smooth that it seems inconceivable that the structure producing processes should be essentially different for SA and SB systems. I do not believe that we can claim to understand these processes until we find one that works for all systems, barred and unbarred.

## References

Freeman, K. C.: 1965, *Monthly Notices Roy. Astron. Soc.* **130**, 183.

Freeman, K. C.: 1966, *Monthly Notices Roy. Astron. Soc.* **134**, 1.

Freeman, K. C.: 1970, *Astrophys. J.*, in press.

Julian, W. and Toomre, A.: 1966, *Astrophys. J.* **146**, 810.

Sandage, A. R.: 1961, *The Hubble Atlas of Galaxies*, Carnegie Institution of Washington Washington.

Vaucouleurs, G. de: 1959, *Handbuch der Physik* **53**, 275.

Vaucouleurs, G. de: 1961, *Astrophys. J.* **133**, 405.

Vaucouleurs, G. de: 1963a, *Astrophys. J. Suppl. Ser.* **8**, No. 74, 31.

Vaucouleurs, G. de: 1963b, *Astrophys. J.* **137**, 720.

Vaucouleurs, G. de: 1963c, *Astrophys. J.* **138**, 934.

Vaucouleurs, G. de, and Freeman, K. C.: 1970, 'Structure and Dynamics of Barred Spiral Galaxies, and in particular of the Magellanic Type', in *Vistas in Astronomy* (ed. by A. Beer), to be published.

Vaucouleurs, G. de, Vaucouleurs, A. de and Freeman, K. C.: 1968, *Monthly Notices Roy. Astron. Soc.* **139**, 425.

## Discussion

*Miller:* Your arguments lean heavily on photometry at light levels well below that of the night sky, and seem to imply much more accurate subtractions of the night sky than can usually be justified.

*Freeman:* Comparison of detailed isophotometry of several objects by independent observers using independent plate material and calibrations indicates that acceptably accurate photographic surface photometry to at least three magnitudes below the night sky level can be achieved.

*Toomre:* In his very nice lecture on barred spirals Ken Freeman also touched on a vital clue to the ordinary spirals. As he said, and as De Vaucouleurs especially has emphasized, it is remarkable that there exists no sharp break between the barred and the obviously disk-like galaxies. Instead, their transition appears to be almost continuous, and upon closer inspection, not only do more and more of the bars, like that of the LMC, turn out to be imbedded in fairly disk-like systems, but even some galaxies as 'normal' as M31 seem to exhibit internal motions suggesting some sort of a bar-like asymmetry.

As a theoretician, I am delighted with this trend of evidence for two reasons: For one thing, I am now convinced that Lin's 'asymptotic' theory of spiral density waves suffers from the serious – but probably non-fatal – difficulty that such already tightly wound waves need continually to be re-plenished, apparently from length scales so large that the present form of that theory ceases to be applicable. (Because of a group velocity inherent to such waves, I might add, this need for replenish-ment is rather more acute than Oort surmised in his opening lecture.) On the other hand, it also seems clear from work beginning with the sheared density waves of Goldreich and Lynden-Bell, that every differentially rotating galactic disk is very *responsive* to excitation. Now this forcing could come tidally from a nearby galaxy, as I discussed briefly in my own talk. Or it could arise from large gas complexes orbiting within the disk itself, as Julian and I once demonstrated. But an even more impressive source of these *forced* spiral waves should be the gravity forces from any slowly rotating bar-like or oval distortion involving a major (and probably central) fraction of the mass of a galaxy.

I do not know, of course, that any given disk is so obligingly distorted. But Freeman's remarks, Lin's dilemma, this spiral responsiveness, and numerical calculations such as those of Kalnajs all suggest that at least a mild case of the bar disease afflicts most of the seemingly ordinary spiral galaxies.

# 69. STRUCTURE AND DYNAMICS OF BARRED SPIRAL GALAXIES WITH AN ASYMMETRIC MASS DISTRIBUTION

G. DE VAUCOULEURS

*University of Texas, Austin, Tex., U.S.A.*

and

K. C. FREEMAN

*Australian National University, Canberra, Australia*

**Abstract.** The asymmetric structure and dynamics of late-type barred spirals is analyzed in terms of a model consisting of a small prolate spheroid (the bar) displaced from the center of a large oblate spheroid (the disk). The non-axisymmetric motions predicted for this model are in remarkable agreement with optical and radio observations of internal motions in several magellanic barred spirals, including NGC 4027 and the Large Cloud. In particular, the model explains the displacement of the apparent center of the rotation curve from the center of the bar.

1. Until recently dynamical models of galaxies rested on the assumption of circular symmetry for ordinary spirals or, at least for barred spirals, of axial symmetry. In general these assumptions appear reasonable for isolated galaxies; for example the luminosity distribution in an ordinary spiral such as M33, type SA(s)cd, is very nearly symmetric about its nucleus at the rotation center (Figure 1a). Nevertheless there is evidence that even in such apparently symmetric systems, a small fraction of the mass, including both stars and gas, is significantly displaced form the center of symmetry. For example in M33 the centroid of the distribution of supergiant stars brighter than $m_B = 20.2$ $(M_B < -4.5)$ (de Vaucouleurs, 1961b) is displaced from the nucleus by some $2'$ to $3' \simeq 500$ pc in position angle 200° (Figure 1b). Similarly, the center of gravity of the integrated 21 cm brightness distribution (Gordon, 1969) is displaced from the optical nucleus by about $4' \simeq 800$ pc in p.a. 220° (Figure 1c). This direction is unrelated to that of M31 which is in p.a. 315°. Here the excess mass involved is a very small fraction of the total mass of the galaxy, probably less than 0.1%.

A more pronounced offset exists in M101 where the distribution of both 21 cm and optical emission is conspicuously asymmetric (Beale and Davies, 1969). Here the mass asymmetry may involve as much as several per cent of the total mass, with 1% in the gas alone, but in this case it may be more a result of tidal interaction with NGC 5474 and other members of the M101 group rather than a manifestation of basic structural asymmetry.

2. In late-type barred spirals, and in particular of the Magellanic type, i.e. for types SB(s)d to SB(s)m of the revised Hubble system (de Vaucouleurs, 1959b), asymmetry is a basic and characteristic property of the mass distribution clearly visible on direct photographs (Figure 2a) and reflected in the surface distributions of practically all components of the system, in particular supergiant stars, neutral hydrogen, luminosity etc. (Figure 2b). Here the asymmetry involves several per cent of the total mass of the

*Becker and Contopoulos (eds.), The Spiral Structure of Our Galaxy, 356–362. All Rights Reserved.*

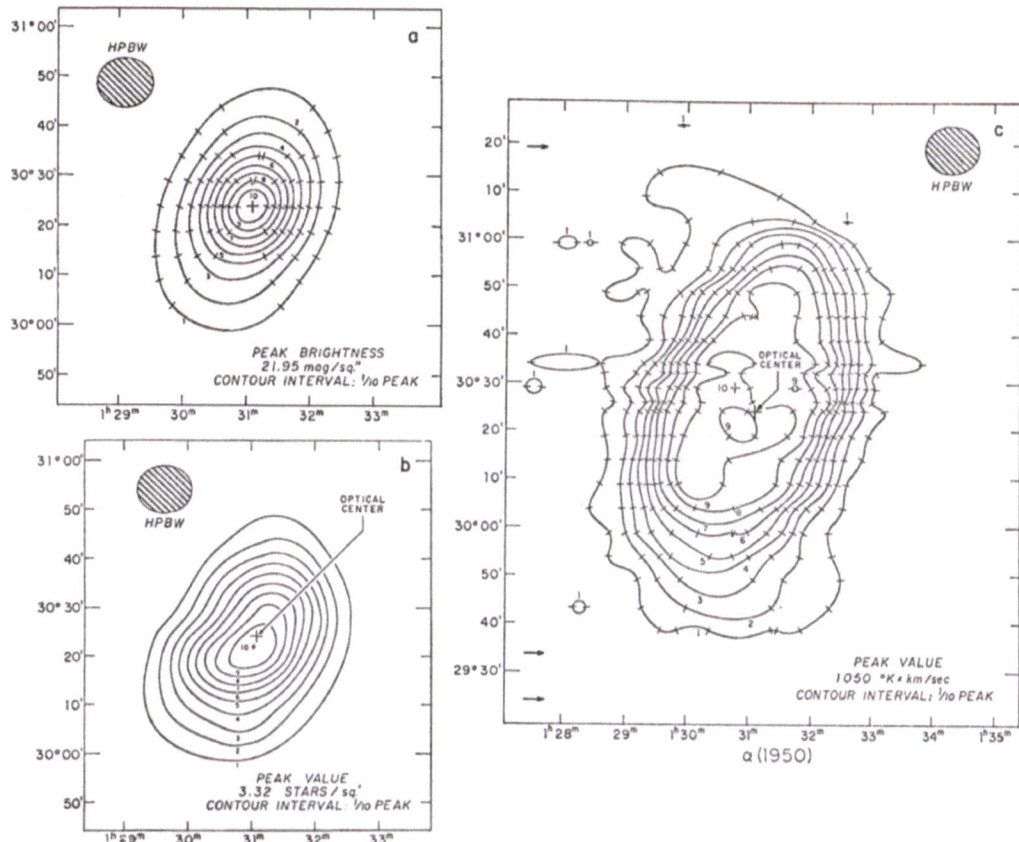

Fig. 1.   Comparison of optical and 21 cm distributions in M33. – (a) Blue light integrated luminosity in B system, after de Vaucouleurs (1959a); – (b) Blue light star counts to $m_B = 20.2$ ($M_B < -4.5$), after de Vaucouleurs (1961b); – (c) 21 cm integrated flux density, after Gordon (1969). (a) and (b) were convolved with 300-foot NRAO antenna beam to match resolution in (c). Note excess of supergiant stars (b) and neutral hydrogen gas (c) in quadrant ssf nucleus which coincides with center of optical isophotes (a).

system and gives rise to the puzzling asymmetry of the rotation curve with respect to the center $C$ of the bar first noted in the Large Magellanic Cloud (Kerr and de Vaucouleurs, 1955; Feast, 1964). Later the same effect was observed in several other systems of closely related types, in particular NGC 55 (de Vaucouleurs, 1960; Robinson and Van Damme, 1966), NGC 4027 (de Vaucouleurs *et al.*, 1968), and NGC 4631 (G. and A. de Vaucouleurs 1963; Roberts, 1968; Crillon and Monnet, 1969). In all these systems the displacement from $C$ of the center of symmetry $C_r$ of the rotation curve is on the order of 0.5 to 1.5 kpc toward the richer side of the system. This is also the side marked by the major spiral arm $B_1$ giving to some systems the superficial appearance of 'one-armed' spirals, although closer inspection always reveals the presence of one or more additional arms. The basic structure of these asymmetric barred spirals is shown in Figure 2a. The so-called 'radio center' $C_r$ of the rotation curve was originally introduced by Kerr and de Vaucouleurs

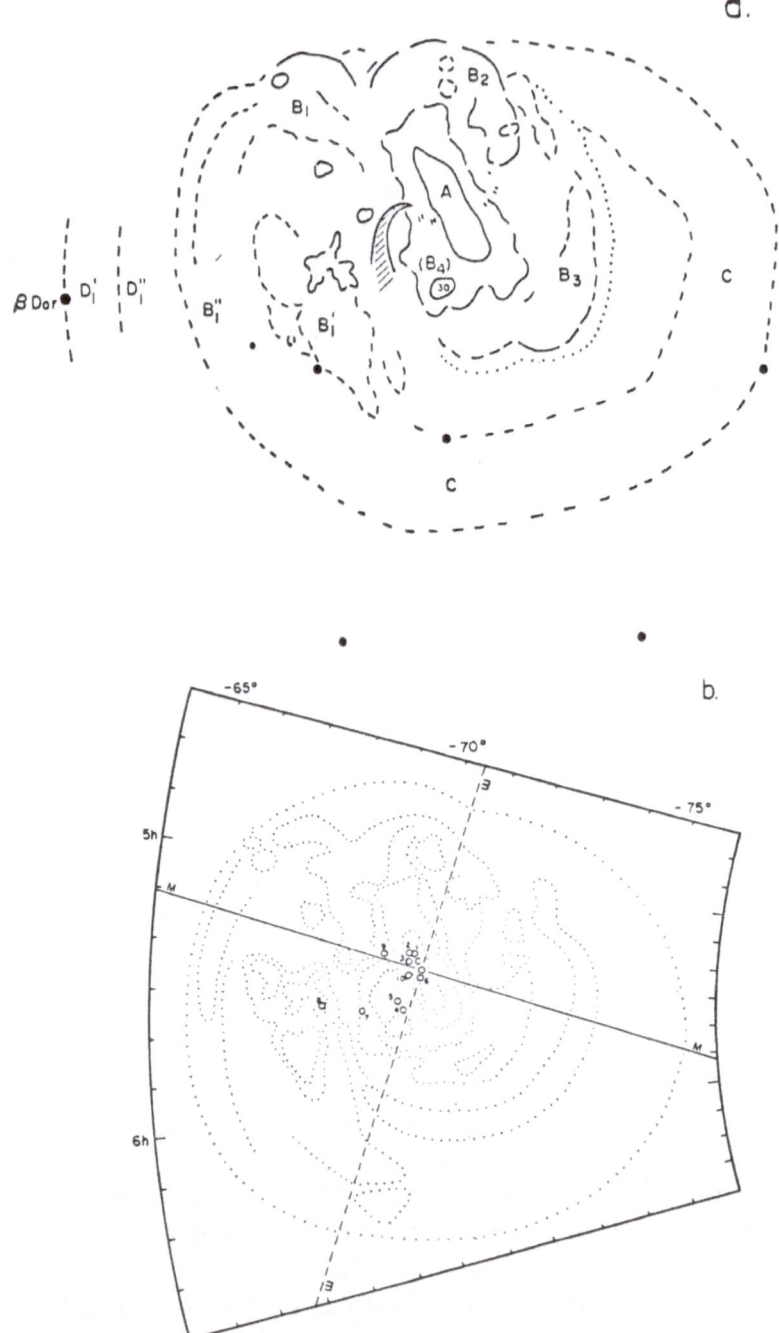

Fig. 2. Asymmetric structure of magellanic-type barred spirals. – (a) Key to barred spiral structure of Large Magellanic Cloud showing axial bar $A$, inner spiral arms $B_1$ to $B_3$, outer loop $C$, and out-lying spiral arcs $D'_1$, $D''_1$. – (b) Locations of various distribution centroids in Large Magellanic Cloud. $C$ = center of symmetry of bar. 1 = surface brightness peak in blue light, 2 = centroid yellow light isophotes, 3 = centroid of planetary nebulae, 4 = geometric center of outer loop, 5 = centroid of outlying clusters, 6 = centroid of all clusters, 7 = centroid of neutral hydrogen, 8 = centroid of bright stars ($M < -5$), 9 = center of symmetry of HI rotation curve ('radio center' $C_r$), 10 = center of symmetry of rotation curve of planetaries.

(1955) simply to replace the actual asymmetric rotation curve by a fictitious, but more or less equivalent symmetric curve to which they could apply the standard analysis based on existing galaxy models which all assumed circular symmetry. The need for a dynamical theory of asymmetric barred systems was clearly recognized, however (de Vaucouleurs, 1960, p. 281).

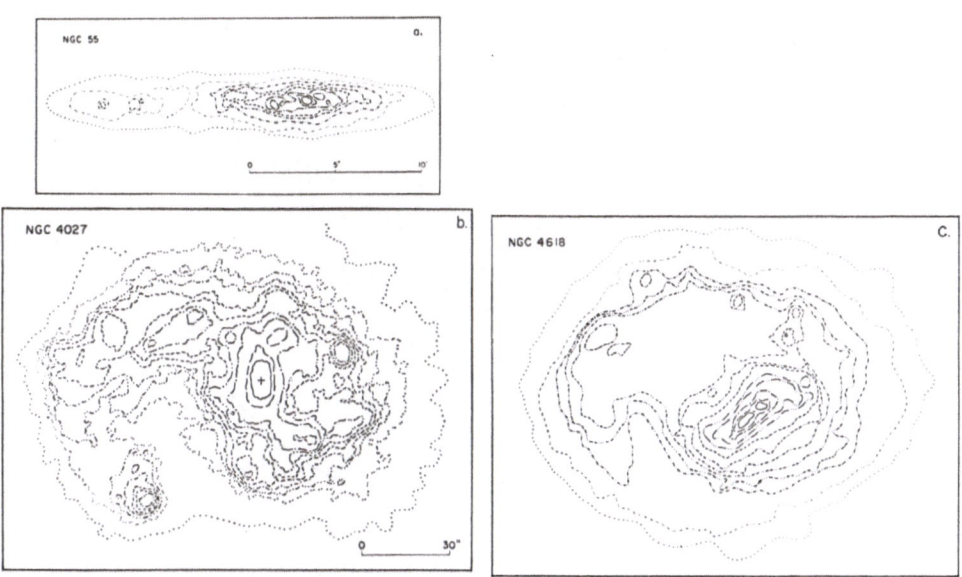

Fig. 3.   Isophotes of three Magellanic barred spirals in blue light. – (a) NGC 55 (de Vaucouleurs, 1961a); – (b) NGC 4027 (de Vaucouleurs *et al.*, 1968); – (c) NGC 4618 (Bertola 1967). Note eccentric location of bar in flat disk component.

**3.** A plausible and still simple model, suggested by the surface photometry of systems such as the Large Cloud, NGC 55, 4027, 4618, etc. (Figures 3a, b, c) is shown in Figure 4; it consists of (1) a large oblate inhomogeneous spheroid of mass $M_d$ with axis ratio $c/a=0.2$ and some suitable density low following one of Perek's (1948, 1962) standard distributions, to approximate the main disk population of the Galaxy, and (2) a small prolate spheroid of axis ratio $c/a=5$ and mass $M_b \simeq 0.1M_d$ approximating the bar.

This bar is displaced from the center of the large spheroid by a small distance $\varLambda$ (Figure 4), typically of the order of 1 kpc in Magellanic spirals. How this situation arises in the first place is not our concern here since observation proves that it is present in nature. One might speculate that when primordial gas masses assemble to form a proto-galaxy there is no particular reason why the assemblage should have spherical or even circular symmetry; the amount of asymmetry may in fact determine the particular galaxy type formed.

**4.** Some of the properties of Magellanic barred spirals can be understood from the properties of test particle orbits in the gravitational field of the configuration described in Section 3. For example, Figure 5 shows the mean rotation curve for NGC 4027

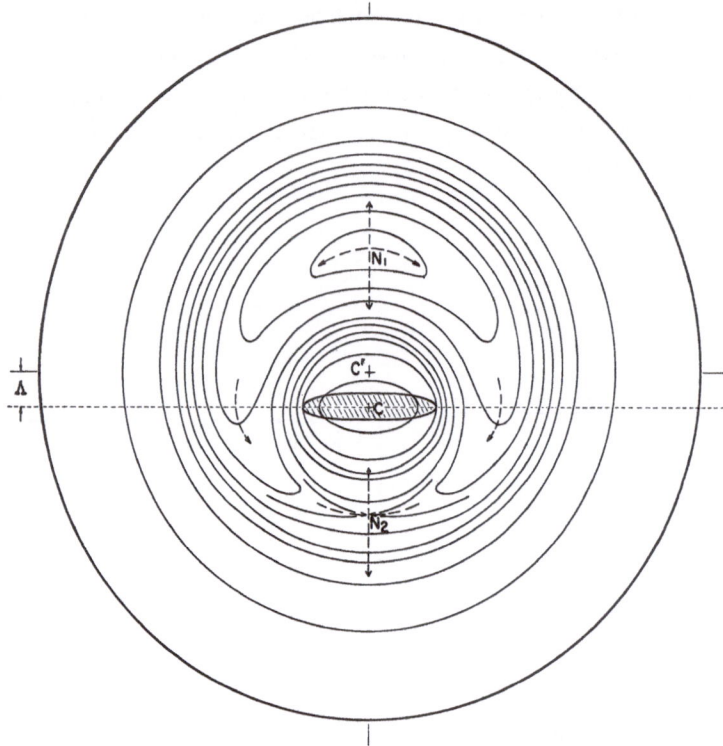

Fig. 4.   Model of asymmetric barred spiral with equipotential lines. The inner prolate spheroid
centered at $C$ (bar) is displaced by a distance $\varLambda$ from center $C'$ of outer oblate spheroid (disk or main
body). Neutral points are at $N_1$ and $N_2$. Compare Figure 3.

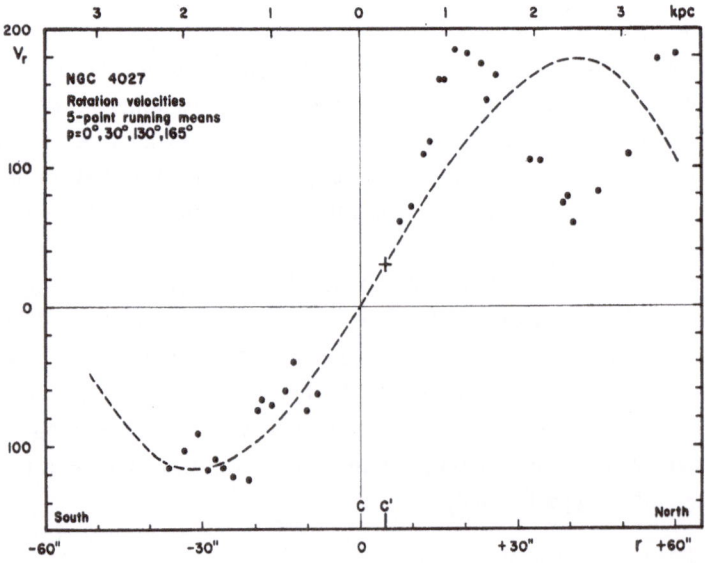

Fig. 5.   The mean rotation curve of NGC 4027. The points are mean points from spectra in four
position angles. The broken line is the best fit cubic to the mean points. $C$ is the bar center, and $C'$
the center of symmetry of the cubic.

(de Vaucouleurs *et al.*, 1968); see also Figure 3b. On the North side of the bar (the major arm side), the mean rotational velocity appears to decrease from 180 km s$^{-1}$ at $R=1.2$ kpc to 70 km s$^{-1}$ at $R=2.4$ kpc, and then increases again to 180 km s$^{-1}$ at $R=3.6$ kpc ($R=0$ is the bar center). There is no corresponding feature in the rotation curve on the other side of the bar.

To explain this feature, we use the model of Figure 4. It is assumed that the bar rotates (synchronously) about the disk center $C'$ in such a way that the potential seen by an observer rotating with the bar is time-independent. We do *not* assume any particular functional dependence of the rotation period on $\Lambda$ and the masses of the bar and the disk, but prefer to estimate the significant dimensionless numbers from observation.

This is an unusual model for the potential field of a galaxy. It cannot be justified dynamically at this stage. We use it because it is simple and includes the relative displacement and orientation of the bar and disk indicated by optical and radio observations. This seems to us to be among the fundamental properties of Magellanic barred spirals.

The equipotentials of the gravitational+centrifugal force field for this model (as seen in a reference frame rotating with the bar) are shown in Figure 4: the dimensionless numbers for this model are appropriate to NGC 4027 (see de Vaucouleurs and Freeman, 1970). Note that (a) there are only two neutral points outside the bar: $N_1$ (stable) and $N_2$ (unstable); (b) the parameters for NGC 4027 locate $N_1$ at about 2 kpc north of the bar center (cf. Figure 5).

There are two families $A$ and $B$ of periodic orbits around the stable neutral point $N_1$ (Figure 6). Calculation of the characteristic exponents for these orbits show that members of these families with amplitudes smaller than $A_1$, $B_1$ respectively (Figure 6)

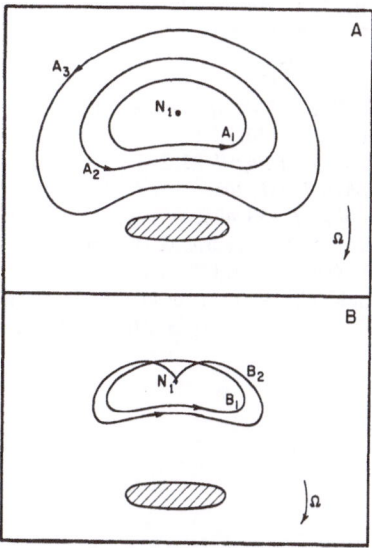

Fig. 6.   Trajectories on the galactic plane for members of two families of periodic orbits near the neutral point $N_1$. The bar, and the sense of the rotation, are also shown.

are stable. Surface of section analysis for values of the Jacobi integral corresponding to these stable orbits shows that they are stable to finite perturbations. This means that it is possible to trap a substantial amount of matter within the region enclosed by the orbit $A_1$; this matter circulates in a retrograde sense around $N_1$.

Now consider the rotation curve that would be observed for this model by an observer located at an infinite distance to the left-hand side of Figure 6; this location is appropriate to NGC 4027 (see de Vaucouleurs et al., 1968). For points in the region enclosed by $A_1$, the matter circulating around $N_1$ would cause an increase over the 'local circular velocity' in the apparent rotational velocity for $R<2$ kpc, and a corresponding decrease for $R>2$ kpc, with approximately the observed amplitude.

The neutral point $N_2$ is unstable. There are no stable periodic orbits corresponding to families $A$ and $B$ and there is no trapping of matter near $N_2$. The disturbance to the rotation curve then appears only on the $N_1$ side of the bar, as observed.

Full details of this work are given by de Vaucouleurs and Freeman (1970).

## Acknowledgements

The research reported here has been supported in part by the U.S. National Science Foundation and by the Research Institute of the University of Texas.

## References

Beale, J. S. and Davies, R. D.: 1969, *Nature* **221**, 531.
Bertola, F.: 1967, *Mem. Soc. Astron. Ital.* **38,** No. 2.
Crillon, R. and Monnet, G.: 1969, *Astron. Astrophys.* **2**, 1.
Feast, M. W.: 1964, *Monthly Notices Roy. Astron. Soc.* **127**, 195.
Gordon, K. J.: 1969, University of Michigan dissertation.
Kerr, F. J. and Vaucouleurs, G. de: 1955, *Australian J. Phys.* **8**, 508.
Perek, L.: 1948, *Contr. Astron. Inst. Masaryk* **1**, No. 6.
Perek, L. 1962, *Adv. Astron. Astrophys.* **1**, 165.
Roberts, M. S.: 1968, *Astrophys. J.* **151**, 117.
Robinson, B. J. and Van Damme, K. J.: 1966, *Australian J. Phys.* **19**, 111.
Vaucouleurs, G. de: 1959a, *Astrophys. J.* **130**, 728.
Vaucouleurs, G. de: 1959b, *Handbuch der Physik* **53**, 275.
Vaucouleurs, G. de: 1960, *Astrophys. J.* **131**, 265.
Vaucouleurs, G. de: 1961a, *Astrophys. J.* **133**, 405.
Vaucouleurs, G. de: 1961b, *Amer. Phil. Soc. Yearbook 1961*, p. 268.
Vaucouleurs, G. and A. de: 1963, *Astrophys. J.* **137**, 363.
Vaucouleurs, G. de and Freeman, K. C.: 1970, *Vistas in Astronomy*, in press.
Vaucouleurs, G. and A. de and Freeman, K. C.: 1968, *Monthly Notices Roy. Astron. Soc.* **139**, 425.

# C. NUMERICAL EXPERIMENTS

# 70. NUMERICAL EXPERIMENTS IN SPIRAL STRUCTURE

R. H. MILLER

*University of Chicago, Chicago, Ill., U.S.A.*

and

K. H. PRENDERGAST and W. J. QUIRK

*Columbia University, New York, N.Y., U.S.A.*

**Abstract.** Results of an *n*-body calculation, containing about 120000 particles, were shown as a motion picture. Some of the particles are treated as 'gas', obeying a special dissipative dynamics, the rest as 'stars'. The system was started as pure 'gas', and 'stars' were made out of the 'gas' in a manner closely mimicking real galaxies. Spiral density wave patterns appear in the 'gas', and last for about 3 'galactic rotations' without substantial change of form. Various experiments are described that have been undertaken in an attempt to learn the roles of various parts of the system in the maintenance of spiral patterns.

The large *n*-body calculation based on the 'game' described earlier (Miller and Prendergast, 1968) has shown spiral patterns which are demonstrated by showing a motion picture of their evolution. This calculation exactly incorporates the essential feature of being microscopically reversible; a feature which corresponds to the very long relaxation times expected in all parts of a galaxy.

Earlier calculations led to very 'hot' stellar populations. In order to look for spiral patterns, a much 'cooler' system seems necessary. We first tried sudden 'cooling'. Immediately after the cooling', gravitational collapse 'heated' the system back to the 'hot' condition it had been in before. Repeated 'coolings' did not help. A continuous 'cooling' scheme that conserves linear and angular momentum while reducing the random velocities in all the particles at each configuration space location at each integration step led to a general mess. The material which is treated this way is something like a 'gas'; it has dissipative properties. The next scheme for 'cooling' is a gradual scheme in which 'stars' are formed from the 'gas'. In this version of the calculation, two populations are represented, that being 'cooled' as in the continuous 'cooling' scheme, being called 'gas', the other, which is not so 'cooled' being called 'stars'. 'Stars' are created at each integration step by changing a certain number of 'gas' particles into 'stars' – shifting them from one constituent population to the other. The number changed is proportional to the square of the number of 'gas' particles at a particular configuration space location. When a model is started, all particles are 'gas'. The constant of proportionality, the 'star creation coefficient', is an important parameter that strongly influences the early evolution of the system.

We have conducted several experiments in an attempt to discover which features of the system are most important to the maintenance of spiral patterns. The value of the computer model is that it gives us a spiral system that may be modified in any way we wish. The problem is to design experiments that will delineate the essential features of systems showing spiral structure; that will show the role of each of the constituent parts. All details of the system are available; we routinely compute all of

*Becker and Contopoulos (eds.), The Spiral Structure of Our Galaxy, 365–367. All Rights Reserved.*
*Copyright © 1970 by the I.A.U.*

the usual galactic parameters such as mean velocities, circular velocity, force, epicyclic frequency, Oort constants, and components of the velocity ellipses at each radius in the system.

Spirals can be made in self-gravitating systems. The additional complications of magnetic fields are unnecessary.

The spiral patterns look like real galaxies. They are similar to real galaxies in that the spiral patterns are delineated in the 'gas' while the 'stars' form a background that shows much less structure. The models differ from real galaxies in that the peculiar velocities of the 'stars' are quite large – of the same order as the circular velocity. The 'star' system is pressure-supported, but the 'gas' moves with nearly the circular velocity. The spiral patterns have some features that agree with expectations from current ideas about spirals: individual mass points move through the patterns, and the patterns extend inward to the inner Lindblad resonance. The patterns become indistinguishable at about the co-rotation point, because of lack of particles.

Several experiments were run to try to determine the role of this 'star' background and the extent to which it is involved in the spiral patterns.

(1) 'Stars' of small peculiar velocity (velocity near the circular) were selected and a density plot was made for one integration step. A pattern was discernible although much less pronounced than in the 'gas'. As 'stars' of larger peculiar velocity were included, the patterns became more difficult to distinguish. These 'stars' show the local minima of the potential field.

The remaining experiments were carried out on a slightly modified system, using a stage of the regular calculation at which the spiral pattern is readily distinguishable as the starting point for a new integration.

(2) The entire 'gas' population (some 16000 particles) was turned into 'stars'. Separate density plots were made for this new 'star' population and for the other particles that were already 'stars' at this integration step. The spiral pattern began to dissipate – it was clearly discernible as a spiral pattern for about one revolution (20 integration steps), but became less and less so as more and more of its 'stars' acquired the large peculiar velocities of the background 'stars'. In a similar experiment in which the 'gas' was turned into 'stars' at an earlier stage, the new 'stars' rather rapidly settled down into patterns reminiscent of barred spirals. A 'bar' formed and 'stars' trailed off its ends to form open spirals. As more and more 'stars' trailed off, the spiral became tighter – a few of the 'stars' fell back toward the 'bar', the 'bar' became smaller and the spiral pattern less distinct. This entire process required about 2–3 revolutions of the bar (40–60 integration steps). These experiments show that the different dynamics of the 'gas' population leads to a somewhat different form for the spiral pattern and helps the pattern to last for longer times.

(3) The 'star' background was frozen into place. This produced a force field that did not evolve further, but was approximately that of a self-consistent static model because of the way in which it was obtained. The 'gas' continued to move in this unchanging background potential. Spiral patterns persisted – perhaps even a little better than they did with the actual 'star' background. In another version of the

experiment, the 'star' background potential was symmetrized (in angle). With this modification, the patterns did not persist as well. These experiments show that the moving structure of the actual 'star' background is not essential to the maintenance of the spiral pattern, but that a grainy or asymmetrical potential is.

(4) A static background potential like that of (3) was obtained from the mass of both the 'gas' and the 'stars'. The 'gas', now treated as having no-self-gravitation, continued to move in this background. The spiral pattern dissipated rather rapidly. This shows that the self-gravitation of the 'gas' is essential to the maintenance of spiral patterns, as might have been expected. When the background potential was symmetrized in angle, as in the second part of (3), the pattern dissipated quickly, showing that while the grainy static background is a little harmful to spiral patterns, an unsymmetrical part is necessary to drive the spiral.

(5) The static background of (3) was once more set up, but now all the mass points that were treated as 'gas' in (3) were treated as 'stars' for this experiment. Again, the spiral pattern persisted about as well as it did in (3). This shows that the special dynamics of the 'gas' component is not necessary to maintain spiral patterns at this stage of evolution. Again, the symmetrization in angle of the second half of (3) caused the patterns to die out.

Taken together, these experiments indicate that: (a) the 'stars' provide a background potential in which the 'gas' moves and must have an asymmetric potential to 'drive' the spiral, but otherwise do not play an essential part in the maintenance of the spiral pattern. High-velocity 'stars' can provide such a potential field without danger of a gravitational collapse. The excess velocity despersion above that necessary to prevent collapse is probably an artifact of our computer model and of the way in which the system was created. (b) Self-gravitation is essential to the maintenance of spiral patterns. The extra dissipative effects in the 'gas' component also help the patterns to survive for a longer time, when the subsystem that shows the spiral pattern can interact with a dynamic background. It helps to 'cool' the population. With a static background, this dissipative effect is less important, but that background cannot be symmetrized.

The method used to establish conditions in which spirals might occur is appealing because of the close analogy with the processes thought to be important in real galaxies, and because it stresses those properties that make spiral patterns stand out in real galaxies.

While we are not yet satisfied with all aspects of the calculation, the fact that spiral patterns have been obtained means that we are on the right track.

These calculations were carried out at the Goddard Institute for Space Studies in New York through the courtesy of Dr. Robert Jastrow, Director. The work has been assisted by grants from the National Science Foundation and by the Atomic Energy Commission.

### Reference

Miller, R. H. and Prendergast, K. H.: 1968, *Astrophys. J.* **151**, 699.

# 71. COMPUTER MODELS OF SPIRAL STRUCTURE

F. HOHL

*NASA Langley Research Center, Hampton, Va., U.S.A.*

**Abstract.** A computer model for isolated disks of stars is used to study the self-consistent motion of large numbers of point masses as they move in the plane of the galactic disk. The Langley Research Center's CDC 6600 computers are used to integrate the equations of motion for systems containing from 50 000 to 200 000 stars. The results are presented in the form of a motion picture.

## 1. The Model

The reported work on computer simulation of disks of stars has so far been confined to infinite doubly periodic systems (Miller and Prendergast, 1968) or to a small numbers of stars (Lindblad, 1960). The model used in the present report simulates the evolution of an isolated disk of stars. Each star in the system moves according to Newton's laws of motion in the self-consistent gravitational field due to the whole system. The motion of the point stars is confined to the plane of the disk. The gravitational potential is calculated on a fixed $N \times N$ array of cells (either $64 \times 64$ or $128 \times 128$) by Fourier transform methods. The model is described in detail by Hohl and Hockney (1969).

The motion of the stars is advanced stepwise in time. One complete cycle for advancing the motion of the system by a small time step consists of the following steps:

(a) The star coordinates are examined to determine the mass density at the center of each of the $N \times N$ array of cells.

(b) From the density distribution over the fixed array of cells, the corresponding gravitational potential at the cell centers is calculated.

(c) The gravitational field at the position of the stars is calculated and is used to advance the position and velocity of the stars by a small time step.

The above three steps represent one cycle and they are repeated until the desired evolution of the system is achieved. The time step is taken sufficiently small and the number of stars sufficiently large such that a decrease in the time step or an increase in the number of stars will not affect the evolution of the system. Results obtained for systems containing 50 000 stars and using 200 time steps per rotational period were found to be qualitatively indistinguishable from those for systems with 200 000 stars and using 400 time steps per rotational period.

## 2. Results

The first systems investigated were initially balanced, uniformly rotating disks of stars with an initial surface density given by $\sigma(r) = \sigma(0) \cdot (1 - r^2/R_0^2)^{1/2}$, where $r$ is the radial coordinate and $R_0$ is the radius of the disk. As predicted by Toomre (1964), the cold (zero velocity dispersion) disk is found to be unstable. The initial star positions

*Becker and Contopoulos (eds.), The Spiral Structure of Our Galaxy, 368–372. All Rights Reserved.*

are obtained by using a pseudo-random number sequence. Depending on the small perturbations caused by the random initial positions, the cold disk breaks up into three to five smaller subsystems during the first rotation. Adding increasing amounts of initial velocity dispersion has the effect of increasing the smallest size of the star condensations that are formed. A velocity dispersion equal to about 27% of the circular velocity at the edge of the cold balanced disk stabilizes the disk against breakup and any local condensations. These results are in agreement with the predictions by Toomre (1964). However, the disks were not stable against long wavelength modes and after about two rotations the disks tend to assume a bar-shaped structure. Similar

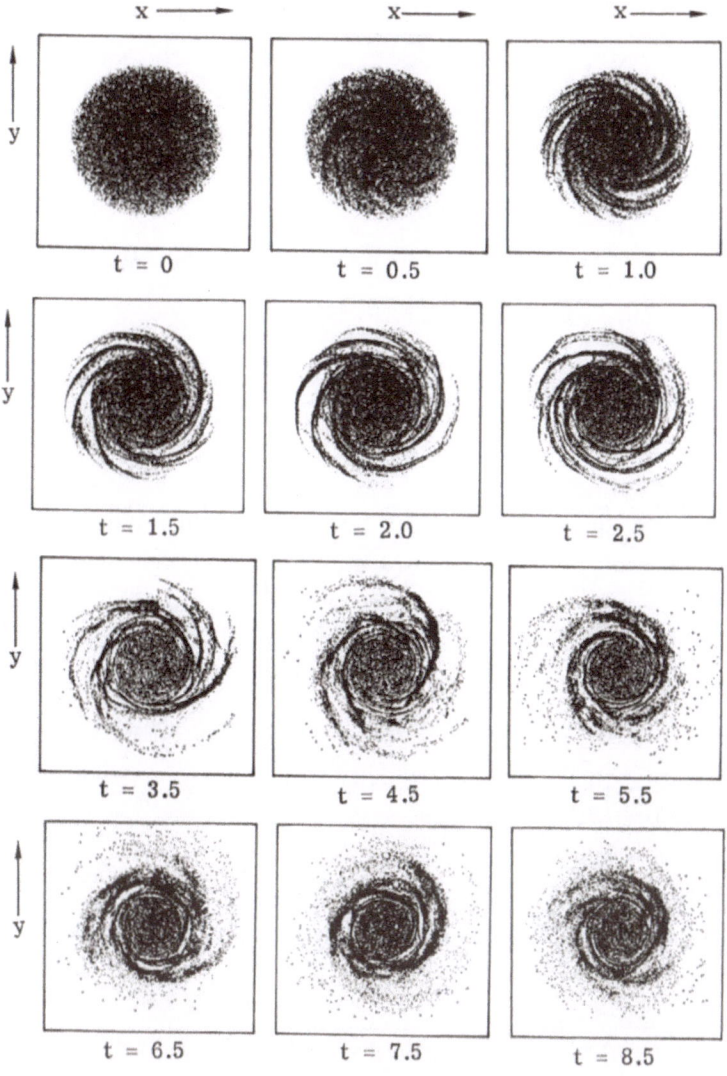

Fig. 1.   Evolution of an initially cold balanced disk of stars under the influence of a fixed radial force similar to the Schmidt model. The initial disk radius is 15 kpc and the time is given in rotation periods at a radius of 10 kpc.

results were obtained for disks with a Gaussian density distribution and with a velocity dispersion equal to (or even greater than) the stabilizing velocity dispersion calculated by Toomre (1964). It is found that rather large central condensations of mass (such as an exponential density distribution) are necessary for a stable disk galaxy.

To study the development of spiral structure, the model was modified to include a fixed central force similar to the Schmidt model of the Galaxy. The mass of the stars in the disk was taken to be from 5 to 50% of the total mass of the Galaxy. The initial distribution of the disk population of stars was a Maclaurin disk with zero radial velocities of the stars and with rotational velocities just sufficient to balance the

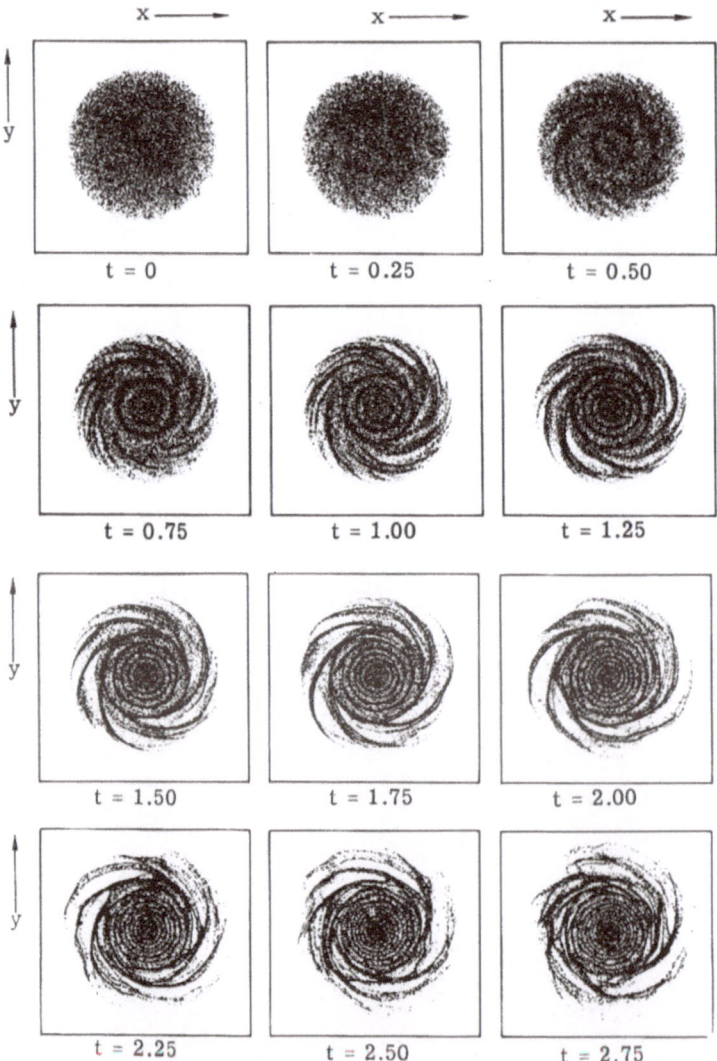

Fig. 2.   Evolution of a disk of stars identical to the system shown in Figure 1, except that the initial star positions have received a small perturbation which is proportional to the distance of the star from the center of the disk.

gravitational attraction. Figure 1 shows the evolution of such a system containing 50000 stars. The disk stars represent 10% of the total mass of the galaxy and the initial radius of the disk is 15 kpc. The time shown is in rotational periods at a radius of 10 kpc. It can be seen that a spiral structure develops and remains even after 8.5 rotations, at which time the calculation was terminated. From a purely kinematical viewpoint, differential rotation should have wound up the spiral structure before 8.5 rotations. In many of the calculations, there appeared to be a tendency to form circular rings of stars in the central region of the disk. To investigate the appearance of the rings more closely, the initial conditions for the system shown in Figure 1 were perturbed by rotating (around the disk center) the initial star positions, about 2° from their equilibrium positions. The resulting evolution is shown in Figure 2. Note that the circular rings originate at the center of the disk and move outward, resulting in circular density waves. These density waves result purely from the epicyclic motion of the stars and no net mass is carried outward by the density waves. Figure 3 is a close-up of the system at $t=0.75$ and shows in greater detail the structure of the system.

### Acknowledgement

The author would like to thank Professor Alar Toomre for numerous discussions and suggestions.

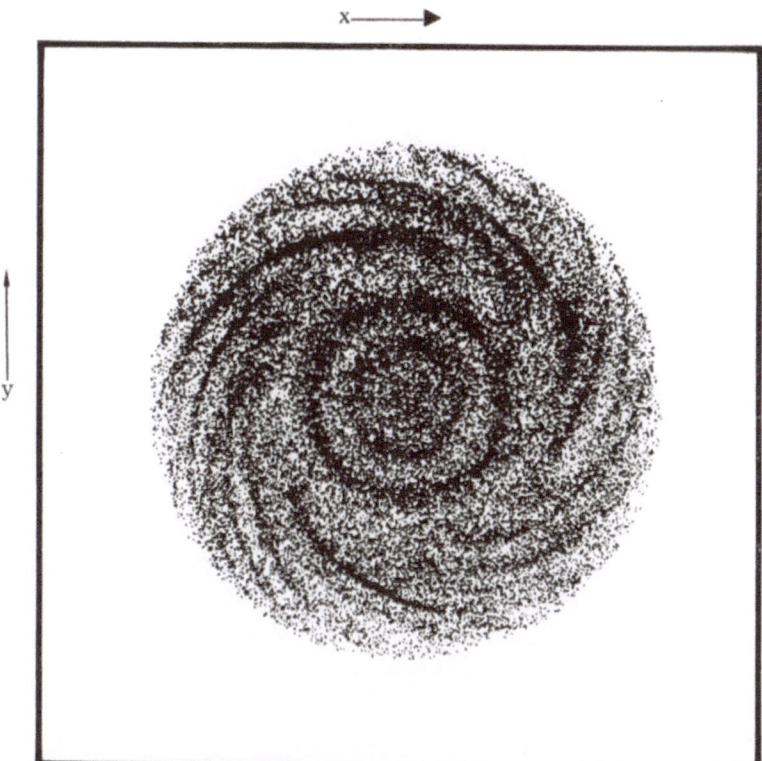

Fig. 3.   An enlargement of the system shown in Figure 2 at $t = 0.75$.

# References

Miller, R. H. and Prendergast, K. H.: 1968, *Astrophys. J.* **151**, 699.
Lindblad, P. O.: 1960, *Stockholm Obs. Ann.* **21**, No. 4.
Hohl, F. and Hockney, R. W.: 1969, *J. Comput. Phys.* **4**, 306.
Toomre, A.: 1964, *Astrophys. J.* **139**, 1217.

PART IV

# COMPARISON OF THEORY AND OBSERVATIONS

# A. GRAVITATIONAL EFFECTS

# 72. INTERPRETATION OF LARGE-SCALE SPIRAL STRUCTURE

C. C. LIN

*Massachusetts Institute of Technology, Cambridge, Mass., U.S.A.*

**Abstract.** The present paper consists of three parts: (1) A general explanation, from a semi-empirical point of view, of the density wave theory and its ramifications, with only a few remarks on those detailed features which have not been stressed before; (2) a statement of our conclusions about the Milky Way System; and (3) a discussion of the problem of the origin of density waves of spiral form.

## 1. Density Wave Theory of Galactic Spirals: A Semi-Empirical Approach

A. EXISTENCE OF 'GRAND DESIGN'

During this conference, several speakers have pointed out that the spiral features in a galaxy are complicated, but there is also no doubt about the existence of a 'grand design', usually two-armed in the inner parts, but often multiple-armed in the outer parts. As properly stressed by Oort (1962) at the Princeton conference, the explanation of the existence of the grand design needed urgent consideration. At the present stage of the development of the theory, now that this grand design is reasonably well understood, it is perhaps more appropriate, if one looks into the future, to pay more attention to the more complicated spiral features on a smaller scale, for here we may learn more about the physical processes in the galaxy. Indeed, Oort has placed slightly greater emphasis on these features in his introductory lecture at this symposium (see p. 1). One might indeed paraphrase his present statement in terms of his 1961 words as follows: Although the general form of the large-scale phenomenon can be recognized in many nebulae, this structure is often hopelessly irregular and broken up.

For the past few years, Shu, Yuan and I myself have been trying to ascertain the grand design of our galaxy by calculating a spiral pattern from the density wave theory (Lin and Shu, 1964, 1966, 1967), in the spirit of the late B. Lindblad, and comparing the theoretical deductions with observational data (Lin *et al.*, 1969; Yuan, 1969). We began our studies by adopting a semi-empirical approach. We followed Oort in dividing the problem of spiral structure into two parts; (a) the persistence of the spiral structure in the presence of differential rotation (or the question of material arms versus density waves), and (b) the origin of this structure. This turns out to be a very fortunate decision, for problem (b) turns out to be a much more difficult theoretical problem than problem (a). Indeed, it is still largely unsolved (cf. Section 3). We shall therefore continue to stress this semi-empirical approach in the present paper.

B. GENERAL FEATURES TO BE EXPLAINED

I list below ten general observational features which one must consider in dealing with spiral features in galaxies. Their explanation in terms of the density wave theory will be considered in the next section.

*Becker and Contopoulos (eds.), The Spiral Structure of Our Galaxy, 377–390. All Reservde.*

(1) Existence of *grand design*

The spiral structure extends as a 'pattern' over the 'whole' galactic disk.

(2) Persistence

(a) The spacing between spiral arms, which is used to classify normal spirals into Sa, Sb, Sc types, is correlated with other physical characteristics of the galaxy.

(b) In particular, a smaller nucleus is associated with wider spacing.

(3) *Two*-armed, trailing

The spiral pattern is in general two-armed and, as far as is known, always trailing.

(4) *Multiple*-armed structure

Multiple-armed structure is often observed in the outer regions of many galaxies.

(5) *Ring* structure possible

A ring structure is often observed in the inner part of many galaxies. This structure is very clearly seen in NGC 5364.

(6) *Slender strings of beads*

The newly formed stars and young H II regions are arranged neatly like a slender string of beads. This indicates star formation in restricted regions, but simultaneously over a wide front. (It is perhaps this feature that led people to think of magnetic containment as a way to pinch the gas along narrow tubes.)

(7) *Dust lanes*

The *principal* dust lanes lie on the *inside* edge of the bright optical arm, although there are other features associated with dust.

(8) *Abundance distribution of* H II

Both the continuum survey of Westerhout and the recombination line observations of Burke, Mezger, and their collaborators Reifenstein and Wilson indicate a marked deficiency of H II region inside of the '3 kpc circle'. This may be compared with the absence of H II regions within the ring of NGC 5364.

(9) *Abundance distribution of* H I

M. Roberts has shown that the peak distribution of H I extends *beyond* the bright spiral structure in M 33 and other galaxies. This agrees in general trend with that known in our galaxy: the peak H I distribution lies outside of the H II peak (Westerhout).

(10) *Magnetic field*

The magnetic field is relatively weak and, in the case of the Orion arm, it lies roughly along the arm (Hiltner).

C. QSSS HYPOTHESIS

It is clear that items (1) and (2) – the persistence feature mentioned above – would strongly suggest the density wave concept. Indeed, based on these features, we postulated the QSSS hypothesis (hypothesis of quasi-stationary spiral structure), which enabled us to explain a large body of observational data. It also raised a number of interesting questions on the dynamical mechanisms in stellar systems; e.g., the behaviors of material arms and of density waves, the origin of density waves, the uniqueness of the spiral pattern, and the behavior of stars, collectively and individually, near the points of co-rotation and Lindblad resonance*.

We can expect the existence of density waves also to have deep implications on the physical processes in the interstellar medium, and in particular on the formation of new stars. We can also expect the success of the theory, as applied to the Milky Way System, to embolden us to apply the theory to external galaxies. These are indeed fruitful avenues for future research work.

We therefore take the existence of density waves, and in particular, the QSSS hypothesis as our central theme. On the one hand, we examine its implications in our own galaxies and in other galaxies. On the other hand, we look for the mechanism for its origin, and use this as a starting point to examine many interesting behaviors of the stellar system.

One important theme to be kept in mind is COEXISTENCE. The complicated spiral structure of the galaxies indicates the coexistence of material arms and density waves, – and indeed of the possible coexistence of several wave patterns. When conflicting results appear to be suggested by observations, the truth might indeed lie in the co-existence of several patterns. Before taking this 'easy way out', one should of course try to examine each interpretation of the observational data as critically as possible.

There is also coexistence in the problem of origin of spiral structure. From our experience with plasma physics, we learned that there are many types of instabilities. Since a stellar system is basically a plasmoidal system, various types of instability can also occur in the problem of the galactic disk. The instability mechanism suggested in Section 3 is therefore only one of several possibilities. It has the advantage of being 'gross' or 'ponderous' instead of being 'delicate'. Such 'gross' instabilities, whenever they exist, usually occur prominently (as our experience in problems of hydrodynamic instabilities tells us). Another 'gross' mechanism for the origin of spiral structure is the existence of a short bar at the center, whose formation may or may not be related to the mechanism discussed in Section 3.

### D. DISCUSSION OF THE GENERAL FEATURES

To go beyond items (1) and the first part of (2), we need to work out the detailed dispersion relationship for density waves. A description of the explanation of the second part of item (2), items (3), (4), and (5) may be found in the paper by Lin *et al.* (1969). We note especially that the ring structure is associated with Lindblad resonance. Items (6), (7), (8), and (9) can be explained in terms of a sudden compression of the gas, as it passes through a density wave, as will be discussed in some detail by W. W. Roberts (this volume, p. 415; see also Roberts, 1969). Clearly, the role of magnetic field (item (10)) is secondary in our picture.

---

* When the stars meet the density wave at a frequency equal to its epicyclic frequency, there is a resonance behavior. This is referred to as Lindblad resonance (cf. the lecture of Contopoulos, this volume, p.303). There is of course also 'resonance' behavior when the stars move with the wave. This is however called specifically co-rotation to avoid confusion.

The detailed discussion, as presented at the conference, will not be given here. I wish only to record the following remarks.

1. The existence of the primary dust lane on the *inner* side of the bright spiral arm is compatible with a *trailing* spiral pattern travelling at a pattern speed *lower* than the material speed (cf., Lynds, this volume, p. 26).

2. The separation between the dust lane and the bright spiral arm should be, according to the theory, a measure of the *difference* between the pattern speed and the material speed (cf., Roberts, this volume, p. 415). It should therefore be *larger in the interior part* of the galaxy (cf., Morgan, this volume, p. 9) where the angular velocity of the material motion is larger, the angular velocity of the pattern being independent of radial distance.

3. As we move away from the center of the galaxy, formation of new stars and hence the bright spiral pattern should terminate where the material speed and the pattern speed become nearly the same. This comment is very useful for determining the pattern speed in external galaxies.

4. External galaxies should be studied in great detail for confirmation of these ideas. In particular, the ring structure deserves special attention, since extraordinary behavior may be expected near resonance.

5. Secondary dust lanes across a spiral arm ('feathers') or on the outer edge of an arm are probably related to the visibility of the dust and other secondary dynamical behavior in an arm. Dust is probably present throughout an arm. These points deserve further attention, as they will cast light on the mechanisms occurring within a spiral arm.

## 2. Spiral Structure of the Galaxy

### A. A BRIEF DESCRIPTION OF THE SPIRAL STRUCTURE

Having stated my general position, I shall proceed directly to a description of our conclusions. In Figure 1, we show an old picture already published in the *Proceedings of the Noordwijk Symposium*. It shows a theoretical pattern of our galaxy that is in reasonable agreement with observations. This diagram gives the *skeleton* of the spiral structure calculated according to the density wave theory, as applied to a basic galactic model given by Schmidt (1965). The pattern is trailing, and rotating around the Galaxy at a pattern speed of $11 \text{ km s}^{-1} \text{ kpc}^{-1}$. Thus, the material motion is *faster* than the wave pattern. According to the calculations of W.W. Roberts (which are qualitatively in agreement with the earlier calculations of Fujimoto for other values of the parameters involved), this leads to a sharp rise of the gas density (on the *inside* of the optical spiral arm) followed by a sudden expansion (towards the *outside* of the spiral arm). We could infer from this that the dust lane would be expected to occur close to the inner edge of the bright optical arm.

The pattern speed of $11 \text{ km s}^{-1} \text{ kpc}^{-1}$ is now considered a lower limit; a more accurate value is perhaps $11.5 \text{ km s}^{-1} \text{ kpc}^{-1}$, if we continue to adopt a model of an infinitesimally thin disk. The detailed calculations were made by Yuan, and will be discussed by him in detail. I shall merely mention here that the spiral pattern derived by him from the observational data differs only slightly from the theoretical pattern. It must be recognized that the agreement should *not* be perfect, since the galactic disk is perhaps not perfectly circular and the actual structure may not be a pure mode in the theory.

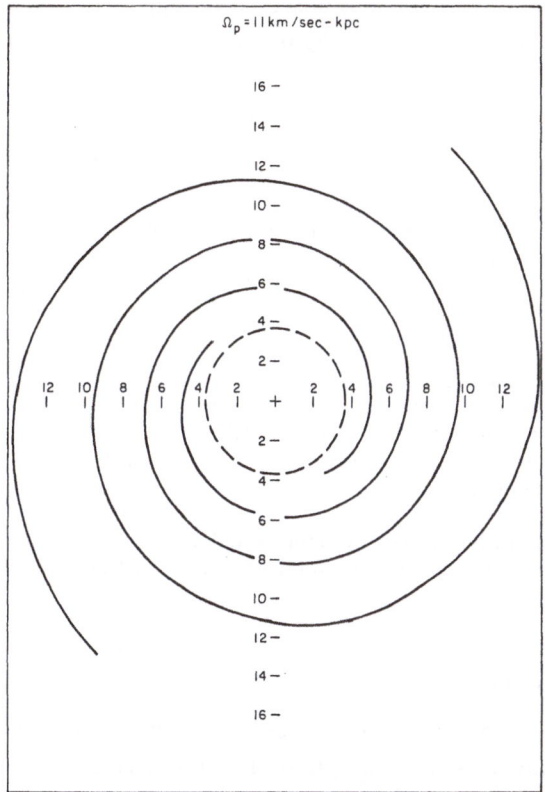

Fig. 1. Spiral pattern for 1965 Schmidt model.

When the thickness of the galactic disk is considered, Shu found that the pattern speed should be approximately 13.5 km s$^{-1}$ kpc$^{-1}$ in order to reproduce the observed spiral pattern. These calculations were made on the basis of his approximate evaluation of the thickness effect. At this symposium, Vandervoort (p. 341) reported further and more careful investigations of the effect of thickness of the galactic disk. The results agree very well with the earlier results obtained by Shu.

Other parameters that characterise the nature of the primary density wave are also shown in Table I. For a complete discussion of the manner in which these parameters are determined, the reader is referred to the paper by Lin *et al.* cited above.

The pattern speed we obtained is in sharp contrast with the value of 32 km s$^{-1}$ kpc$^{-1}$ obtained by Kalnajs. One can reconcile the two different results only if we accept the idea that there may be more than one mode present in the galaxy. The mode we obtained can be compared with a number of observational data, such as (a) the distribution of neutral hydrogen and its correlation with the oscillations in the rotation curve, (b) the distribution of young stars, (c) the origin of the local stars, and (d) the vertex deviation of the velocity distribution of the local stars. The discussion of items (a), (b) and (c) may be found in the paper by Lin *et al.*; item (d) will be discussed in a forthcoming paper by Strömgren, Lin and Yuan. One principal result is that the velocity distribution of the A stars, as observed by Eggen, can be very well reproduced by theoretical calculations based on the parameters shown in Table I.

TABLE I

Certain dynamical parameters

| | |
|---|---|
| 1. rms radial velocity predicted for stars in the solar neighborhood | $\leqslant 37$ km s$^{-1}$ |
| 2. Spiral pattern (primary component), pattern speed | 13.5 km s$^{-1}$ kpc$^{-1}$ |
| 3. In the solar vicinity | |
|    (a) arm spacing | |
|       (between Perseus and Sagittarius arms) | 3.5 kpc |
|    (b) amplitude of spiral gravitational field | 5 % of mean field |
|    (c) amplitude of variation of projected mass density | 10 % of mean |
|       (i) in stars | 5 % |
|       (ii) in gas | 5 % |
|    (d) rms turbulent velocity of the gas (adopted) | 7 km s$^{-1}$ |
|    (e) magnetic field (adopted for dynamical consistency) | 5 $\mu$G |

B. SOME SPECIAL PROBLEMS

I wish now to call attention to three special problems which have been raised from time to time. The answers to these questions are not entirely clear, but I shall offer some opinions and comments.

(1) Do stars and gas move together?

(2) What is the nature of the local arms:
    (i) the Carina arm, (ii) the Orion arm,
    (iii) the Perseus arm, (iv) the Sagittarius arm?

(3) What are the 3 kpc arms?

(1) During this Symposium, Kerr has offered observational evidence that the H I and H II regions move at approximately the same velocity, within a scatter of about 10 km s$^{-1}$. According to the gravitational picture, the stars and gas should move together, in the mean, provided that they are located at the same point in space, and that their velocity dispersions (turbulence in one case, peculiar velocity of stars in the other) exert a similar influence. Neither condition is exactly satisfied when one observes in a given direction. Thus, a small amount of deviation does not give evidence to the existence of other types of forces in action.

(2) The local arms are not necessarily all of the same nature. This is a godsent

opportunity for us to understand the various physical processes, through theoretical considerations and the complementary use of a variety of observations. The Perseus and the Sagittarius arms fit into our large scale structure. The Orion arm is presumably a material arm (possibly formed out of material separated from the Sagittarius arm about $80-100 \times 10^6$ yr ago). The Carina arm could be composed of young objects migrated out of the Sagittarius arm. It may also be a part of a trailing pattern travelling at about $15$ km s$^{-1}$ kpc$^{-1}$. There is also the possibility that it might be a part of a trailing pattern, travelling at a speed *higher* than the local circular speed. Such an arm would have its dust lane on the outside of its bright objects.

(3) The '3-kpc arm' is a deep mystery, as discussed in some detail in the forthcoming paper by W. W. Roberts (1969). One more suggestion might be made here. If indeed the mechanism suggested in Section 3 is correct, then the '3-kpc arm' is a part of a reflected leading wave, of an evanescent type (i.e., one whose amplitude decreases exponentially in space as it leaves the ring of reflection). Such a wave carries no energy flux, the reflected energy being carried outwards by the long waves described in Section 3. This picture is necessarily tentative. A careful study of the reflection mechanism is therefore an urgent matter to prove or disprove the conjecture.

## 3. On the Origin and the Maintenance of Galactic Spirals*

### A. OVERVIEW

Having provided a possible mechanism for a grand design over the whole galactic disk, we must now investigate the origin and the long term maintenance of the density waves that underlie the galactic spiral structure. In his survey lecture at the beginning of this Symposium, Oort raised this question. The consideration of the propagation of a group of density waves (Toomre, 1969) brings the problem even into sharper focus. In the following discussions, we suggest a mechanism for the initiation of trailing spiral waves and for the maintenance of a two-armed spiral pattern in a quasi-stationary manner. In our picture, trailing spiral arms, co-rotating with the general mass motion, are produced by stretching of irregularities at the outer parts of a galaxy through differential rotation. Owing to resonance, the two-armed structure will prevail as the disturbances propagate inwards as a group of waves, which extracts energy from the basic rotation of the galaxy. The pattern speed expected from this mechanism in our Galaxy is thus around $11-13$ km s$^{-1}$ kpc$^{-1}$, consistent with that determined from observational data (Lin *et al.*, 1969).

The reflection of the waves from the central region then stabilizes the wave pattern into a quasi-stationary form by transmitting the signal, via long-range forces, back to the outer regions where the waves originated. Thus, there is necessarily the co-existence of a very loose spiral structure and a tight spiral structure. Population I objects stand out sharply in the tight pattern while stars with large dispersive motion would primarily participate in the very loose pattern.

* The author is grateful to F. Shu and A. Toomre for several helpful discussions.

## B. A STRETCH OF A TRAILING SPIRAL ARM PRODUCED BY DIFFERENTIAL ROTATION

Since the galactic disk is in a state of strong differential rotation, any structural irregularity is likely to be stretched into a part of a trailing spiral arm, moving approximately at the local circular velocity, and inclined to the circular direction at various and varying angles. This material arm exerts its own induced effect and it may be expected that it eventually becomes a roughly self-sustained entity, somewhat like the self-sustained density waves (Lin and Shu, 1964, 1966, 1967) with inherent frequency $v = 0$ (co-rotating waves). The process of such an evolution may be followed by the theoretical studies* of Goldreich and Lynden-Bell (1965) and of Julian and Toomre (1966).

Where would such structural irregularities most likely occur? The answer seems to be: in the outer regions of the galaxy. There is observational evidence for this statement: connecting links between major spiral arms are frequently observed in the outer parts of many galaxies. In our opinion, the Orion arm is such an interarm branch (Lin et al., 1969). There is also a theoretical basis for this conjecture. The Jeans instability of the galactic disk (Toomre, 1964) tends to produce irregularities unless it is stabilized by the peculiar velocities of the stars and the turbulent motion of gas. Since the stellar content is smaller in the outer parts, and the turbulent motion of the gas may be dissipated, these outer parts of a galaxy are likely to be gravitationally unstable. Structural irregularities develop and are stretched out into short spiral arms, co-rotating with the prevailing local circular motion. Such a disturbance $q(\varpi, \theta, t)$, at a distance $\varpi = \varpi_0$ from the galactic center may be represented by a formula of the general form

$$q = F(t/\tau_0, \quad \theta - \Omega_0 t), \tag{1}$$

where $\Omega_0$ is the angular velocity at $\varpi_0$, and $\tau_0$ is a time scale much longer than $2\pi/\Omega_0$. We shall henceforth neglect the slow dependence on time indicated by the argument $t/\tau_0$.

## C. RESONANCE

The disturbance in the form of a piece of a co-rotating spiral arm would naturally exert its influence on other parts of the galaxy, and could induce density waves. But in general its effect would be expected to be limited, unless there is some form of *resonance*. A 'near-resonance' mechanism** becomes apparent when we consider the individual harmonic components of (1) and concentrate on the component with $2\theta$ dependence.

---

* The analysis made by these authors becomes inconclusive for application to waves on a finite disk when the angle of inclination becomes small; for the radial distance of the waves from the galactic center was effectively taken to be infinite. We therefore do not believe that the waves would eventually be wiped out by Landau damping, as indicated by Julian and Toomre.

** As pointed out by G. Contopoulos, the 'near resonance' considered here is, strictly speaking, still quite far away from the resonance condition defined by Equation (5). The term as used here merely indicates that the waves can propagate over a *substantial* part of the disk where $|2(\Omega - \Omega_0)| < \kappa$, which defines the "principal part of the spiral pattern" (Lin and Shu, 1966).

Let us write

$$q = \sum_{m=0}^{\infty} q_m \, e^{im(\theta - \Omega_0 t)} \tag{2}$$

and consider the effect of the individual components. The stars at a galacto-concentric distance $\varpi$ will feel a gravitational field at an angular frequency

$$f = m\left[\Omega(\varpi) - \Omega_0\right] \tag{3}$$

If this angular frequency $f$ is nearly equal to the epicyclic frequency $\kappa(\varpi)$, then there is near-resonance, and we may expect the disturbance (2) to have a large effect. The condition of resonance is

$$\kappa(\varpi) = m\left(\Omega(\varpi) - \Omega_0\right) \tag{4}$$

or

$$\Omega_0 = \Omega(\varpi) - \kappa(\varpi)/m. \tag{5}$$

Clearly, this can be satisfied only for one particular value of $\varpi$. However, near-resonance may be expected for a wide range of values of $\varpi$ if the right-hand side of (5) is nearly constant. In the outer parts of a galaxy, the variations in $\Omega(\varpi)$ and $\kappa(\varpi)$ are small. Thus, the required condition may be nearly satisfied for a substantial range of the radial distance $\varpi$ for a few values of $m$. In the inner parts of a galaxy, the required condition can be satisfied only for $m=2$. (In the case of uniform rotation, $\Omega - \kappa/2$ is exactly equal to zero.) This is a fact noted by B. Lindblad, and it has played an important role in the theory of density waves.

It is thus possible to conclude that the component of (2) with $m=2$ could produce a substantial effect in the whole galaxy, provided it occurs at such a galacto-centric distance that $\Omega_0$ is nearly equal to the nearly constant value of $\Omega(\varpi) - \kappa(\varpi)/2$. In our galaxy, this means that

$$\Omega_0 = 11 - 13 \text{ km s}^{-1} \text{ kpc}^{-1}. \tag{6}$$

and that the disturbances should originate around $\varpi_0 = 15$ kpc. This is indeed a region where the stars are less abundant, and hence the system is gravitationally unstable.

D. PROPAGATION OF THE DISTURBANCE

There is therefore the possibility of a multiple-armed structure at the outer reaches of a galaxy, but all prominent spiral structures are expected to be two-armed. The spiral structure consists of a group of trailing waves with phase velocity (in angle)

$$\Omega_p = 11 - 13 \text{ km s}^{-1} \text{ kpc}^{-1}. \tag{7}$$

Let us now consider how these waves would behave as they propagate in the radial direction.

There are two possible groups of trailing waves (Toomre, 1969): in the range $\Omega > \Omega_p$, they are (i) the short waves with a negative group velocity, and (ii) the long waves with a positive group velocity. Clearly, only the *short trailing* waves will prevail*, if the disturbances are initiated in the manner described above.

It is of course too much to expect that these short trailing waves would naturally produce a quasi-stationary spiral pattern. In the first place, the energy supplied by the disturbances, randomly created at the outer regions, must be limited. Furthermore, it will disperse during propagation. Secondly, the 'slow' dependence on the time scale $\tau_0$ is bound to show up fairly soon, and a well-organized pattern is therefore unlikely to emerge. How can these two difficulties be met?

The first difficulty, we shall see, will be resolved by the feeding of energy from the differentially rotating galactic disk. The second difficulty will be resolved by a reflection of the wave from the central regions of the galaxy, either at a resonance ring or at the galactic center, where a more-or-less bar-shaped structure will be formed.

## E. ENERGY FEEDING DURING PROPAGATION

It is easy to imagine that a *differentially* rotating disk can exchange energy with waves propagating through it, but one must determine whether the waves gain or lose energy. Also, one must consider the associated re-adjustment of the distribution of the rotation of the galaxy; the time scale of this adjustment should be very long. This latter point is easy to settle. A rough estimate of this time scale is $\varepsilon^{-2}$ times a typical period of revolution, where $\varepsilon$ is the fractional variation of the various quantities in the disturbance. If we take $\varepsilon$ to be of the order of 10%, the time scale for this perennial adjustment would be of the order $10^{10}$ yr.

The feeding of energy can be calculated quantitatively. It is known that, in many cases of wave propagation, there is a principle of conservation of density of action (Whitham, 1965; Bretherton and Garrett, 1967). In the case of density waves under consideration, this principle has also been found to be true by Toomre (1969)** based on results obtained by Shu (1968).

Briefly, this principle may be stated as follows. If the energy density of the wave is denoted by $\mathscr{E}$, we define the density of action by

$$\mathscr{A} = \mathscr{E}/f ,\tag{8}$$

where $f$ is the angular frequency seen by an observer co-moving with the general stream. It then follows that

$$\frac{\partial \mathscr{A}}{\partial t} + \nabla \cdot (\mathscr{A} \mathbf{c}_g) = 0 ,\tag{9}$$

where $\mathbf{c}_g$ is the group velocity.

---

* The waves outside the co-rotation point would also be short trailing waves, since they propagate outwards where $\Omega < \Omega_p$. Note that $\Omega(\varpi)$ decreases with increasing values of $\varpi$.
** Toomre's calculations were made for axisymmetric modes, for which the effect of change in frequency is not evident. The extension to non-axisymmetric modes, although natural to anticipate, does involve rather non-trivial calculations. This was completed by Shu. During this Symposium, Kalnajs pointed out that the energy density defined here is that associated with the rotating coordinate system. This does not alter the main line of reasoning here, especially that part related to the vanishing of the group velocity near resonance.

Following the motion of the group of waves, we then have, by (9),

$$\left(\frac{d\mathscr{A}}{dt}\right)_{c_g} = \frac{\partial \mathscr{A}}{\partial t} + (\mathbf{c}_g \cdot \nabla)\,\mathscr{A} = \mathscr{A}\,(\nabla \cdot \mathbf{c}_g), \tag{10}$$

which may also be written as

$$\left\{\frac{\partial}{\partial t} + (\mathbf{c}_g \cdot \nabla)\right\} \log \mathscr{A} = -\,\nabla \cdot \mathbf{c}_g, \tag{11}$$

or

$$\left\{\frac{\partial}{\partial t} + (\mathbf{c}_g \cdot \nabla)\right\} \log \mathscr{E} = \left\{\frac{\partial}{\partial t} + (\mathbf{c}_g \cdot \nabla)\right\} \log f - \nabla \cdot \mathbf{c}_g. \tag{12}$$

If the right-hand side of the above equation is positive, we have an amplification of the wave group during propagation. If it is negative, we have an attenuation. In the present case, we shall find that the right-hand side is indeed *positive*, and we have consequently an amplification of the disturbance energy as the wave propagates inwards, when the waves are stationary ($\partial(\ )/\partial t = 0$). This is very easy to verify since

$$c_g < 0, \quad \text{and} \quad \frac{d\Omega}{d\varpi} < 0. \tag{13}$$

The form (9) yields a more vivid description of the mechanism. In the stationary case, we have

$$\varpi c_g \mathscr{A} = \text{constant}. \tag{14}$$

Near the point of origin, the angular frequency $f$ (Equation (3)) is very small. Consequently (Equation (8)), it requires only a negligible amount of *energy* to produce a substantial amount of *action*. As the group of waves propagates inwards, the angular frequency $f$ steadily increases, and the energy density $\mathscr{E}$ must therefore increase proportionally. In addition, there is further enhancement of $\mathscr{E}$ due to the decrease of $|c_g|$ and $\varpi$. Indeed, if there is a point of inner Lindblad resonance, $|c_g| \to 0$ there*, and the energy density becomes very large. The increase of energy density with decreasing $\varpi$ is of course to be expected from the cylindrical geometry of the galactic system.

We note however that the mechanism will not be operative unless there is a pre-existing mechanism for wave propagation. Furthermore, if the propagation mechanism is imperfect, i.e., if energy may be partly lost by dispersion, the effect of this gradient instability may be somewhat obscured. In the present case, we are fortunate to have a condition of near-resonance, which tends to insure a favorable propagation of the waves.

## F. REFLECTION OF WAVES NEAR THE CENTRAL REGION

From the discussions of the last section, there appear two possibilities for the action of the wave group to become infinite, if we insist on using Equation (14). Either we

---

* Indeed, it vanishes as $(1 - |\nu|)^{3/2}$.

have (i) $\varpi = 0$, or we have (ii) $c_g = 0$. The first condition refers to the galactic center, and the second condition refers to Lindblad resonance. These are the places where the propagation of the group of waves must undergo a change, and it is natural to think of them as places where the waves, having reached a very large (but not infinite) magnitude, are reflected. Let us examine the consequences of this reflection process, beginning with the galactic center.

As the group of waves approaches the galactic center, its greatly increased amplitude will be sufficient to cause the galactic nucleus to be slightly distorted* into a 'short bar'. This 'bar' configuration will be rotating at an angular velocity $\Omega_p$, appropriate to the forcing incoming wave. There is thus an associated gravitational field also rotating at the same angular frequency $\Omega_p$, propagating outwards. Its effect will be particularly strongly felt by the outer rings of the galaxy where the circular velocity is $\Omega_p$; i.e., where the waves originated. Thus, the cycle *does* complete itself and a stationary state may be maintained.

The completion of the cycle must be done with the right phase in time and right distribution in space. An indication for this possibility may be seen by examining the outward propagation. If we persist in applying the results obtained for the short waves to the relatively long waves, we find that the two branches coincide for the condition of co-rotation $v = 0$. Thus, if the energy is carried out by the long trailing waves, it would be forcing the outer parts of the galactic disk with an appropriate 'mode' (i.e., an appropriate distribution in space and variation in time) provided one numerical value characterizing the phase is correctly specified. Since we are dealing with a group of waves whose 'wave length' is quite long, and not really well defined, this is not a critical requirement. At the same time, one cannot expect the re-enforcement to be perfect, but this is again not a critical matter. A *partial feedback* would be able to sustain a relatively stationary pattern, since there is a substantial gain of energy as the short waves move inwards, but not a corresponding loss of energy as the long waves, with length scale on the order of the galactic radius, propagate outwards.

### G. INNER RESONANCE RING

In cases where a sharp Lindblad resonance exists (e.g. in NGC 5364 or the Milky Way), the waves cannot penetrate into the center, but would be reflected already at the resonance ring. The violent response of the stars suggests that they are likely to group themselves into certain orbits, collectively showing an oval structure (cf., Contopoulos, this volume, p. 303), which takes the place of the bar in the above described reflection mechanism; but the general points made above still hold. The precise mechanism of reflection is obviously a very important and very interesting process deserving much future attention.

There is the distinct possibility that the waves would be *absorbed* at the inner resonance ring, and that there is no reflection of the group of waves. If this were to be found to be the case, there would be difficulty in accounting for the regular orga-

---

* In a pure disk model, the 'short bar' thus produced at the center would presumably be represented by a singularity. See Lin and Shu (1964), Rehm (1965), or as quoted by Lin and Shu (1967).

nization of a grand design. It would then be more likely that we would see a rather chaotic superposition of two-armed waves, unless certain nonlinear effects related to resonant stars (Contopoulos, 1970), would prevail.

H. CONCLUSION

We have thus arrived at a picture involving the co-existence of a relatively tightly wound pattern of short waves and a rather loosely wound pattern of long waves. Clearly, objects with small peculiar velocities – particularly Population I objects – would participate in the tight spiral pattern. On the same length scale, the loosely wound pattern, with at most one winding for the whole galaxy would present itself as a distortion of the whole galaxy. It would be less visible to casual optical and radio observations. It could conceivably be mapped out via very careful analysis of optical and radio data.

All the above discussions are still fairly speculative. It would be extremely desirable to follow through the detailed development of these ideas with more careful mathematical analysis.

## 4. Concluding Remarks: A Bird's Eye View of Theoretical Developments

As mentioned before, we use the QSSS hypothesis as the focal point for our theoretical developments. On the side of basic mechanism, we raise various dynamical issues whose clarification may also help us to understand the behavior of electro-magnetic plasmas. For the study of our own galaxy, and perhaps other galaxies as well, there appear to be a better opportunity for the understanding of the physical processes, including such microscopic behavior as the formation of molecules and dust grains, induced by macroscopic compression via density waves. The effect of the density wave can also influence the cosmic ray particles via the galactic magnetic field. Here it is important to keep alive the general concept of density waves, and not to attach it *only* to a two-armed grand design.

We have also the opportunity to explore the external galaxies, since here certain types of observations can be made more easily*, and we have a larger sample of objects for our study. The relationship between normal spirals and barred spirals is a fascinating subject. It is perhaps sufficient, but not necessary, to have a short barred structure at the center of a normal galaxy, to initiate a two-armed grand design. The coexistence of several origins for spiral structure should be stressed. Simultaneously, there is the possibility of the coexistence of several kinds of spiral structures and their actual revelation as observable features. Indeed, it is the feeling of the present writer that relatively loose spiral patterns tend to be obtained by computational or analytical methods directed at the whole galactic disk. The asymptotic method yields relatively tight spirals when near resonance is approached, but it can also yield fairly loose spirals.

---

* One must beware of possibilities of misinterpretation, for we can only measure the velocity in the line of sight. Since the material in a galactic disk has two other components of motion besides that in the circular direction, inferences on the small deviations in circular motion must be drawn with care.

## Acknowledgement

The work reported here has been supported in part by grants from the National Science Foundation and the National Aeronautics and Space Administration.

## References

Bretherton, F. P. and Garrett, C. J. R.: 1967, *Proc. Roy. Soc. London* **A302**, 529.
Contopoulos, G.: 1970, *Astrophys. J.*, in press.
Goldreich, P. and Lynden-Bell, D.: 1965, *Monthly Notices Roy. Astron. Soc.* **130**, 97, 125.
Julian, W. H. and Toomre, A. 1966, *Astrophys. J.* **146**, 810.
Lin, C. C.: 1967, in *Relativity Theory and Astrophysics*, 2: *Galactic Structure*, Amer. Math. Soc., Providence, p. 66.
Lin, C. C. and Shu, F. H.: 1964, *Astrophys. J.* **140**, 646.
Lin, C. C. and Shu, F. H.: 1966, *Proc. Nat. Acad. Sci. U.S.A.* **55**, 229.
Lin, C. C. and Shu, F. H.: 1967, IAU-URSI Symposium No. 31, p. 313.
Lin, C. C., Yuan, C., and Shu, F. H.: 1969, *Astrophys. J.* **155**, 721.
Oort, J. H.: 1962, in *Interstellar Matter in Galaxies* (ed. by L. Woltjer), Benjamin, New York, p. 234.
Roberts, W. W.: 1969, *Astrophys. J.* **158**, 123.
Schmidt, M.: 1965, *Stars and Stellar Systems* **5**, 513.
Shu, F. H.: 1968, Ph.D. Thesis, Harvard University.
Toomre, A.: 1964, *Astrophys. J.* **139**, 1217.
Toomre, A.: 1969, *Astrophys. J.* **158**, 899.
Whitham, G. B.: 1965, *J. Fluid Mech.* **22**, 273.
Yuan, C.: 1969, *Astrophys. J.* **158**, 871, 889.

# 73. THEORETICAL 21-cm LINE PROFILES:
# COMPARISON WITH OBSERVATIONS

## C. YUAN*

*Massachusetts Institute of Technology, Cambridge, Mass., U.S.A.*

**Abstract.** In order to make a direct comparison with observations of the 21-cm line of neutral hydrogen, theoretical profiles based on the ideas of the density-wave theory are constructed for a modified Schmidt model of the Galaxy and its theoretical spiral pattern. The comparison has covered galactic longitudes $l^{II} = 30° - 330°$ with 10° intervals in the galactic plane. Good agreement is found in most of the above directions.

## 1. Introduction

In this report, we shall briefly discuss the comparison of theoretical 21-cm line profiles against observational data. Detailed results of the present study, however, will be published elsewhere (Yuan, 1970). The theoretical profiles are based on a large-scale spiral pattern for the Milky Way System, derived from an application of the density-wave theory of Lin and Shu (1964, 1966). Good agreement has been found in most of the galactic longitudes under study, which range from $l^{II} = 30°$ to $l^{II} = 330°$ at 10° intervals. (Only 10 directions are presented in this paper.) We have thus obtained a spiral pattern for the Milky Way System that is consistent with both observations and the density wave concept.

There are, however, several other important implications in this agreement. The theoretical spiral pattern, at least from the solar vicinity outwards, depends sensitively upon the basic galactic model, composed of two related items, the surface density and the rotation curve. The decision on a basic model for this part of our galaxy, therefore becomes a crucial part of the present study. Furthermore, the construction of the profiles demands, besides the knowledge of the spiral pattern, the detailed information of the interstellar gas, which includes (1) the abundance distribution of neutral hydrogen, (2) the density contrast between the gas in the spiral arm and in the interarm region, (3) the systematic motion of the gas due to the presence of the spiral pattern, and (4) the turbulent speed of gas clouds. Our present study also contains a suggestion on these properties of gas as a whole.

## 2. Theoretical Profiles

The procedure of constructing theoretical profiles is indeed a reverse of the usual method of deducing spiral structure from the 21-cm line observations. In the present study, the construction of the profiles is performed under the following three basic assumptions: (1) The spin temperature of neutral hydrogen is constant throughout the galaxy. (2) The velocity distribution of the gas is Gaussian at each point along

* Present address: City College of New York, N.Y., U.S.A.

*Becker and Contopoulos (eds.), The Spiral Structure of Our Galaxy, 391–396. All Rights Reserved.*
*Copyright © 1970 by the I.A.U.*

the line of sight. (3) The systematic motion and the spiral component of the volume density of the gas have a sinusoidal variation across the spiral arms, as specified by the linear theory of gas dynamics.

The first two assumptions immediately imply the following basic relation

$$T_b(v, l^{II}) \sim \int_0^\infty \varrho(r, l^{II}) \exp\left\{ -\frac{[v - V(r, l^{II})]^2}{2c_t^2} \right\} dr$$

in which $T_b$ is the brightness temperature, $v$ is the line-of-sight velocity and $r$ is the distance measured from the sun. The third assumption specifies the spiral component in the volume density $\rho$ and the systematic deviation of the velocity $V$ along the line of sight from that given by the circular motion in the basic model. Theoretical profiles are obtained by integrating the above equations for various values of $v$, with the turbulent speed $c_t$ and the spiral pattern also assigned. We shall proceed to describe these necessary quantities in the following sections.

## 3. Theoretical Spiral Pattern

Before we can calculate a theoretical spiral pattern, we must have a basic galactic model with reasonable accuracy. We shall adopt essentially the Schmidt model of 1965, but with slight modifications from the solar vicinity outwards.

A. BASIC DISK MODEL

It is well known that the part of the rotation curve beyond the location of the sun could not be accurately obtained from observations. All the existing models have simply extended the rotation curve beyond that point by extrapolation. Such an extrapolation in galactic rotation may cause serious errors in the determination of the mass distribution. The density-wave theory, however, provides an additional check on this extrapolation. On the one hand, the arm spacing according to the theory is linearly related to the surface density. On the other hand, it can be obtained from the 21-cm line profiles with the rotation curve. Thus, the extrapolation has to be so chosen that the corresponding surface density will reproduce an observed arm spacing. We shall apply this idea to the decision on the basic disk model.

First, we notice that the extrapolation adopted in the 1965 Schmidt model is somewhat unsatisfactory in this sense. With all the reasonable pattern speeds, the arm spacing beyond the location of the sun appears always too wide in comparison with observations (Yuan, 1969). A reduction in the surface density, and hence the rotation velocity, seems to be necessary beyond the sun. This may be accomplished by a mass cutoff to the Schmidt model.

We shall only summarize the final results here. With a cutoff at 19 kpc, the adopted model only deviates slightly from the 1965 Schmidt model. The Oort constants are only changed by less than 1%. Local surface density is reduced by 15% and the circular velocity at 15 kpc from the center is lower by 7%.

B. SPIRAL PATTERN

The theoretical spiral pattern is computed at the pattern speed $11.5$ km s$^{-1}$ kpc$^{-1}$. Inside the 10-kpc, this spiral pattern has no essential difference from the one published by Lin *et al.* (1969), but it winds up more tightly beyond that circle (solid curve, Figure 1). In order to account for the existence of the oval or other possible large-scale distortions, some minor adjustment has been made to this theoretical pattern during the calculations of the profiles. The displacements of the arms radially from the original pattern are usually around 300 pc but amount to about 500 pc at a few places (dashed curve, Figure 1).

## 4. The Adopted Model for the Gas Component

A. ABUNDANCE DISTRIBUTION

The abundance distribution adopted in the calculations has no appreciable difference from the familiar diagram of Westerhout (Oort, 1965). The distribution as indicated in Figure 2 is characterized by its double-ring structure. The major ring is located at about 12 kpc, and can be easily recognized as those observed in the external galaxies (Roberts, 1967). The secondary ring is located at about 6 kpc. As we shall see next, this distribution is quite satisfactory as far as the large-scale structure of the gas is concerned.

B. DENSITY CONTRAST AND SYSTEMATIC MOTION

The linear theory of the response in gas provides a useful proportionality among the perturbation of gas density and the amplitudes of the tangential and the radial systematic motion (Lin *et al.*, 1969). Once one quantity is known, the others follow immediately from the proportionality. The tangential systematic motion for the region inside the location of the sun can be directly obtained from the observed rotation curve, whereas the density contrast is relatively easy to estimate for the outer part of the Galaxy, from the observed profiles. Using these values as a guide, we have adopted the following values: The amplitudes of both components of the systematic motion are taken to be 7–9 km s$^{-1}$. The density contrast varies from 2:1 at 5 kpc to 5:1 at 15 kpc.

C. TURBULENT SPEED OF GAS CLOUDS

Turbulent speed is relatively insensitive in our calculation. A simple argument shows that the turbulent speed should roughly vary with the square root of the surface density. In the present study, the local turbulent speed is taken to be 8 km s$^{-1}$. The value is increased to 13 km s$^{-1}$ at 5 kpc and reduced to 6 km s$^{-1}$ at 15 kpc from the center.

D. ORION SPUR AND CARINA BRANCH

In addition to the above-mentioned regular features, we have also included the Orion

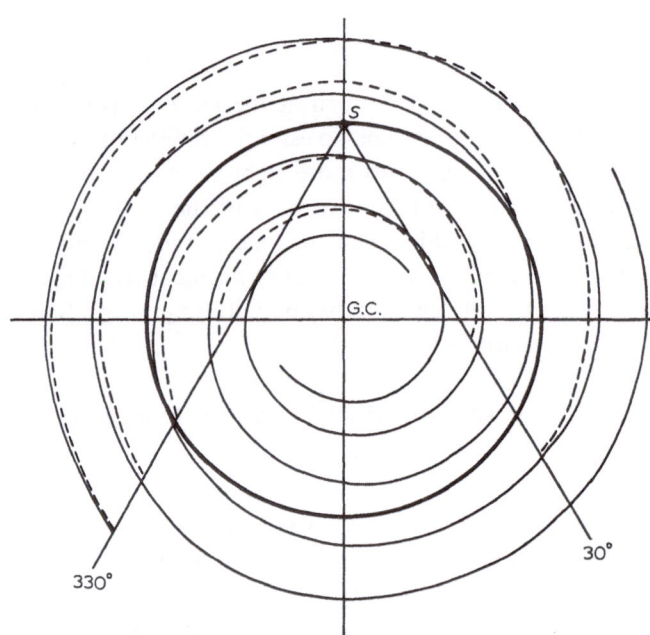

Fig. 1.   Theoretical spiral pattern. The solid curve is the spiral pattern calculated at pattern speed 11.5 km s$^{-1}$ kpc$^{-1}$ and based on the modified Schmidt model. The dashed curve is the spiral pattern actually used in constructing theoretical profiles.

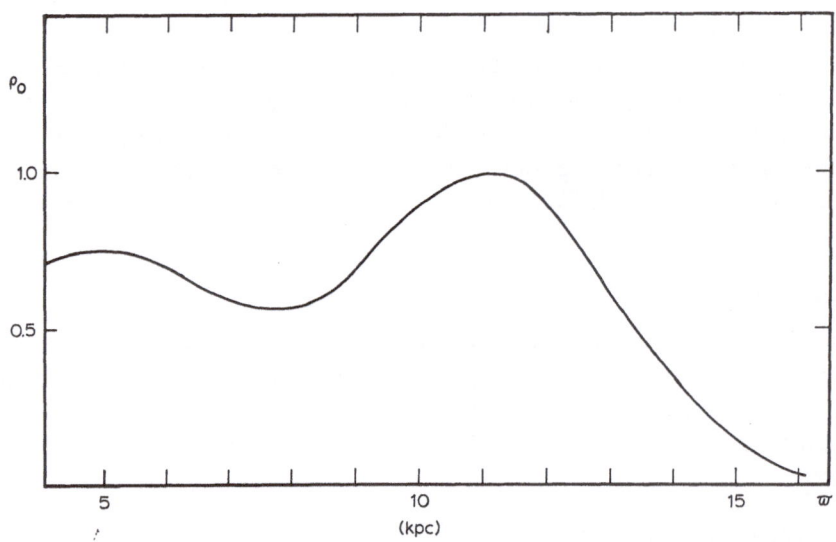

Fig. 2.   The abundance distribution of the gas in the galactic plane.

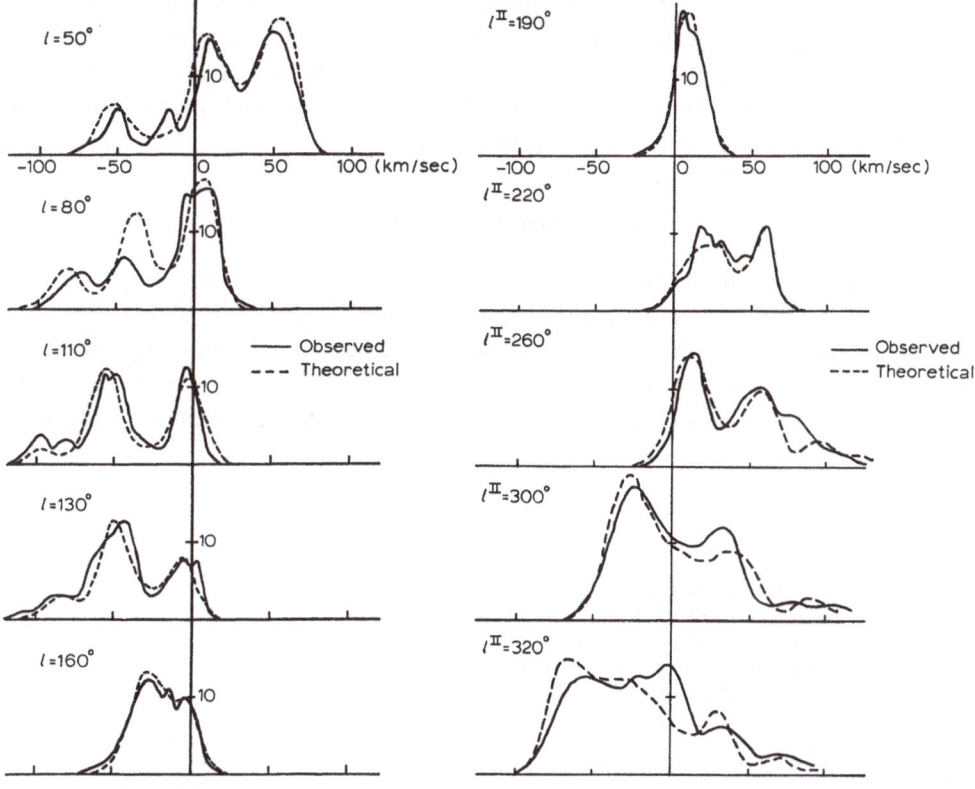

Fig. 3. Comparison with observations
in the northern sky.

Fig. 4. Comparison with observations
in the southern sky.

Spur in our calculation in order to see whether the 'local circular arm' in the northern sky can be reproduced by putting the gas along the optical Orion Spur. The answer is affirmative. In the same manner, we find a minor local feature of the Orion Spur in the southern sky. The gaseous branch in the Carina direction is also included in the study, for the purpose of comparison with Bok's extensive study in that region (1969). We have not found any evidence that the 'Carina branch' has extended into the 10-kpc circle to link itself to the Sagittarius Arm or the local Orion Spur.

## 5. Results

Detailed results in comparison will be published elsewhere (Yuan, 1969). In this paper, only ten directions, uniformly covering the galactic plane from $l^{II}=30°$ to $l^{II}=330°$, are presented (Figures 3 and 4). The locations and the intensities of the peaks and valleys are generally in good agreement with observations. This has greatly encouraged our theoretical investigation and also given us more confidence in the adopted large-scale spiral structure, determined with the guidance of the density-wave theory.

Irregularities such as an interarm branch or spur occur quite commonly on the outer part of the galaxies. In those directions of irregularities, the agreement becomes poor. The Cygnus branch in the northern sky and the Carina branch in the southern sky are these examples. The existence of these branches in the outer part of the Galaxy indicates that other modes, possibly with multiple arms, may also be present in our galaxy.

## Acknowledgements

The observational data in this paper are adopted from (1) Westerhout's Greenbank Survey for $l^{II} = 30°-235°$ (1965), (2) Hindman's data for $l^{II} = 235°-300°$ (1969), and (3) Kerr's data for $l^{II} = 300°-330°$ (1969). I would like to take this opportunity to thank Professor F. Kerr for making these southern data available to me before publication. I also wish to thank Professor C. C. Lin for his constant encouragement during this investigation and many most helpful discussions.

## References

Bok, B.: 1969, private communication.
Kerr, F. J.: 1969, *Australian J. Phys. Astrophys. Suppl.*, No. 9.
Lin, C. C. and Shu, F. H.: 1964, *Astrophys. J.* **140**, 646.
Lin, C. C. and Shu, F. H.: 1966, *Proc. Nat. Acad. Sci. U.S.A.* **55**, 229.
Lin, C. C., Shu, F. H., and Yuan, C.: 1969, *Astrophys. J.* **155**, 721.
Oort, J. H.: 1965, *Trans. IAU* **12A**, 789.
Roberts, M. S.: 1967, IAU Symposium No. 31, p. 189.
Schmidt, M.: 1965, *Stars and Stellar Systems* **5**, 513.
Westerhout, G.: 1966, *Maryland-Greenbank Galactic 21-cm Line Survey*, 1st ed., University of Maryland.
Yuan, C.: 1969, *Astrophys. J.* **158**, 871.
Yuan, C.: 1970, in preparation.

# 74. NEUTRAL HYDROGEN IN THE SAGITTARIUS AND SCUTUM SPIRAL ARMS

W. B. BURTON and W. W. SHANE

*Leyden University Observatory, Leyden, The Netherlands*

**Abstract.** Observations of the neutral hydrogen in the first quadrant of galactic longitude have been analysed. The existence of large-scale streaming motions such as the streaming associated with the Sagittarius arm makes interpretation of the observations in terms of circular galactic rotation unsatisfactory. It is shown that application of the density-wave theory formulated by Lin *et al.* (1969) leads to a more satisfactory interpretation. Using kinematic models based on this theory the distribution and motion of the neutral hydrogen are studied. Failures of kinematic models based on circular rotation are pointed out. A map of the distribution of neutral hydrogen is produced. The Scutum arm is composed of inner and outer arcs both of which seem to be moving outward from the galactic center with velocities of the order of 30 km s$^{-1}$.

## 1. Analysis of the Observations

Profiles of the 21-cm line of H I have been obtained using the 25-m telescope in Dwingeloo, covering the range of (new) galactic longitude $l^{II} = 22°$ to 56° and extending about five degrees on either side of the galactic plane. The observations are generally complete for positive velocities and were made with a bandwidth of 2 or 4 km s$^{-1}$. These observations, which constitute the principal material for the investigations described here, were analysed in detail. We have also made use of additional observations by S.C. Simonson covering some regions closer to the direction of the galactic center and have drawn on a number of other programs in order to complete the survey of the galactic plane from $l^{II} = -3°$ to 70°. The observations have been represented as contour maps showing isophotes of brightness temperature in the velocity-longitude and velocity-latitude planes.

We have found it convenient in analysing the line profiles to decompose them into Gaussian components. Before doing so, however, we applied two significant corrections. Firstly, we reduced the measured brightness temperatures to optical depths assuming a spin temperature $T_s = 135$ K. We did this reluctantly, realizing that the conditions under which such a correction is valid are by no means fulfilled. However the alternative, making no correction before decomposition into additive components, is equivalent to assuming an infinite spin temperature. We argue that the modest correction implied by our choice of a rather high spin temperature is probably an improvement over no correction at all. Secondly, we have, in those regions where it appeared to be of significance, subtracted a contribution which arises from a rather uniform layer of H I with a concentration to the galactic plane somewhat less than that of the major spiral features. This feature, which we call the galactic envelope, will not be discussed in this contribution. Its presence in the uncorrected line profiles led to difficulties in the analysis.

The method of Gaussian analysis was chosen primarily as a matter of convenience.

*Becker and Contopoulos (eds.), The Spiral Structure of Our Galaxy,* 397–414. *All Rights Reserved.*

We were able in this way to describe quantitatively the significant features in the line profiles and to reduce the amount of data by about a half order of magnitude without substantial loss of information.

In most parts of the observed region the Gaussian analysis was performed twice, first for each profile separately in order to identify the structural features and then a second time, introducing the identified features into all profiles at a single longitude, in order to study their dependence upon latitude. At some of the higher longitudes where the line profiles show the least complexity, this double analysis was not required.

It was now possible to determine the dependence of optical depth, velocity, and dispersion in velocity, upon latitude. The first of these could be represented by one or more Gauss functions and the others by polynomials. After a slight further simplification it was possible to describe all of the observations at a single longitude by a sum of bivariate normal functions. Comparison of this distribution with the corresponding contour map of brightness temperature as a function of latitude and velocity shows that all of the features visible on the latter are represented quantitatively by the former.

The large number of components required to give an acceptable representation of the data, as many as twenty at one longitude, reflects the complexity of the detailed structure. Even more detail is visible on high resolution surveys. In order to derive a picture of the most general features of spiral structure it was necessary to combine those components which appeared to be related to a single feature of the order of size of a spiral arm. In order to do this we examined in detail the behavior of each component as a function of longitude as well as the latitude distribution, which also provided the principal criterion for resolution of the distance ambiguity. The high resolution Maryland-Green Bank 21-cm Line Survey was indispensable in following the structure through the complex region near the plane at lower longitudes. In this step, as elsewhere, frequent reference was made to the observed line profiles in order to avoid being misled through the complexity of the analysis. In the higher longitude region, where the profiles were less complex, this part of the analysis presented little difficulty.

Each feature studied in this way can be described in detail, but description through the parameters derived from the above analysis is more useful. These parameters represent maximum optical depth, mean position in latitude and in velocity, dispersion in latitude and in velocity, and rate of change of mean velocity with latitude. Of these, the mean velocity is the most significant, and is plotted, for all the major features identified, in the velocity-longitude diagram (Figure 1). The shaded bands represent the observed features, as described in the caption. The figure is seen to be dominated by two major loops, which we suppose to represent the Sagittarius and Scutum spiral arms. But before we can discuss the space distribution of the neutral hydrogen, we must derive a suitable model on the basis of which we can determine kinematic distances.

## 2. Interpretation of the Velocity-Longitude Diagram

Interpretation of the velocity-longitude diagram requires answers to two questions, which cannot be considered separately:

(i) what are the characteristics of the velocity field of the Galaxy; and (ii) how is the hydrogen distributed in space throughout the Galaxy.

As a first step we assume that the hydrogen rotates in circular orbits according to a velocity field described by a smooth rotation curve, $\Theta_c(R) = 250.0 + 4.05 \, (10 - R) - 1.62 \, (10 - R)^2$, where $R$ is distance from the galactic center expressed in kpc. This

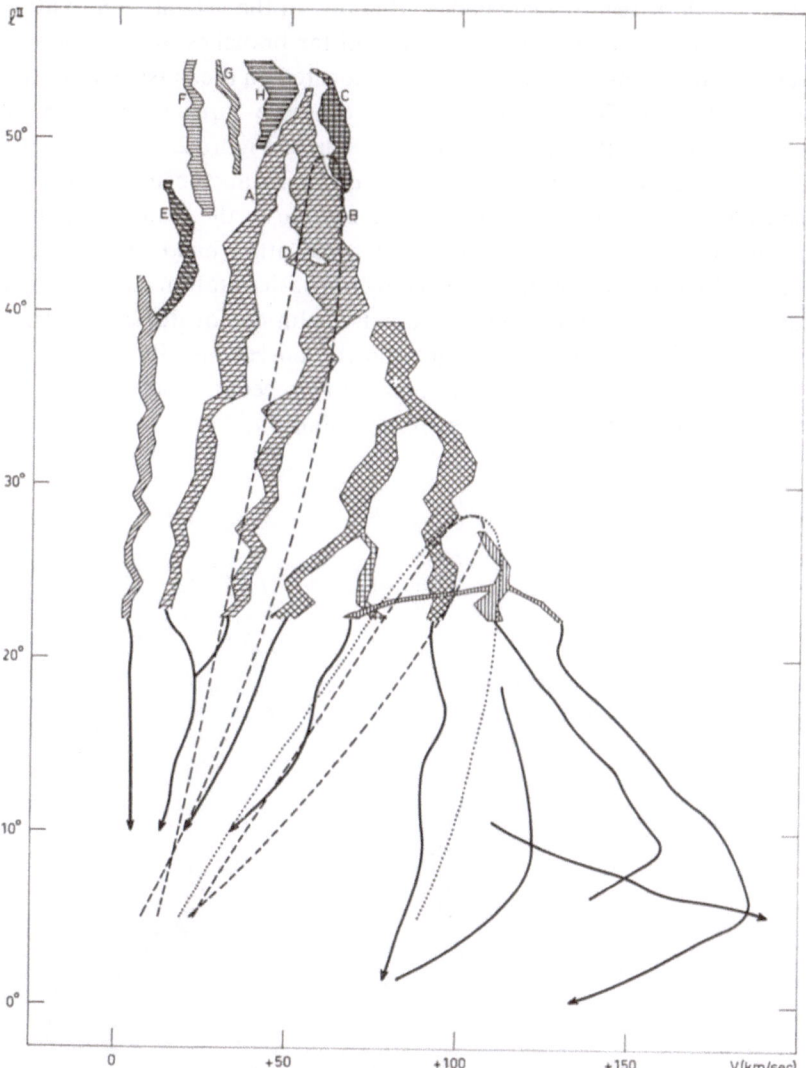

Fig. 1.    Observed velocity-longitude diagram for $22° \leqslant l^{II} \leqslant 56°$. Separate features are identified by different sorts of shading and, in the upper range of longitude, by letters. The breadth of a feature in velocity is proportional to the integrated HI content along a line of sight at its central latitude. The diagram has been extended schematically to lower longitudes on the basis of a preliminary analysis by S. C. Simonson. The broken curves represent the axes (density maxima) of the Sagittarius and Scutum arms based on model I. The dotted curve represents the same for the Scutum arm based on model IV.

basic rotation curve differs by only a few km s$^{-1}$ from that of Kwee *et al.* (1954). Furthermore we suppose a one-to-one correspondence between the velocity-longitude patterns in Figure 1 and concentrations of hydrogen in space.

First step: $\Theta_c(R)$ & $V(l^{II}) \Rightarrow$ Map.

To illustrate some of the problems encountered using these simple assumptions we consider the three features labelled A, B and C in the velocity longitude diagram. Features A and B are respectively the near and far branches of the Sagittarius arm, and feature C is the high-velocity stream of gas located on the outside of the Sagittarius arm (Burton, 1966a). The central velocities of these three features are extracted from Figure 1 and replotted as dots in the left side of Figure 2. If the velocities of the Sagittarius arm patterns A and B are converted to distance using the basic rotation curve then the vicinity of the subcentral-point circle in the resulting map is empty, which is certainly not a realistic situation. On the other hand, the velocities of the streaming feature C are larger than allowed by the basic rotation curve, so that following the assumptions of the first step there is no solution for distance to this feature.

In order to avoid forbidden velocities and the unrealistic situation of the empty subcentral-point region, we depart from the first procedure and introduce flexibility into both the velocity field and the density distribution.

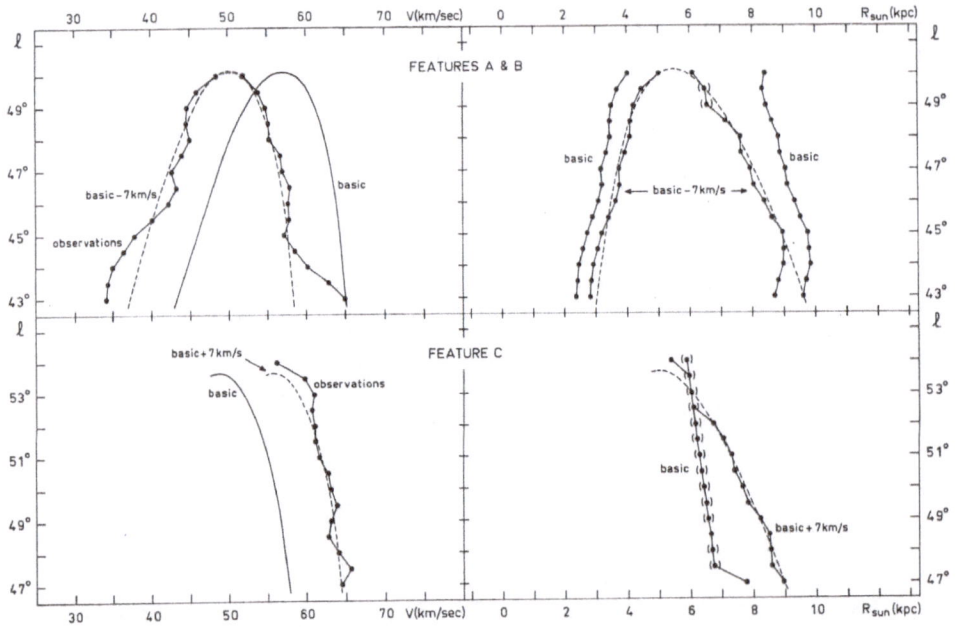

Fig. 2.    The observed velocities of features A, B and C, and kinematically derived distances from the Sun. The velocities are taken from Figure 1. Distances are derived by using the basic rotation curve or a shifted rotation curve. Dots in parentheses correspond to forbidden velocities and are plotted at the distance of the subcentral point. The smooth loops show the velocities and distances, based on the indicated rotation curve, for exponential spirals chosen to fit the observations.

In the second step we abandon circular rotation. We use the same basic rotation curve but postulate for each feature additional motion perpendicular to galactic radii, superimposed on the basic rotation.

Second step: $\Theta_c(R)$ + perturbation & $V(l^{II}) \Rightarrow$ Map.

The streaming is adjusted until the space distribution is reasonable. In the case of the Sagittarius arm features A and B, the subcentral-point region is not empty if a streaming of 7 km s$^{-1}$ against galactic rotation is superimposed on the basic rotation. In the case of feature C, by postulating a streaming in the direction of the galactic rotation of 7 km s$^{-1}$ superimposed on the basic rotation, the velocities are not forbidden and the distance of the feature from the Sun can be calculated.

The fifteen year old Kootwijk map was drawn assuming pure circular rotation. In the case of the Kootwijk map it was the density distribution which was adjusted when necessary; the velocity-field assumptions were not relaxed. For example, intensities found at velocities greater than allowed by the assumed rotation curve were smeared along 2 kpc of line of sight, symmetric with respect to the subcentral point. Accordingly, large-scale streaming motions may appear on the resulting map as spurious spiral arms or spiral-like fragments.

A map of the region $43° \leqslant l^{II} \leqslant 56°$ drawn following the procedures of the second step is shown in Figure 3. Such a map is necessarily composite since different perturbations to the basic rotation curve will in general be necessary to acccount for the behavior of the various patterns in the velocity-longitude diagram. There are also patterns for which there is no reason to adopt any deviation from the basic rotation. When the distance ambiguity has to be resolved, as in preparing Figures 2, 3 and 12, the distribution in latitude serves as the primary criterion. Where available, absorption measurements are also helpful.

In a third step we will try to interpret the observed streaming motions in terms of the first-order density-wave theory as formulated by Lin et al. (1969).

Third step: $\Theta_c(R)$ + density-wave perturbations & $V(l^{II}) \Rightarrow$ Models & Map.

## 3. Model Calculations

A contour map of hydrogen in the galactic plane is shown in Figure 4. This map was produced in cooperation with S. C. Simonson. The models may be compared with this reference map, although our interpretation is based on analysis of many more measurements. It is evident from the map that the run of terminal velocities with longitude shows irregularities. The bump around $l = 52°$ is associated with the Sagittarius spiral arm; the other main bump, around $l = 32°$, is associated with the Scutum arm. A more or less continous spiral arm appears at negative velocities, and an arm outside the Sagittarius arm appears at velocities near zero, superimposed on local gas. Regions of minimum intensity in the contour map can be just as important in the analysis as regions of high intensity. The minimum intensity regions between the

Sagittarius arm and the Scutum arm, and between the Sagittarius arm and the arm
outside it (at velocities near zero), are both evident in the reference map.

Model I is based on the first-order density-wave theory and is represented as a
contour map in Figure 5. The axis of the Sagittarius arm is seen tangentially at $l^{II} = 50°$
in the model and the axis of the Scutum arm is seen tangentially at $l^{II} = 29°$. Non-circular
arms located closer to the galactic center than 10 kpc will appear as loops in velocity-
longitude space, but resolution of the near and far branches of these arms is generally
prevented in the models by the large contour interval.

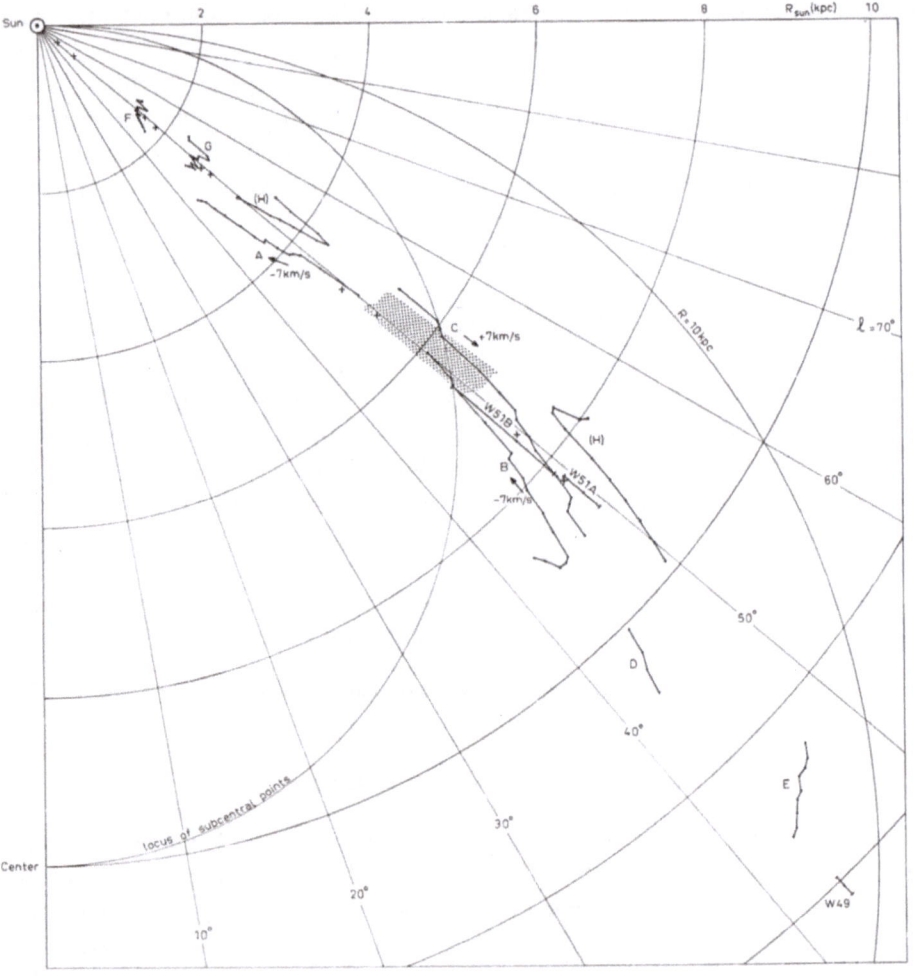

Fig. 3.   Map of H I distribution near the galactic plane in the region $43° \leqslant l^{II} \leqslant 56°$. The kinematic
distances to the Sagittarius arm features A and B, and to the high-velocity feature C, are based on
the basic rotation curve plus a perturbation of the amount and direction indicated. The distances
to the other features follow from the basic rotation curve alone. The distance ambiguity has not been
resolved for feature H. The distances to the H II regions indicated are based on recombination line
velocities (Mezger and Höglund, 1967). The absorption spectrum of the H II region W51 (Burton,
1966b) shows dips at velocities corresponding to the distances indicated by the crosses.

Irregularities in the terminal velocities are evident due to streaming in the sense of the density-wave theory and associated with the Sagittarius and Scutum arms. Two arms outside of the Sagittarius arm are included in the model; these two arms are too far from the galactic center to be seen tangentially so that only one branch appears on the model. One arm inside the Scutum arm is included in the model.

The parameters which are adjusted in constructing the model are the tilt angle of the two-arm exponential spiral and the density contrast of the gas, which is expressed as the ratio of the surface density at the axis of an arm to surface density in the interarm region.

The most important observations to be respected when adjusting the model parameters are the locations and amplitudes of the irregularities in the terminal velocities.

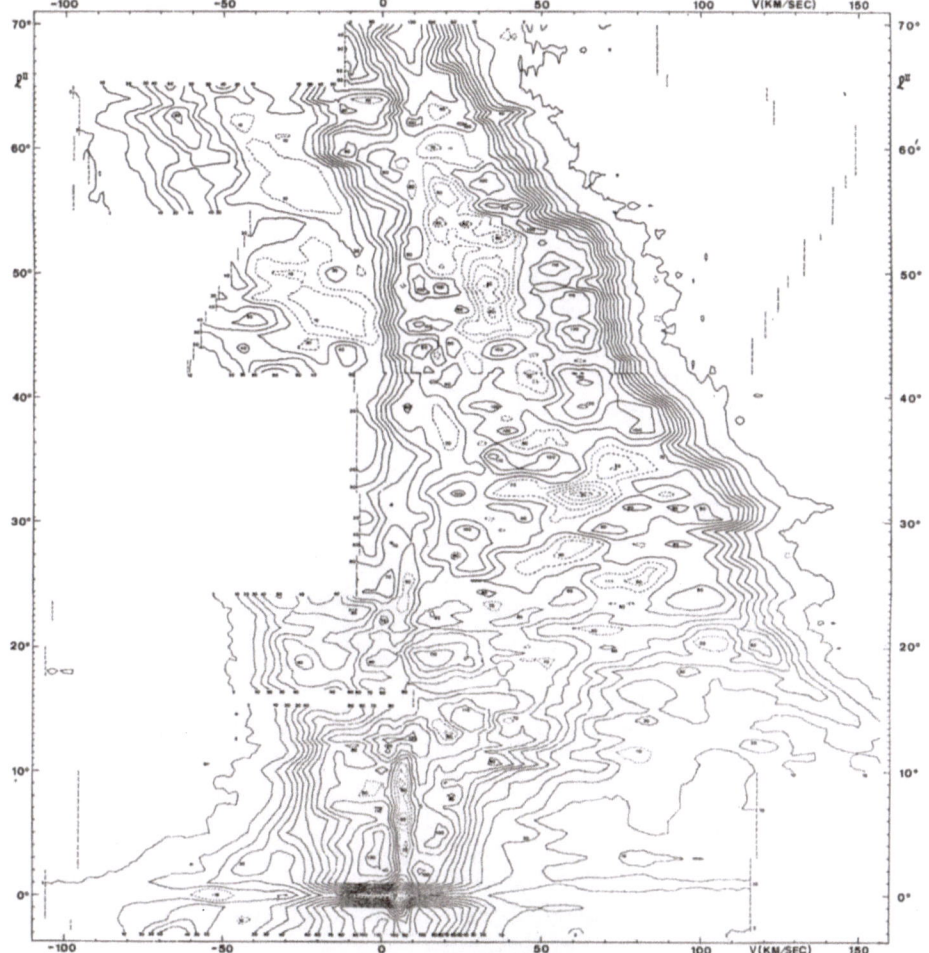

Fig. 4.   Contour map of H I brightness temperature in the galactic plane. The bandwidth is 2 km s$^{-1}$ above $l^{II} = 24°$ and 4 km s$^{-1}$ below. Contours were drawn at 2 K, 10 K, 20 K, etc. Broken-line contours enclose regions of low brightness temperature.

A scale factor puts the bump associated with the Sagittarius arm at the correct longitude of 50°. The tilt angle determines the longitude of the bump associated with the Scutum arm. If a constant value for the tilt is chosen, then the shape of the bumps is wrong and the body of the map agrees poorly with the observations, primarily because the other arms are not located in the correct positions. A better fit is obtained by taking the tilt angle to be a linear function of distance from the galactic center, such that the outer arms are more circular than the inner arms.

Once the tilt angle is specified, the amplitude of the bumps in the terminal velocities is determined by the gas density contrast parameter. The gas density contrast also

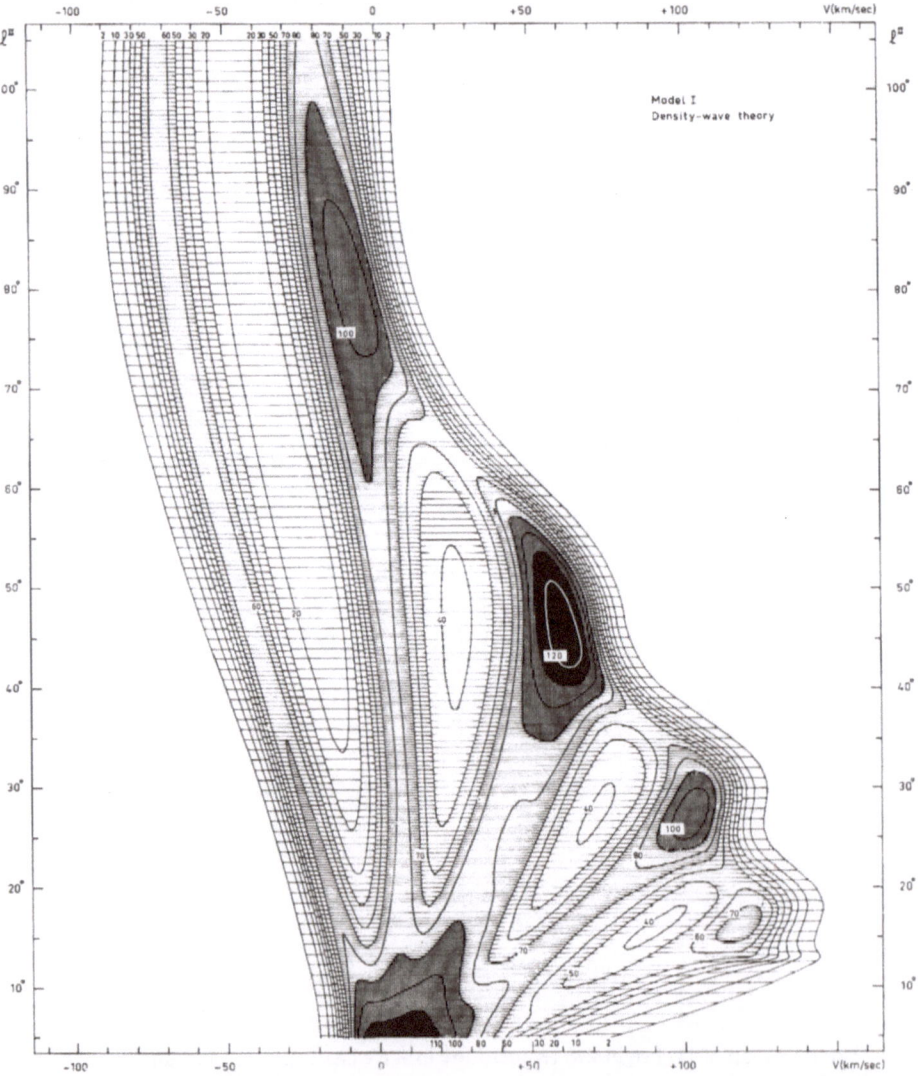

Fig. 5.   Model I contour map for the galactic plane based on first-order density-wave theory.

determines the relative intensities of peaks and valleys deep in the contour map. In the first-order density-wave theory the gas arms coincide with the stellar arms, and the gas density varies in a smoothly oscillating manner along galactic radii.

The tilt used in model I varies linearly from 8° (with respect to a circle) at 5 kpc from the galactic center to 5° at 10 kpc. The contrast between gas density on the axis of an arm and interarm density varies linearly from 3:2 at 5 kpc from the center to 3:1 at 10 kpc. The maximum streaming velocity predicted by the density-wave theory using this tilt and density contrast is about 8 km s$^{-1}$. The density-wave pattern speed is held constant at 13 km s$^{-1}$ kpc $^{-1}$. Formally the tilt angle is directly related to the pattern speed through the dispersion relationship of the density-wave theory. However, in constructing the model it was convenient to retain the tilt angle as a free parameter in finding a simple model with density-wave streaming in reasonable agreement with observations. Comparison of the pattern derived here with that

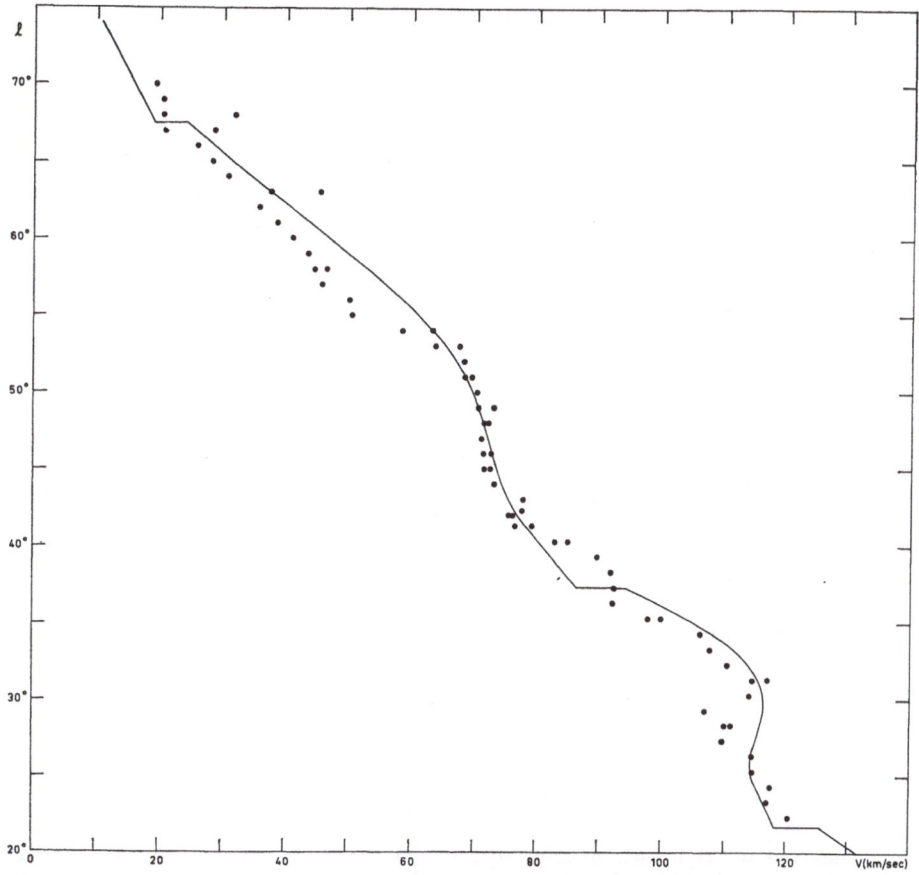

Fig. 6.   Comparison of terminal velocities derived from observations and from model I. The dots are the observed terminal velocities measured by Shane and Bieger-Smith (1966). The line shows the terminal velocities calculated from model I.

formally derived from the constant pattern speed of 13 km s$^{-1}$ kpc$^{-1}$ showed that
they agreed well, especially in the region under discussion.

The line profiles employed in the model contour map are smoothed with a velocity
dispersion $\sigma_v = 9.0 - 0.4$ $R$ (km s$^{-1}$). The average gas density is constant at $<n_H> =$
0.4 atoms cm$^{-3}$ for $R < 10$ kpc and decreases linearly toward a cut-off at $R = 25$ kpc.
A constant spin temperature $T_s = 135$ K was assumed.

The run of observed terminal velocities calculated by Shane and Bieger-Smith
(1966) is compared in Figure 6 with the terminal velocities calculated in the same way
from the profiles used in the model I contour map. The agreement is fairly good. The
poorest agreement is in the region of the secondary bump at $l^{II} = 38°$, which does not
appear in the model, and near $l^{II} = 55°$ where the observed velocities drop by more than
10 km s$^{-1}$ in one degree of longitude. Abrupt changes in velocity or in density can
not be explained by the first order density-wave theory. Sudden drops in terminal
velocity derived from the model are due to the fact that at longitudes just below the
drop the highest-velocity maxima are contributed by gas on the outside of one arm
whereas at longitudes just above the drop they are contributed by gas streaming, in
the opposite sense, along the inside of the next arm.

A comparison of the observed rotation curve with the unperturbed basic rotation
curve and with that derived from model I is given in Figure 7.

The geometry through which the space density distribution is transformed into the
velocity-longitude relationships which we observe is complicated. The general ridge
of intensities at high velocites in model I illustrates the effect of an extended path-

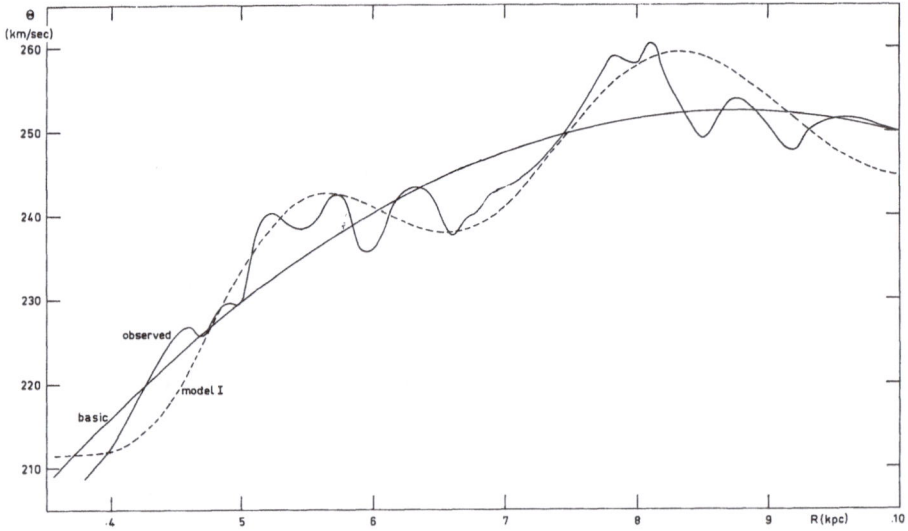

Fig. 7. Observed and computed apparent rotation curves. The irregular curve is that derived by
Shane and Bieger-Smith (1966). The smooth curve is the basic rotation curve; the dashed one shows
the basic rotation curve perturbed by the density-wave theory streamings of model I. In the presence
of deviations from circular motion the maximum velocity might not come from the subcentral point;
nevertheless the figure is drawn assuming that it does.

length over which the velocity changes only slowly. Another example of the same situation is the very strong pattern in the model between $l^{II} = 65°$ and $l^{II} = 90°$, near zero velocity. The volume-density distribution is smooth along all the arms in the model; the changes as one moves along the axis of an arm in the velocity-longitude space of Figure 5 are due to crowding in velocity. One of the most important large-scale characteristics of the observations shown in the reference map in Figure 4 is the general ridge of high velocity hydrogen. Furthermore, the observations show a strong pattern beginning at $l^{II} = 65°$ and continuing to higher longitudes. These observed high intensities are due to real hydrogen, but not necessarily to a spatial concentration of hydrogen. This path-length effect must be kept in mind when converting the observed velocity-longitude patterns into a map of the Galaxy, in order to avoid spurious features on the map. The pattern labelled H in Figures 1 and 3 is a case in point. Above $l \approx 53°$ this pattern appears as the high-velocity ridge on the observed contour maps. It is possible that these intensities are contributed by interarm gas which is not concentrated into a spiral-like fragment.

In reality, there are bound to be irregularities in the radial velocities as well as in the spatial structure. Large-scale irregularities of only a few km s$^{-1}$ in the radial velocities will have consequences in the profiles comparable to those due to irregularities in the structure.

In order to demonstrate the effects of streaming, model II has been calculated using the same density distribution as used in model I, but using the basic rotation curve alone without the density-wave theory perturbations. The contour map of this circular rotation model is shown in Figure 8. Model II has a strikingly different appearance from model I, even though the input space distribution is exactly the same in both models. Of course the run of terminal velocities in model II is quite smooth. The slope of the major patterns in the model contour maps is partly determined by the amount of streaming. The slopes of the patterns in the density-wave theory model agree better with the observations than the slopes in the circular rotation model.

Model III, shown in Figure 9, is based on the assumptions of circular rotation and no gas between arms. The distribution of the average density $<n_H>$ is the same in this model as in the previous ones. The hypothesis of no gas between arms is unlikely. Furthermore the model contour map shows that these assumptions also cannot represent the observations, the major failures being that the shape of the irregularities in the run of terminal velocities is all wrong, and that the high-velocity ridge of hydrogen, so characteristic of the observations, is missing in model III.

The sort of insights which can be gained from model calculations into the complicated relationships between space distribution and velocity-longitude diagrams leads us to suggest that it might prove useful to calculate model line profiles whenever one produces a map of the spiral structure. These profiles, which would be based on the velocity field used to derive the kinematic distances in the map, would then be compared with the observations to test if the spiral structure map is a reasonable one.

It is clear that attempts to derive the velocity field must be based on simultaneously derived solutions for the distribution of the gas. In the first step outlined above, the emphasis was on the hydrogen distribution. In the second and third steps the emphasis was on the velocity field. The density-wave theory provides a velocity field in better agreement with the observations than that provided by circular rotation.

With a better velocity field, can we draw a better map of the Galaxy? The velocity field derived from fitting the density-wave theory to the observations already implies a mass distribution. Since the theoretical model clearly does not represent the observations perfectly, and in some regions it fails grossly, a map based on the velocity-field

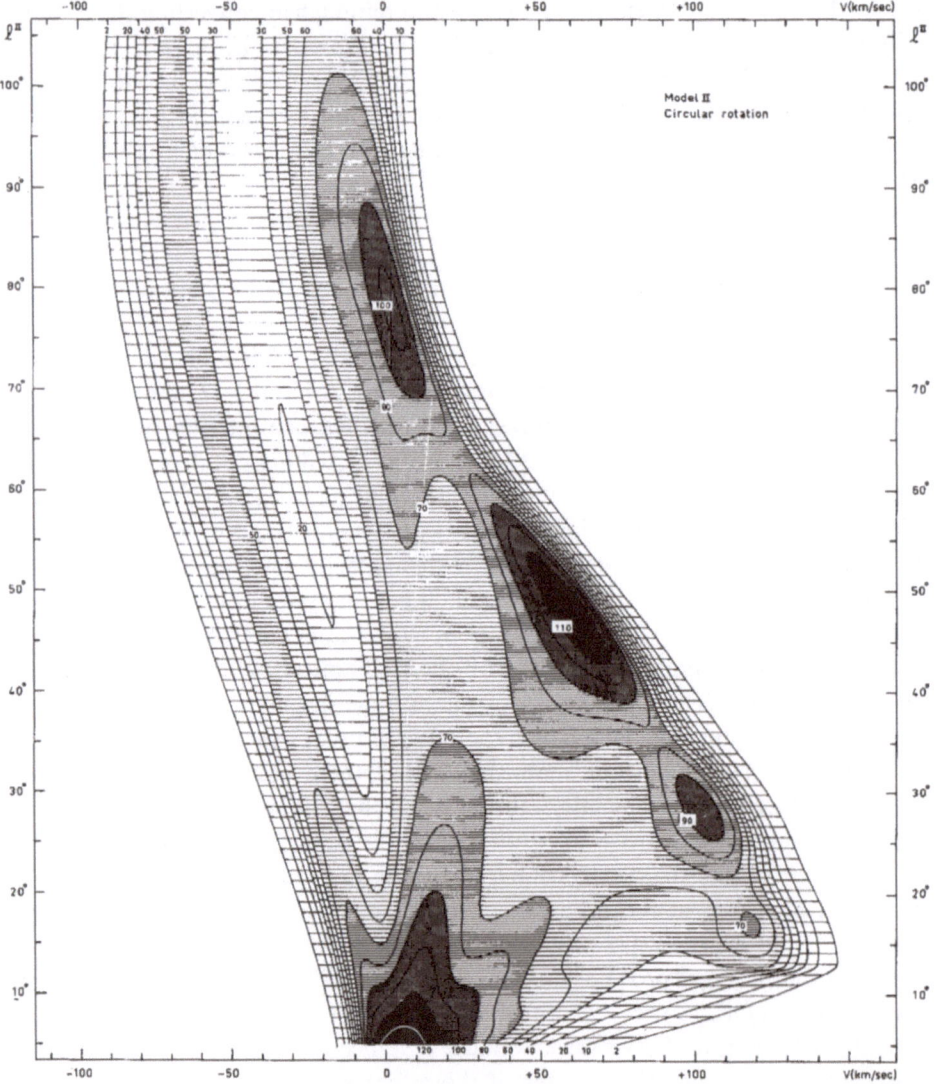

Fig. 8.   Model II contour map based on circular rotation. The same density distribution is used in this model as in model I.

of the model will place some features in space where they contradict the space distribution used as input in deriving the velocity field.

It seems best as a next step to take the insights we have won into the velocity field and to go back to the observations, but now with the emphasis again on the hydrogen density distribution.

## 4. The Velocity Field in the Scutum Arm

In order to study more closely the properties of the models it is useful to compare the velocity-longitude diagram observed with that corresponding to the density

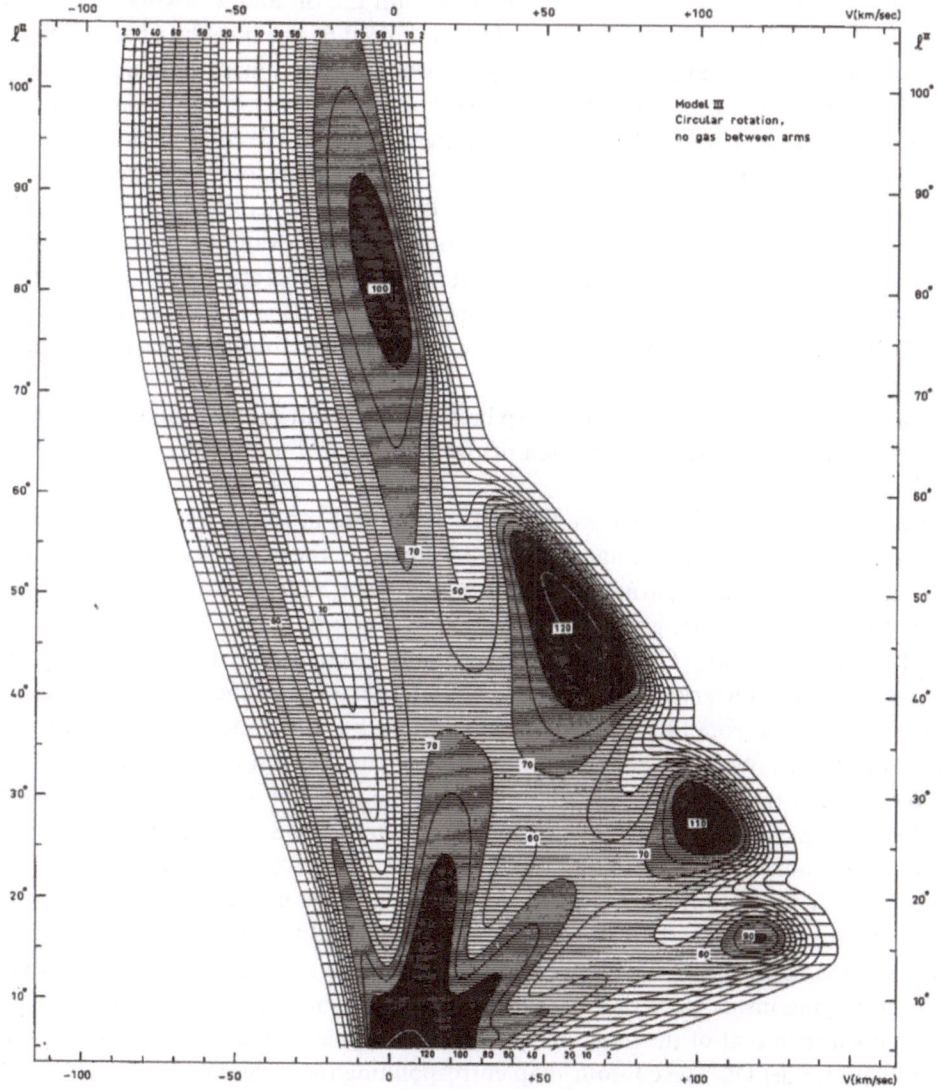

Fig. 9.    Model III contour map based on circular rotation and no gas between arms.

maxima of the model. This method has the advantage that the most important features of the whole observational material are presented in a simple way. The comparison must be made with caution, however, as dissimilar features are being plotted; for example, the persistent intensity maximum at high velocity which is such a prominent feature of all the contour maps is absent from the model results as shown in this way.

In Figure 1 we have plotted the axes of the Sagittarius and Scutum arms computed for model I as broken curves. The agreement with the observations is not very satisfactory. Considering the Scutum arm, we see that the loop of the arm derived from the model is far too closed and is located at too low a longitude. In order to open the loop we may try to increase the tilt of the arm. This is an inefficient process, since the density-wave streaming velocities will increase with the tilt and will work to close the loop, so that a tilt of more then 20° would be required to produce the observed figure. Such high tilts are not expected in the neighbourhood of the inner Lindblad resonance, and it would be necessary to introduce a very rapid increase in tilt inward from the Sagittarius arm and probably a corresponding increase in the multiplicity of arms.

An alternative possibility is to introduce a radial expansion in the region of the Scutum arm. Such an assumption introduces dynamical difficulties and it is questionable if the first-order density-wave theory could have any validity in the presence of such an expansion. The hypothesis, however, serves to represent these observations and is supported by three independent observational results; these are the expansion of the 3-kpc arm as seen in absorption, the rapid increase in the observed maximum velocity which sets in for longitudes lower than about $l^{II} = 22°$ and which could be explained otherwise only by a very sharp break in the rotation curve, and the extension of the far (i.e. higher velocity) branch of the Scutum arm to $l^{II} = 0°$ at the velocity of $+70$ km s$^{-1}$.

For the reasons given above we favor the expansion model, which we call number IV, over one with rapidly changing tilt, and have made the corresponding calculation. We have used the same parameters as in model I but have introduced an additional axisymmetric expansion, given, as a function of distance from the galactic center, in the table in Figure 10. The behavior assigned at distances less than 2 kpc from the center is quite arbitrary. The velocity-longitude diagram of the axis of the Scutum arm as calculated from this model is shown in Figure 1 as a dotted curve. The Sagittarius arm, lying outside the region of expansion, is unchanged from model I.

We see that this model represents quite well the opening of the Scutum arm loop and the extension to $l^{II} = 0°$. The assumed location of the Scutum arm in the model, however, is unaltered, and it lies several hundred parsecs inward from the observed location. We cannot alter the position of the arm in the model without spoiling the fit to the rotation curve data (Figure 6), so we conclude that the main concentration of gas in the Scutum arm may lie outside the axis of the mass concentration. A secondary arc, lying inside the axis of the arm, is also seen in Figure 1. This loop is even more open than that of the main arm, implying a higher expansion velocity, in agreement with model IV. The contour map corresponding to model IV is shown in Figure 10. Comparison of this map with Figure 4 reveals not only the similarity in behavior

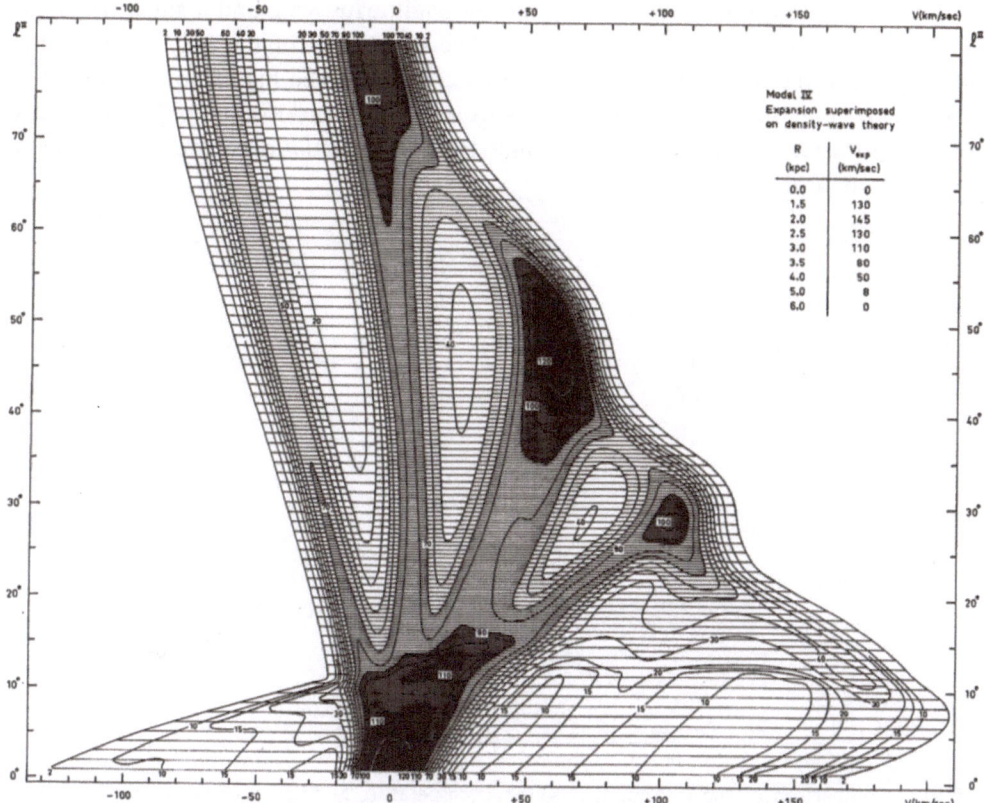

Fig. 10.  Model IV contour map based on the same assumptions as model I but with the radial
expansion, given in the table, superimposed on the other motions.

of the Scutum arm but also general agreement of the maximum velocity at $l^{II} < 22°$
and a small feature at negative velocity close to $l^{II} = 0°$ which may correspond to the
3-kpc arm.

In order to examine further the doubling of the Scutum arm, we have compared
the maximum brightness temperatures at high velocity as observed and as computed
from model IV. These are plotted in Figure 11 for the longitude range $22° \leqslant l^{II} \leqslant 42°$.
The rise toward the maximum of the Sagittarius arm is accurately expressed by the
model, but the broad maximum of the Scutum arm shown in the model is clearly
split in the observations. In this representation the inner and outer parts appear to
be of about equal strength.

## 5. The Distribution of the Neutral Hydrogen

Using the velocity field of model IV we have constructed a map (Figure 12) of the
inner region of the Galaxy in the range $l^{II} = 22°3$ to $42°3$. We see the arcs belonging to
the Scutum arm as well as a small feature which may be associated with the far

branch of the outer arc, both branches of the Sagittarius arm, and a small fragment
which may belong to the next outer arm. The extensions of the Scutum arm arcs
outward along the locus of subcentral points are due to the high-velocity ridge discussed above and do not necessarily represent a spatial concentration of H I. The axes
of the Sagittarius and Scutum arms assumed for the calculation of the velocity field
are shown as broken curves.

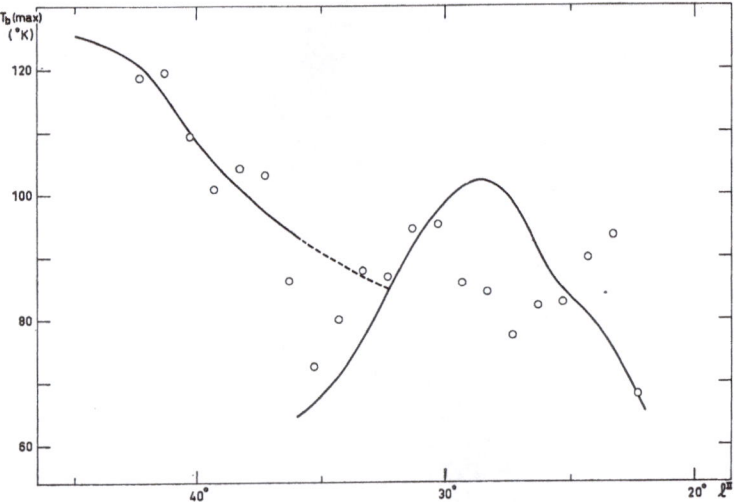

Fig. 11.   Comparison of the maximum brightness temperature at high velocity observed (circles)
and calculated from model IV (curve); the broken part of the curve represents an alternative choice
for the velocity of the maximum.

It is tempting to compare the location of the H I as plotted in Figure 12 with that
of the arms as introduced, mainly from dynamical considerations, into the model
calculations. For two reasons one must be very cautious about drawing conclusions
from such a comparison. In the first place, the position of the arms assumed for the
model calculations is determined largely from the evidence of the rotation curve
which pertains only to the neighbourhood of the subcentral points, and from which
the location throughout the region is inferred. A perturbation near this locus may
result in a substantial shift of the assumed position of the whole arm. In the second
place, the first-order density-wave theory does not pretend to give an accurate picture
of either the density or the velocity field within an arm. Recent calculations by
Roberts (1969) have, in fact, shown that substantial deviations from the first-order
theory may be expected. Thus kinematic distances derived from the first-order
theory may contain systematic errors of the order of the dimensions of an arm.
Furthermore the first-order theory can hardly be expected to be valid under the
conditions of expansion assumed here. The presence of the deformation in the rotation
curve in the neighbourhood of the Scutum arm indicates, however, that the velocity

field is in fact perturbed in a manner similar to that predicted by the first-order theory, which then serves as a means of extrapolation.

Bearing in mind the above limitations, we can nevertheless draw some conclusions about the distribution of HI in the region considered. It appears that the Scutum arm is divided into two arcs separated by almost 1 kpc. The gas in both arcs appears to be moving outward from the galactic center with velocities approximately equal to those given in the table in Figure 10. The appearance of the Scutum arm in the velocity-

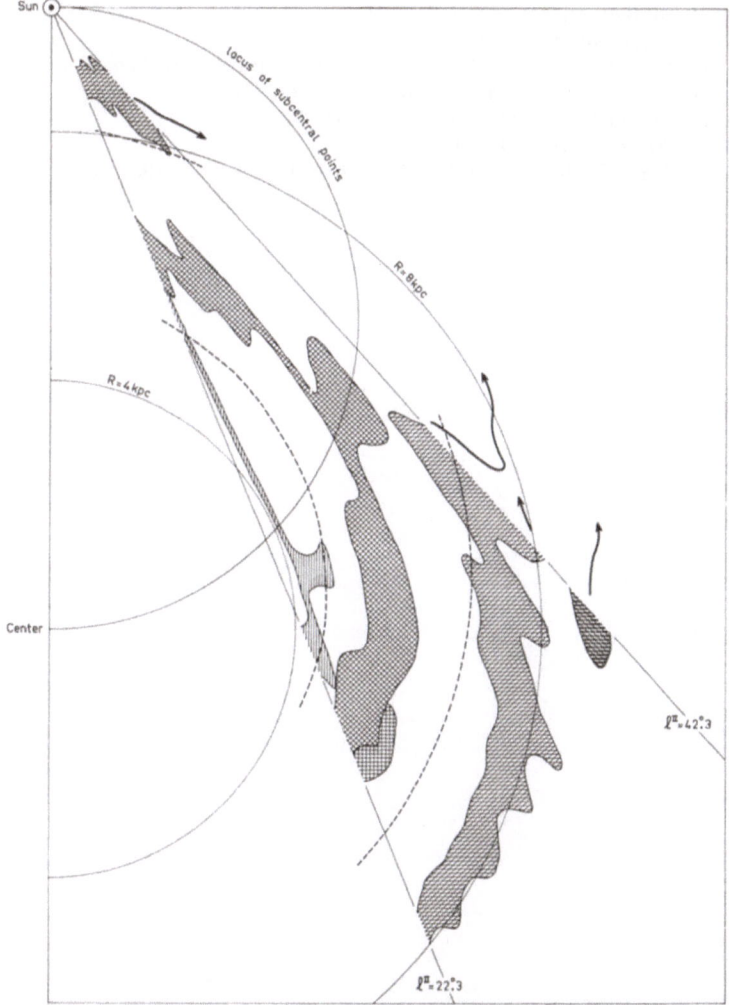

Fig. 12. Map of the HI distribution near the galactic plane in the region $22°.3 \leq l^{\mathrm{II}} \leq 42°.3$. The kinematic distances are based on model IV. The breadth of each feature represents the HI content (not its spatial extent), $\sigma_v \tau = 5$ km s$^{-1}$ being represented by 1 kpc along the line of sight. The shading is the same as that used in Figure 1. The extension into the higher longitude region is indicated schematically. The broken curves indicate the locations of the axes of the Sagittarius and Scutum arms adopted in all of the model calculations.

longitude diagram supports this conclusion. The Sagittarius arm as plotted does not closely resemble a spiral, the far branch being almost circular and the near branch quite sharply inclined. This suggests a large scale asymmetry in either the velocity field or the density distribution or both, but in view of the limitations discussed in the preceding paragraph such a conclusion cannot be drawn with any confidence on the basis of the present model calculations.

### References

Burton, W. B.: 1966a, *Bull. Astron. Inst. Netherl.* **18**, 247.
Burton, W. B.: 1966b, *Astron. J.* **71**, 848.
Kwee, K. K., Muller. C. A., and Westerhout, G.: 1954, *Bull. Astron. Inst. Netherl.* **12**, 211.
Lin, C. C., Shu, F. H., and Yuan, C.: 1969, *Astrophys. J.* **155**, 721.
Mezger, P. G. and Höglund, B.: 1967, *Astrophys. J.* **147**, 490.
Roberts, W. W.: 1969, *Astrophys. J.* **158**, 123.
Shane, W. W. and Bieger-Smith, G. P.: 1966, *Bull. Astron. Inst. Netherl.* **18**, 263.

# 75. LARGE-SCALE GALACTIC SHOCK PHENOMENA AND THE IMPLICATIONS ON STAR FORMATION

W. W. ROBERTS, JR.

*University of Virginia, Charlottesville, Va., U.S.A.*

**Abstract.** The possible existence of a stationary two-armed spiral shock pattern for a disk-shaped galaxy, such as our own Milky Way System, is demonstrated. It is therefore suggested that large-scale galactic shock phenomena may very well form the large-scale triggering mechanism for the gravitational collapse of gas clouds, leading to star formation along narrow spiral arcs within a two-armed grand design of spiral structure.

We have been reminded throughout this symposium that spiral galaxies have many irregular features, including multiple spiral arms; yet *two* principal arms making up a large scale grand design of spiral structure can often be traced to the central region and sometimes to the very center of these galactic systems. Often the spiral arms show a very high resolution into *knots* that are generally interpreted as H II regions and associations of stars. While the young stars practically always appear in stellar associations, the young stellar associations with their corresponding brilliant H II regions often occur in chains and spiral arcs similar to strings of beads within the larger grand design of spiral structure. In view of these observational studies, a basic problem stands out: namely, what physical mechanism could trigger star formation along a two-armed grand design of spiral structure in this orderly fashion.

The density wave theory has been suggested to account for the grand design feature. Yet, the presence of a marked grand design of spiral structure in many disk-shaped galaxies signifies an even more marked and narrow region for the initiation of star formation within the gaseous spiral arms of the grand design. In the density wave theory, this is indeed a very urgent problem to be explained; for, the gas stays inside the spiral arm for a relatively short period of time. Furthermore, in the linear theory, the gas concentration in a density wave extends over a broad region and could not be expected to provide a sufficiently rapid triggering mechanism to produce narrow spiral strips of newly-born luminous stars. We are therefore led to infer the presence of large-scale 'galactic shocks' that would be capable of triggering star formation in such narrow spiral strips over the disk, and we must therefore further refine the linear density wave theory by including nonlinear effects. In fact, we might expect self-sustained density waves in the gaseous component of the galactic disk to grow and develop in the course of time into disturbances with shock-like nature. Thus, the self-sustained density waves obtained by Lin and Shu in a linear theory (1964, 1966) might be expected in a nonlinear theory to develop into large-scale shocks with a regular two-armed grand design of spiral structure over the disk. Therefore, it is important to show that two periodically-located spiral shock waves present on the large scale

*Becker and Contopoulos (eds.), The Spiral Structure of Our Galaxy, 415–422. All Rights Reserved.*
*Copyright © 1970 by the I.A.U.*

throughout the galactic disk are compatible with the general nature of a stationary nonlinear gas flow about the disk.*

The motion considered is that of the continuum of turbulent gas composing the gaseous disk, moving in a gravitational field consisting of a two-armed spiral field superposed on the Schmidt model for the Milky Way System. Suppose we now look at the asymptotic nonlinear solution for the gas motion we have obtained (Roberts, 1969). This nonlinear solution making up the gas flow picture over the galactic disk describes isothermal gas flow in closed, nearly concentric, and twice-periodic stream-tube bands that pass through a Two-Armed Spiral Shock pattern. Figure 1 shows this

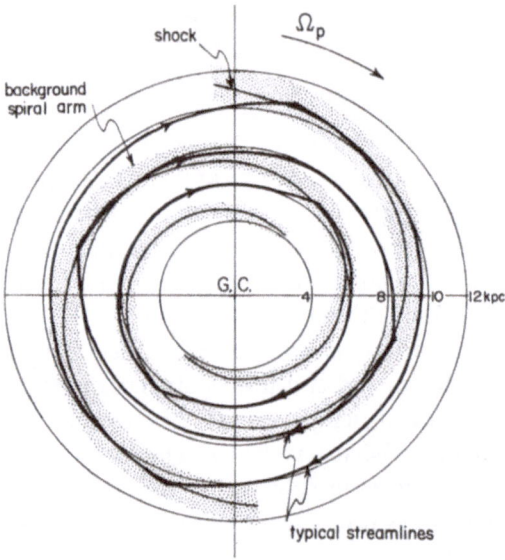

Fig. 1.   Shock and background spiral pattern in the galaxy. Each arrowed gas streamline appears as a sharp-pointed oval with a sharp turning point at each shock.

simple model of gas flow, which I would like to refer to as the *TASS* picture – the *Two-Armed Spiral Shock picture*. Numerical calculations for regions between the radii of 3–4 kpc and 12 kpc in the Schmidt model confirm the compatibility of two periodically-located shocks lying along and within the imposed two-armed background spiral pattern. The background pattern may be regarded as the composite pattern of all the moderately-old stars of ages greater than perhaps 30 million years and therefore does not stand out in observational studies. Arrowed gas streamlines which turn sharply at each shock are sketched for several typical radii.

---

* Fujimoto actually carried out some nonlinear calculations on the gaseous component of the galactic disk as early as 1966. At that time it was difficult to construct a large-scale nonlinear gas picture for our Milky Way System; however his results nevertheless provided important insight for the initial stages of this work. A comparison of the basic differences between Fujimoto's work and the present investigation is given in Roberts (1969).

Suppose we focus our attention momentarily on a given streamtube. There are five basic independent parameters that govern the nature of the streamtube and the gas flow along it:

(1) $i$  the angle of inclination of a spiral arm to the circumferential direction, taken as about 8°;

(2) $\Omega_p$ the angular speed of the spiral pattern, taken as 12.5 km s$^{-1}$ kpc$^{-1}$;

(3) $F$  the amplitude of the spiral gravitational field taken as a fixed fraction, 5%, of the smoothed axisymmetric gravitational field;

(4) $\varpi$  the average radius of the streamtube; and

(5) $a$  the mean turbulent dispersion speed of the gas along the streamtube, taken in the range between 4 km s$^{-1}$ and 10 km s$^{-1}$.

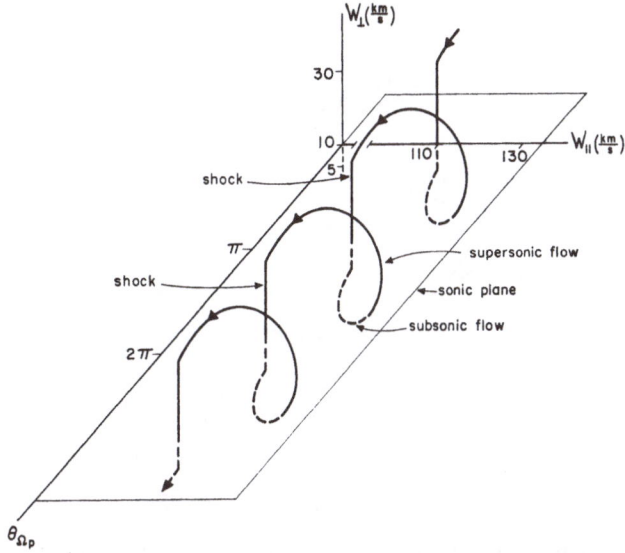

Fig. 2. The nature of gas flow along a typical streamtube.

Once these five parameters are specified the shock location with respect to the background spiral arm is automatically determined. Figure 2 illustrates the nature of the gas flow along a typical streamtube. $W_\perp$ is the velocity component of gas across a spiral arm, and $W_\parallel$ is the velocity component of gas along a spiral arm. The gas begins the same cycle again at every successive large-scale shock. The motion of the gas along a typical streamtube may be visualized in simplified terms as the nonlinear counterpart of epicyclic motion, as modified by gaseous 'pressure'.

This simplified interpretation is shown in Figure 3, which illustrates the radial motion of a gas particle in its epicyclic potential well centered at a fixed radius, $\varpi_0$. Since the gas and the trailing spiral pattern travel about the disk at different angular speeds, where the angular speed of the gas may be much larger than the pattern speed, an observer travelling with the gas (at $\varpi_0$), would see one spiral arm, after another spiral arm, travel past him toward the galactic center. At time $t=0$, the gas particle, the dark blob, is feeling the tendency to be dragged along by the gravitational field of the spiral arm, which is moving inwards faster than the blob. At a time roughly equal to half the period, the gas blob has been dragged inwards as far as its potential well will allow. After breaking away from spiral arm 1 and returning back across its potential well, it meets spiral arm 2 and shock 2. At this shock the gas particle begins the cycle once again.

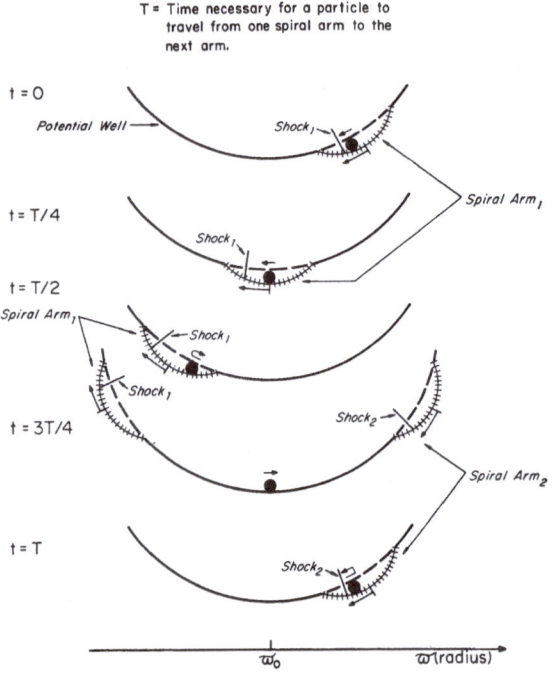

Fig. 3.   Simplified interpretation of the radial motion of a gas particle in its epicyclic potential well.

Spitzer (1968) has considered a number of typical gas clouds and has concluded that the large clouds may not be far from the critical condition where gravitational collapse becomes possible. On the basis of Spitzer's investigation and the present results, it is now suggested that galactic shock waves may very well form the triggering mechanism for the gravitational collapse of gas clouds, leading to star formation. Since newly-born stars give rise to H II regions, galactic shocks may be visualized as the necessary forerunners of the prominent H II regions as well as the newly-born stars. In Figure 4, we view this possible star formation mechanism. Gas moves from left to right. Before reaching the shock, some of the large clouds and cloud complexes may be on the verge of gravitational collapse. A sudden compression of the clouds in the shock to perhaps five to ten times their original density could conceivably trigger the gravitational collapse of some of the largest gas clouds. As the gas leaves the shock region, it is rather quickly decompressed, and star formation ceases.

Attention may be focused on the narrowness of the gas density peak located adjacent to the shock. Such a narrow peak in density and pressure indicates appreciable star formation may take place only over the narrow spiral region lying just behind the large-scale galactic shock. Over the time period of 30 million years necessary for the formation and evolution of the relatively massive stars initiated at the shock, gas traverses a distance normal to the spiral arm of only about ⅛ of the total wavelength separating successive arms. Therefore, the relatively massive and luminous newly-born stars initiated in the peak of gas concentration at the shock are confined to the inner side of the observable gaseous spiral arm of HI; for, when they pass outside this region, they no longer are newly-born or luminous. Since dust and gas travel together,

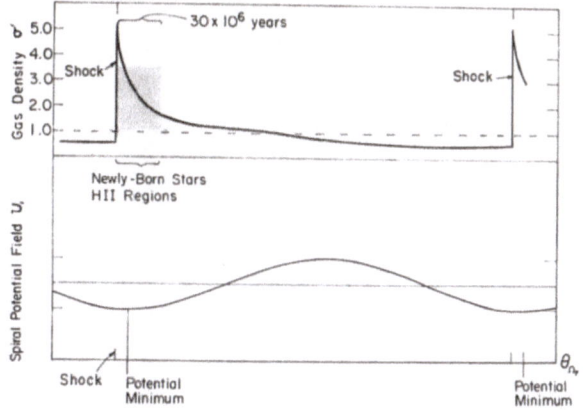

Fig. 4.    Gas density distribution along a typical streamtube in the TASS picture. The potential minimum corresponds to the density maximum and the center of the background spiral arm.

the region for most prominent dust concentration also lies at the sharp HI peak adjacent to the shock. As the dust leaves the shock region, it is decompressed, and some of it may even be evaporated away by the newly-born stars and the HII regions, leaving lesser concentrations of dust to be found within and outside of the HII regions.

We can now visualize how this model galaxy may appear with large-scale shocks present. When observations of the galactic disk are made, the moderately-old star background distribution making up the imposed two-armed spiral pattern is not seen. Basically, what we do see are:

(1) the observable gaseous spiral arms of HI, and

(2) the newly-born luminous stars and brilliant HII regions.

Figure 5 illustrates this observable physical picture according to the shock predictions. A shock together with a sharp HI peak and a prominent dust lane occur on the inner edge of the observable gaseous spiral pattern of HI. On the inside of this observable spiral pattern (but just outside the shock and the sharp HI peak) lie the regions of newly-born stars and the HII regions. I think we can find general agreement here between this theoretical picture and observational studies of the Milky Way System by Morgan (this volume, p. 9), who showed some beautiful photographs of various galaxies in which the locations of the newly-born stars, the luminous H II regions, and the prominent dust lanes occurred for the most part along narrow spirals. I should also like to mention the convincing agreement here with the observational studies of Beverly Lynds, presented at this symposium, who found in her studies of 17 galaxies that the most prominent dust lanes generally lie inside of the spirals of newly-born stars and the brilliant H II regions. There is also general agreement with the observational studies of Kerr *et al.* (1968), also presented at this symposium, who found that the luminous H II regions seem to lie just slightly off the ridge lines of H I in the sense

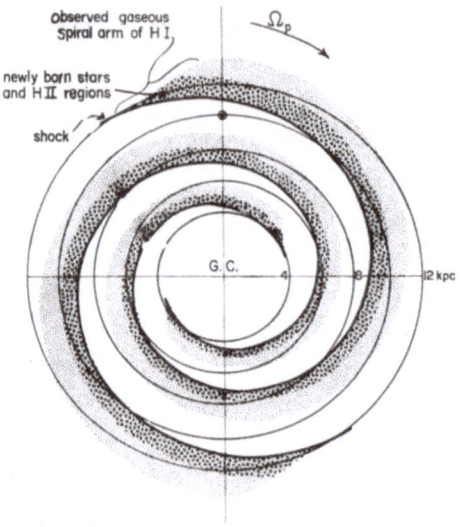

Fig. 5.   The observable spiral pattern in the model galaxy.

of slightly larger radial velocity, or just slightly outside the sharp H I peak in our interpretation.

Suppose we now examine how the strength of the TASS pattern varies with radius from the galactic center. Figure 6 sketches the behavior of the gas density for stream-tubes at various radii. We first might note the peak representing the effective compression of the gas by the shock. This effective gas compression produced by the TASS pattern rises monotonically with decreasing radius from about 5 in the 11 kpc region to 8–9 in the 3–4 kpc neighborhood. It is the maximum value of gas density that likely suggests the actual degree of gas cloud compression and possibly the relative amounts of star formation and H II concentrations at various radii. Therefore, the TASS picture indicates a tendency toward greater star formation in the 4 kpc neighborhood. This feature of enhanced star formation and H II concentration toward the inner regions of the TASS picture shows general agreement with the observational evidence of Kerr and Westerhout (1965) and Kerr *et al.* (1968) for the mean H II distribution plotted with respect to radius over the galactic disk with the highest concentration of H II found between 4 kpc and 6 kpc.

Numerical calculations have been carried out over the disk out to the outer bound of 12–13 kpc where the TASS pattern terminates.

The TASS pattern rapidly diminishes in strength as we follow a spiral arm outwards from 11 kpc. Further out along the arm in the 12 to 13 kpc neighborhood, we visualize the TASS picture to merge into the density wave picture. Observational evidence for an outer cutoff of the TASS pattern appears in the observational studies of M. S. Roberts who found for a number of Sc-type galaxies that the circumferential bands with highest H I distribution do not coincide with and, in fact, lie significantly outside

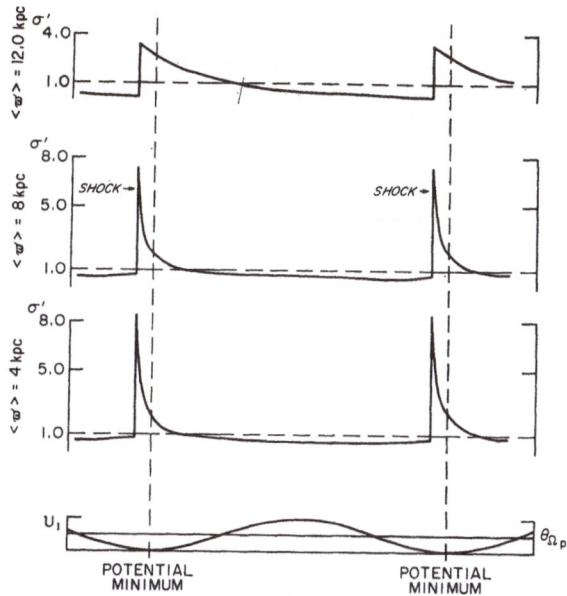

Fig. 6. Variation in the effective gas compression of the TASS pattern with respect to radius.

of the circumferential bands containing the most prominent newly-born stars and H II regions. For our own Milky Way System the same phenomenon seems to occur as illustrated by the Westerhout (1958) and van Woerden (Oort, 1965) results for the H I and H II distributions plotted with respect to radius over the disk, with the H II lying much interior to the H I. These observational results then seem to indicate a strong triggering mechanism that is capable of affecting the production of H II according to the mechanism's own pattern, considerably independent of the distribution of H I. This evidence then adds further support for a large-scale TASS pattern that can control to a large extent the star formation process as well as the radial distribution of H II over the galactic disk. Indeed, the reason why there may be no newly-born stars and no H II regions coinciding with the regions of highest H I distribution in our own Galaxy and in many Sc-type galaxies may be that there is no large scale shock pattern present at such large radii where the H I distribution is maximal.

I would like to end on one final note. The Two-Armed Spiral Shock picture seems to be capable of accounting for a good many striking features associated with the grand design in many spiral galaxies, our own Galaxy included. On the other hand, we must not forget the many types of irregularities that often exist: such as the inter-arms branching across the grand design or the multiple sub-branches sometimes extending from a single large-scale spiral arm. There are no doubt many other types of irregularities as well, and perhaps not all are secondary effects. Yet at this stage, there is optimism about the TASS picture and the possibilities for its extension and refinement.

## References

Fujimoto, M.: 1966, IAU Symposium No. 29, p. 453.
Kerr, F. J. and Westerhout, G.: 1965, *Stars and Stellar Systems* **5**, 167.
Kerr, F. J., Burke, B. F., Reifenstein, E. C., Wilson, T. L., and Mezger, P. G.: 1968, *Nature* **220**, 1210.
Lin, C. C. and Shu, F. H.: 1964, *Astrophys. J.* **140**, 646.
Lin, C. C. and Shu, F. H.: 1966, *Proc. Nat. Acad. Sci. U.S.A.* **55**, 229.
Oort, J. H.: 1965, *Trans. IAU* **12A**, 789.
Roberts, W. W.: 1969, *Astrophys. J.* **158**, 123.
Spitzer, L.: 1968, *Stars and Stellar Systems* **7**, 1.
Westerhout, G.: 1958, *Bull. Astron. Inst. Netherl.* **14**, 215, 261.

# 76. DEVIATION OF THE VERTEX OF THE VELOCITY ELLIPSE OF YOUNG STARS AND ITS CONNECTION WITH SPIRAL STRUCTURE

SIR R. WOOLLEY

*Royal Greenwich Observatory, Hailsham, Sussex, England*

**Abstract.** The deviation of the vertex of the velocity ellipse of stars of spectral class A is shown clearly in a diagram published by Strömberg (1946); no such clear deviation is shown by the velocity ellipses of stars of other spectral types in his diagrams. This deviation of the vertex or *tilt* of the velocity ellipse can have some connexion with spiral structure, and there seem to be two possibilities, (a) that the tilt is confined to young stars only, and has something to do with an initial condition of their formation; and (b) that the tilt is a manifestation of the nature of the attracting field in the neighbourhood of the sun, which is not central on account of the attractions of spiral arms.

The present note examines data taken from the velocities of nearby stars, and uses the classification of the late type main sequence stars put forward by Wilson, based on observations of the reversals in the Ca$^+$ H and K lines.

From this material it is concluded that the vertex deviation *is* confined to young stars, and that hypothesis (a) rather than hypothesis (b) is correct – or at least strongly indicated by available observations. The paper goes on to show that the deviation of the vertex can be explained by supposing that the stars were formed comparatively recently in a thin strip of the galaxy more or less at right angles to the direction of the galactic centre – in fact, in something like a spiral arm.

**1.** The velocities discussed in this note are constructed from *trigonometrical* parallaxes and proper motions, and from radial velocities. They are taken from Gliese's (1957) catalogue and its revision (Gliese, 1969), and from an extension of Gliese's catalogue which uses trigonometrical parallaxes down to 0″.040 now being compiled at the Royal Greenwich Observatory. The transverse and radial velocities are solved along rectangular ($U$, $V$, $W$) axes of which $U$ is directed towards the centre of the galaxy. $V$ is in the galactic plane in the direction in which rotation takes place and $W$ is perpendicular to the galactic plane and positive northwards. These velocities have been referred to an adopted standard of rest which has the velocities

$$U = -10 \text{ km s}^{-1}$$
$$V = -10 \text{ km s}^{-1}$$
$$W = - 7 \text{ km s}^{-1}$$

relative to the sun.

**2.** The vertex deviation may be examined quite readily by inspection of a simple plot of the $U$ and $V$. Three figures are shown. Figure 1 shows all the A stars in Gliese's catalogue (filled circles) and all the very young late type main sequence stars according to Wilson's criterion (open circles). (These are stars to which he assigns the Ca$^+$ emission index $+3$. A paper discussing this criterion has been prepared by Wilson and Woolley.) The figures representing old stars are left separate to avoid crowding and Figure 2 shows the velocities of stars to which Wilson assigns zero (crosses) or negative

*Becker and Contopoulos (eds.), The Spiral Structure of Our Galaxy,* 423–432. *All Rights Reserved.*
Copyright © 1970 *by the I.A.U.*

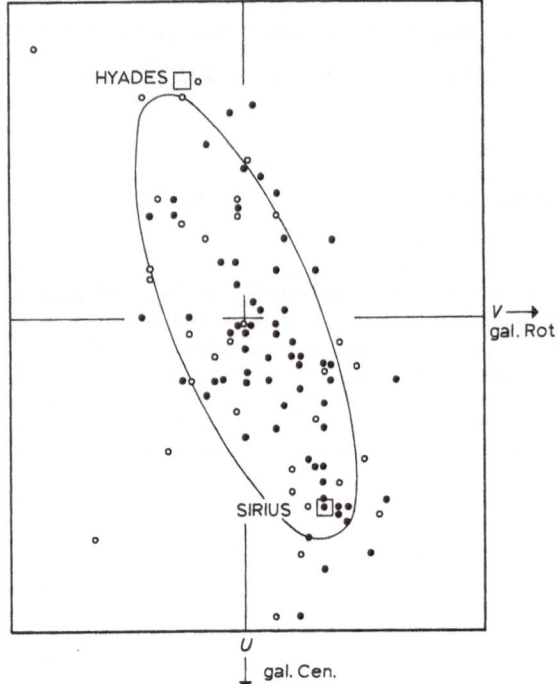

Fig. 1.   Velocities of A-type stars in the solar neighbourhood (●) and of late type main sequence
stars to which Wilson assigns Ca⁺ emission intensities of $+3$ or greater (○).

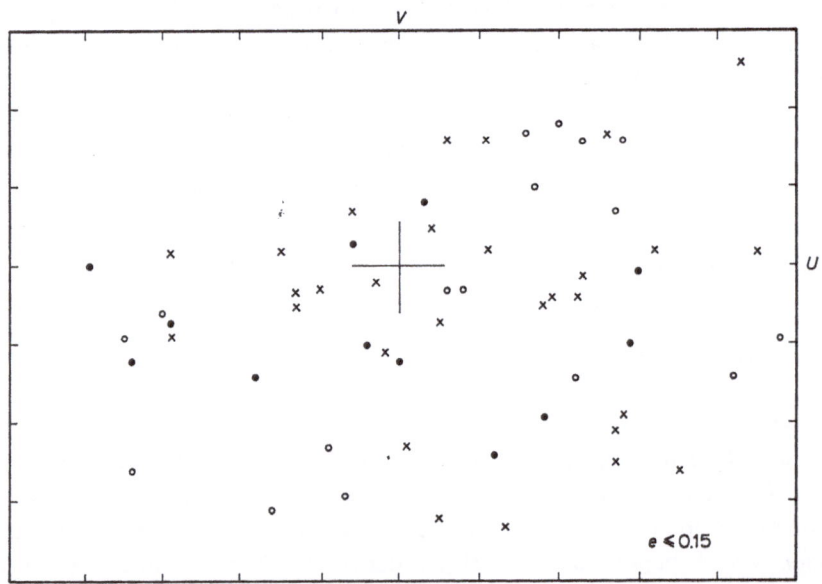

Fig. 2.   Velocities of late type main sequence stars with Ca⁺ emission zero or negative: Wilson
No. zero (×), Wilson No. $-1$, (○), Wilson No. $-2$, $-3$, $-4$ (●).

calcium emission (open circles). Figure 3 shows the corresponding diagram for evolved stars in Gliese's catalogue, that is stars of MK luminosity classes II, III and IV and of spectral type later than F5, the different symbols representing different spectral types in this diagram.

In Figure 1 the vertex deviation is obvious and in both Figures 2 and 3 show little sign, if any, of it.

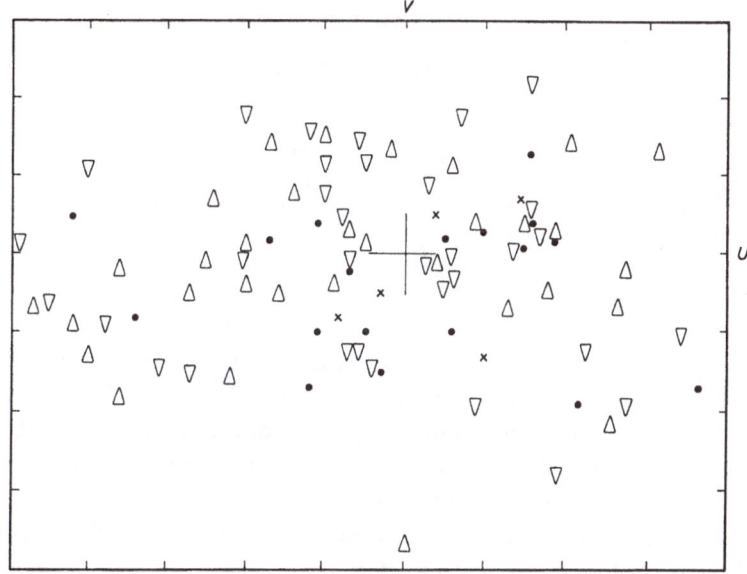

Fig. 3. Velocities of late type evolved stars: F ($\bullet$), G ($\triangle$), K ($\triangledown$), M ($\times$).

A numerical analysis can be made as follows. The quadratic $AU^2 + 2FUV + BV^2$ can be referred to its principal axes so that it becomes $\mathfrak{A}u^2 + \mathfrak{B}v^2$ by the substitution

$$u = lU + mV \qquad v = -mU + lV$$

where $l = \cos\theta$ and $m = \sin\theta$ and

$$\tan 2\theta = -2F/A - B.$$

Now if the velocity function is elliptic, i.e.

$$N(U, V) = \text{const} \times \exp\{-(AU^2 + 2FUV + BV^2)\},$$

the ratios of the mean values of $U^2$, $UV$ and $V^2$ are given by

$$\overline{U^2} : \overline{UV} : \overline{V^2} :: B : -F : A,$$

so that the deviation of the vertex $\theta$ can be found from

$$\tfrac{1}{2}\tan 2\theta = \frac{\sum UV}{\sum U^2 - \sum V^2}.$$

Once $\theta$ has been found we can compute $l = \cos\theta$ and $m = \sin\theta$ and then find the ratio of the principal axes from

$$\mathfrak{A}/\mathfrak{B} = \frac{(Al^2 - Bm^2)}{(Bl^2 - Am^2)} = \frac{(l^2 \sum U^2 - m^2 \sum V^2)}{(l^2 \sum V^2 - m^2 \sum U^2)}.$$

Applied to the stars selected for illustration the results are given in Table I.

TABLE I

| Class of stars | No. of stars | Deviation | Ratio of principal axes |
|---|---|---|---|
| | | $\theta$ | $(\mathfrak{A}/\mathfrak{B})^{1/2}$ |
| Sp. Type A | 66 | $+22°.3$ | 3.25 |
| Ca$^+$ emission $+3$ | 33 | $+19°.2$ | 2.85 |
| Ca$^+$ emission zero and minus | 65 | $+\ 3.0$ | 1.41 |
| Late type evolved stars | 69 | $+\ 3.1$ | 1.82 |

In all cases a ceiling has been imposed, namely only those stars for which the eccentricity of the galactic orbit does not exceed $e = 0.15$ have been kept. This has been done partly to avoid loading the products $\sum U^2$, $\sum UV$ and $\sum V^2$ with very high entries and partly to avoid confusing changes in deviation due to stellar age with changes due to the eccentricity of the orbit – since the young stars are all stars with low eccentricity. A breakdown of the data for the old stars confirms the result that the low tilt is *not* a matter of eccentricity. Notice that the ratio of the principal axes is quite different for the young and old stars.

3. The initial conditions which give rise to the appearance of a particular velocity ellipse in the neighbourhood of the sun can be investigated with the help of Bok's (1934) Equations (1). They are of course very similar to equations used by Lindblad.

We consider motion in two dimensions only (that is, in the plane of the galaxy).

If $(x, y)$ are rectangular coordinates in fixed axis in the galaxy $(R + \xi, \eta)$ are rectangular coordinates in axes rotating with a constant angular speed as in the galaxy; then at an instant when the $(R + \xi)$ axis makes an angle $\theta$ with the $x$-axis

$$
\begin{aligned}
x &= (R + \xi)\cos\theta - \eta\sin\theta & y &= (R + \xi)\sin\theta + \eta\cos\theta \\
\dot{x} &= (\dot{\xi} - \eta\omega)\cos\theta & \dot{y} &= (\dot{\eta} + (R + \xi)\omega)\cos\theta \\
&\quad - (\dot{\eta} + (\xi + R)\omega)\sin\theta & &\quad + (\dot{\xi} - \eta\omega)\sin\theta \\
\ddot{x} &= (\ddot{\xi} - 2\dot{\eta}\omega - (\xi + R)\omega^2)\cos\theta & \ddot{y} &= (\ddot{\eta} + 2\dot{\xi}\omega - \eta\omega^2)\cos\theta \\
&\quad - (\ddot{\eta} + 2\omega\dot{\xi} - \eta\omega^2)\sin\theta & &\quad + (\ddot{\xi} - 2\eta\omega - (R + \xi)\omega^2)\sin\theta
\end{aligned}
$$

and when $\cos\theta = 1$ and $\sin\theta = 0$ (i.e. in axes rotating with angular velocity $\omega$)

$$
\begin{aligned}
x &= R + \xi & y &= \eta \\
\dot{x} &= \dot{\xi} - \eta\omega & \dot{y} &= \dot{\eta} + (R + \xi)\omega \\
\ddot{x} &= \ddot{\xi} - 2\dot{\eta}\omega - (R + \xi)\omega^2 & \ddot{y} &= \ddot{\eta} + 2\dot{\xi}\omega - \eta\omega^2.
\end{aligned}
$$

If $r$ is the distance from the centre of the galaxy, or $r^2 = x^2 + y^2 = (R + \xi)^2 + \eta^2$, the forces $F(x)$ and $F(y)$ directed to the centre of the galaxy are related to the force $F(r)$ by

$$F(x) = \frac{R + \xi}{x} F(r), \quad F(y) = \frac{\eta}{r} F(r)$$

and if we approximate to the galactic attraction by setting

$$F(r) = -\alpha r^{-2} - \beta r^{-3},$$

then, ignoring quantities of the second order,

$$F(x) = -\alpha R^{-2} - \beta R^{-3} + \xi(2\alpha R^{-3} + 3\beta R^{-4})$$
$$F(y) = -\eta(\alpha R^{-3} + \beta R^{-4})$$

Now

$$\ddot{x} = F(x) \quad \text{and} \quad \ddot{y} = F(y).$$

Choose $\omega$ so that

$$R\omega^2 = \alpha R^{-2} + \beta R^{-3}. \tag{1}$$

Then

$$\ddot{\xi} - 2\dot{\eta}\omega - \xi(3\omega^2 + \beta R^{-4}) = 0$$
$$\ddot{\eta} + 2\dot{\xi}\omega = 0. \tag{2}$$

These are Bok's equations for the case $F(r) = \alpha r^{-2} + \beta r^{-3}$. These equations admit of a solution, namely

$$\xi = A \cos(pt + \gamma) + \xi_0, \quad \eta = -B \sin(pt + \gamma) + \lambda t + \eta_0 \tag{3}$$

if

$$-p^2 A + 2\omega p B - A(3\omega^2 + \beta R^{-4}) = 0 \tag{4}$$

$$2\omega\lambda + \xi_0(3\omega^2 + \beta R^{-4}) = 0 \tag{5}$$

$$pB - 2\omega A = 0. \tag{6}$$

Eliminating $A/B$ from (4) and (6) we have

$$p^2 = \omega^2 - \beta R^{-4}. \tag{7}$$

Given $R$, $\omega$ is fixed by (1) so that $p$ is fixed by (7). Since $A/B$ is fixed by (6) and $\lambda/\xi_0$ by (5), we are left with four disposable constants

$$A, \gamma, \xi_0, \eta_0$$

to be determined by the initial conditions, i.e. the values of $\xi$, $\eta$, $\dot{\xi}$, $\dot{\eta}$ at $t=0$.

4. Consider a set of stars at $\xi=0$ and $\eta=0$ at $t=0$ (i.e. the stars now in the solar neighbourhood) and consider a set of these whose velocities lie on an ellipse (centred on the circular velocity). If $\xi=0$ and $\eta=0$ at $t=0$,

$$0 = A \cos\gamma + \xi_0, \quad 0 = -B \sin\gamma + \eta_0,$$
$$\dot{\xi}_0 = -pA \sin\gamma, \quad \dot{\eta}_0 = -pB \cos\gamma + \lambda.$$

Hence

$$\dot{\xi}_0 = -\frac{p^2}{2\omega}\,\eta_0\,,$$

$$\dot{\eta}_0 = \xi_0\left(\frac{pB}{A} + \frac{\lambda}{\xi_0}\right),$$

using Equations (5) and (7). Hence if the velocities $\dot{\xi}_0$ and $\dot{\eta}_0$ lie on an ellipse so do $\xi_0$ and $\eta_0$ (which are in fact the coordinates of the epicentre relative to the local standard of rest). Let this ellipse be

$$a\xi_0^2 + 2f\,\xi_0\eta_0 + b\eta_0^2 = c^2\,.\tag{8}$$

But

$$\xi = \xi_0 + A\cos(pt + \gamma) = \xi_0(1 - \cos pt) - \frac{A}{B}\,\eta_0\sin pt\,,$$

$$\eta = \eta_0 + \lambda t - B\sin(pt + \gamma) = \eta_0(1 - \cos pt) + \xi_0\left(\frac{\lambda t}{\xi_0} + \frac{B}{A}\sin pt\right),$$

so that $\xi, \eta$ are given by the relations

$$\xi = h\xi_0 + k\eta_0\,,$$
$$\eta = l\xi_0 + h\eta_0\,,$$

where at any given time $h$, $l$ and $k$ are constants given by

$$h = 1 - \cos pt\,,\quad k = -\frac{p}{2\omega}\sin pt\,,\quad l = \frac{\lambda}{\xi_0}t + \frac{2\omega}{p}\sin pt\,.$$

Solving for $\xi_0$, $\eta_0$ in terms of $\xi$ and $\eta$, we obtain

$$\xi_0 = (h\xi - k\eta)/(h^2 - kl)$$
$$\eta_0 = (h\eta - l\xi)/(h^2 - kl)\,.$$

Accordingly, if $\xi_0$, $\eta_0$ satisfy (8), $\xi$, $\eta$ satisfy

$$a(h\xi - k\eta)^2 + 2f(h\xi - k\eta)(h\eta - l\xi) + b(h\eta - l\xi)^2 = c^2(h^2 - kl)^2$$

which can be written as

$$\mathfrak{A}\xi^2 + 2\mathfrak{F}\xi\eta + \mathfrak{B}\eta^2 = c^2(h^2 - kl)^2\,,\tag{9}$$

where

$$\mathfrak{A} = ah^2 - 2fhl + bl^2\,,\quad \mathfrak{B} = ak^2 - 2fkh + bh^2\,,$$
$$2\mathfrak{F} = -2ahk + 2f(h^2 + kl) - 2bhl\,.$$

Accordingly $(\xi, \eta)$ lies on an ellipse also centred on the local standard or rest and given by (9). It also follows that every star whose velocity at $t=0$ lies within the original velocity ellipse, so that its epicentre $(\xi_0, \eta_0)$ lies within (8), lies inside the ellipse (9) at any time $t$. (It lies on an ellipse similar to (9) but with a smaller value of $c^2$.)

**5.** It is convenient to work in units in which $R=1$ and $\omega=1$. These imply $\alpha+\beta=1$. In these units

$$p^2 = \alpha, \quad A/B = \sqrt{\alpha/2}, \quad 12\,\lambda = -\xi_0(3+\beta).$$

In other investigations currently being pursued at the Royal Greenwich Observatory we use $\alpha = 1.76245$. With this value of $\alpha$,

$$
\begin{aligned}
p &= 1.327\,57, \\
B/A &= 1.50651, \\
\lambda &= -1.11878\,\xi_0, \\
\dot\xi_0 &= -1.88122\,\eta_0, \\
\dot\eta_0 &= +0.88122\,\xi_0.
\end{aligned}
$$

The ellipse which contains most of the velocities of the A-type stars and young K stars (with strong $Ca^+$ reversals) is

$$2.297U^2 + 6.320UV + 8.704V^2 = 900 \quad (\text{km s}^{-1})^2 \tag{10}$$

with the convention that $U$ is directed away from the centre of the galaxy, i.e. $U=+\dot\xi$; and $V$ is positive in the direction in which rotation takes place. Then

$$8.704\xi_0^2 - 6.320\xi_0\eta_0 + 2.296\eta_0^2 = 0.018544 \tag{11}$$

in the figure describing the limiting epicentres. At $t=-3$ and $t=-4$ we have the following limiting ellipse:

|  | $t=-3$ | $t=-4$ |
|---|---|---|
| $\mathfrak{A}$ | 117.2 | 92.6 |
| $2\mathfrak{F}$ | $-23.46$ | $+11.32$ |
| $\mathfrak{B}$ | 3.296 | 1.54 |
| $c^2/(h^2-kl)^2$ | 0.4625 | 0.2055 |
| Tilt | $-5°.8$ | $+3°.5$ |
| Ratio of principal axes | 7.49 | 7.76 |
| Length of semi-axis major | 0.47 | 0.41 |

At $pt=-2\pi$, or $t=-4.74$, the ellipse degenerates to a straight line along the $\eta$-axis whose half length is 0.244 units.

**6.** For ellipses as large as this it is not really accurate to ignore the square of the eccentricity, which is implied in using the Bok equations; but the errors involved can be ascertained in any particular case by running a computer programme. The analytical and approximate solution gives a qualitative idea of the events; and the young stars now seen in the solar neighbourhood must have occupied areas very like the ellipses given by the approximate theory.

To see the significance of these areas, we enquire what are the figures corresponding to those velocities which we do *not* see in the young stars in the solar neighbourhood, and we examine the velocity ellipse.

$$2.296U^2 + 6.320UV + 8.704V^2 = 900 \quad (\text{km s}^{-1})^2, \tag{12}$$

that is an ellipse similar to (10) but tilted in the opposite sense.

The ellipse for $t = -3$ corresponding to (12) is

$$22.85\xi^2 - 16.38\xi\eta + 13.27\eta^2 = 0.4625 \tag{13}$$

and it is shown in Figure 4 plotted together with the ellipse

$$117.2\xi^2 - 23.46\xi\eta + 3.30\eta^2 = 0.4625 \tag{14}$$

which corresponds to (10). The shaded area is *forbidden* in the sense that if any stars had been found at $t = -3$ in this area, and had got to the solar neighbourhood at $t = 0$, then their velocities would have been outside (10) (but inside (13) instead). The semi-axis minor of (14) is 0.063 units or 630 parsec. This means that the star forming arm had a width of 1260 parsec, which seems reasonable.

A direct computation was made of the positions of each of the stars shown in Figure 1, at $t = -4$, assuming only the force field

$$F(R) = -1.76245R^{-2} + 0.76245R^{-3}$$

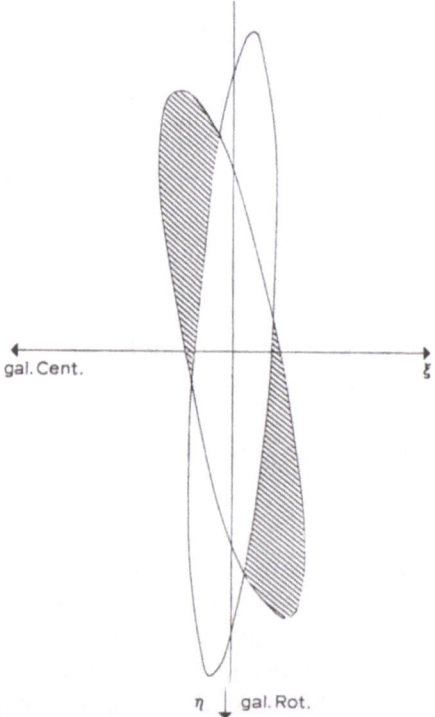

Fig. 4.   Space distribution of stars, at $t = -4$, whose present position is the solar neighbourhood and whose velocities lie inside ellipses tilted at $+22°$ and $-22°$ (shaded).

and the points are shown in Figure 5, in axes fixed in the galaxy at $t = -4$ and the origin being the position of the present centre of rest carried back to that time – i.e. the point occupied at $t = -4$ by an object now in the solar neighbourhood and having the circular velocity, or $U = 0$ and $V = 0$. This figure shows the points occupied in space, at that time, of objects which are all *now* in the solar neighbourhood, but are distributed in *velocity* space as we see in Figure 1.

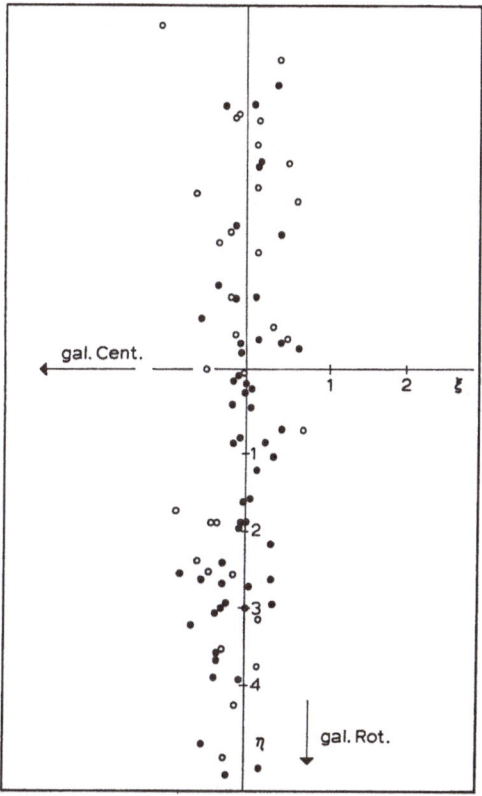

Fig. 5. Actual points in space, at $t = -4$, of the stars now in the solar neighbourhood having velocities (relative to the circular velocity) shown in Figure 1.

7. Figure 5 suggests that the young Gliese stars were formed in a region resembling a spiral arm at $t = -3$ or $t = -4$, or that is to say between 1.0 and $1.3 \times 10^8$ years ago. We may test this by enquiring how the stars compare with those in the Pleiades and Hyades clusters, whose ages Sandage (1957) placed at $2 \times 10^7$ and $4 \times 10^8$ years respectively.

This could be readily examined if we were able to draw a satisfactory HR diagram for the Gliese A stars. Unfortunately the only means of doing this is by constructing absolute magnitudes of the Gliese stars from their trigonometrical parallaxes, some of which are poorly determined, and it seems better to refer to spectral types, of which the counts are given in Table II

## TABLE II
### Number of stars of a given spectral type

| Sp. | Pl. | Gl. | Hy | Sp | Pl. | Gl. | Hy | Sp | Pl. | Gl. | Hy |
|-----|-----|-----|----|----|-----|-----|----|----|-----|-----|----|
| B5 | – | 1 | – | A0 | 1 | 13 | – | A5 | 2 | 4 | 3 |
| B6 | 3 | – | – | A1 | 4 | 2 | – | A6 | 2 | 1 | 3 |
| B7 | 3 | 1 | – | A2 | 1 | 12 | 1 | A7 | 2 | 10 | 3 |
| B8 | 5 | 1 | – | A3 | 4 | 8 | 0 | A8 | 1 | 0 | 3 |
| B9 | 3 | 1 | – | A4 | 2 | 8 | 0 | A9 | ? | 2 | 4 |

Pl. = Pleiades;   Gl. = Gliese;   Hy = Hyades.

These figures indicate strongly that the Gliese stars are much younger than the Hyades, and support the age of 1 to $1.5 \times 10^8$ years for the majority of the Gliese stars.

## Acknowledgement

I am particularly indebted to Miss Rosalind Johnston, who put the programme described through the computer, and who has read and discussed with me the manuscript of this paper.

## References

Bok, B. J.: 1934, *Harvard Circular*, No. 384.
Gliese, A.: 1957, *Astron. Rechen-Inst. Heidelberg, Mitt. A*, No. 8.
Gliese, A.: 1969, *Veröff. Astron. Rechen-Inst. Heidelberg*, No. 22.
Sandage, A.: 1957, in *Proc. Vatican Conference on Stellar Populations*, p. 41.
Strömberg, G.: 1946, *Astrophys. J.* **104**, 12.

# 77. POSSIBLE INFLUENCE OF THE GALACTIC SPIRAL STRUCTURE ON THE LOCAL DISTRIBUTIONS OF STELLAR RESIDUAL VELOCITIES

M. MAYOR

*Observatoire de Genève, Genève, Switzerland*

The residual velocity distribution, observed in the solar neighbourhood, exhibits several peculiarities for the stars of low epicyclic energy: vertex deviation, two-stream distribution, etc.

There are mainly two interpretations of these observational facts: direct influence of initial conditions or irregularities of the gravitational potential. Samples of stars (such as the stars closer than 20 pc from the sun, or samples of giant stars) having practically all performed several galactic revolutions do present the aforementioned peculiarities for low epicyclic energy. It seems hard to admit that the initial conditions of these 'old' stars could produce such peculiarities.

We have worked out an explanation of these distribution anomalies as a result of the self-sustained density waves considered by Lin and Shu (1964). We adopted, for the parameters of spiral structure, the values given by Lin *et al.* (1969), viz.

$$\Omega_{\text{p}} \approx 12 \text{ km s}^{-1} \text{ kpc}^{-1}; \quad \frac{\text{spiral field}}{\text{mean field}} \approx 5\%; \quad i \approx 6°, \tag{1}$$

where $\Omega_{\text{p}}$ is the pattern angular velocity and $i$ the pitch angle, at the sun's position, of the trailing arm.

The linearized theory of a stellar population's response to the density wave has led to a satisfactory qualitative agreement with the observations about the vertex deviation and its dependence on the velocity dispersion.

However, a detailed calculation of the residual velocity distribution function, perturbed by a spiral field, is out of the range of validity of the linear theory if the stellar populations have small velocity dispersions. In order to compare the calculated and observed loci of equal density in the plane $(c_{\varpi}, c_{\theta})$ of residual velocity components, we have obtained the non linear response to the spiral potential.

The Liouville equation for the relative perturbation $\psi$ due to the spiral potential reads symbolically

$$L_0 \{\psi\} = L_1 \{Q_0\} + [\psi L_1 \{Q_0\} - L_1 \{\psi\}] \tag{2}$$

where $\Psi = \Psi_0 (1 + \psi)$ is the perturbed distribution function and $\Psi_0$ the equilibrium distribution, and $Q_0 = -\ln \Psi_0$. $L_0$ and $L_1$ are differential operators defined by Lin (1967).

The complete Equation (2), including non linear perturbation terms within the brackets, has been solved by an iterative process.

The response of various stellar populations moving across the density wave defined by (1) has been calculated. Locating the sun on the inner edge of a spiral arm, it is possible to reach a satisfactory agreement with the observed distributions; the computed distributions allow for a correct vertex deviation in sign as well as in magnitude; they also reveal a two stream structure.

It therefore seems possible to explain the 'anomalies' of residual velocity distribution for low epicyclic energy in terms of a spiral density wave.

A paper containing the above results has been submitted for publication in *Astronomy and Astrophysics*.

## Acknowledgements

I wish to express my gratitude to Prof. P. Bouvier and Dr. L. Martinet for many stimulating and useful discussions during this study.

## References

Lin, C. C.: 1967, in *Relativity Theory and Astrophysics*, 2 *Galactic Structure* Amer. Math. Soc., Providence, p. 66.
Lin, C. C. and Shu, F. H.: 1964, *Astrophys. J.* **140**, 646.
Lin, C. C., Yuan, C., and Shu, F. H.: 1969, *Astrophys. J.* **155**, 721.

# 78. ADDITIONAL EVIDENCE FOR THE CATACLYSMIC ORIGIN OF SPIRAL STRUCTURE IN GALAXIES

K. RUDNICKI

*Warsaw University Observatory, Warsaw, Poland*

**Abstract.** The photograph of the galaxy NGC 3486 shows its spiral structure in statu nascendi. The evident explanation is that of a cataclysmic origin of the visible spiral arms.

The photograph of the galaxy NGC 3486 (right ascension $= 10^h57^m6$, declination $= +29°15$ according to Zwicky and Herzog (1963) taken with the 125 cm Palomar Schmidt Telescope is presented (Figure 1). Spiral arms of this galaxy have plenty of splits which form a secondary spiral structure. Split spiral arms in galaxies are well known. Many of them are shown in Arp's Atlas 1966 (e.g. Nos: 7, 9, 25, 26, 29, 41, 108, 113, 114). In NGC 3486, however, the secondary arms form in many cases almost right angles with the primary arms at the connection points. Thus any explanation of the origin of such spiral structure by a gravitational instability or by an electromagnetic field is rather unlikely. The secondary arms seem to be jets curved subsequently due to the rotation. Since no basic difference is visible in appearance of both the primary and the secondary spiral arms, this can serve as an additional argument in favour of a cataclysmic origin of the spiral structure of galaxies in general.

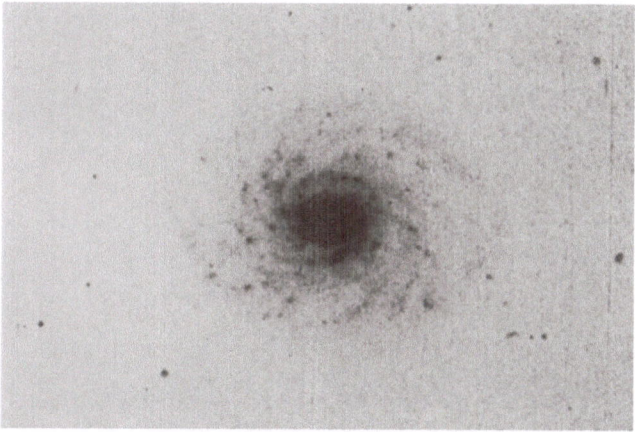

Fig. 1. A photograph of NGC 3486 obtained with the 125 cm Palomar Schmidt telescope. Emulsion Kodak 103a-0, without filter. Date 1962, April 1, $6^h25^m$ UT; exposure time 9 min. North is at the top east to the left. Scale: 6.5 mm = 1 min of arc.

Even if we interpret the complicated structure of NGC 3486 as a superposition of two spiral structures of independent origins, the cataclysmic explanation of them seems again the most probable one.

# References

Arp, H.: 1966, *Astrophys. J. Suppl. Ser.* **14**, 1.
Zwicky, F. and Herzog, E.: 1963, *Catalogue of Galaxies and Clusters of Galaxies*, Vol. II, California Institute of Technology.

# B. MAGNETIC EFFECTS

# 79. MAGNETIC FIELDS AND SPIRAL STRUCTURE

L. WOLTJER

*Dept. of Astronomy, Columbia University, New York, U.S.A.*

The most valuable new information on galactic magnetic fields is derived from the study of Faraday effects in extragalactic radio sources and pulsars. The rotation of the plane of polarization of an electromagnetic wave with a wavelength of 1 m – the so called Rotation Measure RM – which traverses a medium with electron density $n_e$ (cm$^{-3}$) and magnetic field intensity **H** (in $\mu$G) is given by

$$\text{RM} = 0.81 \int n_e \mathbf{H} \cdot \mathbf{dl}, \tag{1}$$

where **dl** is the element of length along the line of sight expressed in pc. In the case of the pulsars, the delayed arrival of the pulses at the longer wavelengths immediately yields the integrated electron density along the line of sight, that is

$$\int n_e \, dl. \tag{2}$$

If we make the *assumption* that the fluctuations in **H** and $n_e$ are uncorrelated the average field strength along the line of sight can be obtained by comparing expressions (1) and (2). Results so far obtained range between 1 and 3 $\mu$ Gauss for the mean field parallel to the line of sight. From an analysis of the distribution of pulsars we conclude that the mean interstellar electron density is about 0.05 cm$^{-3}$ in a layer which extends to no less than 200 pc from the galactic plane. In such a layer a mean line of sight field of 2 $\mu$ Gauss also suffices to account for the observed Faraday rotations in intermediate latitude extragalactic objects. On the basis of these data we provisionally conclude that the mean value of the large scale ($L > 100$ pc) magnetic field near the sun amounts to 3 or 4 $\mu$ Gauss. Small scale random fields do not contribute much to the Faraday effects and could be stronger. The intensity of the synchrotron radiation from the galactic disk indicates that such fields may exist. The strength of the systematic field derived here is compatible with the Zeeman measurements reported by Dr. Verschuur, which indicate that stronger fields may occur in dense clouds.

The energy density associated with a field of 3.5 $\mu$ Gauss is $5 \times 10^{-13}$ erg cm$^{-3}$. For comparison a flow of interstellar gas with a hydrogen density of 0.5 cm$^{-3}$ would have the same energy density at a velocity of 10 km s$^{-1}$ and this indicates the order of magnitude of the motions that may be substantially influenced by the magnetic field. The systematic deviations from the circular velocity caused by this field are much smaller ($\simeq 0.1$ km s$^{-1}$) because a small change ($\varDelta\theta$) in the circular velocity $\theta$ leads to a large change in kinetic energy ($\varDelta E \approx \rho\theta\varDelta\theta$). The low energy density of the large scale field and its inability to affect the circular motions of the gas make it virtu-

*Becker and Contopoulos (eds.), The Spiral Structure of Our Galaxy, 439–440. All Rights Reserved.*
*Copyright © 1970 by the I.A.U.*

ally certain that the magnetic field cannot be the main cause of spiral structure. Of course this does not mean that it is also negligible when we consider the detailed structure of a spiral arm.

The structure of the field is still uncertain; in the future it may well give valuable information on the motions in the interstellar gas. Matthewson has shown that the interstellar polarization data (radio and optical) may be interpreted on the basis of a rather tightly wound helical field in the general neighborhood of the sun. The axis of the field would be directed towards $l = 90°$. It is noteworthy that such a field would show no obvious relation to the local hydrogen structure. The origin of a helical component also poses problems; perhaps one could think of a torsional oscillation in a uniform unidirectional field, but the amplitude would have to be large to make a tight helix.

On a larger scale one would expect the field to follow the fluid motion. Hence in a density wave picture of spiral structure the magnetic field should be strongest in the arms, where the density is highest, and the field should be oriented more or less parallel to the arm. The winding-up problem is not affected because the field is tied to the fluid and not to the density wave as such. If the field had been originally uniform in the galactic disk and if it has been steadily wound up for $10^{10}$ years the thickness of each winding near the sun amounts to only 100 pc; from one winding to the next the field would change sign. The presently available Faraday rotation data do not support such a picture, although the low latitude data are too few for a definite conclusion. One simple way out is to have a basically toroidal field in which each field line is more or less circular. But it is difficult to judge the likelihood of such a field in the absence of a theory for the origin of magnetic fields in galaxies.

# 80. ON THE CIRCULATION OF GAS NEAR
# THE GALACTIC CENTER

E. A. SPIEGEL

*Columbia University, New York, New York, U.S.A.*

**Abstract.** A model of a quasi-steady circulation of gas near the galactic center is considered to explain the outward motion of the 3-kpc arm. A hydromagnetic wind from the galactic nucleus reaches the 3-kpc arm, where a shock is formed; then the gas moves out of the plane and eventually returns to the nucleus. The secular behavior of the model is discussed.

In a discussion of the spiral structure of our galaxy it is natural to inquire briefly into the origin of the striking outward motion of the 3-kpc arm. The first question that one might ask is whether the observed motion represents an outward swing of the gas which has a large oscillating radial component of velocity. In this case, if the 3-kpc arm represents the compressional phase of a wave motion, we might expect that at some galactic longitudes, gas should be falling toward the galactic nucleus. Certainly, this possibility is difficult to rule out observationally, but the situation should be highly dissipative since the flow must have a large component normal to the wave, and it is difficult to imagine what might maintain such large radial motions. Nevertheless, this is a possibility that remains open.

The other possible source of the radial motions that has been widely considered is a flow from the galactic center. Here again, opinion is divided on the time dependence of the flow. One school of thought postulates an explosion in the galactic center as the cause of the outflow, while the other possibility that has been considered is a quasi-steady wind from the galactic center. Both points of view have their difficulties, and I shall not attempt to summarize them here. Certainly the two possibilities need further discussion, and definite models are required for the purpose. Here, I wish to outline a possible model for a steady flux of gas from the galactic center.

The simplest possible case is an axisymmetric, steady flow, of an isothermal, nonturbulent gas, with no magnetic fields. It is then straightforward to treat a disk-like outflow of this kind from the galactic center (Moore and Spiegel, 1968) if it is further assumed that the gravitational field is specified. The discussion is much like that of the solar wind and the solution which is most compatible with observations accelerates through a sonic point, producing large radial motion. This radial motion is arrested by a rather complicated shock (or compressible hydraulic jump) which ionizes the gas and lifts much of it from the galactic plane. If it may be supposed that the gas that leaves the plane rains back into the galactic center the difficulty of a source of mass for the flow is alleviated. It is also tempting to associate the shock with the 3-kpc arm.

The model described has two serious deficiencies. First, it does not provide a cause for the outflow so that the flow is effectively taken as an initial condition. Second, it leads to angular-momentum conservation, hence to $v \propto r^{-1}$, where $v$ is the azimuthal velocity component and $r$ is the distance from the galactic center. This result is not compatible with the observed rotation law. To remove these deficiencies we might try to introduce a nonradial force, and a magnetic field seems to be suitable for this purpose (Moore and Spiegel, 1966).

Mestel, Moore, and I have recently re-examined the picture of a hydromagnetic wind from the galactic center, adding to the above-listed assumptions that the gas is cold. We have also adopted the artifice of a cylindrical geometry. In that case the governing equations can be solved completely as is known from numerous works on stellar wind theory and galactic hydromagnetics (Woltjer, 1965). In this model we have not included a gravitational $z$-force and we have assumed for simplicity a monotonic variation of gravitational potential with distance from the axis of rotation. We further assume the existence of a central gaseous object which is relatively massive (say $10^8 M_\odot$) and fairly compact (tens of parsec). The model is encouraging in two respects.

First, there is the conclusion that gas just at the edge of the central object, rotating with it, but with no radial velocity, is accelerated outward. This outward acceleration arises from the tendency of a poloidal field to cause the gas to corotate with the central body. Thus, the gas will rotate more rapidly than the circular velocity so that its centrifugal force surpasses the gravitational force. We conclude then that a moderately strong poloidal field near the galactic center ($\lesssim 10^{-4}$ G) drives a moderately strong wind from the galactic center. Unfortunately, there are several parameters in the problem (the angular velocity of the central object, the magnetic flux, the mass flux, and the energy of the flow) in addition to those of the gravitational field, so that many solutions are possible. The type that seems preferable gives a continued radial acceleration which, at a certain distance outward, becomes infinite, and we anticipate a shock wave, much as in the nonmagnetic case. This shock would inhibit outflow beyond it, so that the flow is largely decoupled from that in the outer galaxy. In a disk-like flow of this kind, we believe that material could again return to the central object, and that the magnetic field lines would also reconnect in this way.

The second interesting consequence of the model is that the rotation curve of the gas resembles the observed one near the galactic center, irrespective of the details of the mass distribution. For the solution we have outlined, with increasing $r$, the rotational velocity rises to a maximum, drops to a relative minimum, and then rises again for a short distance, where it then reaches the singular acceleration just mentioned. The locations of the maximum and the minimum depend on the unknown parameters mentioned above, so that in this model, the rotation curve cannot be used to infer the mass distribution in the central regions of the galaxy.

We conclude therefore, that a moderately strong field near the galactic center leads us to expect a flow from the galactic center and such a flow would seem to be a good candidate for the cause of the radial motion of the 3-kpc arm. Indeed, I feel that the shock in the outflow should be associated with the arm. On the other hand, I must stress that the study of such models is far from complete. All the ways we have considered of reconnecting the field lines produce magnetic neutral points. Thus, we have to expect instabilities in the model, perhaps of an explosive kind. Other instabilities must be investigated as well, such as non-axisymmetric gravitational stabilities.

A crucial unsolved problem is the secular behavior of the model. Radiative losses in the shock require an energy source, and this must come ultimately from the gravitational store of the galactic nucleus. If the energy source is the potential of the rotating central gas mass postulated above, then the situation I have described can last only a few galactic rotation periods. But if non-axisymmetric instabilities can permit gravitational coupling between the stars in the galactic center and the gas, the model can be maintained for perhaps a galactic lifetime. Of course, it must ultimately lead to the collapse of the central region with a large increase of the central angular velocity. The shocked region too, with its plasma and magnetic fields, would also spin very rapidly, and this may provide an avenue of the much-discussed dramatic evolution of galactic nuclei with accompanying large luminosities. But this phase of the problem cannot yet be discussed adequately.

## References

Moore, D. W. and Spiegel, E. A.: 1966, in *Proceedings of the Summer School on Interstellar Gas Dynamics* (ed. by D. E. Osterbrock), Madison.
Moore, D, W. and Spiegel, E. A.: 1968, *Astrophys. J.* **154**, 863.
Woltjer, L.: 1965, *Stars and Stellar Systems* **5**, 531.

## Discussion

*Oort:* Referring to your remarks concerning the determination of the gravitational forces in the central part of the galaxy we have of course realized from the beginning that in the region where the large radial motions of the gas occur the gravitational forces could not be derived from the motion of the gas. However, within the rapidly rotating disk of 800 pc radius the rotation may well correspond approximately to the gravitational force. Rougoor and I have indicated that the observed run of the rotational velocity in this disk corresponds roughly to what one would expect if the density distribution in the nuclear region of our Galaxy would be similar to that in the Andromeda nebula (as inferred from the distribution of the light). Just outside the disk the transverse velocity of the gas seems to drop to quite low values.

Observations by Van der Kruit (in press) suggest that gas is ejected from the galactic nucleus under a large angle with the galactic plane. It is conceivable that at the radius where this gas falls into the disk this gives rise to the streams with low angular momentum.

*Schmidt-Kaler:* The dip in the rotation curve of the gas is very well shown in the Andromeda Nebula at 10′ from the centre according to very recent unpublished work of Rubin and Ford. However, a dip at this point appears also in the rotation curve of M31 as derived by Babcock (*Lick Obs. Bull.* (1939), No. 498) from the stars, and these are probably pretty old. How does that fit into your theory?

*Spiegel:* The rotation curve of the stars can of course be used to determine the mass distribution, assuming that the velocity dispersion in the stars is small enough and that nonradial gravitational forces are not large. The relevance of my remarks depends on whether there are appreciable differences between the rotation curves of the gas and the stars. If such differences exist, then I would claim that they are probably due to dynamical pressures or magnetic forces and the theory discussed may be relevant. This would be especially true if a pronounced relative minimum appears in the rotation curve of the gas. As I do not have the data of Rubin and Ford, I am not able to give here the results of such a comparison.

# 81. MAGNETOHYDRODYNAMICAL MODELS OF HELICAL MAGNETIC FIELDS IN SPIRAL ARMS

M. FUJIMOTO

*Nagoya University, Nagoya, Japan*

and

M. MIYAMOTO

*Tokyo Astronomical Observatory, Tokyo, Japan*

**Abstract.** A circular arm with elliptical cross-section is used as a model of the spiral arm. It is demonstrated magnetohydrodynamically that interstellar gas may flow in a helical path along the axis of the arm and interstellar helical magnetic lines of force can spiral around it. It is pointed out that though the rolling motions of gas do not always indicate the existence of the helical magnetic field, the latter cannot be in a stationary state without the former. Some observational supports to the present model are given.

## 1. Motions of Gas in a Spiral Arm

A circular arm is assumed as a model of the spiral arm, whose density $\varrho_s$ is uniform and whose cross-section is elliptical with semi-major axis, $a$, in the galactic plane and semi-minor axis, $b$, perpendicular to it. As shown in Figure 1, a local Cartesian coordinates system $(x, y, z)$ is introduced with the origin at the arm axis which is situated at galactocentric distance $r_0$. This local coordinate system rotates in the direction of the $y$-axis with the angular velocity $\Omega$ of galactic rotation at $r=r_0$. Since $r_0 \gg a, b$, in general, we can restrict our discussions only to a small portion of the galaxy near the arm axis. In deriving equations of motion of gas, we expand the pressure and gravitational forces in Taylor series around the arm axis, and take the first two terms. The first term balances the centrifugal force due to the rotation of the local coordinate

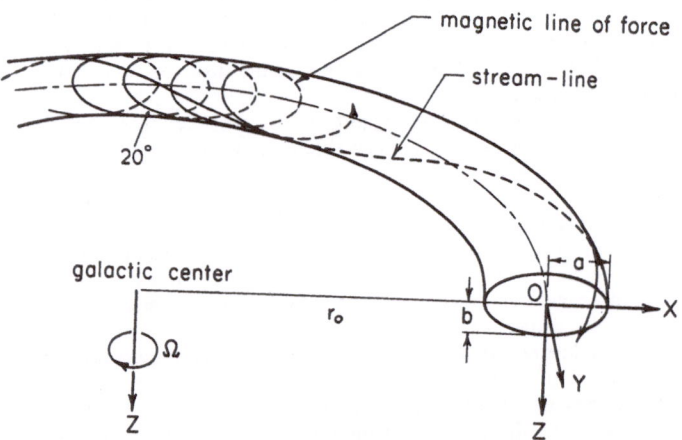

Fig. 1.   Helical magnetic fields and rolling motions of the gas.

*Becker and Contopoulos (eds.), The Spiral Structure of Our Galaxy, 444–447. All Rights Reserved.*
*Copyright © 1970 by the I.A.U.*

system, and the second term is linear in $x$ and $z$. Then the $y$- and time-independent equations of motion are written as follows,

$$U_x \frac{\partial U_x}{\partial x} + U_z \frac{\partial U_x}{\partial z} = 2\Omega U_y - C \cdot x + F_1 ,$$ (1)

$$U_x \frac{\partial U_y}{\partial x} + U_z \frac{\partial U_y}{\partial z} = -2\Omega U_x \quad + F_2 ,$$ (2)

and

$$U_x \frac{\partial U_z}{\partial x} + U_z \frac{\partial U_z}{\partial z} = \quad -D \cdot z + F_3 ,$$ (3)

where $C$ and $D$ are constants evaluated at $r=r_0$ and $F_1$, $F_2$, and $F_3$ are accelerations due to the Lorentz force. It has been shown by Fujimoto and Miyamoto (1969) that the following motion and helical field satisfy Equations (1)–(3),

$$\mathbf{U} = (Lz, Mx, Nx)$$ (4)

$$\mathbf{B} = (B_* Lz, B_0 + B_* Mx, B_* Nx), \quad \text{for} \quad (x^2/a^2) + (z^2/b^2) \leqslant 1 ,$$ (5)

and

$$\mathbf{B} = 0, \quad \text{for} \quad (x^2/a^2) + (z^2/b^2) > 1 ,$$ (6)

where $L$, $M$, $N$ and $B_*$ are constants still to be determined by Equations (1)–(3), and $B_0$ denotes a uniform field parallel to the arm. In an inertial frame, the motion $U_x = Lz$, $U_y = Mx$ and $U_z = Nx$ gives a helical motion along the arm. Note that we have $\mathrm{rot}(\mathbf{U} \times \mathbf{B}) = 0$ and $\mathbf{U} \times \mathbf{B} \neq 0$ for $\mathbf{U}$ and $\mathbf{B}$ in Equations (4)–(6). Substituting Equations (4)–(6) in Equations (1)–(3), we have, in the case of the tightly wound helical fields,

$$L = \pm \frac{a}{b} \sqrt{D} (1 - E_{\mathrm{Mg}}/E_{\mathrm{K}})^{-1/2}$$ (7)

$$M = -2\Omega (1 - E_{\mathrm{Mg}}/E_{\mathrm{K}})^{-1} ,$$ (8)

and

$$N = \mp \frac{b}{a} \sqrt{D} (1 - E_{\mathrm{Mg}}/E_{\mathrm{K}})^{-1/2} ,$$ (9)

with the constraint on the model,

$$D - C = 4\Omega^2 (1 - E_{\mathrm{Mg}}/E_{\mathrm{K}})^{-1} .$$ (10)

which implies that stream lines are closed in the local coordinate system. We have the following identities,

$$\frac{E_{\mathrm{Mg}}}{E_{\mathrm{K}}} = \frac{B_*^2}{4\pi \varrho_s} = \frac{\text{Energy density of the poloidal components of the helical field}}{\text{Energy density of the poloidal components of the rolling motion of the gas}}.$$

## TABLE I

Velocities of rolling motions of gas and helical magnetic field for various values of $E_{Mg}/E_K$

| | $E_{Mg}/E_K$ | | | | |
|---|---|---|---|---|---|
| | 0.0 | 0.1 | 0.3 | 0.5 | 0.7 |
| $U_x$ | $\pm 8.0z$ km s$^{-1}$ | $\pm 10.0z$ | $\pm 15.5z$ | $\pm 25.1z$ | $\pm 46.6z$ |
| $U_y$ | $-5.0z$ | $-5.5x$ | $-7.1x$ | $-10.0x$ | $-16.7x$ |
| $U_z$ | $\mp 0.9x$ | $\mp 1.1x$ | $\mp 1.7x$ | $\mp 2.8x$ | $\mp 5.2x$ |
| $B_x$ | 0 | $\pm 2.1 \times 10^{-6}z$ G | $\pm 5.6 \times 10^{-6}z$ | $\pm 1.2 \times 10^{-5}z$ | $\pm 2.6 \times 10^{-5}z$ |
| $B_y$ | 0 | $-1.1 \times 10^{-6}x$ | $-2.6 \times 10^{-6}x$ | $-4.6 \times 10^{-6}x$ | $-9.1 \times 10^{-6}x$ |
| $B_z$ | 0 | $\mp 2.3 \times 10^{-7}x$ | $\mp 6.2 \times 10^{-7}x$ | $\mp 1.3 \times 10^{-6}x$ | $\mp 2.8 \times 10^{-6}x$ |
| $\sqrt{\langle v_2 \rangle}$ | 9.5 km s$^{-1}$ | 9.5 | 9.1 | 8.1 | 5.2 |

(1) $x$ and $z$ are measured in 100 pc; (2) $\sqrt{\langle v^2 \rangle}$ is the root mean square molecular (or random) velocity of the gas at the arm axis; (3) the constant $B_*$ is taken as positive. When it is negative, the signs in $B_x$, $B_y$ and $B_z$ are to be reversed.

Table I lists numerical solutions for the characteristics of the models of the arm at $r_0 = 10$ kpc in our Galaxy, in which we have assumed $\varrho_s = 2$ hydrogen atoms cm$^{-3}$ and $b/a = 1/3$ with $a = 250$ pc. The uniform component $B_0$ can be estimated from the observed pitch angle of the helical field. They are about $B_0 = 3 \times 10^{-7} \sim 2 \times 10^{-6}$ G, corresponding to $E_{Mg}/E_K = 0.1 \sim 0.7$ in Table I. We find from Table I that the velocities of the rolling motion and the magnitudes of the helical field amount to $10 \sim 20$ km s$^{-1}$ and $10^{-6} \sim 10^{-5}$ G, respectively, near the arm surface.

## 2. Conclusions and Comparisons with Observations

Magnetohydrodynamical treatments have been made on interstellar helical magnetic fields in a spiral arm. It is concluded that interstellar gas can flow in a helical path around the arm axis and that a stationary helical magnetic field can be maintained with such rolling motions. The helical magnetic field as well as the velocity field in our stationary model are demonstrated schematically in Figure 1. Equations (7)–(9) show that rolling motions in the gas do not always indicate the existence of a helical field.

It is possible to compare the present models with observations of the peculiar motions of the gas and with the local magnetic fields in spiral arms in our galaxy. The neutral hydrogen gas in the 4-kpc arm exhibits a rolling motion, as first found by Oort. This same type of motion has been found in the Perseus arm by Westerhout (1970) and Kerr (1970). The velocity difference in the line-of-sight component amounts to $|\delta v_r| \simeq 20$ km s$^{-1}$ above and below the galactic plane in the region $l^{II} = 110° \sim 130°$. Velden (1970) has observed radial velocities of neutral hydrogen gas in intermediaté latitudes ($|b^{II}| = 10° \sim 30°$) toward the direction of the galactic anti-center. The radial velocities take the values $\mp 4.5$ km s$^{-1}$ for $b^{II} = 10° \sim 30°$ and

$b^{II} = -10° \sim -30°$, respectively; this implies the existence of rolling motions, at least, in the solar neighbourhood. The sense of this motion is that of a right-handed helix in both the Orion and the Perseus arms. The velocities of these rolling motions are compatible with those derived in the present models (see Table I).

Mathewson (1968) has combined his polarization data on 1400 stars with those measured in the northern hemisphere, to deduce a magnetic field model for the solar vicinity. His model construction was carried out by trial and error with different configurations until a good approximation was obtained to the observed $E$-vector distribution for polarized starlight. The best model, in which magnetic field lines are tightly wound around the spiral arm, can explain not only the distribution of the $E$-vectors, but also sign-reversals of rotation measures for polarized extragalactic radio sources outside the galactic plane and polarization planes for the non-thermal galactic radio continuum. Since Mathewson has already made extensive comparisons between his helical field models and the observations, and has obtained good fits, we need not discuss these in more detail. We shall, however, point out some differences between these models and those discussed in the present paper.

The helical pattern in the Mathewson's models is sheared through an angle of 40° in the galactic plane, in a counter-clockwise sense when viewed from the north pole, and it is vertical to the galactic plane. In the present models, on the other hand, the helical pattern is not sheared and it is tilted with respect to the galactic plane by an angle of 20° (Figure 1). Another discrepancy between observations and the results of the present model is observed in the positions of sign-reversal for the rotation measures. Actual sign reversals are observed at $l^{II} \sim 40°$ and $\sim 220°$ (Gardner and Davies, 1966; Berge and Seielstad, 1967; Gardner et al., 1967), while the theory presented here predicts directions of $l^{II} \sim 0°$ and $\sim 180°$.

Many other observational characteristics of the local magnetic field can be explained by the present model.

A more extended paper with the same title has been published in *Publ. Astron. Soc. Japan* (Fujimoto and Miyamoto, 1969).

### References

Berge, G. L. and Seielstad, G. A.: 1967, *Astrophys. J.* **148**, 367.
Fujimoto, M. and Miyamoto, M.: 1969, *Publ. Astron. Soc. Japan* **21**, 194.
Gardner, F. F. and Davies, R. D.: 1966, *Australian. J. Phys.* **19**, 129.
Gardner, F. F., Whiteoak, J. B., and Morris, D.: 1967, *Nature* **214**, 371.
Kerr, F. J.: 1970, IAU Symposium No. 38, p. 95.
Mathewson, D. S.: 1968, *Astrophys. J.* **153**, 447.
Velden, L.: 1970, IAU Symposium No. 38, p. 164.
Westerhout, G.: 1970, IAU Symposium No. 38, p. 122.

# 82. LOCAL MAGNETIC FIELD AND ITS RELATIONS TO THE LOCAL SPIRAL ARM STRUCTURE

G. D. VERSCHUUR

*National Radio Astronomy Observatory, Charlottesville, Va., U.S.A.*

No paper was submitted.

# 83. POLARIZATION OF SOUTHERN OB-STARS

G. KLARE and T. NECKEL

*Landessternwarte and Max-Planck-Institut für Astronomie, Heidelberg-Königstuhl, Germany*

The polarization data of 1421 southern OB-stars of the Heidelberg catalogue have been measured and plotted in a galactic $l^{II}$, $b^{II}$-diagram. For some longitude intervals the relationship of the standard deviation of the electric vector alignment and the galactic longitude was computed.

In 1966 a survey of OB-stars in the southern Milky Way has been completed at Heidelberg (Klare and Szeidl, 1966). 1660 objects between $l^{II} = 230°$ and $l^{II} = 20°$ could be identified with the use of 90 objective prism plates of the small Hamburger Schmidt-Spiegel at Boyden. In earlier publications we have shown, that these OB-stars depict the spiral structure of the Milky Way up to distances of 4 kpc from the sun, as shown in Figure 1 (Klare and Neckel, 1967). Also 21 OB-groups were found (Klare, 1967).

In January 1968 we began polarization- and UBV-magnitude measurements of all the OB-stars of the Heidelberg catalogue. We are using a 50-cm cassegrain reflector, which was built in Heidelberg and has been operating since autumn 1967 at the Boyden Observatory, South Africa. For polarization measurements a rotating polarization foil in the light path was used. Polarized light modulates the photo-current from the

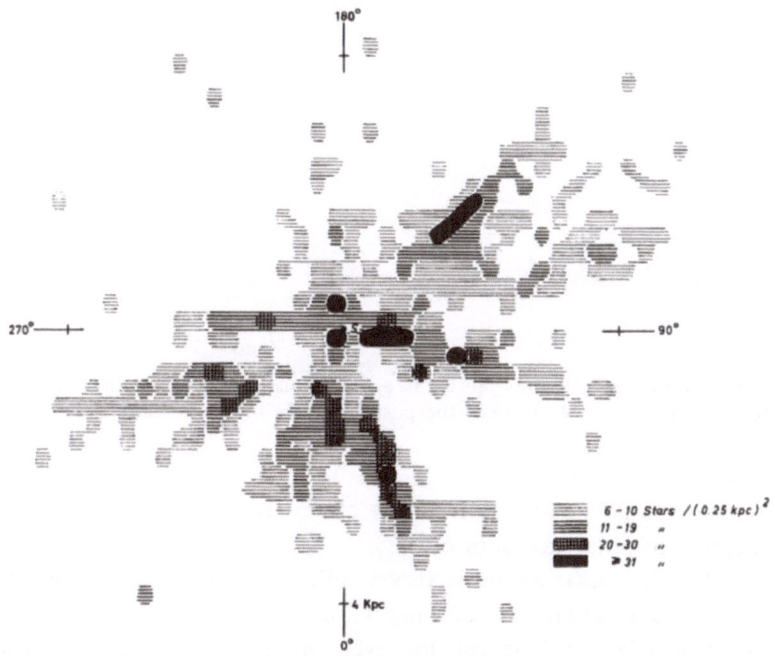

Fig. 1.   Distribution of 5083 OB+- and OB⁰-stars, and 1090 OB-stars with known MK-spectral type in the galactic plane.

multiplier. The photo-current then goes through a voltage-to-frequency converter whose output impulses are counted by 3 counters. Two of the counters work only half a period and they are separated by a quarter of a period (Leinert *et al.*, 1967).

To date we have measured the polarization of 1421 stars; a total of 2500 individual measurements were made. The accuracy of our measurements generally is 1 to $2^0/_{00}$ in the amount of polarization and $5°$ in the polarization angle. We would like to point out, that the few stars, which have been observed as well by Hiltner as by us agree very well with each other.

We hope to be able to complete the polarization program within a few months. I will now discuss only the preliminary results.

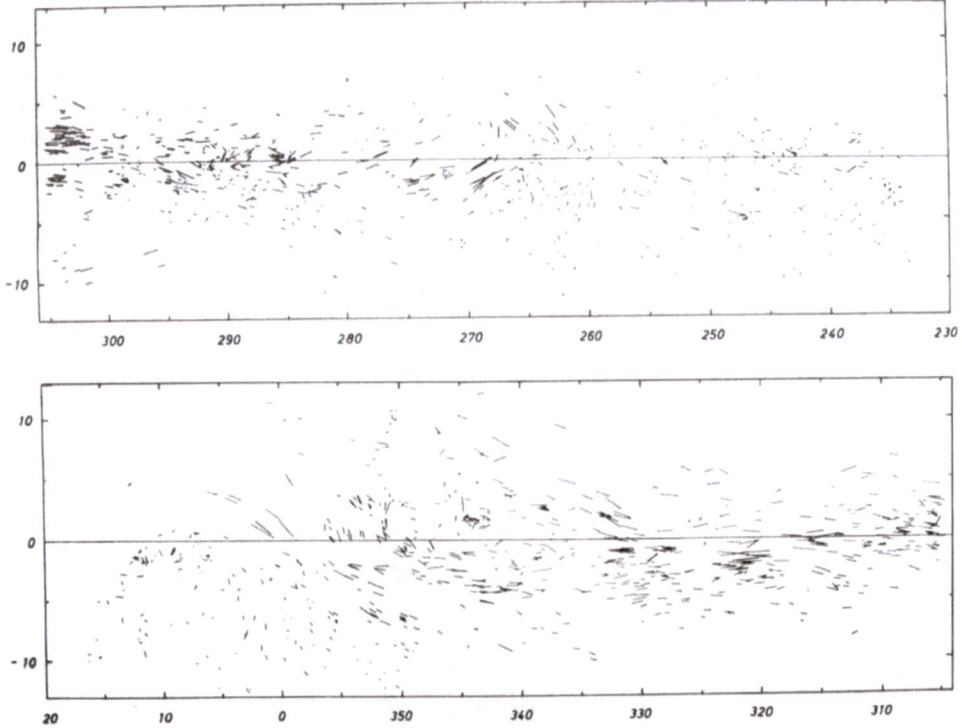

Fig. 2.    The polarization of OB-stars plotted in a $l^{II}$, $b^{II}$-diagram. The length of the lines indicates the relative amount of polarization, the position angle indicates the plane of vibration.

Figure 2 contains all polarization values of the 1421 OB-stars. The longest lines represent degrees of polarization of 6 to 7%.

Up to $l^{II} = 250°$ the polarization is very small, which is not surprising, because in this direction we have very little absorbing matter. In the adjacent region to $l^{II} = 300°$, the amount of polarization is larger, however, the directions of polarization show a large scatter. Between $l^{II} = 300°$ and $l^{II} = 345°$ the degree of polarization is very large and the direction very uniform and parallel to the galactic equator. From $l^{II} = 345°$

to $l^{II} = 20°$ we observe a whirling-like distribution, the directions of the polarization are partly vertical to the galactic equator, often they are inclined by 45°.

In Figure 3 we see the relationship of the standard deviation of the polarization angles from the mean value in some longitude intervals and the galactic longitude. Crosses represent the observations of Hiltner, and points represent our observations. The dot and the cross near $l^{II} = 0°$, which show a large discrepancy, cannot really be compared, due to a selection effect. The dot corresponds to many stars, distributed over a rather extended area, while the cross (Hiltner) corresponds only to a few stars concentrated on a small area, where the alignment of the electric vectors is quite uniform.

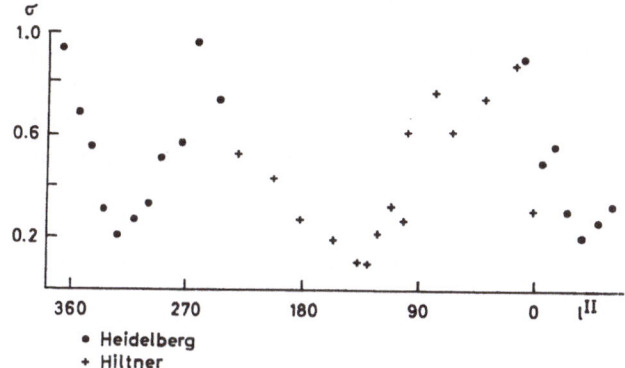

Fig. 3. The relationship of the standard deviation, $\sigma$, of the electric vector alignment and the galactic longitude.

According to the interpretation of Hiltner the maximum at $l^{II} = 80°$ corresponds to the tangential direction of the local spiral arm in the solar neighbourhood; the minimum at $l^{II} = 140°$ corresponds to that direction, in which the line of sight is vertical to the magnetic field.

In the southern Milky Way we observe a maximum in the standard deviation at $l^{II} = 260°$. In this direction we expect to find – due to spiral arm indicators – the tangential direction of the local spiral arm, opposite to $l^{II} = 80°$.

Also the minimum of the standard deviation in the southern sky at $l^{II} = 320°$ lies opposite to the minimum in the northern sky, at $l^{II} = 135°$.

## References

Klare, G.: 1967, *Z. Astrophys.* **67**, 249.
Klare, G. and Szeidl, B.: 1966, *Veröff. Landessternwarte Heidelberg-Köningstuhl* **18**, 9.
Klare, G. and Neckel, T.: 1967, *Z. Astrophys.* **66**, 45.
Leinert, C., Schmidt, T., and Schwarze, B.: 1967, *Veröff. Landessternwarte Heidelberg-Königstuhl* **20**, 23.

# 84. ON A COROLLARY TO THE MAGNETIC DIPOLE THEORY
# OF THE ORIGIN OF SPIRAL STRUCTURE

PARIS PİŞMİŞ

*Universidad Nacional Autónoma de México, Mexico City, Mexico*

**Abstract.** Some observable features of our Galaxy are explained in terms of a theory of the origin of spiral structure, whose basic feature is a gradually contracting gaseous subsystem with a magnetic dipole of which the axis is very close to the galactic plane.

## 1. Introduction

The aim of this communication is to present a few consequences (a corollary) of a theory, which was advanced some years ago by this author, to account for the origin and maintenance of the spiral structure in a galaxy. A series of ideas and arguments leading to the formulation of the theory are discussed in a number of papers published since 1960 (Pişmiş, 1960; Huang and Pişmiş, 1960; Pişmiş, 1961) while the outline of the theory itself with an application to the galaxy M31 has appeared in 1963 (Pişmiş, 1963). Here a brief sketch of this theory will be given prior to the presentation of a corollary to it.

## 2. The Theory

At present it is generally accepted that the stellar subsystems are formed at successive epochs in the course of the evolution of a galaxy. In a galaxy where the subsystems, from halo down to disk, are already formed, we assume that there still remains an organized gaseous subsystem, a flat spheroid, concentric with the galaxy at large and that this gaseous subsystem possesses, and is held together by, a magnetic field which, as a rough first approximation, is taken to be a dipole. We postulate that *the magnetic dipole is centrally located and its axis is perpendicular to the axis of rotation of the galaxy.* Such an orientation for the dipole field is intuitively quite an unorthodox assumption but herein lies the novelty of our approach.

It is expected that due to infinite conductivity of the interstellar gas the spheroid will rotate as a rigid body. No distortion of the dipole field inside the spheroid should thus occur. Next, it is plausible that the gaseous spheroid will suffer a gradual contraction and during this process matter and field will contract together. The only regions from where gas may leak out are those of the magnetic poles. The locus of the gas thus left behind will be a spiral due to its slower rotation speed with respect to the contracting spheroid. (In a forthcoming paper a quantitative treatment of the loss of mass and angular momentum is given.)

An application of this mechanism to M31 using the observed rotation curve of this galaxy has yielded a pattern of two intertwined spirals, with 1.5 turns each, in $3 \times 10^9$ years (for details see Pişmiş, 1963).

The gas may also leave the polar regions with a radial outward velocity, in other words it may be ejected. It is easy to show that, be it with leakage or with ejection, a spiral pattern will ensue by the mechanism sketched above. It may thus be stated that as a consequence of this theory:

(a) A spiral gravitationed potential field is created and it is superposed on a smooth axis-symmetrical potential field of the previously existing stellar component of a galaxy.

(b) The bi-symmetry of the spiral pattern is accounted for.

(c) The magnetic field lines are clearly along the spiral pattern.

(d) Two consecutive arms in the grand spiral design are of opposite polarity.

It should be emphasized that the main function of the magnetic dipole field is to funnel the gas through two diametrically opposite points of the bulge and that, unlike previous magnetic theories, no account is taken of the magnetic force in the maintenance of the spiral forms. Although much remains to be done on the physics of the theory, interesting consequences emerge as a corollary to our proposition (Pişmiş, 1968).

## 3. Corollary

If the magnetic dipole is not exactly perpendicular to the rotation axis but deviates slightly from it, the magnetic field will change sign while crossing the plane of symmetry – the galactic plane – and two consecutive arms will be above and below the plane respectively.

To fix ideas assume, now, that the positive pole is slightly above the plane of symmetry of the Galaxy and the negative pole below the plane. The two sets of spirals engendered by the rotating dipole will no more be co-planar. The spiral produced by the positive pole will always remain above the plane and the diametrically opposite arm will always remain below the plane; therefore along a radius vector in the plane the arms will be seen alternately above and below the plane and their distance from the plane will increase as the edges of the galaxy are approached. This result seems to be supported by recent optical data and by the 21 cm surveys (Henderson, 1967). Moreover the sign of the magnetic field will be positive above and negative below the plane, a result also in line with recent observations showing a general reversal of the magnetic field while crossing the galactic plane.

One other consequence of the deviation from orthogonality of the dipole may be mentioned here. This is the increase of the overall width of the galactic stratum from the center towards the edge of a galaxy. According to recent knowledge (Henderson, 1967) the thickness of the neutral hydrogen layer in the Galaxy appears to vary monotonically from the center outwards; the thickness is about 250 pc at the position of the sun and increases to about 500 pc at the edge of the Galaxy. Taking account of the natural width of the arms and their irregularities it is not difficult to see that the combined width of the two sets of intertwined spirals may appear as a layer of neutral hydrogen which widens outwards.

An estimate of the angle of inclination of the magnetic dipole to the average galactic plane using the above estimates of the variation of thickness of the hydrogen layer has yielded a value of 45′ for this angle.

In concluding we may state that a simple corollary to our magnetic dipole theory for the origin of spiral structure appears to account for the following three phenomena suggested by observation, but apparently quite unrelated to one another, at first sight.

(1) The reversal of the magnetic field while crossing the galactic plane.

(2) The alternately up and down position of the consecutive spiral arms.

(3) The thickening of the neutral hydrogen layer towards the edges of the galaxy.

## References

Henderson, A. P.: 1967, Ph. D. Thesis, University of Maryland.
Huang, S. S. and Pişmiş, P.: 1960, *Bol. Obs. Tonantzintla Tacubaya* 2, 7.
Pişmiş, P.: 1960, *Bol. Obs. Tonantzintla Tacubaya* 2, 3.
Pişmiş, P.: 1961, *Bol. Obs. Tonantzintla Tacubaya* 3, 3.
Pişmiş, P.: 1963, *Bol. Obs. Tonantzintla Tacubaya* 3, 127.
Pişmiş, P.: 1968, *Bol. Obs. Tonantzintla Tacubaya* 4, 229.

# PART V

# 85. SUMMARY AND OUTLOOK

B. J. BOK

*Steward Observatory, Tucson, Ariz., U.S.A.*

## 1. Preamble

Our Symposium has truly been held in the spirit of IAU Commission 33. This Commission has traditionally provided the opportunity for optical and radio astronomers to meet with theorists interested in the dynamics of stellar systems. Our good relations with Commission 34 assure us that in our deliberations we shall not lose sight of the physical implications of our theoretical and observational developments. The past and present Presidents of Commission 33 are proud of the fact that they – and their Organizing Committees – took the initiative for holding this Symposium in Basel. We need not hide the fact that the plan to go to Basel came because many of us wished to honor Wilhelm Becker and his associates for the beautiful and basic researches they have done relating to the spiral structure of our Galaxy.

After Shapley's major discovery of the distant center of our Galaxy, with our sun assigned a position about 10 kpc from the center, the scene was set for the developments to follow. The outstanding new approaches came through the Oort-Lindblad theory of galactic rotation and Trumpler's proof for the presence of a general galactic absorption of the order of half a magnitude per kpc or greater in visual light for positions along the galactic equator. However, we seemed stymied in our attempts to study the spiral structure of our Galaxy in the 1930's and early 1940's, until in the late forties there came Baade and Mayall's propositions, based upon studies of M31, which showed that H II Regions and O to B2 stars are probably the best spiral tracers. Morgan took up Baade's challenge to determine accurate absolute magnitudes and absorption-corrected distances, and, between Christmas and the New Year of 1951, he presented the world with the Morgan-Sharpless-Osterbrock picture of the basic spiral structure near the sun. To Wilhelm Becker and his associates goes the credit for having put our local structure in order.

At the time when the first optical pictures began to appear, the discovery of the 21-cm line led Oort and his many associates – of whom I need only mention here Van de Hulst, Müller, Schmidt and Westerhout – to use the 21-cm profiles for probing the spiral structure of the Milky Way within reach of the Netherlands Radio Telescope at Dwingeloo; and Kerr, Hindman and others covered the parts of the Milky Way observable from Sydney, Australia.

Theory was slow in catching up with observations. Until half a decade ago, most of us in the field were of the opinion that the magnetic fields near the galactic plane were probably as strong as 20 or 30 $\mu$G, and fields of this stength would probably have proved sufficiently strong to hold the spiral arms together as magnetic tubes. This was an illusion – for it is now clear that the fields are no stronger than two or three

*Becker and Contopoulos (eds.), The Spiral Structure of Our Galaxy, 457–473. All Rights Reserved.*

microgauss at most. During all of the magnetic period basic research on the purely gravitational approach was being done by one person: Bertil Lindblad. His work on the theories of epicyclic orbits and on resonance phenomena laid the foundation for many current theoretical developments and Lindblad's name has been frequently mentioned in our deliberations. It is a pity that he was taken away from us rather early in life. He would have thoroughly enjoyed the present Symposium, and we ask his distinguished son, Per Olof, to bring our greetings to his mother. Theory took a new turn about five years ago, when Lin and Shu entered the field with the density wave theory. Whereas Lindblad concentrated on the orbits of stars and on waves in systems of stars, Lin placed the emphasis on density waves of stars and especially in the gaseous interstellar medium. The Lin-Shu theory is now in full bloom, but we must not fool ourselves and think that all is done except the mopping up. We have heard it repeatedly said at this Symposium that the Lin-Shu theory is only a first-order theory and that there is no guarantee that it will not require major modification before all is said and done. There is controversy aplenty even within the MIT-Harvard family and this is all to the good. Observers are following the theoretical debates with care and more and more shall we be able to present decisive observational data capable of proving or disproving the tenets of the theorists.

It is wonderful indeed to have had with us at this Symposium so many of the Prophets of the past. However, we must not overlook the danger that their formidable presence in our midst may serve to intimidate the young astronomers, physicists and mathematicians present who have fresh ideas on spiral structure problems. We must not let THE ESTABLISHMENT guide our future course. As far as I am concerned, my sympathy and attention have gone to the unconventional young astronomers with new approaches – and I for one will stick with them until their ideas are definitely shown to be wrong.

It is indeed a formidable task for me to try and present in one hour a summary of our deliberations of the past six busy days and also to discuss future trends. This Symposium has produced several ground-breaking papers, a large number of new and solid contributions to our knowledge and understanding of spiral structure, and there have been also some excellent peripheral papers. But we must admit that there were also a dozen or more poor and irrelevant papers, which took valuable time. In my survey and comments, I shall not deal with even the best of the peripheral papers – which include the fine papers by Vera Rubin and Ford, and by Ejnasto. I shall also not speak about those researches reported by our Russian friends which mostly did not deal specifically with spiral structure. I beg your forgiveness for not mentioning some papers fully deserving of comment, for I am only human and the time for preparation of my text was short – as was the time for presentation. The harvest is a rich one, but it had to be completed and stored before our departure from Basel!

## 2. Spiral Arms

We should ask ourselves first of all: *What is a spiral arm?* The beautiful photographs of typical spiral galaxies beyond our own – if there is such a thing as 'a typical spiral galaxy' – show us spiral features that are primarily gaseous in nature. The spiral arms that concern the optical and the radio astronomer and also the theorist in the field are most neatly demarcated as long connected streamers of hydrogen gas, many of which have shapes that resemble logarithmic spirals. The gas associated with a specific spiral feature is principally neutral atomic hydrogen, but inside these H I streamers there are generally many concentrations of more than average density, in which star formation has recently taken place and in which OB stars with ages of the order of $10^7$ years and less are found. These OB stars come either alone or in OB associations or clusters, and near them we find most of the hydrogen ionized and observable as H II regions by both optical and radio methods. With reference to the Local Standard of Rest in our rotating Galaxy, these OB stars have not had time to have moved more than 50 to 100 parsec from their places of origin, which means that features at distances of 2 or 3 kiloparsec from the sun are observed by us within 1, or at most 2, degrees of their places of origin. At this Symposium we were reminded of many facts that we might otherwise overlook. For example, Weaver remarked that the temperature of the H I in spiral arms is probably close to 100 K, whereas in the interarm regions – with gas densities of the order of one-tenth that in the arms – we may have average temperatures of 1000 K and more; and we always bear in mind that the H II regions have temperatures of the order of 10000 K.

Apart from the OB stars, alone or in associations and clusters, the long-period cepheids and possibly also the WR stars and the early Be stars, are about the only decent spiral tracers. As McCuskey made clear in his thorough paper, the common stars are generally too old to be of much help. They may define 'fossil' spiral features, but most of the common stars have moved too far away from their places of origin to suit us. Moreover, the original regions of their births will probably be no longer in troughs of spiral potential. By now some of these positions of origin are likely to be in inter-arm regions!

To provide a basis for the beginning of a theory for the spiral structure of our Galaxy the optical and radio observers are asked to provide answers to two difficult varieties of fundamental questions. These are: (1) what are the principal characteristics of a spiral feature, be it a section of a major spiral arm, a connecting link, or one of the wisps or feathers that we observe emanating from spiral arms; (2) what appears to be the best pattern of overall spiral structure applicable to our Galaxy? The optical astronomer who studies our Galaxy is mostly limited to providing answers to the first question, whereas the giving of answers to questions relating to overall structure and pattern is primarily the province of the radio astronomer. The optical astronomer helps to provide guidelines for answers to questions relating to patterns and to overall structure mostly through the study of galaxies beyond our own stellar system which show spiral features of one sort or another. The theorist blends it all, constantly showing impatience at the inefficiency of optical and radio astronomers alike, who struggle with their basic data and who, for the taste of the theorist, are not providing fast enough simple and comprehensive answers to the basic questions which the theo-

rists wish to have disposed of once and for all. Most of the theorists at our Symposium
were applied mathematicians with commendable profound interests in gas and stellar
dynamics. I only wish that we might have included a bigger representation of physicists
and astrophysicists, for the great problems of our Galaxy are basically problems of
physics.

### 3. Spiral Galaxies of the Local Group and Beyond

Our Symposium began with two fine presentations; one by Oort, the other by Morgan,
who both stressed the terrific importance of studies of galaxies of the Local Group
and beyond that show spiral features.

Morgan showed us some beautiful H$\alpha$ photographs taken mostly by Garrison with
the Yerkes 40-inch Reflector fitted with a reducing camera of Meinel design and
appropriate interference filters. Many comparable H$\alpha$ photographs were shown, by
Courtès, Cruvellier, the Georgelins and Monnet, by Mme Pronik and by Khachikian.
I understand that more could have been shown – by Bertola for example – if there had
been time. It has been wonderful to have seen these excellent samples of photography
for some of our prized neighbors, M31, M33, M51, M83, M101, NGC 628, NGC
1232, NGC 4254, NGC 4643, NGC 6943, and others. To supplement the material
presented at our Symposium, we now have Hodge's fine photographs published earlier
this year (*Astrophys. J. Suppl.* **18**, 73, 1969) and there is the promise of an extensive
study of H$\alpha$ photographs to be published by Kristian and others at Mount Wilson and
Palomar Observatories.

While it is essential that we should stress in all theories of spiral structure that the
prevailing type of structure shows two principal arms, we should bear constantly in
mind that there are many exceptions to this rule. Oort stressed in his Opening Address
that in half of the spiral galaxies we have evidence for more than two arms. Often we
find features between the arms, or secondary branches, or links between arms, or
feathers and bifurcations in major arms, some so strong that it may be difficult to follow the basic pattern. Morgan and Vorontsov-Velyaminov considered the problems
in their addresses. Vorontsov-Velyaminov considers the lack of agreement between the
available spiral plots of our Galaxy a very serious matter and he presented evidence
to show that there may be no traceable spiral structure at distances greater than 10 kpc
from the center. Bortchkhadze's paper drew renewed attention to the presence of
several separate spiral arms in the multiple-arm system NGC 1232. He finds also as
many as nine sections of arms in one quadrant of NGC 5247. Rudnicki added NGC
3486 to the list of notable multiple arm systems. It is worth stressing that multiple
arms prevail in the outer parts of spiral galaxies, but that close to the nucleus two
distinct arms are generally shown.

Many of the participants to the Symposium were bewildered by the variety of spiral
structures and vaguely spiral-like structures for the neighbor galaxies for which
photographs were shown. It was reassuring to hear de Vaucouleurs state unequi-
vocally that all of these can be sorted according to about 10 classes, with *openness* of

spiral features, the presence of a *ring* and of a *bar* as the first items to be judged for classification.

The difficulties of classification by appearance alone were stressed in Morgan's address. He showed how relatively straightforward it is to classify irregular and spiral galaxies by integrated spectra ranging from A to K for Irr to Sc to Sb to Sa to S0, with the giant ellipticals naturally following after that.

To supplement the plentiful H$\alpha$ photographs, we need now extensive radial velocity studies of the types reported by Courtès, Cruvellier, the Georgelins and Monnet and by Mrs. Simkin. And we are of course all waiting anxiously for the results of the promised high resolution 21-cm surveys of galaxies; Baldwin promised results with resolutions of the order of 1 min of arc, or better! It is likely that from studies of nearby galaxies with spiral features, we shall be able to derive in the near future firm data on the relative widths of the spiral arms outlined in H II and compare these with the widths of the H I arms. This observational information is urgently needed for the study of *where* inside a spiral feature star formation has most likely taken place. Is it near the inner or near the outer rim of a spiral arm, or does the backbone of star formation run down the middle of the H I arm? How wide is the band of star formation and how does its width compare to the width of the H I feature? Is star formation limited – as I have long thought it is – to regions of abnormally high gas density, say 10 times the average for the arm as a whole?

The excellent paper by Beverly Lynds taught us much that is new about the distribution of dust in spiral galaxies. In the first phase of her work she has studied the distribution of dust in 17 Sc galaxies, all of them seen practically face-on. Her sketches confirm and extend our earlier knowledge on the subject. The 100-inch and 200-inch Reflector photographs, made available to her by Sandage, confirm Sandage's and Baade's assertions that dust lanes appear first close to the nucleus, to be followed by luminous spiral arms, with the dust arms in the inner parts at the insides of the luminous arms. Lynds finds that young stars and dust go together and that the brightest H II regions are generally found at the edges of dark lanes. Dark 'feathers' are often traced across a luminous arm and these must be considered a common feature, according to Lynds. An interesting result of her studies is that in several galaxies, notably in NGC 628, the luminous outer arms are 'sandwiched' between obscuration on either side. The continuation and extension of this work is obviously important and urgent.

## 4. Gravitational Theory

We are fortunate indeed that the theorists attended our Symposium in force. The theoretical keynote address was brilliantly delivered by Contopoulos, who gave us a full introduction to the gravitational approaches to the dynamics of spiral structure. At the start he intimated with his customary firmness that discussions of possible magnetic effects would be taboo. And so they were! We had to wait until the end of our third day, when I took it upon myself to refer to the fact that not so long ago spiral arms were tentatively explained as being held together by magnetic forces.

There were some mighty attractive features to the theories of a decade back in which spiral arms were viewed as magnetically bound tubes of ionized gas. Unfortunately the observed galactic magnetic fields fell short in strength by an order of magnitude compared to the required fields, but no one should overlook the celestial message reaching us through maps of optical polarization, which show the most orderly alignment when one observes across a galactic spiral feature. The magnificent work of the MIT group loosely headed by C. C. Lin has made the pendulum of interpretation swing toward Bertil Lindblad's gravitational approach, and this is wonderful indeed. But I look forward to the next decade in which the pendulum may favor a middle course with shared emphasis on gravitational and magnetic effects. I hope that the members of our Symposium who will attend the Symposium next week in the Crimea will remind the physicists present there that the astronomers and applied mathematicians are urgently in need of consultants in physics to help them resolve some of the big problems of spiral structure.

At the start of his remarks, Contopoulos stressed that the basic contribution of Lin and Shu was that they apparently removed the dilemma of the winding up of spiral arms. Lin, Shu and Yuan introduced and developed the concept of a spiral wave pattern of potential, moving through our Galaxy at a rate of rotation less than that given by the Oort-Lindblad rotation. All workers in the field give deserved great credit to Bertil Lindblad's concept of a gravitational theory which he began to develop more than a quarter of a century ago. The spiral arms are not viewed as permanent entities with some stars and some atoms and nebulae belonging to a spiral feature, others not, but rather do we see spiral arms now as loci of star concentration coinciding with spiral-shaped potential troughs. The stars and the gas move through the spiral arms, but on the average they prefer to linger in the arms and speed through the relatively thin inter-arm voids. Once a spiral system of two trailing arms has been established, the system will according to Lin and Shu apparently be a gravitationally self-perpetuating one. By the proper selection of the rate of rotation for the spiral potential field, with a value for the pattern speed $\Omega_p = 13.5$ km s$^{-1}$ kpc$^{-1}$ now favored by Lin, Shu and Yuan, a tightly wound near-circular system of two trailing spiral arms is produced, which shows many features known to exist in our own and neighboring galaxies.

As expected, grave doubts were expressed about some of the theoretical interpretations and different approaches were suggested. Contopoulos referred to these new ideas, many of them quite different from the Lin-Shu-Yuan approach. Later in the program the relevant papers were presented and we heard about Fujimoto's hydrodynamical approach, about Kalnajs's work on very open spiral systems with pattern speeds very much greater than those of Lin and Yuan, about Toomre's modification of Lin's theory, in which the waves show inward motions, and about Hunter's work on large-scale oscillations of galaxies. It is useful to have these approaches developed further since they may come in very handy if and when observers obtain evidence for axisymmetrical radial inflow or outflow of gas in our Galaxy. As of now, the observed humps in the basic observed curves giving circular velocity versus distance from the galactic center, one applicable to the first quadrant of galactic longitude, the other to the fourth quadrant, can be best understood in terms of the theoretical spiral structures and fields developed by Lin and by Yuan. But we should not forget that 10 years ago many of us looked with favor upon Kerr's suggestion of a radial outflow of gas at the sun's position of 7 or more km per sec. Hunter's modes would help understand such results.

The basic kinematical problem was discussed in a thorough paper by Blaauw, who finds that the Local Standard of Rest – as defined by objects associated with the halo and the nucleus of our Galaxy – has velocity components of the order of 5 to 10 km s$^{-1}$ with respect to the gas and the youngest stars. He stressed in his presentation that the gas in our vicinity is surprisingly quiescent, but that the youngest stars move in turbulent fashion in the solar neighborhood. He noted that any observer located in the Perseus Arm would have difficulties in sorting out the kinematics of motions of stars within a few hundred parsec of the sun.

Throughout the theoretical presentations, the limitations of Lin's first-order theory were stressed. First-order theory fails near the Lindblad resonance points, especially near the critical distance where rotational and epicyclic frequency become commensurable. Four-arm spirals then become a possibility! Lin has concluded that spiral-shaped gravitational waves of potential cannot exist within the inner Lindblad resonance circle.

On the fourth day of our Symposium Lin and Yuan spoke about recent developments in the density wave theory. The observational astronomer is especially pleased to learn about the interest our theoretical colleagues are showing in observations, and it is a source of regret to the observers, optical and radio alike, that we cannot agree as yet on the full outlines of spiral structure for our Galaxy. Give us a few more years, and we shall be able to tell you all right! Lin and Yuan have fitted their theoretical pattern primarily to the Kerr diagram of spiral structure. The theoretical pattern will have to be revised if the spiral diagram of Weaver proves to be the more acceptable one. In all observed radio diagrams the Orion Feature looks increasingly more like a spur than a major arm. The most spectacular optical feature, the Carina Feature, remains a source of embarrassment to radio astronomers and theoreticians alike!

Yuan has undertaken the time-consuming, but very useful, task of calculating 21-cm profiles on the basis of his modified Lin model. This approach is to be highly recommended for the future, for it is only in this manner that we shall be able to decide in the end which theoretical model fits best with our observations.

The paper by Yuan gave us some useful basic tests of the Lin-Shu theory. A further test – to which we should refer at this time – is that provided by the analysis of Burton and Shane. Their first observational analysis, announced 3 years ago, gave evidence already for streaming effects very much in line with those to be expected from the Lin-Shu density wave theory.

Lin made some very interesting suggestions regarding the outer-arm structure. Multiple arm systems become quite possible in the outer parts, but the spiral arms will probably not extend beyond the place where the pattern speed equals the rate set by normal galactic rotation. As a matter of fact, he suggests that the pattern speeds of external galaxies can be found in principle by noting the distance from the center of the galaxy at which the spiral arms seem to end. There is considerable doubt about the fate of density waves in the outermost parts of galaxies.

## 5. Origins of Spiral Patterns

The Lin-Shu-Yuan theory of spiral structure suggests how, in all likelihood, the spiral potential field, once formed, may be automatically maintained, but the authors have not yet suggested a well-developed mechanism for the original formation of the basic spiral potential field. Toward the end of the Symposium, Lin made a suggestion, which was developed as a result of discussions with Shu and Toomre. Trailing spiral arms, co-rotating with the general mass motion, are produced through the stretching of irregularities in the outer parts of a galaxy because of effects caused by differential rotation. Resonance produces a situation in which only a two-armed structure will prevail when the disturbances propagate inwards as a group of waves. These waves should extract their energy from the basic rotation of the galaxy, as sketched by Lynden-Bell and others.

Most of us present at the Basel Symposium felt quite pessimistic about the present status of our knowledge *why* it is that spiral structure seems to prevail on such a wide scale in galaxies. The exception was Oort, who, in his Opening Address said: "Indeed, any mechanism of formation of a rotating galaxy is likely to produce a system with large-scale initial asymmetry in the plane of rotation, which will almost automatically develop into a two-armed spiral structure." Let us hope Oort is right. On Saturday afternoon we were exposed to a number of widely different theoretical approaches by Toomre, Lynden-Bell, Shu, and Kalnajs, and on Monday morning we heard further on the subject from Vandervoort, and indirectly from Marochnik.

Toomre put some of the possible blame for starting our Galaxy's spiral structure on a close passage of the Large Magellanic Cloud to the plane of our Galaxy. But on further consideration he was not very certain of this approach, for he 'shuddered' at the resulting violent effects that would be produced if this were to happen, effects so turbulent and disruptive in character that they would hardly help to give birth to the majestic and all-pervading spiral potential field of Lin.

Lynden-Bell sketched an approach in which a ring would form at the distance from the center of the galaxy at which the rotation period would be in resonance with the pattern speed of the density wave. If the star density decreases outward, then there would be more stars pushing just inside of the ring, stars with greater rotational speed, than there would be stars holding back the ring – laggards Lynden-Bell called them – at distances just outside. This mechanism would provide the energy to start a density wave on its track outward from the inner ring. Thus we may expect a trailing spiral pattern outside the resonance ring. The main energy of the spiral pattern is thus derived from galactic rotation. Lynden-Bell pointed out that the Lin-Shu theory in its present form does not consider the transfer of momentum from one part of a galaxy to another, something which he considers basic for our understanding of the starting and maintaining of a spiral pattern of potential.

Shu was quite pessimistic. He pointed out that resonances primarily affect stars and *not* the interstellar gas, the atoms of which are constantly interacting. This conclusion applies especially to the Lindblad resonances, in which we deal with stars moving in neat and relatively constant epicyclic orbits. Stars do not interact violently in short times. Shu expressed the opinion that resonances have a tendency to hinder spirality rather than help it and he sees no evidence for instabilities general enough and sufficiently powerful to start a natural disturbance in the interstellar gas on the scale required by the prevalence of spiral structure in galaxies.

Vandervoort showed that spiral-like streamers of finite amplitude will occur naturally if a thin gaseous rotating disk – think of Population I – rotates in the midst of a gravitational field produced by a not-so-flat disk and a halo of old stars – think of Population II. Inside the possibly very thin first disk spiral features will naturally develop. They are like thin spiral frosting on a very large cake!

To initiate a spiral potential field, one obviously requires some feature that is non-axisymmetrical in character. Following Freeman's comprehensive and elegant treatment of Barred Spirals, Toomre advanced the suggestion that we may have overlooked the most promising spiral feature of all: *the Bar*. This suggestion is very much in line with that made on several past occasions by Pikel'ner. Toomre noted that Freeman and de Vaucouleurs had stressed the smoothness of the transition from extreme barred spirals, to spiral systems with minor bars, to true spiral systems – if there are such things. Thin gaseous disks in rotation are eager to be excited and any disturbance will almost naturally produce a spiral potential field. The Lin-Shu theory, or one of its modifications, probably provides the machinery for keeping the two-arm spiral potential field intact once it is formed.

One approach to the formation of a Bar comes from Kalnajs's work. He showed that flat-disk galaxies, which are stabilized by random motions to withstand violent axisymmetric instabilities, are still capable of gradual evolution. In the inner part of a rotating galaxy some sort of collapse may take place, which would result in a density wave which would in turn produce a feature looking very much like a bar with two little trailing streams. Angular momentum and energy must be disposed of and these may be transmitted to the outer parts. Kalnajs requests observers to provide him with data on the velocity fields for the whole of a galaxy, not just for sections of arms, since he wishes to know the observed deviations from circular motion and fit these into his theory.

Our theorists should be encouraged to continue their search for basic mechanisms to produce spiral potential fields on a large scale in rotating gaseous systems – and barred spirals should by all means be included in all such studies. The observational astronomers interested in studies of galaxies should make increased efforts to study the structural and kinematic properties of barred spirals in the expectation that the dynamicists will find much use for the results derived from such observational studies. The properties of cosmic dust in bars should be studied. During our discussions, there was brief mention of some hydrodynamical laboratory experiments that might throw further light on problems of the formation and stability of spiral features. This is an approach that deserves to be encouraged.

Prendergast and Miller presented the Symposium with the results – in a film – of some spectacular numerical experiments. The results of similar work by Hohl were presented by Toomre. It is now practicable to study the evolution of star systems with tens or hundreds of thousand of stars, and follow their patterns of motion and distribution under the inverse square law of attraction. In Prendergast and Miller's experiment clouds of gas are gradually turned into stars, the rate of star formation being held proportional to the square of the density. In the film we saw trailing spirals, bars and rings, forming and disintegrating before our eyes. Density wave effects were especially noticeable when about 25% of the galaxy was still in gaseous form. These striking patterns were found in the gas rather than in the star pictures.

Miller commented that these experiments are not done to provide at great expense spectacular films, but that we now have a tool for actually experimenting with spiral galaxies and studying problems of internal structure, motion, stability, and evolution.

In judging these films one should bear in mind that they show basically information on the motions of mass points or mass units in a self-perpetuating gravitational field. Care must be exercised when direct comparisons are made with photographs of spiral galaxies – which show basically brightness distributions. We should, furthermore, bear in mind that the underlying gravitational potential field is mostly controlled by the stars, rather than by the gas. The gaseous spiral patterns must be viewed against this background of a distribution of stellar masses.

So much for galaxies other than our own, and for pure theory. Now let us turn to our Galaxy and discuss its observed radio and optical properties, bearing in mind that this is a Symposium on The Spiral Structure of *our* Galaxy!

## 6. Radio Spiral Structure of Our Galaxy

The *Baseler Nachrichten* referred to our Symposium as the first international gathering ever of optical astronomers, of experts in galactic dynamics, and of the world's greatest radio *astrologers*. I viewed with pleasure the horoscopes of galactic spirality, which our distinguished radio colleagues had prepared for the occasion. Some of these were skillfully animated in the best of the early Mickey Mouse tradition. It seems as though Mezger, Kerr, Westerhout and Weaver are doing as well by our Galaxy as Kepler did by Wallenstein!

The major radio astronomical papers at our Symposium were those of Mezger and others at NRAO, MIT and the CSIRO on distribution of H II regions, and of Kerr, Weaver, Westerhout, Varsavsky and others on H I features of our Galaxy. This Symposium will become known as the occasion on which major steps forward were made in our knowledge of the radio astronomical evidence for spiral structure in our Galaxy.

Mezger's presentation of the H II data by himself, Wilson, Gardner and Milne showed that we now have information on the radial velocities of H II regions for practically all of our Galaxy. The interpretation of these data is not a simple matter, since there still is much confusion about near and far distances corresponding to radial velocities for H II Regions observed within the circle with a radius of 10 kpc centered upon the center of our Galaxy. Morgan commented that many ambiguities can be resolved in the years to come by careful optical work of the variety that is now being done for the transparent sections of the Milky Way in Carina. The H109-$\alpha$ velocities, when interpreted by themselves alone and as best we can, do not yield spiral patterns that make sense. But these data should be invaluable when sorted in conjunction with H I-21 cm profiles and with optical data. All of us who work on the spiral structure of our Galaxy will want to have the NRAO-CSIRO-MIT Catalogue within easy reach at all times.

Two outstanding facts are to be noted. The first is that there is a striking apparent total absence of H II regions within a central circle with a radius of 4 kpc. The second item to be recorded is that there exists a class of giant H II regions, all with flux densities equal to 4 times that of the Orion Nebula or greater, which show their highest concentration within the band between 4 and 6 kpc from the galactic center. The bulk of the H II regions peaks at distances between 4 and 8 kpc from the center, which is in marked contrast to the H I distribution, which peaks between 8 and 15 kpc from the galactic center.

Conflicting H I results were reported by Kerr and by Weaver.

Kerr presented the basic picture reproduced in his 1969 *Annual Review* article, one that is based on the work of himself, Hindman, Westerhout and Henderson. The Perseus Arm, the Cygnus-Carina Arm (with the sun at the inside and the Orion Spur emanating from it), the Sagittarius Arm and the Norma-Scutum Arm are basic features of Kerr's patterns of spiral structure. These arms appear to be near-circular and tightly wound with pitch angles of the order of 5° to 7°. Kerr notes that the H II Regions are generally found close to the ridges of greatest H I intensity. In the outer parts of our Galaxy, the contrast between arm and interarm gas density runs as high as 10 to 1, probably less in the central regions of our Galaxy. The dipping of the spiral arms over long stretches below and above the galactic plane is quite striking, features that stand out wholly apart from possible oval distortions and the bending of the galactic plane in the outer parts.

This may be a good place to mention in passing a very important result announced by Oort, who spoke of the work of Miss Kepner. She finds marked concentrations of H I at heights above or below the plane as great as 1 to 3 kpc. The observed velocities show this hydrogen gas to be clearly related to some of the major spiral features, like the Perseus Arm. It must have been expelled with terrific speeds – possibly explosive speeds – from its parent spiral arm to have reached such heights. Similar results were announced by Weaver from his northern survey.

From the observations made at Hat Creek in California, supplemented by the Kerr and Hindman data for the southern Milky Way, Weaver derives a spiral pattern for our Galaxy very different from that favored by Kerr. His pitch angles for the arms are of the order of 12° to 14° and the principal arms that are shown are the Perseus and Sagittarius Arms. The sun is in the Orion Spur – which is not a major feature. The Carina Arm is almost wholly incorporated in the great Sagittarius Arm, and a residual feature becomes a very minor local loop of the H I pattern. The Norma-Scutum Arm is not clearly shown as in other diagrams. Weaver pays much attention in his analysis to the directions from the sun in which the velocity vs. galactic longitude diagrams exhibit loops which show that for that direction one looks tangentially along a major spiral arm. One such loop is at galactic longitude 284° (in Carina), the other at longitude 50° approximately.

Every effort must be made to find with minimum delay the reasons for the differences between the Kerr and Weaver patterns. The procedures used by Kerr have been described in detail, but Weaver's results are so recent that we have only a general notion about his procedures of analysis. Weaver stressed that his results are still highly provisional. Both analyses suffer from relative incompleteness of data in the southern hemisphere, where there are many gaps in the coverage of profiles and where there is a need for tighter latitude coverage. Thus far Weaver has admittedly not paid sufficient attention to integrated intensities of his profiles and he has not used effectively the distribution in latitude for disentangling the near and far H I in directions for which the radial velocity resolution is poor. There has been no time to call in radio H109-α velocities and data on optical H II regions to adjust his 21-cm results. In a way we are stymied until Weaver and Kerr get together and arrive either at more nearly compatible spiral patterns, or until one shows that the other's pattern has serious deficiencies. It is most desirable that the available data be published in full – and, as I understand it, this is being done. Careful independent analysis of the basic material by others seems desirable, but this is not to be undertaken lightly. Perhaps such analysis should wait until the southern profiles catch up in quality and coverage with the northern ones.

Varsavsky reported on the progress of the work now under way in Argentina. He confirms many of the features shown by Kerr. The low-latitude work of Miss Garzoli should contribute especially to our knowledge in the critical longitude range 270° to 310°.

We should not fail to refer to the beautiful results shown in the form of films by Westerhout and by Weaver. Westerhout concluded from his observations with the NRAO 300-foot antenna that H I spiral arms are really conglomerations of H I clouds, with little background in between, that the H I density between arms is often quite low, that the Orion Arm or Spur is far less impressive than the Perseus Arm, and that we have now good evidence for a Distant Arm, well beyond the Perseus Arm.

Burton and Shane presented some new results for the inner arm structure. Their analysis relates beautifully the expanding 4 kpc Arm to inner and outer arms of which sections are viewed on both sides of 56° galactic longitude. The ratio of arm to inter-arm density seems to be different here from the outer parts, with 1 to 3 being a favored average for interior structures. Their values for the pitch angles of the arms are 8° for the inside features, 5° for the outside ones.

To summarize: We now have information on spiral features in the range of distance from 4 to 5 kpc from the center to 14 or 15 kpc, but the present controversies will have to be resolved before we can say that we have a spiral pattern that is generally agreed upon as the more or less overlying pattern for our Galaxy. Observers are not yet ready to present the theorists with the 'grand design' of C. C. Lin's dreams.

## 7. Optical Spiral Structure of Our Galaxy

The discussion of optical studies relating to the spiral structure of our Galaxy opened with a paper by McCuskey on the stellar component of our Galaxy.

McCuskey concludes that (1) the $OB^+$ and $OB^0$ stars, WR stars, early Be stars, O Associations, some S and N (carbon) stars, are linked to spiral structure; (2) the B5 and B8 to A3 main sequence stars, classical cepheids of longest periods, early M giants (including variable stars) are possibly loosely associated; (3) the remaining main sequence stars to F8, the yellow-red giants, the M giants later than M5, as well as the remaining S and N stars, are not related to spiral structure.

Next comes a comprehensive paper by two of our hosts, Becker and Fenkart. This paper extends the results given half a decade ago by Becker, Fenkart and Steinlin and it confirms the existence of three optical spiral features, the Perseus, the Orion and the Sagittarius Arms, while adding a fourth, the Norma-Scutum Arm. Fenkart pointed out that the possible connection between the Cygnus and Carina Arms, which I have advocated in the past in the longitude section 280–300°, is erroneous since there is a void of 1.5 kpc between the sun and Carina in which there are no HII regions and OB clusters in association. The Becker-Fenkart spiral pattern fits well into the overall spiral structure in the outer parts of NGC 1232.

Courtès, Cruvellier, the Georgelins and Monnet presented their beautiful work on optical HII regions. With the aid of radial velocities and optical distances for exciting stars, they have been able to outline an optical pattern, which includes the Perseus, Orion, Sagittarius and Norma-Scutum Arms. The French astronomers conclude that the OB stars, the long-period cepheids and the HII regions all have precisely the same velocity characteristics. We note that the pitch angles of the spiral arms traced by Courtès et al. are about 20°. The Swiss and the French astronomers place the sun at the inside of the Orion Arm.

I am happy to let these results stand unchallenged, and yet I feel that the last word may not yet have been said on this subject. Even with the best efforts, optical astronomy is by its very nature interstellar-absorption limited and, while we may for some longitudes reach to 6 or 8 kpc distance from the sun, we shall more likely on the average be limited to distances half that great. The radio astronomer is not limited in his analysis by absorption. Once we know the kinematics of our Galaxy, we can analyse our 21-cm profiles and H109-α velocities and examine the overall pattern of structure. The optical astronomers are all agreed where are the principal concentrations in spiral arms visible from the sun's position in our Galaxy. The connecting of these clumps and sections of spiral arms into an overall spiral pattern should come from a blending of optical and radio information. I hope to see the day when we shall be able to hand to the theoretical astrophysicists and applied mathematicians a trustworthy model of spiral distribution and kinematics ready for comprehensive theoretical analysis; we are not yet at this goal.

Tammann presented a paper on the galactic distribution of long-period cepheids, those with ages of the order of $3 \times 10^7$ years and less, periods 11 days and greater. Color observations in UBV alone can yield good absorptions and distances for stars that are far more distant than the faintest OB stars within reach. The discovery rate of these cepheids must be stepped up, for here is an area in which optical astronomy stands to make major new contributions. Tammann's paper deserves careful attention.

Increasing emphasis is being given to detailed optical studies of sections of the Milky Way in which we are clearly observing specific spiral features. Mrs. Dickel, jointly with Wendker and Bieritz, has made a special study of the nebulosities in the Cygnus X region, longitudes 70° to 90°. Not only have they found structural patterns which appear to be related to the local magnetic field, but they have obtained new and interesting data on the geometry and internal structure of the spiral feature. The paper by Dickel *et al.* sets a fine pattern for future research.

During the early days of the study of spiral structure by optical means the Carina-Centaurus section, between longitudes 270° and 300°, did not receive the attention it deserved. This is no longer so. At our Symposium there were five papers devoted to this section and those immediately preceding and following it. The days of Carina neglect are past.

Optical data, supported by radio evidence from Argentina and Australia, show that the Carina feature is sharply bounded at galactic longitude 283° to 285° and less sharply at about 295° to 300°. The lower longitude limit is shown beautifully by Graham's work which proves the presence of OB giants and supergiants along the sharp edge over distances between 2 and 8 kpc from the sun. Velghe and Denoyelle have studied the thickly absorbed section with longitudes 260° to 283°. The absorption is very great in spots but, in spite of this, the absence of OB giants and supergiants between longitudes 265° and 275° seems definite. The absence of radio continuum found at these longitudes gives further proof that between 283° and 285° we are truly observing along an edge of a major spiral feature. Lyngå has shown that the absorption at the far-longitude edge is not exceptionally great. He finds a total absorption $A_V$ of about 2 magnitudes at $l^{II} = 298°$ applicable to a distance of 3 kpc and nothing much beyond to a distance of 6 kpc. Obviously the inside of the arm is not very rich in cosmic dust.

Because of these low absorptions it becomes possible to outline the detailed internal structure of the Carina Spiral Feature. Together with my associates Alice Hine and Ellis Miller, making use of the standard UBV sequences established jointly with Priscilla Bok, we have been able to study the 'tree-rings' for two sections through the Carina spiral feature, one at 2 kpc, the other at 4 kpc from the sun. Each cross-cut in the section between longitudes 283° and 295° is rich in OB stars, H II regions and cepheids. These are spread in depth. E.g. Seggewiss, has determined the distance to one OB concentration, which is about 2 kpc. The point to stress is that the ridge of

H II concentrations and of young OB clusters and associations is narrow compared to the broad band of H I.

The width of the ridge of H II features is about 800 pc, that of the H I spiral arm 1500 pc. For the Carina Spiral Feature we have now definitely located the position of the ridge of young stars and ionized hydrogen within the broad spiral feature. It should be possible to do in the future much detailed structural analysis on the internal affairs of spiral arms. A purely theoretical paper by Roberts bears directly upon these problems. On the basis of the Lin-Shu density-wave theory, he has calculated where, inside a broad H I feature, the strongest gas concentration will occur. He visualizes the formation of a shock wave, which will remain stationary and coincident with the background spiral arm, and which will lead to the piling up of much gas along a narrow ridge within the broad H I feature. Near the ridge dense gas clouds will be naturally condensed to such an extent that star formation takes place. Roberts' Two-Arm-Spiral Shock-Pattern provides good check lines of positions where, according to theory, observers should find bands of star formation. It is most encouraging that Roberts' work suggests the formation of narrow ridges of star formation within broad neutral hydrogen features which is precisely what we observe in the Carina-Centaurus Feature.

One result of the studies of the Carina-Centaurus Section by both optical and radio techniques is that the major spiral features are here found concentrated at galactic latitudes $-1°$ to $-3°$. At 5 kpc from the sun the principal concentrations lie 200–300 parsecs below the standard reference flat galactic plane.

The listing of OB stars at Warner and Swasey Observatory by Stephenson and Sanduleak promises to yield a wealth of new material for future studies of the southern Milky Way. Our Symposium voiced strong support for the suggestion that funds be made available for the preparation of first-class identification charts to show the positions of the southern OB stars.

I should mention that a potentially new spiral tracer was added to our lists. Mills gave evidence to show that pulsars are found mostly in or close to the galactic plane and they seem to show a remarkable preference for being located in spiral arms!

## 8. Influence of Magnetic Fields

I shall be brief in my comments on the final morning session. Woltjer's opening remarks set the tone for the analyses of magnetic fields and their effects. The average large-scale magnetic field can be determined by now quite well from Faraday rotation observations, since values of $n_e$ (of the order of 0.05 per cm³) from pulsar data are now available and these appear to be quite precise. The fields are of the order of a few microgauss, probably no greater than 3 microgauss. Such magnetic fields are so weak that they will largely be dragged along with the gas and, if the gas goes through a density wave, the field may be expected to be bent accordingly. From Woltjer's paper and the ones that followed it, it became clear that magnetic fields do not contribute effectively to the maintenance of spiral structure. However, they do have some significant side-effects. Their influence on the rolling of spiral arms may become a very active topic for future observational and theoretical analysis. And we may expect strong magnetic fields in condensing gas clouds.

Mathewson's discovery of a localized helical field complicates all analyses of polarization data. Fortunately optical polarization measurements are now becoming available in abundance, and it should be possible to disentangle effects caused by the local field from those produced by spiral features. Polarization data were especially needed for the long-neglected southern Milky Way. Verschuur reported that Mathewson has ready for publication a catalog with several thousand measurements, and Klare and Neckel showed their results for over 1400 stars, all of them along the southern Milky Way. These data should be exceedingly informative, especially when combined with good data on magnitudes, colors and absolute magnitudes of all stars with polarization measurements. Let us hope that the helical field may not unduly complicate the analysis and interpretation!

The final morning session was a very rich one. We had Spiegel's presentation of the conditions inside the 4 kpc arm. He visualizes a system of hydromagnetic flow acting upon gas emitted by the galactic nucleus and thrown out from the galactic nucleus into the central region, where he assumes the streaming to be guided by a magnetic field. At about 4 kpc from the center, this gas is ejected from the galactic plane, some to the north, the rest to the south. Supposedly, most of it returns in graceful arcs toward the galactic nucleus.

The final paper of our marathon series was one by Paris Pişmiş. She spoke about her series of papers on the Origin of Spiral Structure published in Tonantzintla and Tacubaya Bulletins Nos. 19, 21, 23, and 30. Time for presentation was short – so we were all asked to read these papers with care. Which is one reason why I shall now sit down and go home!

### Notes Added December 1, 1969:

The readers of the Basel-Symposium Volume should be aware of two papers about to appear in print, which were not available to us at Basel.

(1) At the Symposium much indirect attention was given to barred spirals as a possible intermediate stage in the formation of normal spiral galaxies. Following the Basel Symposium, Margaret Burbidge prepared a paper summarizing the published observational data. This paper will be published in the January/February issue of *Comments on Astrophysics and Space Physics*. She describes a variety of models of barred spirals, in which the morphological trend is from subclass *c* toward subclass *a*, and in which barred spirals will naturally tend to evolve into normal spirals. More observational material on absorption lines in barred spirals is urgently required. Some years ago A. D. Code reported, at a Colloquium held at Steward Observatory, large velocity gradients exhibited by absorption features in bars, but no published data are available.

(2) The December, 1969, issue of *Sky and Telescope* contains an article by H. C. Arp: 'On the Origin of Arms in Spiral Galaxies'. Arp explores in this article the suggestion

that spiral arms represent the tracks of material ejected from galaxies. This possibility was not discussed at Basel and deserves careful further attention by theoretical astrophysicists, optical and radio astronomers. The approach represents truly an "explosive theory for the origins of spiral arms" quite different from the theoretical approach of the Lin-Shu theory of gravitational density waves.

# DESIDERATA FOR FUTURE WORK

*Oort:* For the spiral structure in our own Galaxy the most important desideratum would seem to be to get a picture of the spiral arms based on direct distance criteria, independent of the radial velocity. Such a picture might be obtained in the future if a further study of giant H II regions would provide data for a determination of their distances from optical, infrared or radio observations.

Further extension of the so successful Basel work on distant galactic clusters, possibly by searches in the far infrared, may also contribute in a quite significant way.

A more accurate location of the spiral arms observed in the 21-cm line may be obtained by a large extension of the work on 21-cm absorption lines in the radiation of distant galactic sources. This would appear to be an extremely important undertaking.

Important further information on the dynamics of spiral arms can undoubtedly be obtained from the accurate determination of the birth places of stars by the methods developed by Strömgren. The significance of this was clearly illustrated by the results described by Lin and Yuan during this Symposium.

The greatest advances for the spiral problem in the next few years are likely to come from higher-resolution radio surveys of external galaxies. With large synthesis instruments beamwidths of the order of $\frac{1}{3}$ to $\frac{1}{2}$ min of arc will soon be realized, both for continuum observations and for 21-cm line work. For one or two dozens of the nearest large spirals this will make it possible to obtain accurate charts of the arms of neutral hydrogen as well as of those of ionized hydrogen, and of the distribution of the non-thermal radiation relative to the gas arms. It will enable astronomers to obtain fairly reliable data on the relative density of the gas between arms, on the widths of the H I arms and on the relative situation of dust, gas and stars. The line observations will furnish a wealth of data on rotation, on radial components of the gas motion, and on internal motions in the arms. Study of the central regions should permit the investigation of features like the 3-kpc arm in other galaxies, and perhaps to obtain an insight into the relation between such expanding features and the general spiral patterns.

Evidently, more refined multicolour tracings through spiral arms may add greatly to the understanding of their birth process.

It may be desirable in the further development of the theory of spiral structure to take account, on one hand, of the 'events' which apparently cause part of the gas to be carried up to large distances from the galactic plane, and, on the other hand, of the inflow of gas from intergalactic space, which might have sensible effects on the dynamics of the arms.

*Becker and Contopoulos (eds.), The Spiral Structure of Our Galaxy, 474–478. All Rights Reserved.*
*Copyright © 1970 by the I.A.U.*

*Becker:* Allow me some words about our future work at the Basel Observatory concerning the galactic disc and the spiral structure.

Although our investigations of galactic clusters and H II regions will be continued, they are approaching their natural limits, set by the power of medium size telescopes and by the fact that the photometric determination of distances over 4 kpc is no longer precise enough for reliable results.

Among the roughly thousand galactic clusters in the catalogues of Alter and Rupprecht, some 400 may be found with distances under 4 kpc. About 250 have already been observed. Among the remaining 150 clusters, maybe 30 to 40 will be young enough to help to define the spiral pattern according to our experience.

We hope to complete the picture above all in three points. We expect a better definition of the arm following the Perseus arm and we think it might be interesting to know whether the local arm joins really the Scutum-Carina arm in a distance of about 3 to 4 kpc in the direction $l^{II} = 30°$ to $55°$, as it might be expected according to the radio-astronomical observations of Weaver. The large pitch angle of about $25°$ of the local arm is a strong point in favour of this possibility. Finally it would be of considerable interest to find out, whether there exists a connection between the Sagittarius-Scutum arm and the Carina arm in the directions between $300°$ and $330°$. If this should be the case, the whole feature would have to be called optically Scutum-Sagittarius-Carina arm.

Unfortunately the number of clusters is not large in the two regions mentioned above. Between $30°$ and $55°$ there are only 3, between $300°$ and $330°$ 21 clusters, 7 of which being already observed. Only the observations can decide, whether there are sufficiently young clusters among these objects.

The situation is similar with respect to the H II regions. Optically only those objects are of interest, whose exciting stars are known or can be identified with certainty and localized with optical methods. In the two interesting regions mentioned above, there are 10 and 7 H II regions respectively, the latter ones ($l^{II} = 300°$ to $330°$) being rather small features mostly.

Their exciting stars have not yet been observed, but a corresponding programme might be executed probably in a short time.

Should all these proposed observations of galactic clusters and H II regions not be able to decide the open questions, then these questions will probably remain unanswered by optical means, since the other indicators of spiral structure cannot be localized individually with sufficient precision.

I want to add some remarks about the possibilities of practical stellar statistics as far as they concern the spiral structure and as far as they are based on photometric measurements. Stars should not be forgotten completely in this context.

Dr. McCuskey has made clear that the important and detailed investigations carried out at the the Warner and Swasey Observatory with stellar-statistical methods of spectral classification and photometry yield no indications of spiral structure in the solar neighbourhood. Their importance lies in another direction.

The gravitational theory of spiral structure leaves no doubt that its findings refer to the interstellar matter and to the young objects being born there. However it is not impossible that there exists a certain large scale structure outlined by older stars.

In connection with the problems of spiral structure practical stellar statistics has to take a new attitude. Without any doubt only such a programme can be successful which reaches considerably larger distances than do the spectral classifications. That can be done with a three-colour photometry, which, comprising stars of solar luminosity and brighter than about 20th magnitude, should reach to the limits of the galactic system on our side of its centre. The practically effective limit is set by interstellar absorption in each direction.

A programme covering the local arm and the inter-arm region between the local and the Scutum-Sagittarius arms has been started several years ago in Basel. It is guided by the following consideration: without any doubt there exist discrete star clouds in the Milky Way as well as in other spiral systems. Nor can it be denied that these clouds follow the spiral arms to a large extent.

Since, according to existing results, the stellar clouds do not consist exclusively of interstellar matter and young objects, but also of older stars, the stellar-statistical findings of a three-colour photometry might contribute to a certain extent to the resolution of questions of the spiral structure. In addition to that, one can collect experience concerning the different composition of star clouds regarding the spectral types. This experience might gain some relevance with respect to the connection between spiral structure and evolution. However, it would lead too far in the frame of this symposium, to consider the results of this programme in a more detailed way.

In any case we hope to get additional points of view from the theory of spiral structure, which would allow us to shape the observation programmes in close contact with it.

*Lin:* I would like to make three short comments from the theoretical point of view.

## 1. *Mechanism and the Origin of Density Waves*
The observed spiral structure consists of a 'grand design' and many other secondary features. There is presumably a co-existence of material arms and density waves. There may be co-existence of several wave patterns. One of our principal interests in the near future will be the understanding of the mechanisms involved, especially those that can give rise to the wave patterns.

From our experience with plasma physics, we know that there is a variety of mechanisms for instabilities in plasmas. We must therefore not close our eyes to the possibility of several mechanisms giving rise to similar spiral patterns. We must, however, guard against invoking physically non-existent components as a cause for instability. Personally, I am still convinced of the importance of gravitational instability in the outer reaches of the galaxies, as I described in my talk.

As I also mentioned there, attention should be directed to resonant stars that play an important role in the mechanism of maintaining density waves in stellar systems.

We wish to associate the ring structure in external galaxies and the 3-kpc arm in the Milky Way with Lindblad resonance.

## 2. *Application to External Galaxies*

Besides continuing to push for better comparison between theory and observations within our own Galaxy, we should now turn our attention to external galaxies. There are at least three lines of research that one could pursue.

(a) *Normal spirals: determination of the pattern.* After the rotation curve and the mass model are established for an external galaxy, one can get a first approximation to its spiral pattern, if the pattern speed is known. We now have a way of estimating this approximately, since the outer edge of the spiral pattern is roughly associated with co-rotation. It is my understanding that Frank H. Shu is pursuing this approach in connection with the galaxy M51.

(b) *Barred spirals: a conjecture.* The arrangement of dust lanes and prominent H II regions in a galaxy such as NGC 1300 suggest that they might also be caused by gas undergoing a sudden compression while flowing through a density wave pattern. Such a picture would also be compatible with the dust lanes along the bar. We therefore propose the re-examination of the theory of barred galaxies from this point of view. This view is compatible with the gradual transition between normal spirals and barred spirals, as strongly advocated by de Vaucouleurs.

(c) *Ring structure: high resolution observation in external galaxies.* We know that the 3-kpc arm separates the part of the galaxy abundant with H II regions from that deficient in H II regions, and we believe that there is a ring structure similar to that seen in external galaxies, such as NGC 5364. I would therefore like to recommend the careful examination of the behaviour of the gas and stars near the ring of NGC 5364 for possible anomalous behaviour similar to that associated with the 3-kpc arm. Consultation with Van den Bergh indicates that the optical observations needed are within the realm of observational possibilities. For the observation of neutral hydrogen, much higher resolution is needed than that available at the present time.

## 3. *Physical Processes in the Interstellar Medium*

Having gone this far with the understanding of the grand design, we can turn our attention to the next smaller scale, to consider the rolling motion (torsional oscillation?) of a spiral arm and its implications on the magnetic field, as discussed by Woltjer. We have noted that we can account for the wider separation between the dust lane and the H II arm in the inner part of M51 (and the narrower separation in the outer parts), as emphasized by Morgan. In general, we can look for a deeper understanding of the physical processes in the interstellar medium once we know there are density waves.

Since the several spiral arms nearby are presumably not all of the same nature, a comparative study of the physical processes in them would be very helpful to our understanding. Strong compression due to density waves triggers the formation of bright stars and H II regions. This same dynamical process would have its effect on

H$_I$ regions, on the magnetic field (and its consequent effect on cosmic ray particles), on the formation of dust lanes, and possibly on the formation of molecules as well. We can indeed look forward to an exciting time in the study of interstellar medium.

## Discussion

*Van den Bergh:* It is the purpose of this remark to draw attention to a rather striking feature of spiral structure in galaxies that has not been discussed at this Symposium. Observations of late-type galaxies show that the strength of the spiral phenomenon is a function of the luminosity (or mass) of the galaxy in which it occurs. Supergiant galaxies tend to have long and well developed spiral structure. Normal giant galaxies usually exhibit only rudimentary spiral structure. Finally the majority of late-type dwarf galaxies exhibit no spiral structure at all, i.e. they are Magellanic type irregulars. The dependence of the ratio of spirals to irregulars on luminosity is illustrated by the data given in Table I.

TABLE I

Percentage of all intermediate- and late-type galaxies which are irregular ($H = 100 \, \text{km s}^{-1} \, \text{Mpc}^{-1}$ assumed)

| $\langle M_{pv} \rangle$ | No. galaxies classified | Per cent irregular |
|---|---|---|
| − 20.1 | 113 | 0 |
| − 19.2 | 201 | 0.5 |
| − 18.2 | 93 | 7.5 |
| − 16.6 | 88 | 36 |
| − 15.0: | 104 | 62 |